Modelling and Identification with Rational Orthogonal Basis Functions

Peter S.C. Heuberger, Paul M.J. Van den Hof
and Bo Wahlberg (Eds.)

Modelling and Identification with Rational Orthogonal Basis Functions

With 78 Figures

Peter S.C. Heuberger
Delft Center for Systems and Control
Delft University of Technology
Delft, 2628 CD
The Netherlands

Paul M.J. Van den Hof
Delft Center for Systems and Control
Delft University of Technology
Delft, 2628 CD
The Netherlands

Bo Wahlberg
S3 - Automatic Control Group
Royal Institute of Technology
Stockholm, 100 44
Sweden

British Library Cataloguing in Publication Data
Modelling and identification with rational orthogonal basis functions
1. Functions, Orthogonal 2. Linear systems - Mathematical models
I. Heuberger, Peter S. C. II. Hof, Paul M. J. van den III. Wahlberg, B. (Bo)
515.5′5

Library of Congress Control Number
2005925762

ISBN 978-1-84996-976-5
Springer Science+Business Media
springeronline.com

Softcover reprint of the hardcover 1st edition 2005

Typesetting: Electronic camera-ready by editors

69/3830-543210 Printed on acid-free paper SPIN 11364665

Preface

This book deals with the construction and use of rational orthogonal basis functions in modelling and identification of linear dynamical systems. It gathers results of the research efforts of nine specialists in the field, who have been very much intrigued by this subject over the past 10 to 15 years.
Since the early 1990s, the theory and application of non-conventional basis functions have attracted more and more attention in the fields of systems and control theory and signal processing. This research extends the systems theory of Laguerre functions and Kautz functions, which were introduced several decades ago. However, the important contributions to efficiently solving problems of system identification and approximation using these functions were only clarified about a decade ago. Since then, generalizations of the classic basis functions have been worked out, which incorporate a huge amount of freedom for design and adaptation.

Decomposing dynamical systems in terms of orthogonal expansions enables the modelling/approximation of the system with a finite length expansion. By flexibly tuning the basis functions to the underlying system characteristics, the rate of convergence of these expansions can be drastically increased. This leads to highly accurate models (small bias) being represented by few parameters (small variance). Additionally, algorithmic and numerical aspects are more favorable. The basis functions are applied in problems of identification, approximation, identification for control (uncertainty modelling), and adaptive filtering. Besides powerful algorithms, they also provide useful analysis tools for understanding the underlying identification/approximation algorithms.

Independent seeds for this work were planted around 1990 starting with the Ph.D. thesis of Peter Heuberger in Delft, under supervision and with impetus of Professor Okko Bosgra, on the use of basis function models in identification and transformation. Around the same time, Bo Wahlberg, in Sweden, published his work on the identification of Laguerre models. His objective was to generalize the theory of Lennart Ljung on identification of high-order finite impulse response models. This work started in a discussion with Graham Goodwin (when Bo Wahlberg was a post-doc in Newcastle, Australia) who pointed to the work by Guy Dumont on control using Laguerre models. A few years later, in 1993, Brett Ninness completed his Ph.D. thesis in Newcastle, Australia, on deterministic and stochastic modelling. Since then, developments within the field, as well as international cooperation, have grown steadily. Sabbatical visits of Bo Wahlberg and József Bokor to Delft, of Paul Van den Hof and Håkan Hjalmarsson to Newcastle, Australia, and of Tomás Oliveira e Silva and Brett Ninness to Stockholm were all very fruitful and generated, besides new research results, also an increasingly growing enthusiasm for this problem field, of which the interesting and relevant fields of applications were growing steadily. New Ph.D. students entered the area, among them Zoltan Szabó in Budapest and Thomas de Hoog in Delft, who both finished their theses in this area and became contributors to this book. Besides this group of people, many more colleague researchers and students were involved in developing the insights and results that are reported here. This is witnessed in the extensive bibliography that is added.

Discussion on publishing a book on orthogonal basis functions dates back to 1995, when the three editors met for a 2-day session in the Netherlands and produced a first table of contents. Later, the project was extended, the scope widened, and international collaboration became more pronounced. The first joint result of this collaboration was the organization of a pre-conference tutorial workshop at the 1997 IEEE Conference on Decision and Control, in San Diego, CA, which was followed by a similar workshop at the 1999 IFAC World Congress in Beijing, P.R. China, and the organization of a tutorial session at the 12th IFAC Symposium on System Identification, 2000, in Santa Barbara, CA. The written material that was produced for these events served as a starting point for the book that is now completed.

Nine international authors have contributed to this project, which culminated in a carefully edited and coordinated volume of 13 distinctive chapters, authored by different groups of authors. The various chapters can be read independently, but the material has been carefully selected and structured to make up a comprehensive text with a logical line of reasoning:

- Construction and analysis of generalized orthogonal basis function (GOBF) model structures (Chapters 1–3)
- System identification in a time domain setting, and related issues of variance, numerics, and uncertainty bounding (Chapters 4–7)
- System identification in the frequency domain (Chapters 8–9)
- Design issues and optimal basis selection (Chapters 10–11)
- Transformation and realization theory (Chapters 12–13)

The book is written as a research monograph with a survey focus. It is meant to be interesting for a broad audience, including researchers and graduate students in systems and control, as well as in signal processing. It also offers comprehensive material for specialized graduate courses in these areas. Control engineers may appreciate the abundance of technical problems that can be solved efficiently by the tools developed here.

We enjoyed working on this book. We hope that you as a reader enjoy reading it, and we welcome any feedback that you might have.

Delft, The Netherlands; Stockholm, Sweden
November 2004

Peter Heuberger
Paul Van den Hof
Bo Wahlberg

Contents

5 Variance Error, Reproducing Kernels, and Orthonormal Bases

6 Numerical Conditioning

List of Abbreviations and Symbols

arg min	Minimizing argument
ARX	Autoregressive exogenous
ARMAX	Autoregressive moving average exogenous
A	State matrix of state space model
B	Input-to-state matrix of state space model
C	State-to-output matrix of state space model
D	Direct feedthrough matrix of state space model
(A, B, C, D)	State space realization
$\beta(\omega)$	Phase function of inner function
BJ	Box-Jenkins
c_k	Sequence of coefficients in scalar-valued basis functions
$\mathbb{C}^{p \times m}$	Set of complex-valued matrices of dimension $p \times m$
$\text{Cov}\{\}$	Covariance
$\delta(t)$	Discrete-time pulse function
δ_{ij}	Kronecker delta function
$d_n(S; B)$	n-width of S in the Banach space B
$\mathbb{D}$	The unit disc: $\{z \mid \|z\| < 1\}$
e_i	Euclidean basis vector
$e(t), e_t$	White noise process
ε	Prediction error
ε_n	Squared error of the GOBF with n sections
$\mathbb{E}$	The exterior of the unit circle
$\mathsf{E}\{\}$	Expectation (expected value of a random variable)
$\bar{\mathsf{E}}\{\}$	Generalized expectation
FIR	Finite impulse response
$F_k(z)$	General form of scalar-valued basis function in $\mathcal{H}_2$
$f_k(t)$	General form of scalar-valued basis function in ℓ_2
$\varphi(t), \varphi_t$	Regression vector
Φ	Matrix of regression vectors $[\varphi(1) \cdots \varphi(N)]^T$
$\varphi_n(z, \mu)$	Reproducing kernel
$\Phi_x(e^{i\omega}), \Phi_x(\omega)$	Spectral density of x

$\Phi_{xy}(e^{i\omega}), \Phi_{xy}(\omega)$	Cross-spectral density of x and y
G_b	Inner function generating the basis
G	Input/output part of system model
$g(t), g_t$	Pulse response sequence of G
G_0	Input/output system
$g_0(t)$	Pulse response sequence of G_0
$\mathcal{G}$	Model set (deterministic part only)
Γ_k^f, Γ_k^b	Reflection coefficients
$\Gamma_n(z)$	Vector of basis functions $[F_1(z) \cdots F_n(z)]^T$
GOBF	Generalized orthogonal basis function
$\mathbf{H}$	Hankel matrix
$\mathcal{H}_2^{p\times m}, \mathcal{H}_2^{p\times m}(\mathbb{E})$	Set of real $p\times m$ matrix functions that are squared integrable on the unit circle and analytic on $\mathbb{E}$ (superscripts omitted if $p = m = 1$)
$\mathcal{H}_2(\lvert z\rvert > R)$	Hardy space of square integrable functions on $\lvert z\rvert = R$, analytic in the region $\lvert z\rvert > R$
$\mathcal{H}_2(\lvert z\rvert > R; j)$	jth orthogonal subspace of $\mathcal{H}_2(\lvert z\rvert > R)$ induced by G_b
$\mathcal{H}_2^\perp$	Orthogonal complement of $\mathcal{H}_2$ in $\mathcal{L}_2$
$\mathcal{H}_{2-}$	Hardy space of functions in $\mathcal{H}_2$, that are zero at infinity
$\mathcal{H}_\infty, \mathcal{H}_\infty(\mathbb{E})$	Set of functions that are bounded and analytic on $\mathbb{E}$
H	Noise-to-output transfer of system model
H_0	System noise filter
i	Complex number, $i = \sqrt{-1}$
I_n	$n \times n$ identity matrix
Im[ ]	Imaginary part
$\ell_2^{m\times n}[0,\infty)$	Space of matrix sequences $\{F_k \in \mathbb{C}^{m\times n}\}_{k=0,1,2,\cdots}$ such that $\sum_{k=0}^{\infty} tr(F_k^* F_k)$ is finite
$K_n(z,\mu)$	Reproducing kernel
K_k	Sequence of coefficients in block-grouped basis functions
K_k^f	Corresponding coefficients in a 'forward' linear prediction
K_k^b	Corresponding coefficients in a 'backwards' linear prediction
K_k^o	Coefficients in a orthonormal series expansion using the x_k^o signals
$\ell_2[0,\infty)$	Space of squared summable sequences on the time interval $\mathbb{Z}_+$
LTI	Linear time-invariant
$L_2^{p\times m}$	Set of real $p\times m$ matrix functions that are squared integrable on the unit circle (superscripts omitted if $p = m = 1$)
$L_k(z)$	Laguerre orthogonal (scalar) basis functions in $\mathcal{H}_2$
$l_k(t)$	Laguerre orthogonal (scalar) basis functions in ℓ_2
λ	Complex indeterminate resulting from λ-transform
M_k	Markov parameter
$\mathcal{M}_k$	Extended Markov parameter
$\mathcal{M}$	Model set (deterministic and stochastic part)

n	State-space dimension; model order; dimension of parameter vector
n_b	Number of poles in block-repeated basis functions
$N(\lambda)$	Multi-variable inner function derived from scalar basis generation inner function G_b
ω	Radial frequency
OBF	Orthogonal basis function
OE	Output error
p	Number of outputs
p_k	Pole
p_k^0	Poles of a nominal plant
$p_{j,k}^0$	Poles of the jth nominal plant
plim	Probability limit
P	Controllability Gramian; covariance matrix
Q	Observability Gramian
q^{-1}	Backward shift or delay operator
q	Forward shift operator
$\rho(\xi)$	Asymptotic rate of decay of the squared error
$\rho(\xi; j)$	Asymptotic rate of decay of the squared error for the jth nominal model
r_k, R_k	Correlation function (scalar and vector valued)
$\mathbb{R}^{p\times m}$	Set of real-valued matrices with dimension $p \times m$
$\mathcal{RH}_2^{p\times m}$	Subspace of $\mathcal{H}_2^{p\times m}$ of real rational functions
$\mathcal{RH}_{2-}^{p\times m}$	Subspace of $\mathcal{H}_{2-}^{p\times m}$ of real rational functions
Re[ ]	Real part
s	Complex indeterminate
$\mathcal{S}$	Data generating system
σ	Singular value
σ_e^2	Variance of stochastic process e
Σ_k^f, Σ_k^b	Covariance matrix of the forward and backward predictions
Sp	Span
θ	Parameter vector
Θ	Set of parameter vectors
$\mathbb{T}$	The unit circle, $\{z \mid \lvert z\rvert = 1\}$
$T_{k,l}^{[\theta]}$	Matrices used to compute partial derivatives of x_k w.r.t. θ
u	Input signal
Var{}	Variance
$V_k(z)$	Block-repeated orthogonal basis functions in $\mathcal{H}_2$
$V_N(\theta)$	Sum of squares loss-function
V_N	Vector of noise samples $[v(1) \cdots v(N)]^T$
$v_k(t)$	Block-repeated orthogonal basis functions in ℓ_2
$v(t)$	Output noise
$W_k(z)$	Block-repeated orthogonal (dual) basis functions in $\mathcal{H}_2$
$w_k(t)$	Block-repeated orthogonal (dual) basis functions in ℓ_2

x	State
x_k	The state of the kth section of the GOBF model
x_k^f	'Forward' signals in a lattice structure of the GOBF model
x_k^b	'Backward' signals in a lattice structure of the GOBF model
x_k^o	Orthonormal signals in a lattice structure of the GOBF model
ξ_k	Poles of basis functions
x	ℓ_2-signal
X	Complex function (Z-transform of x)
$\widetilde{X}$	System/operator transform (function) of X
$\mathcal{X}(k)$	(Column vector) coefficient sequence in (block-grouped) signal decomposition
$\breve{x}$	Coefficient in basis function expansion
$\breve{X}$	Hambo signal transform
X_o	Observability Gramian
X_c	Controllability Gramian
y	Output signal
Y_N	Vector of output samples $[y(1)\cdots y(N)]^T$
$\mathbb{Z}_+$	Set of non-negative integers
z	Complex indeterminate
z_k	Zero
$(\cdot)^T$	Transpose of a matrix
$\overline{(\cdot)}$	Complex conjugate for scalars; elementwise complex conjugate for matrices
$(\cdot)^*$	Complex conjugate for scalars; complex conjugate transpose for matrices; (Hilbert space) adjoint of an operator: $G^*(z) = G^T(z^{-1})$
$\lVert\cdot\rVert_2$	Induced 2-norm or spectral norm of a constant matrix, *i.e.* its maximum singular value; ℓ_2-norm; $\mathcal{H}_2$-norm
$\lVert\cdot\rVert_\infty$	$\mathcal{H}_\infty$-norm
$Vec(\cdot)$	Vector-operation on a matrix, stacking its columns on top of each other
$\otimes$	Kronecker matrix product
$\langle F, G\rangle$	Inner product of F and G in $L_2^{p\times m}(\mathbb{T})$: $\frac{1}{2\pi i}\int_0^{2\pi}\text{Trace}\{F^T(e^{i\omega})\overline{G(e^{i\omega})}\}d\omega$.
$\langle x, y\rangle$	Inner product of x and y in $\ell_2^n(J)$: $\sum_{k\in J} x^T(k)\overline{y(k)}$.
$[\![x, y]\!]$	ℓ_2 cross-Gramian or matrix 'outer product' $\sum_{k\in J} x(k)y^T(k)$, with $x\in\ell_2^{n\times p}(J)$, $y\in\ell_2^{m\times p}(J)$.
$[\![X, Y]\!]$	L_2 cross-Gramian or matrix 'outer product' $\frac{1}{2\pi i}\oint X(z)Y^*(1/z)\frac{dz}{z}$, with $X\in L_2^{n\times p}(\mathbb{T})$, $Y\in L_2^{m\times p}(\mathbb{T})$, where $Y^*(1/z) = \sum_k y(k)^* z^{-k}$.

List of Contributors

József Bokor
Computer and Automation Research Institute
Hungarian Academy of Sciences
Kende u. 13-17, 1111 Budapest XI, Hungary
bokor@sztaki.hu

Peter Heuberger
Delft Center for Systems and Control
Delft University of Technology
Mekelweg 2, 2628 CD Delft, The Netherlands
p.s.c.heuberger@dcsc.tudelft.nl

Håkan Hjalmarsson
S-3 Automatic Control, KTH
SE-100 44 Stockholm, Sweden
Hakan.Hjalmarsson@s3.kth.se

Thomas de Hoog
Philips Research Laboratories,
Prof. Holstlaan 4, 5656 AA Eindhoven, The Netherlands
Thomas.de.Hoog@philips.com

Brett Ninness
School of Electrical Engineering and Computer Science
University of Newcastle
University Drive, Callaghan, NSW 2308, Australia
brett@ee.newcastle.edu.au

Tomás Oliveira e Silva
Departamento de Electrónica e Telecomunicações
Universidade de Aveiro
3810-193, Aveiro, Portugal
tos@det.ua.pt

Zoltan Szabó
Computer and Automation Research Institute
Hungarian Academy of Sciences
Kende u. 13-17, 1111 Budapest XI, Hungary
szaboz@daedalus.scl.sztaki.hu

Paul Van den Hof
Delft Center for Systems and Control
Delft University of Technology
Mekelweg 2, 2628 CD Delft, The Netherlands
p.m.j.vandenhof@dcsc.tudelft.nl

Bo Wahlberg
S-3 Automatic Control, KTH
SE-100 44 Stockholm, Sweden
bo.wahlberg@s3.kth.se

Introduction by the Editors

A common way to represent the transfer function of linear time-invariant dynamical systems is to employ a series expansion representation of the form $G(z) = \sum_{k=1}^{\infty} g(k)z^{-k}$, where $\{g(k)\}$ refers to the impulse response of the system. Finite expansion representations are known as finite impulse response (FIR) models, and they are very useful dynamical models in many areas of engineering, ranging from signal processing, filter design, communication, to control design and system identification.

FIR models have many attractive properties, such as the fact that the coefficients $g(k)$ appear linearly in $G(z)$, leading to many computational and analytical advantages. However, a problem can be that for systems with high sampling rates and dominant high- and low-frequent dynamics, a very high number of terms in the expansion may be required to obtain an acceptable approximation of the true system dynamics.

As an alternative, generalized basis function models are considered of the form $G(z) = \sum_{k=1}^{\infty} c(k)F_k(z)$, for a predefined set of rational basis functions $\{F_k(z)\}$.

In this book, a full presentation is given of the rich theory and computational methods that are available for generalized rational basis function models, and in particular, in modelling of dynamical systems and in system identification. When estimating dynamical models of processes on the basis of measured input and output signals, appropriate model structures are required to parameterize the models. A central problem is to select model structures that are sufficiently rich to contain many models so to be able to find accurate approximations of our process to be modelled (small bias). On the other hand, the model structures should be sufficiently parsimonious to avoid a large variance. This bias/variance trade-off is at the very heart of the model structure selection problem and therefore central to the problem area considered in this book. A wide variety of other issues will also be treated, all related to the construction, use, and analysis of model structures in the format of rational basis function expansions.

In the first introductory chapter, Bo Wahlberg, Brett Ninness, and Paul Van den Hof provide the basic motivation for the use of basis function representations in dynamic modelling through some simple examples. Starting from finite impulse response (FIR) representations and Laguerre models, generalized forms are shown, incorporating a level of flexibility – in terms of a selection of poles – that can be beneficially used by the model builder. Additionally, the development of basis function structures is set in a historical perspective.

Bo Wahlberg and Tomás Oliveira e Silva continue this setting in Chapter 2, *Construction and Analysis*, where the general theory and tools for the construction of generalized orthonormal basis functions (GOBF) are explained. Parallel developments are shown either through Gram-Schmidt orthonormalization of first-order impulse responses or through concatenation of balanced state space representations of all-pass systems. The standard tapped-delay line, as a simple representation and implementation for linear time-invariant systems, is shown to be generalized to a concatenation of all-pass systems with balanced state readouts.

In Chapter 3, *Transformation Analysis*, authored by Bo Wahlberg, an alternative construction of the basis functions is given, resulting from a bilinear mapping of the complex indeterminate z in the system's Z-transform $G(z)$. Similar in spirit to the bilinear mapping of the complex plane that occurs when relating continuous-time and discrete-time systems, these mappings can be used to arrive at attractive representations in transform domains, as *e.g.* representations that have fast decaying series expansions. By choosing the appropriate mapping, pole locations of the transform system can be influenced.

The application of orthonormal basis function model structures in system identification is addressed in Chapter 4, *System Identification with Generalized Orthonormal Basis Functions*. Paul Van den Hof and Brett Ninness outline the main results in a tutorial chapter on time domain prediction error identification. The benefits of linear-in-the-parameters output-error structures are discussed, and bias and variance results of the basis function model structures are specified.

In Chapter 5, *Variance Error, Reproducing Kernels, and Orthonormal Bases*, Brett Ninness and Håkan Hjalmarsson give a detailed analysis of the variance error in system identification. Relations between GOBF model structures and fixed denominator model structures are discussed. Existing (asymptotic) expressions for variance errors are improved and extended to finite model order expressions, showing the instrumental role of orthogonal basis functions in the specification of variance errors for a wide range of model structures, including the standard OE, ARMAX, and BJ model structures.

The same authors focus on numerical conditioning issues in Chapter 6, *Numerical Conditioning*, where the advantages of orthogonality in model structures for identification are discussed from a numerical perspective. A relation is also made with improved convergence rates in adaptive estimation algorithms.

In Chapter 7, *Model Uncertainty Bounding*, Paul Van den Hof discusses the use and attractive properties of linear-in-the-parameters output-error model structures when quantifying model uncertainty bounds in identification. This problem has been strongly motivated by the recent developments in identification for control and shows that linear model structures as GOBF models are instrumental in the quantification of model uncertainty bounds.

The material on system identification is continued in Chapters 8 and 9, where József Bokor and Zoltan Szabó present two chapters on frequency domain identification. In Chapter 8, *Frequency-domain Identification in* $\mathcal{H}_2$, results from a prediction error approach are formulated. Asymptotic (approximate) models are shown to be obtained from the unknown underlying process by an interpolation over a particular (basis-dependent) frequency grid, which is a warped version of the equidistant Fourier grid. In Chapter 9, *Frequency-domain Identification in* $\mathcal{H}_\infty$, the prediction error framework is replaced by an identification framework in a worst-case setting, relying on interpolation techniques rather than stochastic estimation. Results for GOBF models are presented in this setting, and extensions to (non-rational) wavelets bases are considered.

In Chapter 10, *Design Issues*, Peter Heuberger collects practical and implementation issues that pertain to the use of GOBF models in identification. Aspects of finite data, initial conditions, and number of selected basis poles come into play, as well as iterative schemes for basis poles selection. Also, the handling of multi-variable systems is discussed, including the use of multi-variable basis functions.

Optimal basis selection is presented in Chapter 11, *Pole Selection in GOBF Models*, where Tomás Oliveira e Silva analyses the question concerning the best basis selection for predefined classes of systems. 'Best' is considered here in a worst-case setting. The classical result that an FIR basis (all poles in zero) is worst-case optimal for the class of stable systems is generalized to the situation of GOBF models.

In Chapter 12, *Transformation Theory*, authored by Peter Heuberger and Thomas de Hoog, the transform theory is presented that results when representing systems and signals in terms of GOBF models with repeating pole structures. The resulting Hambo transform of systems and signals is a generalization of the Laguerre transform, developed earlier, and can be interpreted as a transform theory based on transient signal/systems' responses. This in contrast to the Fourier transform being a transform based on a systems' steady-state responses. Attractive properties of representations in the transform domains are shown.

The transformation theory of Chapter 12 appears instrumental in solving the realization problem as presented by the same authors in Chapter 13, *Realiza-*

tion Theory. The topic here is the derivation of minimal state space representations of a linear system, on the basis of a (finite or infinite) sequence of expansion coefficients in a GOBF series expansion. It appears that the classical Ho-Kalman algorithm for solving such problems in FIR-based expansions can be generalized to include the GOBF expansions.

This book contains many results on the research efforts of the authors over the past 10 to 15 years. In this period the work on rational orthogonal basis functions has been very much inspiring to us, not in the least by the wide applicability of the developed theory and tools.
Finite impulse response models are extensively, and often routinely, used in signal processing, communication, control, identification, and many other areas, even if one often knows that infinite impulse response (IIR) models provide computational and performance advantages. The reason is often stability problems and implementation aspects with conventional IIR filters/representations. However, this is, as shown in this book, not a problem if one uses orthogonal basis function representations. Our mission has been to provide a framework and tools that make it very easy to evaluate how much one gains by using orthogonal basis function models instead of just FIR models.
We hope that with the material in this book, we have succeeded in convincingly showing the validity of the statement:

> *Almost all of what you can do using FIR models, you can do better for almost the same cost using generalized orthonormal basis function (GOBF) models!*

Delft and Stockholm
November 2004

Peter Heuberger
Paul Van den Hof
Bo Wahlberg

1

Introduction

Bo Wahlberg[1], Brett Ninness[2], and Paul Van den Hof[3]

[1] KTH, Stockholm, Sweden
[2] University of Newcastle, NSW, Australia
[3] Delft University of Technology, Delft, The Netherlands

1.1 Why Orthonormal Basis Functions?

By approximating the impulse response of a linear time-invariant system by a finite sum of exponentials the problem of modelling and identification is considerably simplified. Most often, finite impulse response (FIR) approximations are used. However, by instead using orthonormal infinite impulse response (IIR) filters as basis functions, much more efficient model structures can be obtained.
Over the last decade, a general theory has been developed, which extends the work on Laguerre filters by Wiener in the 1930s, for the construction and analysis of more general rational orthonormal basis functions for the class of stable linear systems. The corresponding filters are parameterized in terms of pre-specified poles, which makes it possible to incorporate *a priori* information about time constants in the model structure.
The main applications are in system identification and adaptive signal processing, where the parameterization of models in terms of finite expansion coefficients is attractive because of the linear-in-the-parameters model structure. This allows the use of linear regression estimation techniques to identify the system from observed data. Orthonormality is often associated with white noise input signals. However, the special shift structure of generalized orthonormal basis functions gives a certain Toeplitz structure for general quasi-stationary input signals, which can be used to construct efficient algorithms and to derive statistical performance results. The theory of reproducing kernels can be used to obtain reliable variance expressions for estimation of even more complex models.
The application potentials of orthogonal basis functions go beyond the areas of system identification and adaptive signal processing. Many problems in circuit theory, signal processing, telecommunication, systems and control theory, and optimization theory benefit from an efficient representation or parameterization of particular classes of signals/systems. A decomposition of signals/systems in terms of orthogonal (independent) components can play an

important role in corresponding optimization algorithms where the choice for particular orthogonal structures can be made dependent on prior knowledge of the object (signal/system) that has to be described.
In this book, main attention will be focused on the construction and use of orthogonal basis functions model structures in modelling and system identification.

1.2 Preliminaries; Notation

1.2.1 Transfer Function, Impulse Response, and Frequency Response

Consider a discrete time linear system

$$y(t) = G(q)u(t), \tag{1.1}$$

where $u(t)$ is the input signal and $y(t)$ the corresponding output signal. The system is represented by the transfer operator $G(q)$, where q is the forward shift operator, $qu(t) = u(t+1)$, and q^{-1} is the delay (backward shift) operator, $q^{-1}u(t) = u(t-1)$. For stable systems, we have the impulse response representation

$$y(t) = \sum_{k=0}^{\infty} g_k u(t-k), \quad G(q) = \sum_{k=0}^{\infty} g_k q^{-k}.$$

By $G(z)$, $z \in \mathbb{C}$ (the complex plane), we denote the corresponding transfer function

$$G(z) = \sum_{k=0}^{\infty} g_k z^{-k},$$

and by $G(e^{i\omega})$, $\omega \in [-\pi, \pi]$, the frequency response. We will often assume strictly proper transfer functions,

$$\lim_{|z|\to\infty} G(z) = 0,$$

i.e., at least one delay between the input and the output signal. It implies that $g_0 = 0$. This is common in digital control applications where zero-order hold circuits are often used. It is, however, easy to modify the results to be presented to include a direct term in the transfer function.

1.2.2 State-space Model

A discrete time state-space model is defined by

$$x(t+1) = Ax(t) + Bu(t)$$
$$y(t) = Cx(t) + Du(t).$$

Here $x(t) \in \mathbb{R}^n$ is the state vector, $u(t) \in \mathbb{R}^m$ is the input signal, and $y(t) \in \mathbb{R}^p$ is the corresponding output signal. We will most often only consider the single-input single-output (SISO) case $m = p = 1$. The corresponding transfer function equals

$$G(z) = D + C(zI - A)^{-1}B.$$

Signals will generally be denoted by small characters, *e.g.* $x(t)$, while Z-transforms will be denoted by capitals, *e.g.*

$$X(z) = \sum_{k=0}^{\infty} x(k)z^{-k}.$$

1.2.3 Function Spaces and Inner Products

Let $\mathbb{D}$ denote the unit disc: $\{z, \ |z| < 1\}$, $\mathbb{E}$ the exterior of the unit disc including infinity: $\{z, \ |z| > 1\}$, and $\mathbb{T}$ the unit circle: $\{z, \ |z| = 1\}$. By $\mathcal{H}_2(\mathbb{E})$ we mean the Hardy space of square integrable functions on $\mathbb{T}$, analytic in the region $\mathbb{E}$. We denote the corresponding inner product of $X(z), Y(z) \in \mathcal{H}_2(\mathbb{E})$ by

$$\langle X, Y \rangle := \frac{1}{2\pi} \int_{-\pi}^{\pi} X(e^{i\omega})Y^*(e^{i\omega})d\omega = \frac{1}{2\pi i} \oint_{\mathbb{T}} X(z)Y^*(1/z^*)\frac{dz}{z},$$

where * denote complex conjugation, *i.e.* $Y(z) = \sum_k y_k z^{-k} \Rightarrow Y^*(1/z^*) = \sum_k y_k^* z^k$, which for real valued $\{y_k\}$ equals $Y(1/z)$. [1]
Sometimes we will consider more general functions spaces then $\mathcal{H}_2(\mathbb{E})$, *e.g.* the Hilbert space of complex matrix functions of dimension $p \times m$ that are square integrable on the unit circle denoted by $L_2^{p\times m}(\mathbb{T})$.

1.2.4 Cross-Gramian

The transpose of a matrix A is denoted by A^T and the complex conjugate transpose (Hermitian transposition) by A^*, *e.g.* if

$$A = \begin{pmatrix} a_{11} & a_{12} \\ a_{21} & a_{22} \end{pmatrix} \quad \Rightarrow \quad A^* = \begin{pmatrix} a_{11}^* & a_{21}^* \\ a_{12}^* & a_{22}^* \end{pmatrix}.$$

[1] A more strict definition of the inner product in $\mathcal{H}_2(\mathbb{E})$ is based on a limit when the radius $r \to 1^+$. However, because the radial limits of functions in $\mathcal{H}_2(\mathbb{E})$ exist almost everywhere in $\mathbb{T}$, it is usual to evaluate the inner product using these boundary values. The contour integral form of the inner product should only be used if both $X(z)$ and $Y(z)$ do not have singularities on $\mathbb{T}$.

The notation $\bar{A}$ is sometimes used to denote pure complex conjugation (without transpose), *i.e.* $\bar{A} = [A^*]^T$.
For matrix-valued functions $X(z) \in \mathcal{H}_2^{n \times p}(\mathbb{E})$, $Y(z) \in \mathcal{H}_2^{m \times p}(\mathbb{E})$, we will use the cross-Gramian

$$[\![X, Y]\!] := \frac{1}{2\pi i} \oint_{\mathbb{T}} X(z) Y^*(1/z^*) \frac{dz}{z}.$$

Notice that if $|z| = 1$ then $1/z^* = z$. For transfer functions with real valued impulse responses $Y^*(1/z^*) = Y^T(1/z)$. The cross-Gramian will sometimes be called the matrix 'outer product'.
For vector-valued functions $X(z), Y(z) \in \mathcal{H}_2^{n \times 1}(\mathbb{E})$ the vector inner product equals $\langle X, Y \rangle = [\![X^T, Y^T]\!]$.

1.2.5 Norm and Orthonormality

The norm of $X(z) \in \mathcal{H}_2(\mathbb{E})$ equals

$$||X|| := \sqrt{\langle X, X \rangle}.$$

Two transfer functions $X_1(z)$ and $X_2(z)$ are called *orthonormal* if the following conditions hold

$$\langle X_1, X_2 \rangle = 0, \quad ||X_1|| = ||X_2|| = 1.$$

1.2.6 Impulse Response

The functions z^{-k}, $k = 1 \ldots n$ belong to $\mathcal{H}_2(\mathbb{E})$, and are orthonormal

$$\langle z^{-k}, z^{-l} \rangle = \frac{1}{2\pi i} \oint_{\mathbb{T}} z^{-k} z^{l} \frac{dz}{z} = \begin{cases} 1, & l = k \\ 0, & l \neq k. \end{cases}$$

The Laurent series expansion (in z^{-1}) of $G(z) \in \mathcal{H}_2(\mathbb{E})$ equals

$$G(z) = \sum_{k=1}^{\infty} g_k z^{-k}, \tag{1.2}$$

where the impulse response coefficients are given by

$$g_k = \langle G, z^{-k} \rangle = \frac{1}{2\pi i} \oint_{\mathbb{T}} G(z) z^k \frac{dz}{z}.$$

1.2.7 Convergence and Completeness

Convergence in mean of the infinite sum (1.2) holds if

$$||G|| = \sqrt{\sum_{k=1}^{\infty} |g_k|^2} < \infty.$$

This implies that the functions $\{z^{-k},\ k = 1,2\ldots\}$ are complete in $G(z) \in \mathcal{H}_2(\mathbb{E}),\ G(\infty) = 0$. However, we will often consider the more restrictive class of input-output stable systems for which

$$\sum_{k=1}^{\infty} |g_k| < \infty.$$

This implies that the frequency response $G(e^{i\omega})$ is a continuous function and that

$$|G(e^{i\omega}) - \sum_{k=1}^{n} g_k e^{-i\omega k}| \leq \sum_{k=n+1}^{\infty} |g_k| \to 0, \quad \text{as } n \to \infty. \tag{1.3}$$

The convergence is then uniform in ω.

1.3 Motivating Example from System Identification

Let us first study the following simple example. Consider a stable first-order discrete time system with transfer function,

$$G(z) = \frac{b}{z-a}.$$

The objective is to estimate $G(z)$ from sampled observations of the input signal $u(t)$ and output signal $y(t)$ of the system.

1.3.1 ARX Model Estimation

The simplest and most common used system identification approach is to write the input-output relation (1.1) in an equation error form (ARX model),

$$y(t) = ay(t-1) + bu(t-1).$$

Notice that the output is linear in the unknown parameters a and b. This means that any linear regression estimation procedure, *e.g.* the standard least square method, can be used to estimate the unknown parameters from measured data. The story would end here if the following problems with this approach did not exist:

- The estimated parameters will be biased in case of measurement noise, which cannot be modeled as an auto-regressive (AR) process with the same poles as the system transfer function.
- The performance can be improved by using higher order auto-regressive exogenous (ARX) models, but then one may run into numerical problems in case of lack of input signal and/or noise excitation, resulting in possible unstable pole/zero cancellations.

- In case of under-modelling, the bias error is determined by the cost function

$$\frac{1}{2\pi}\int_{-\pi}^{\pi}|G_0(e^{i\omega})-\frac{B(e^{i\omega})}{A(e^{i\omega})}|^2\Phi_u(e^{i\omega})|A(e^{i\omega})|^2d\omega + \frac{1}{2\pi}\int_{-\pi}^{\pi}\Phi_v(e^{i\omega})|A(e^{i\omega})|^2d\omega, \quad (1.4)$$

  where $G_0(z)$ is the true transfer function, $B(z)/A(z)$ is the ARX transfer function estimate, $\Phi_u(e^{i\omega})$ is the power spectral density of the input signal, and $\Phi_v(e^{i\omega})$ the power spectral density of additive noise.
  The factor $|A(e^{i\omega})|^2$ will act as a frequency weighting function, which in fact will emphasizes the fit at higher frequencies.

There exist many more or less ad hoc modifications of system identification using an ARX structure, which will mitigate these effects.

1.3.2 OE Model Estimation

One common approach is to use an output error (OE) model structure. For the first-order example, we then use the input-output model

$$y(t) = bu_f(t), \quad u_f(t) = au_f(t-1) + u(t-1).$$

In case the pole a is known, $u_f(t)$ can be directly calculated by just filtering the input signal $u(t)$ by $1/(q-a)$. The output is then linear in the unknown parameter b, which thus can be estimated using linear regression techniques. We will call this a fixed denominator model.
However, for the general case the OE model output is a non-linear function of the pole parameter a, because $u_f(t)$ then is a non-linear function of a. Hence, numerical optimization techniques have to be used to determine the optimal OE estimate of a.
The good aspects of OE modelling are

- There is no bias due to noise, as long as the input signal is uncorrelated with the noise.
- The OE structure will allow robust estimation of high-order models, as will be shown below.
- The asymptotic frequency domain bias error is governed by the cost function

$$\frac{1}{2\pi}\int_{-\pi}^{\pi}|G_0(e^{i\omega})-\frac{B(e^{i\omega})}{A(e^{i\omega})}|^2\Phi_u(e^{i\omega})d\omega,$$

  where $B(z)/A(z)$ now equals the OE transfer function estimate. Notice that the $|A(e^{i\omega})|^2$ weighting factor in (1.4) has been removed.

However, the main limitation with this approach is that it involves iterative calculations to find the optimal estimate. Because the cost-function may have local minima, except for the special case with no under-modelling and rich input signal, one has to be careful in finding good initial model estimates and using validation procedures in order not to get stuck in a local minima.

1.3.3 FIR Model Estimation

The standard approach to obtain an OE structure, which is linear-in-the-parameters, is to parameterize the transfer function in terms of the impulse response $\{g_k\}$ of the system,

$$G(z) = \sum_{k=1}^{\infty} g_k z^{-k}. \tag{1.5}$$

The impulse response for the first-order example equals

$$g_k = b\,a^{(k-1)}, \quad k \geq 1.$$

To obtain a linear-in-the-parameters model, the sum (1.5) can be truncated after $k = n$, to give the so-called finite impulse response (FIR) approximative model

$$y(t) = \sum_{k=1}^{n} g_k u(t-k).$$

The idea is now to directly estimate the impulse response parameters g_k, $k = 1 \ldots n$.
As in (1.3), the frequency domain approximation error can be bounded as

$$|G(e^{i\omega}) - \sum_{k=1}^{n} g_k e^{-i\omega k}| \leq \sum_{k=n+1}^{\infty} |g_k| \tag{1.6}$$

and is thus proportional to the rate of which the impulse response tends to zero as k tends to infinity. For the first-order example, the error is proportional to $|a|^n$. This means that for $|a|$ close to zero, good low-order FIR approximations can be obtained, while if $|a|$ is close to one, the sum (1.6) converges slowly to zero and one has to use a large-order n for an acceptable FIR approximation.

1.3.4 Summary

The conclusion of the above discussion is the obvious observation that a high-order FIR approximation must be used to approximate systems with long (relative to the sampling interval) impulse responses. Because the variance of the transfer function estimate is proportional to the number of estimated parameters, this implies large variance errors for short data records. Our objective is to derive more compact approximative models!

We refer to [164] or [278] for thorough discussions of system identification in general, and the above observations in particular. See also Chapter 4.

1.4 From Pole Selection to Basis Functions

Let us now study how to obtain infinite impulse response (IIR) output error (OE) model structures, which are linear in the unknown parameters.

1.4.1 Fixed Denominator Model

Assume that the poles, ξ_k, $k = 1 \ldots n$, of the system to be modeled have been specified. The corresponding transfer function is then rational and of order n, *i.e.*

$$G(z) = \frac{B(z)}{A(z)},$$

where

$$B(z) = b_1 z^{n-1} + \ldots + b_n,$$

$$A(z) = z^n + a_1 z^{n-1} + \ldots + a_n = \prod_{k=1}^{n} (z - \xi_k).$$

The input-output relation can then be written

$$y(t) = B(q) q^{-n} u_f(t), \quad A(q) q^{-n} u_f(t) = u(t).$$

Because $u_f(t)$ now can be pre-calculated, the model output signal depends linearly on the coefficients, $b_1 \ldots b_n$, of $B(q)$, which thus can be estimated using standard linear regressions estimation techniques. This is often called a fixed denominator model.

1.4.2 Rational Basis Functions

From a basis functions point of view, define

$$\bar{F}_k(z) = \frac{z^{k-1}}{A(z)}, \quad k = 1 \ldots n.$$

Then the fixed denominator transfer function model can be written

$$G(z) = \sum_{k=1}^{n} b_k \bar{F}_k(z).$$

Notice that the basis transfer functions, $\bar{F}_k(z)$, have infinite impulse responses, and hence we have no problems describing systems with long impulse responses (memory).

However, in most applications it is unrealistic to assume perfect knowledge of the time constants or the poles of system. Hence, we would like to have more flexible models, which take the uncertainty of the *a priori* information about pole locations into account.

It should also be remarked that the fixed denominator basis functions are not orthogonal (unless $A(z) = z^n$, *i.e.* the FIR case). This implies that the corresponding estimation problem can be numerically ill-conditioned. Notice the transformed input signal $u_f(t)$ is prefiltered by a low-pass filter, $1/(z^{-n} A(z))$, and hence less exciting than $u(t)$.

1.4.3 Pole Locations

The key question when using fixed denominator (poles) models is of course how to choose the poles. From an engineering point of view it seems reasonable to choose them as close as possible to the true poles. However, in order to have approximations that converge to the true system as the number of used basis functions tends to infinity, one has to be more precise. The necessary and sufficient condition for completeness of rational basis functions is that

$$\sum_{k=1}^{\infty}(1-|\xi_k|)=\infty.$$

The interpretation is that the sequence of poles cannot converge (if at all) too fast to the unit circle. See Section 2.2 for more details.
Examples of useful pole locations are:

FIR:	$\xi_k = 0,$	$k = 1, 2, \dots$
Laguerre:	$\xi_k = a,$	$k = 1, 2, \dots$
Two-parameter Kautz:	$\xi_{2k-1} = \xi_1,\ \xi_{2k} = \xi_2,$	$k = 1, 2, \dots$
GOBF:	$\xi_{j+(k-1)nb} = \xi_j,$	$j = 1, \dots n_b,\ k = 1, 2, \dots$
Dyadically spaced lattice:	$\xi_{k,j} = (1-2^{-k})e^{2\pi i j/2^k},$	$j = 0, \dots, 2^k - 1,$ $k = 1, 2, \dots$

The poles determine the subspace wherein we would like to find the optimal transfer function estimate. Basis functions are just tools to describe the subspace, and will not change the basic structure of the subspace.

1.4.4 Continuous Time

Almost all results in this book will be presented for discrete time systems. There is, however, a completely analogous theory for continuous time systems, where we work with Laplace transforms instead of Z-transforms. The exterior of the unit circle $\mathbb{E}$ is replaced by the right half of the complex plane $\mathbb{C}^+$. The corresponding scalar inner product is defined as

$$\langle \mathsf{X}, \mathsf{Y} \rangle := \frac{1}{2\pi i}\int_{i\mathbb{R}} \mathsf{X}(s)\mathsf{Y}^*(-s^*)ds,$$

where integration is over the imaginary axis.
It is possible construct an explicit isomorphism between $\mathcal{H}_2(\mathbb{E})$ and $\mathcal{H}_2(\mathbb{C}^+)$ using the bilinear transformation

$$z \mapsto s = a\frac{z-1}{z+1}, \quad a > 0.$$

Assume that $F_k(z) \in \mathcal{H}_2(\mathbb{E})$, and that

$$F_k(z) \mapsto \mathsf{F}_k(s) = \frac{2a}{s+a} F_k\left(\frac{a+s}{a-s}\right) \qquad k = 1, 2.$$

Then $\mathsf{F}_k(s) \in \mathcal{H}_2(\mathbb{C}^+)$, and

$$\langle F_k, F_l \rangle = \langle \mathsf{F}_k, \mathsf{F}_l \rangle, \quad k, l = 1, 2.$$

This means that the bilinear transformation preserves orthonormality. See [234] for further details.

Example 1.1. The orthonormal 'shift' functions $F_k(z) = z^{-k}$, $k = 0 \dots n$ in $\mathcal{H}_2(\mathbb{E})$, map to the so-called continuous time Laguerre functions

$$\mathsf{F}_k(s) = \frac{2a}{s+a}\left(\frac{a-s}{a+s}\right)^k, \quad a > 0,\ k = 0, \dots n,$$

which via the isomorphism are orthonormal in $\mathcal{H}_2(\mathbb{C}^+)$.

1.5 Historical Overview

Having provided some discussion as to how the ideas of rational orthonormal bases may be applied to certain current system identification problems, we close with a brief survey of the pedigree of these ideas by tracing their genealogy.

1800s Although rational functions are considered in these notes, the consideration of polynomials seems to be involved in the genesis of the ideas, wherein during the late 19th century it was noted that differential equations defined operators and that these operators often have an integral representation that is 'self-adjoint' and hence defines an orthonormal basis that the solution is most simply expressed with respect to. Nowadays such differential equations are termed 'Sturm-Liouville' systems, well-known cases being Schrödinger's equation and the so-called Wave equation [43]. In these instances, the orthonormal bases are defined in terms of Laguerre and Legendre orthonormal functions that themselves are defined in terms of Laguerre and Legendre orthogonal polynomials.

These are polynomials that are orthogonal on the real line with respect to certain weights, and they have been studied since the time of Gauss in relation to efficient methods of numerically evaluating integrals via so-called quadrature formulae. Nowadays they are still being evaluated for this purpose and others related to studying problems of approximation theory.

The relationship between these Laguerre and Legendre polynomials and the Laguerre and Legendre bases mentioned in modern day signal processing and control literature is via Fourier and Bilinear transformation of the so-called Laguerre and Legendre functions, which are the corresponding polynomials multiplied by exponential functions. See [215] for details on these points, together with discussion of the relationship of other classical 19th and early 20th century orthogonal polynomials (Chebychev polynomials for example) to rational orthonormal bases that are relevant to modern day signal and control theory problems.

1925–1928 The first mention of rational orthonormal bases seems to have occurred in this period with the independent work of Takenaka [296] and Malmquist [180]. The particular bases considered were those that will be denoted as 'generalized orthonormal basis functions (GOBF)' in Chapter 2. The context of this early work was application to approximation via interpolation, with the ensuing implications for generalized quadrature formula's considered.

1930s–1940s Nowadays, the work of Malmquist is more remembered than that of Takenaka because Walsh credits Malmquist with first presenting the formulation of GOBFs that formed the genesis of Walsh's wide ranging work that further studied the application of these bases for approximation on the unit disk (discrete time analysis in modern day engineering parlance) and on the half plane (continuous time). Walsh wrote dozens of papers on this topic, but the major work [320] collects most of his results over the years.

At the same time, Wiener began examining the particular case of continuous time Laguerre networks for the purpose of building the optimal predictors that he began thinking about at MIT, and later developed during the war for anti-aircraft applications. Early work due to his student Lee appeared as [157], which was later presented more fully in [158]. Wiener's own work was finally declassified in 1949 and appeared as [329], although there are mentions of his thoughts on the utility of orthonormal parameterizations in [328].

Also of fundamental importance in this period was that Szegö provided a unified discussion and analysis of polynomials orthogonal on various domains, including the unit circle. At the same time in Russia, Geronimus was also working on these ideas, but they were not to appear until 1961 [99].

1950s This was a very active period. For example, the classic work [108] (first edition) appeared in which the earlier orthogonal polynomial ideas of Szegö were applied via the consideration of the associated Toeplitz matrices, to various problems in function approximation, probability theory and stochastic processes. In particular, the famous Levinson recursions for the solution of Wiener's optimal predictors were cast in terms of orthonormal bases and the associated 'three term recurrence' formulas that may be used to calculate the bases.

Also in this period, Kautz [143, 144] revisited the formulation of GOBFs and their continuous time version for the purposes of network synthesis, as did the work [122]. From a mathematical point of view, a standard reference on orthogonal polynomials was produced [261], and the genesis of examining orthonormally parameterized approximation problems (which is system identification) via the ideas of 'reproducing kernels' was formulated [19].

1960s In this period, work on orthonormal parameterizations for systems and control applications includes that of [37,186,258], which again revisited the GOBF bases (although using a different formulation, that of [54]), which examined issues of completeness, that of [56] that was specific to Laguerre bases, and the famous book [137] where Chapter 12 provides discussion of orthonormal parameterizations for control theory applications. The classic engineering monograph [75] also provided an overview of the use of certain orthonormal parameterizations for system analysis purposes.
From a mathematical perspective, the book [3] appeared that explored for the first time in monograph form the application of Szegö's orthonormal polynomial bases to various approximation problems. As well, the seminal works [51, 62, 86] appeared, which contain superb expositions of the role of orthogonality and reproducing kernels in an approximation theory context.

1970s This period saw the genesis of the application of orthonormal parameterizations for the purpose of VLSI implementation of digital filter structures. Fettweiss's so-called wave digital filter formulation [88–90] seems to have begun the area, with the monograph [257] and paper [200] being the most readable account of the ideas. Allied to these ideas was the work [78] examining the use of the GOBFs for the purposes of optimal prediction of stationary processes and that of [141] providing further examination of orthogonal polynomial bases. Laguerre bases were examined for the purposes of system identification in [148, 149].

1980s–1990s This period has seen, at least in the engineering literature, an explosion of interest in the use of orthonormal parameterizations. In a signal processing setting see [60,72,210,249,331], in a control theory oriented system identification setting see [16, 25, 27, 29, 55, 57, 58, 132, 214, 225, 306, 313,315], in a model approximation setting see [175, 176, 245, 319], for applications to adaptive control problems see [82, 338, 339], for applications to VLSI implementation of digital filter structures see [74, 304], and for applications to modelling and predicting stationary stochastic processes see [76–78, 318]

2000s The first draft of this book was presented at the *36th IEEE CDC Preconference Workshop no. 7, San Diego, CA, 1997,* and the second version at the *14th IFAC World Congress, Workshop no. 6, Beijing, PRC, 1999.* The summary above presents more or less the development up to mid-1999. In order to summarize the main progress after that date, a literature search on relevant work has been performed.

- There has been quite a lot of work in the approximation properties of orthonormal bases functions, see *e.g.* [5–15, 41, 93, 156, 174, 223, 288,

333]. New methods for finding optimal approximations can be found in [26, 45, 47, 260, 263, 264, 297–299].

- Realization theory is another area where there has been major progress since 1999. Examples of relevant work are [1, 65–68, 127, 128, 130, 207, 238, 268, 291, 310, 312, 316, 317, 341, 342].
- Fast algorithms for adaptive signal processing have recently been derived in [187–191].
- System identification using orthonormal basis function models are still an active area of research. Recent results can be found in [4,30,53,94, 104,202,224,290]. In particular, variance properties of identified models have been analysed in [208, 218, 219, 332].
- Application of orthonormal basis function in control and signal processing is still a a very active area. In particular, orthonormal basis function models have been very useful in predictive control. Interesting applications include [21, 46, 59, 85, 138, 142, 150, 151, 172, 183, 205, 206, 227, 228, 239, 243, 321, 323, 334, 336].

2

Construction and Analysis

Bo Wahlberg[1] and Tomás Oliveira e Silva[2]

[1] KTH, Stockholm, Sweden
[2] University of Aveiro, Aveiro, Portugal

2.1 Objectives

In this chapter, we introduce some basics methods to construct and analyze rational orthonormal functions. Two different techniques are described in detail:

- The Gram-Schmidt procedure applied to transfer functions.
- A matrix factorization technique based on state-space models.

2.2 Rational Orthogonal Functions

The theory of orthogonal functions is indeed classical and one of the most developed fields of mathematics. See *e.g.* [40,61,96,139,201,204,261,294] for introductions to and overviews of this area.

2.2.1 Notation

We will use the notation introduced in Chapter 1 and will start by repeating the most important definitions: The Hardy space of square integrable functions on the unit circle, $\mathbb{T}$, and analytic outside the unit circle including infinity, $\mathbb{E}$, is denoted by $\mathcal{H}_2(\mathbb{E})$. The corresponding inner product of $X(z), Y(z) \in \mathcal{H}_2(\mathbb{E})$ is defined by

$$\langle X, Y \rangle := \frac{1}{2\pi i} \oint_{\mathbb{T}} X(z) Y^*(1/z^*) \frac{dz}{z},$$

where * denotes complex conjugate. Notice that if $|z| = 1$ then $1/z^* = z$. The squared norm of $X(z)$ equals $||X||^2 = \langle X, X \rangle$. Two transfer functions $F_1(z)$ and $F_2(z)$ are orthonormal if $||F_1|| = ||F_2|| = 1$ and $\langle F_1, F_2 \rangle = 0$. For matrix valued functions we will use the cross-Gramian or matrix 'outer product'

$$[\![X, Y]\!] := \frac{1}{2\pi i} \oint_{\mathbb{T}} X(z) Y^*(1/z^*) \frac{dz}{z},$$

where * denotes complex conjugate transpose for matrices. For vector valued functions $X(z), Y(z) \in \mathcal{H}_2^{n\times 1}(\mathbb{E})$, the vector space inner product equals $\langle X, Y\rangle = [\![X^T, Y^T]\!]$. An import class of functions are functions with real valued impulse responses, *i.e.* $Y(z) = \sum_k y_k z^{-k}$, where $\{y_k\}$ are real numbers. Then $Y^*(1/z^*) = Y^T(/z)$, and the cross-Gramian is a real valued matrix satisfying $[[\![X, Y]\!]]^T = [\![Y, X]\!]$.

2.2.2 The Gram-Schmidt Procedure

Consider two linearly independent functions $\bar{F}_1(z), \bar{F}_2(z) \in \mathcal{H}_2(\mathbb{E})$[1], which form a basis for their span. An orthonormal basis for this subspace can easily be obtained using the Gram-Schmidt procedure,

$$\begin{aligned} K_1(z) &= \bar{F}_1(z), \\ F_1(z) &= \frac{K_1(z)}{||K_1||}, \\ K_2(z) &= \bar{F}_2(z) - \langle \bar{F}_2, F_1\rangle F_1(z), \\ F_2(z) &= \frac{K_2(z)}{||K_2||}. \end{aligned}$$

Then $||F_1|| = ||F_2|| = 1$ and

$$\langle F_2, F_1\rangle = \frac{1}{||K_2||}\left(\langle \bar{F}_2, F_1\rangle - \langle \bar{F}_2, F_1\rangle ||F_1||^2\right) = 0,$$

because $||F_1|| = 1$. Hence, $F_1(z)$ and $F_2(z)$ are orthonormal, and they span the same space as $\bar{F}_1(z)$ and $\bar{F}_2(z)$. Notice that $K_2(z)$ is the projection of $\bar{F}_2(z)$ onto the orthogonal complement to the space spanned by $F_1(z)$.
In general, $F_i(z)$ can recursively be determined by normalizing the projection of $\bar{F}_i(z)$ onto the orthogonal complement of $\mathrm{Sp}\{F_1(z)\ldots F_{i-1}(z)\}$. Hence, it is in principle easy to construct orthonormal functions from a given set of functions. Also notice that the orthonormal functions are by no means unique. For example, changing the ordering of the functions will affect the corresponding basis, but the span of the basis functions will remain the same. The following example illustrates this idea for rational functions with real poles.

Example 2.1. Consider the functions $\bar{F}_1(z) = 1/(z-a)$ and $\bar{F}_2(z) = 1/(z-b)$, $-1 < a, b < 1$, $a \neq b$. These functions are linearly independent but are not orthonormal in the space $\mathcal{H}_2(\mathbb{E})$. Using the Gram-Schmidt procedure, we find

[1] Overbar is used here to mark functions that are not orthonormal, and should not be confused by complex conjugate (denoted by * in this chapter).

$$F_1(z) = \frac{1}{z-a}/(1/\sqrt{1-a^2}) = \frac{\sqrt{1-a^2}}{z-a},$$
$$K_2(z) = \frac{1}{z-b} - \frac{\sqrt{1-a^2}}{1-ab}\frac{\sqrt{1-a^2}}{z-a} = \frac{b-a}{1-ab}\frac{1}{z-b}\frac{1-az}{z-a},$$
$$F_2(z) = \frac{\sqrt{1-b^2}}{z-b}\frac{1-az}{z-a}.$$

The Cauchy integral formula can be used to calculate the scalar products, *e.g.*

$$\langle \bar{F}_2, F_1 \rangle = \frac{1}{2\pi i}\oint_{\mathbb{T}} \frac{F_1(1/z)}{z-b}\frac{dz}{z} = F_1(1/b)/b,$$

because the only pole of the integrand inside the unit circle is $z = b$. Note that $z = 0$ is not a pole because $F_1(\infty) = 0$. The functions $F_1(z)$ and $F_2(z)$ are now by construction orthonormal.
The function $F_2(z)$ has a special structure. It can be factored as

$$F_2(z) = \frac{\sqrt{1-b^2}}{z-b}G_b(z),$$

with

$$G_b(z) = \frac{1-az}{z-a}.$$

The transfer function $G_b(z)$ is called a first-order all-pass transfer function, and satisfies $G_b(z)G_b(1/z) = 1$.
These results can be extended by including a third function:

$$\bar{F}_3(z) = \frac{1}{z-c}, \quad -1 < c < 1, \ c \neq a, b.$$

The Gram-Schmidt procedure then gives

$$K_3(z) = \bar{F}_3(z) - \langle \bar{F}_3, F_2 \rangle F_2(z) - \langle \bar{F}_3, F_1 \rangle F_1(z),$$
$$F_3(z) = \frac{K_3(z)}{||K_3||},$$

where $\langle \bar{F}_3, F_1 \rangle = F_1(1/c)/c$ and $\langle \bar{F}_3, F_2 \rangle = F_2(1/c)/c$. Further calculations give

$$F_3(z) = \frac{\sqrt{1-c^2}}{z-c}\frac{1-az}{z-a}\frac{1-bz}{z-b},$$

which, by the Gram-Schmidt construction, is orthogonal to $F_1(z)$ and $F_2(z)$. Projecting $\bar{F}_3$ onto the space orthogonal to the first two basis functions corresponds to multiplication by an all-pass function $G_b(z)$ with the same poles as the previous two basis functions,

$$G_b(z) = \frac{(1-az)(1-bz)}{(z-a)(z-b)}, \quad G_b(z)G_b(1/z) = 1.$$

The normalization is the same as for the first basis function, since $||G_b|| = 1$.

This example can be extended to a set of n first-order transfer functions with distinct real poles.

2.2.3 The Takenaka-Malmquist Functions

In the previous section, we showed how to construct orthonormal basis functions using the Gram-Schmidt procedure. For the general case with possibly multiple complex poles, this will result in the so-called Takenaka-Malmquist functions

$$F_k(z) = \frac{\sqrt{1-|\xi_k|^2}}{z-\xi_k} \prod_{i=1}^{k-1} \left[\frac{1-\xi_i^* z}{z-\xi_i} \right], \quad k = 1,2,\ldots, \qquad \xi_i \in \mathbb{C}, \ |\xi_i| < 1. \quad (2.1)$$

This structure can be anticipated from the Cauchy integral formula,

$$\langle F_k, F_1 \rangle = F_k(1/\xi_1^*)\sqrt{1-|\xi_1|^2} = 0, \quad \forall k > 1,$$

which forces $F_k(z)$ to have a zero at $1/\xi_1^*$. Using this argument for $\langle F_k, F_j \rangle$, $j = 2, \ldots k-1$, leads to the zero structure $F_k(1/\xi_j^*) = 0$, $j < k$.
The projection property of all-pass function

$$G_b(z) = \prod_{i=1}^{k-1} \left[\frac{1-\xi_i^* z}{z-\xi_i} \right],$$

i.e.

$$\langle F_j, FG_b \rangle = 0, \ j \leq k-1, \quad \forall F(z) \in \mathcal{H}_2(\mathbb{E}),$$

is known in a somewhat more abstract setting as the Beurling-Lax theorem, [254,319].
The Takenaka-Malmquist functions will in general have complex impulse responses. The all-pass function $G_b(z)$ will then satisfy the condition

$$G_b(z) G_b^*(1/z^*) = 1.$$

2.2.4 Error Bounds

Let the poles of of the basis functions $\{F_k(z)\}_{k=1,\ldots n}$ be $\{\xi_i \in \mathbb{C}, |\xi_i| < 1, i = 1 \ldots n\}$, and define the corresponding all-pass transfer function

$$G_b(z) = \prod_{i=1}^{n} \left[\frac{1-\xi_i^* z}{z-\xi_i} \right].$$

Decompose $G_0(z) \in H_2(\mathbb{T})$ as

$$G_0(z) = G_n(z) + E_n(z) G_b(z),$$

where $G_n(z) \in \mathrm{Sp}\{F_k(z)\}_{k=1,\ldots n}$, and $E_n(z) \in H_2(\mathbb{T})$. Then $E_n(z)G_b(z)$ is orthogonal to $G_n(z)$ because of the projection property of $G_b(z)$ and hence $G_n(z)$ is the optimal approximation of $G_0(z)$ in the space spanned by the given basis functions. Now make a (partial fraction expansion) decomposition of

$$G_0(z)[G_b(z)]^{-1} = D_n(z) + E_n(z),$$

where the poles of $D_n(z)$ equals the zeros of $G_b(z)$, which are outside the unit circle, and $E_n(z)$ has the same poles as $G_0(z)$, which are inside the unit circle. Hence, we have decomposed $G_0(z)[G_b(z)]^{-1}$ into its stable part and its completely unstable part. From the Cauchy integral formula we then have

$$E_n(z) = \frac{1}{2\pi i}\oint_{\mathbb{T}} \frac{G_0(\zeta)}{\zeta - z} G_b^*(1/\zeta^*)d\zeta, \quad |z| \geq 1.$$

Here we have used $[G_b(z)]^{-1} = G_b^*(1/z^*)$. Hence

$$G_0(z) - G_n(z) = \left[\frac{1}{2\pi i}\oint_{\mathbb{T}} \frac{G_0(\zeta)}{\zeta - z} G_b^*(1/\zeta^*)d\zeta\right] G_b(z).$$

If

$$G_0(z) = \sum_{j=1}^{n_0} \frac{b_j}{z - p_j} \quad \Rightarrow \quad E_n(z) = \sum_{j=1}^{n_0} \frac{b_j\, G_b^*(1/p_j^*)}{z - p_j}$$

and thus

$$G_0(z) - G_n(z) = \left[\sum_{j=1}^{n_0} \frac{b_j\, G_b^*(1/p_j^*)}{z - p_j}\right] G_b(z). \tag{2.2}$$

Example 2.2. Let us illustrate the calculations for the a first-order system and the FIR basis

$$G_0(z) = \frac{1}{z - p}, \quad F_k(z) = z^{-k},\ G_b(z) = z^{-n}.$$

Here

$$G_0(z) = z^{-1} + \ldots + p^{n-1}z^{-n} + p^n z^{-n}\frac{1}{z - p},$$

$$D_n(z) = \frac{z^{n-1} + \ldots + p^{n-1}}{z^n}, \quad E_n(z) = \frac{p^n}{z - p},$$

and thus

$$G_0(z) - G_n(z) = \frac{1}{z - p} p^n z^{-n} = G_0(z)G_b(1/p)G_b(z).$$

2.2.5 Completeness

The above expressions for the approximation error have previously been derived in *e.g.* [224]. They are in particular useful to bound the approximation error on the unit circle, *i.e.*

$$|G_0(e^{i\omega}) - G_n(e^{i\omega})| = \left| \sum_{j=1}^{n_0} \frac{b_j}{e^{i\omega} - p_j} G_b^*(1/p_j^*) \right| \leq \sum_{j=1}^{n_0} \left| \frac{b_j}{e^{i\omega} - p_j} \right| |G_b^*(1/p_j^*)| , \tag{2.3}$$

because $|G_b(e^{\omega})| = 1$. The rate of convergence to zero of the right-hand side of the error expression (2.3) as n tends to infinity is closely coupled to the size of $|G_b^*(1/\zeta^*)|$, $|\zeta| < 1$. In [16] the following upper bound is derived

$$|G_b^*(1/\zeta^*)| \leq \exp\left[-\frac{1}{2}(1 - |\zeta|) \sum_{j=1}^{n} (1 - |\xi_j|) \right], \quad |\zeta| < 1.$$

If

$$\sum_{j=1}^{\infty} (1 - |\xi_j|) = \infty \tag{2.4}$$

this implies that $|G_b^*(1/\zeta^*)| \to 0$ as $n \to \infty$ and the approximation error will tend to zero. It turns out that the so-called Szász condition (2.4) implies that the Takenaka-Malmquist functions are complete in $H_2(\mathbb{T})$ (and also in $H_p(\mathbb{T})$, $1 \leq p < \infty$, and in the disc algebra of functions analytic outside and continuous on the unit circle). This means that the approximation error measured by the corresponding norm can be made arbitrarily small by choosing a large enough value for n. If the Szász condition (2.4) is not satisfied, the Blascke product $G_b(z)$ will converge to an $H_2(\mathbb{T})$ function, which will be orthogonal to $F_k(z)$ for all k. Hence, the Takenaka-Malmquist functions are then not complete, and thus the condition (2.4) is both necessary and sufficient. The Szász condition has a long history and was already known by Takenaka [296]. The original paper, [292], by Szász contains a rather complete analysis of the closure of certain sets of rational functions in various spaces. Recent work in this area includes [16].

2.2.6 Real Impulse Response

The Z-transform $F(z)$ of a real sequence satisfies $F(z) = F^*(z^*)$. Hence, complex poles (and zeros) will appear in complex conjugate pairs. As previously mentioned, the Takenaka-Malmquist functions will in general not have real impulse responses. If $\xi_{j+1} = \xi_j^*$ it is possible to form linear combinations of $F_j(z)$ and $F_{j+1}(z)$ to obtain two orthonormal functions with real impulse responses, which span the same space.

Deeper insights into the construction of orthonormal functions can be gained by using state-space models and matrix algebra. This is the topic of the next section.

2.3 State-space Theory

2.3.1 Introduction

The basic idea of this chapter is to construct and analyze state-space models

$$x(t+1) = Ax(t) + Bu(t),$$

where $u(t)$ is the input signal, and $x(t) \in \mathbb{R}^n$ is the state vector, for which the transfer functions from the input to the states,

$$V(z) = [F_1(z) \dots F_n(z)]^T = (zI - A)^{-1}B,$$

are orthonormal. This means that

$$[\![V, V]\!] = \frac{1}{2\pi i} \oint_{\mathbb{T}} V(z)V^T(1/z)\frac{dz}{z} = I \text{ (the identity matrix).}$$

Here we will specify the eigenvalues of A, which are the poles of $\{F_k(z)\}$. We will only study state-space models with real A, B-matrices, *i.e.* with real valued impulse response.

To understand the basic ideas of using state-space theory to find rational orthonormal basis functions, consider the following example.

Example 2.3. Returning to Example 2.1, consider the following state-space model

$$\begin{aligned} \bar{x}_1(t+1) &= a\bar{x}_1(t) + u(t) \\ \bar{x}_2(t+1) &= b\bar{x}_2(t) + u(t). \end{aligned}$$

The transfer functions $\bar{F}_1(z) = 1/(z-a)$ and $\bar{F}_2(z) = 1/(z-b)$ are then the transfer functions from the input $u(t)$ to the states $\bar{x}_1(t)$ and $\bar{x}_2(t)$, respectively. Define the state vector $\bar{x}(t) = [\bar{x}_1(t)\ \bar{x}_2(t)]^T$ and the corresponding state-space matrices

$$\bar{A} = \begin{bmatrix} a & 0 \\ 0 & b \end{bmatrix}, \quad \bar{B} = \begin{bmatrix} 1 \\ 1 \end{bmatrix}.$$

The input-to-state vector transfer function

$$\bar{V}(z) = (zI - \bar{A})^{-1}\bar{B}$$

here equals

$$\bar{V}(z) = \begin{bmatrix} \bar{F}_1(z) \\ \bar{F}_2(z) \end{bmatrix} = \begin{bmatrix} \frac{1}{z-a} \\ \frac{1}{z-b} \end{bmatrix}.$$

2.3.2 State Covariance Matrix

Assume that the input signal $u(t)$ to the system is a zero mean white noise process with variance 1, *i.e.*

$$\mathsf{E}\{u(t)u(t+k)\} = \begin{cases} 1, \; k=0 \\ 0, \; k \neq 0 \end{cases}.$$

The state covariance matrix

$$\bar{P} = \mathsf{E}\{\bar{x}(t)\bar{x}^T(t)\}, \tag{2.5}$$

of a stable state-space model

$$\bar{x}(t+1) = \bar{A}\bar{x}(t) + \bar{B}u(t),$$

satisfies the Lyapunov equation

$$\bar{P} = \bar{A}\bar{P}\bar{A}^T + \bar{B}\bar{B}^T. \tag{2.6}$$

The state covariance matrix $\bar{P}$ also equals the so-called *controllability* Gramian of the state-space model. The reason why we are interested in the state covariance matrix is that $\bar{P}$ corresponds to the matrix 'outer product' of the state vector transfer function $\bar{V}(z)$. That is:

$$\begin{aligned} \bar{P} &= \frac{1}{2\pi i}\oint_{\mathbb{T}} \bar{V}(z)\bar{V}^T(1/z)\frac{dz}{z} \\ &= [\![\bar{V},\bar{V}]\!] = \begin{bmatrix} \langle \bar{F}_1, \bar{F}_1 \rangle & \langle \bar{F}_1, \bar{F}_2 \rangle \\ \langle \bar{F}_2, \bar{F}_1 \rangle & \langle \bar{F}_2, \bar{F}_2 \rangle \end{bmatrix} \end{aligned}$$

This follows from

$$\bar{P} = \frac{1}{2\pi i}\oint_{\mathbb{T}} (zI - \bar{A})^{-1}\bar{B}\bar{B}^T(z^{-1}I - \bar{A}^T)^{-1}\frac{dz}{z} = \sum_{k=0}^{\infty} \bar{A}^k\bar{B}\bar{B}^T(\bar{A}^T)^k,$$

where the last equality follows using the power series expansion of $(zI - \bar{A})^{-1}$, satisfies Equation (2.6).

Example 2.4. For Example 2.3 we have

$$\bar{P} = \begin{bmatrix} \dfrac{1}{1-a^2} & \dfrac{1}{1-ab} \\ \dfrac{1}{1-ab} & \dfrac{1}{1-b^2} \end{bmatrix}. \tag{2.7}$$

2.3.3 Input Balanced State-space Realization

The basic idea is now to find a new state-space realization for which the state covariance matrix equals the identity matrix, $P = I$. The corresponding input

to state transfer functions will then be orthonormal. Furthermore, they will span the same space as the original functions, as only linear transformations are considered. A state-space realization for which $\bar{P} = I$ is called *input balanced*, [198].
Let $x(t) = T\bar{x}(t)$, where T is a square non-singular transformation matrix, and denote the transformed state covariance matrix by

$$P = \mathsf{E}\{x(t)x^T(t)\} = T\bar{P}T^T.$$

The orthonormalization problem now corresponds to finding a transformation matrix T such that $T\bar{P}T^T = I$, where $\bar{P}$ is a given symmetric positive definite matrix. This is a standard problem in linear algebra. The solution is to take T equal to a square (matrix) root of $\bar{P}^{-1}$, *i.e.* $\bar{P}^{-1} = T^T T$. This follows from

$$T\bar{P}T^T = T[\bar{P}^{-1}]^{-1}T^T = T[T^T T]^{-1}T^T = I.$$

Notice that we have to assume that $\bar{P}$ is nonsingular. This is the same as assuming that the state-space model is controllable, as $\bar{P}$ also equals the controllability Gramian. This corresponds to the assumption that original functions are linearly independent.

The square root of a matrix is by no means unique. Let Q be a orthogonal matrix, $Q^T Q = I$. Then QT is also a square root of $\bar{P}^{-1}$, and thus a permissible transformation. We refer to *e.g.* [103,155] or any other standard book in matrix computations for a detailed discussion of this topic.

An interesting option is to apply the Gram-Schmidt QR-algorithm to a permissible transformation, $T = QR$, where Q is orthogonal and R is upper triangular, and then use the transformation $x(t) = R\bar{x}$. By working on permuted rows and columns of $\bar{P}$, it is also possible to find a lower triangular matrix L (using the Gram-Schmidt QR-algorithm), such that $x(t) = L\bar{x}$ implies that $P = I$. All this is more or less the same as applying the Gram-Schmidt procedure directly to the transfer functions.

Example 2.5. For Example 2.3 we can take

$$L = \begin{bmatrix} \sqrt{1-a^2} & 0 \\ -\dfrac{\sqrt{1-b^2}(1-a^2)}{a-b} & \dfrac{\sqrt{1-b^2}(1-ab)}{a-b} \end{bmatrix},$$

which is obtained by identifying the implicit transformation used in the Gram-Schmidt procedure in Example 2.1. It is easy to verify that $L\bar{P}L^T = I$, where $\bar{P}$ is given by (2.7).

2.3.4 Construction Method

Summarizing the observations in the above example leads to the following method for construction of orthonormal functions:

1. Consider the set spanned by the n linearly independent rational transfer functions
$$\{\bar{F}_1(z) \dots \bar{F}_n(z)\}.$$
2. Determine a controllable state-space model
$$\bar{x}(t+1) = \bar{A}\bar{x}(t) + \bar{B}u(t)$$
with input-to-state transfer functions
$$\bar{V}(z) = [\bar{F}_1(z) \dots \bar{F}_n(z)]^T = (zI - \bar{A})^{-1}\bar{B}.$$
The eigenvalues of $\bar{A}$ equal the poles of $\{\bar{F}_1(z) \dots \bar{F}_n(z)\}$.
3. Determine the state covariance matrix (controllability Gramian) $\bar{P}$ by solving the Lyapunov equation
$$\bar{P} = \bar{A}\bar{P}\bar{A}^T + \bar{B}\bar{B}^T.$$
This is a linear equation in the elements of $\bar{P}$, for which there exists efficient numerical algorithms.
4. Find a matrix square root T of the inverse of $\bar{P}$, *i.e.* $T\bar{P}T^T = I$. The solution is not unique. There exist several numerically robust algorithms for solving this problem.
5. Make a transformation of the states, $x(t) = T\bar{x}(t)$. The transformed state-space model is
$$x(t+1) = Ax(t) + Bu(t), \quad A = T\bar{A}T^{-1},\ B = T\bar{B}.$$
The new input-to-state transfer function equals
$$V(z) = (zI - A)^{-1}B = T\bar{V}(z).$$
The components of $V(z) = [F_1(z) \dots F_n(z)]^T$ then form an orthonormal basis for $\text{Sp}\{\bar{F}_1(z) \dots \bar{F}_n(z)\}$.

2.4 All-pass Transfer Function

As first noted by Mullis and Roberts [257], all-pass transfer functions provide a powerful framework for factorization of state covariance matrices. In this section, we will summarize this theory.

Consider a single input single output asymptotically stable all-pass transfer function $H(z)$ of order m with real valued impulse response,

$$H(z)H(1/z) = 1, \tag{2.8}$$

completely determined by its poles $\{\xi_i \in \mathbb{C},\ |\xi_i| < 1\}$,

$$H(z) = \prod_{i=1}^{m} \left[\frac{1 - \xi_i^* z}{z - \xi_i} \right]. \tag{2.9}$$

Such a transfer function is often called an inner function, and the representation (2.9) is called a Blaschke product. We will use the notation $H(z)$ for an arbitrary all-pass transfer function, and the notation $G_b(z)$, as in the previous section, for all-pass filters directly connected to certain orthogonal basis functions.
Let (A, B, C, D) be a minimal input balanced state-space realization of $H(z)$. The corresponding state covariance matrix satisfies $P = APA^T + BB^T$ for unit variance white noise input. Also recall that we can ensure that $P = I$, by a proper choice of state vector.

2.4.1 Orthogonal State-space Models

The impulse response of the all-pass system can be written in terms of the state-space matrices,

$$h_k = \begin{cases} D, & k = 0 \\ CA^{k-1}B, & k \geq 1 \end{cases}.$$

The all-pass property (2.8) implies

$$\sum_{k=1}^{\infty} h_k h_{k+j} = \begin{cases} 1, & j = 0 \\ 0, & j \geq 1 \end{cases},$$

which can be re-expressed as

$$DD^T + CPC^T = 1,$$
$$CA^{j-1}(BD^T + APC^T) = 0, \quad j \geq 1.$$

Here P is, as before, the controllability Gramian. The last equation implies that $BD^T + APC^T$ is in the null-space of the observability matrix of $H(z)$. Because the state-space realization of $H(z)$ is assumed to be minimal, and hence observable, this implies that $BD^T + APC^T = 0$. Using $P = I$, we then obtain the three conditions

$$AA^T + BB^T = I,$$
$$DD^T + CC^T = 1,$$
$$BD^T + AC^T = 0,$$

or more compactly

$$\begin{bmatrix} A & B \\ C & D \end{bmatrix} \begin{bmatrix} A & B \\ C & D \end{bmatrix}^T = I, \tag{2.10}$$

i.e. the corresponding matrix is orthogonal. A state-space realization

$$\begin{bmatrix} x(t+1) \\ y(t) \end{bmatrix} = \begin{bmatrix} A & B \\ C & D \end{bmatrix} \begin{bmatrix} x(t) \\ u(t) \end{bmatrix},$$

of $H(z)$ which satisfies (2.10) is called orthogonal.

The above characterization of SISO all-pass filters also holds for multi-input multi-output (MIMO) all-pass filters. A MIMO all-pass system with transfer function $H(z) = D + C(zI - A)^{-1}B$ is orthogonal if and only if it has an orthogonal realization. Note that $H(z)$ is all-pass if and only if $H(z)H^T(1/z) = I$.

2.4.2 Connections of All-pass Filters

Given a set of poles, we can construct an orthogonal state-space realization of the corresponding all-pass transfer function $H(z)$. Different orthogonal realizations and factorizations of

$$\begin{bmatrix} A & B \\ C & D \end{bmatrix} \tag{2.11}$$

will lead to different orthonormal basis functions. Notice that the matrix (2.11) is by no means unique, as a unitary state-space transformation does not change the orthonormality property.

The key observation of Roberts and Mullis, [257], is that orthogonality and the all-pass property are preserved through several different connections of all-pass filters (some or all of these filters may have more than one input and/or more than one output). We will be most interested in cascade connections, for which the following key result holds.

Theorem 2.1. *Consider two orthogonal all-pass filters $H_1(z)$ and $H_2(z)$ with state vectors $x_1(t)$ and $x_2(t)$, respectively. Then the cascade (serial) connection $H_2(z)\,H_1(z)$ is also all-pass and orthogonal with the state vector:*

$$x(t) = \begin{bmatrix} x_1(t) \\ x_2(t) \end{bmatrix}.$$

The result can of course be generalized to an arbitrary cascade of all-pass filter as shown in the figure below.

The proof of Theorem 2.1 is constructive. Define the orthogonal state-space realizations of $H_1(z)$ and $H_2(z)$ as

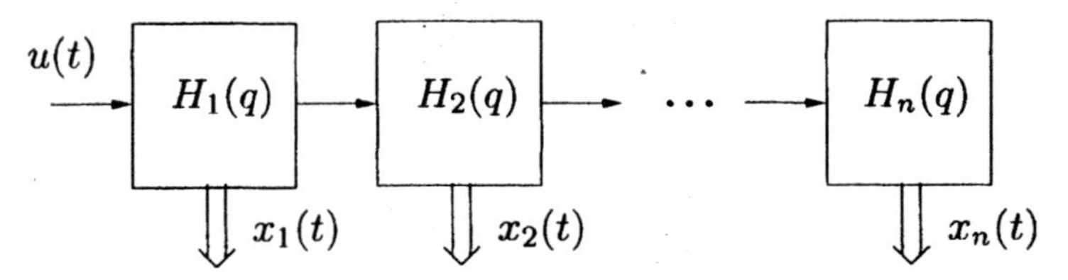

Fig. 2.1. Cascade all-pass filter network. Notice that the transfer functions from input-to-states are orthonormal, and hence the factorization provides a parameterization of different classes of orthonormal basis functions.

$$\begin{bmatrix} A_1 & B_1 \\ C_1 & D_1 \end{bmatrix}, \quad \begin{bmatrix} A_2 & B_2 \\ C_2 & D_2 \end{bmatrix},$$

with input signals $u_1(t)$ and $u_2(t)$ and output signals $y_1(t)$ and $y_2(t)$, respectively. By combining the states, we obtain

$$\begin{aligned} \begin{bmatrix} x_1(t+1) \\ x_2(t+1) \\ y_2(t) \end{bmatrix} &= \begin{bmatrix} I & 0 & 0 \\ 0 & A_2 & B_2 \\ 0 & C_2 & D_2 \end{bmatrix} \begin{bmatrix} x_1(t+1) \\ x_2(t) \\ y_1(t) \end{bmatrix} \\ &= \begin{bmatrix} I & 0 & 0 \\ 0 & A_2 & B_2 \\ 0 & C_2 & D_2 \end{bmatrix} \begin{bmatrix} A_1 & 0 & B_1 \\ 0 & I & 0 \\ C_1 & 0 & D_1 \end{bmatrix} \begin{bmatrix} x_1(t) \\ x_2(t) \\ u_1(t) \end{bmatrix} \end{aligned}$$

Hence, the total matrix for the cascaded system

$$\begin{bmatrix} A & B \\ C & D \end{bmatrix} = \begin{bmatrix} I & 0 & 0 \\ 0 & A_2 & B_2 \\ 0 & C_2 & D_2 \end{bmatrix} \begin{bmatrix} A_1 & 0 & B_1 \\ 0 & I & 0 \\ C_1 & 0 & D_1 \end{bmatrix} = \begin{bmatrix} A_1 & 0 & B_1 \\ B_2C_1 & A_2 & B_2D_1 \\ D_2C_1 & C_2 & D_2D_1 \end{bmatrix}$$

is orthonormal, as it is the product of two orthonormal matrices. This proves the result.

2.4.3 Feedback Connections of All-pass Filters

Feedback connections in an orthogonal all-pass filter also give rise to orthogonal all-pass filters. To show this, consider Example 10.4.3 in the book by Roberts and Mullis, [257], where feedback is introduced in the orthogonal all-pass filter described by the state variable equations

$$\begin{bmatrix} x(t+1) \\ y_1(t) \\ y_2(t) \end{bmatrix} = \begin{bmatrix} A & B_1 & B_2 \\ C_1 & D_{11} & D_{12} \\ C_2 & D_{21} & D_{22} \end{bmatrix} \begin{bmatrix} x(t) \\ u_1(t) \\ u_2(t) \end{bmatrix} \tag{2.12}$$

via the connection $u_2(t) = y_2(t)$. This connection implies that D_{22} is a square matrix. In the above formulas delay free loops will be present unless $D_{22} = 0$. Assuming that $I - D_{22}$ is invertible, elementary algebra yield

$$\begin{bmatrix} x(t+1) \\ y_1(t) \end{bmatrix} = \left\{ \begin{bmatrix} A & B_1 \\ C_1 & D_{11} \end{bmatrix} + \begin{bmatrix} B_2 \\ D_{12} \end{bmatrix} (I - D_{22})^{-1} \begin{bmatrix} C_2 & D_{21} \end{bmatrix} \right\} \begin{bmatrix} x(t) \\ u_1(t) \end{bmatrix}.$$

The matrix inside curly braces is the DC gain of another orthogonal all-pass filter, with state-space realization

$$\left[\begin{array}{c|cc} D_{22} & C_2 & D_{21} \\ \hline B_2 & A & B_1 \\ D_{12} & C_1 & D_{11} \end{array}\right].$$

But the DC gain of orthogonal all-pass filters must be an orthogonal matrix because

$$H(z)H^T(1/z) = I.$$

Thus, feedback connections of the type described above give rise to orthogonal all-pass filters. The same result holds if the feedback connection is made via a general orthogonal all-pass filter (with the appropriate number of input and output signals). Writing the state-space realizations of the two filters together, and using an obvious notation for the (feedback) orthogonal filter, yields

$$\begin{bmatrix} x(t+1) \\ y_1(t) \\ y_2(t) \\ x_f(t+1) \\ y_f(t) \end{bmatrix} = \begin{bmatrix} A_1 & B_1 & B_2 & 0 & 0 \\ C_1 & D_{11} & D_{12} & 0 & 0 \\ C_2 & D_{21} & D_{22} & 0 & 0 \\ 0 & 0 & 0 & A_f & B_f \\ 0 & 0 & 0 & C_f & D_f \end{bmatrix} \begin{bmatrix} x(t) \\ u_1(t) \\ u_2(t) \\ x_f(t) \\ u_f(t) \end{bmatrix}$$

with the connections $u_f(t) = y_2(t)$ and $u_2(t) = y_f(t)$. These connections are of the type used initially, with the D_{22} of the first example replaced by the direct sum of D_{22} and D_f of the second. Thus we have proved the following result, which will be used in Subsection 2.6:

Theorem 2.2. *If $I - D_{22}$, where D_{22} is defined in Equation (2.12), is invertible, then the feedback connection of an orthogonal all-pass filter made via another orthogonal all-pass filter is also orthogonal and all-pass.*

2.4.4 Orthogonal Filters Based on Feedback of Resonators

Define

$$\begin{bmatrix} x(t+1) \\ y(t) \end{bmatrix} = \begin{bmatrix} A_r & B_r \\ C_r & 0 \end{bmatrix} \begin{bmatrix} x(t) \\ e(t) \end{bmatrix}.$$

Now apply the feedback $e(t) = u(t) - y(t)$ to obtain the closed loop system

$$\begin{bmatrix} x(t+1) \\ y(t) \end{bmatrix} = \begin{bmatrix} A_r - B_r C_r & B_r \\ C_r & 0 \end{bmatrix} \begin{bmatrix} x(t) \\ u(t) \end{bmatrix}.$$

This will be an orthogonal realization if

$$\begin{bmatrix} A_r - B_rC_r & B_r \\ C_r & 0 \end{bmatrix}^T \begin{bmatrix} A_r - B_rC_r & B_r \\ C_r & 0 \end{bmatrix} =$$

$$\begin{bmatrix} A_r^TA_r - C_r^TB_r^TA_r - A_r^TB_rC_r + C_r^TB_r^TB_rC_r + C_r^TC_r & A_r^TB_r - C_r^TB_r^TB_r \\ B_r^TA_r - B_r^TB_rC_r & B_r^TB_r \end{bmatrix}$$

$$= \begin{bmatrix} I & 0 \\ 0 & 1 \end{bmatrix},$$

with the simple solution

$$B_r^TB_r = 1, \quad C_r = B_r^TA_r, \quad A_r^TA_r = I.$$

Let $G_r(z)$ be the transfer function of the open loop system, $G_r(z) = C_r(zI - A_r)^{-1}B_r$, and $H(z)$ be the transfer function of the corresponding closed loop all-pass system. Then, using some manipulation, we can write the open loop transfer function as:

$$G_r(z) = \frac{H(z)}{1 - H(z)}.$$

Hence, the eigenvalues of A_r (the poles of $G_r(z)$) satisfy $H(\lambda_r) = 1$ and thus $|\lambda_r| = 1$. This can also be seen from the orthonormality condition $A_r^TA_r = I$. Such a system, with all poles on the unit circle, is called a *resonator*. Here we have assumed a delay-free closed loop, *i.e.* no direct term in $G_r(z)$. This will introduce a pure delay in $H(z)$, *i.e.* a pole in $z = 0$, which is a restriction which may have to be removed before construction of the corresponding input-to-state orthogonal transfer functions $V(z)$. This approach has been extensively studied in [240, 241].

2.4.5 Basic Orthogonal All-pass Filters

The simplest all-pass building blocks are:

- First-order all-pass filters with transfer functions

$$H_i(z) = \frac{z - a_i}{1 - a_iz}, \qquad -1 < a_i < 1,$$

  with orthogonal state-space realizations

$$\begin{bmatrix} A_i & B_i \\ C_i & D_i \end{bmatrix} = \begin{bmatrix} a_i & \sqrt{1 - a_i^2} \\ \sqrt{1 - a_i^2} & -a_i \end{bmatrix}. \tag{2.13}$$

- Second-order all-pass filters with transfer functions

$$H_j(z) = \frac{-c_jz^2 + b_j(c_j - 1)z + 1}{z^2 + b_j(c_j - 1)z - c_j}, \qquad -1 < b_j < 1, \ -1 < c_j < 1,$$

  with orthonormal state-space realizations

$$\begin{bmatrix} A_j & B_j \\ C_j & D_j \end{bmatrix} = \left[\begin{array}{cc|c} b_j & c_j\sqrt{1-b_j^2} & \sqrt{1-b_j^2}\sqrt{1-c_j^2} \\ \sqrt{1-b_j^2} & -b_jc_j & -b_j\sqrt{1-c_j^2} \\ \hline 0 & \sqrt{1-c_j^2} & -c_j \end{array}\right]. \quad (2.14)$$

It should be noted that the realization is not unique.

- First-order all-pass section with a pole at the origin and two input and output signals

$$\begin{bmatrix} A_j & B_j \\ C_j & D_j \end{bmatrix} = \left[\begin{array}{c|cc} 0 & 0 & 1 \\ \hline \sqrt{1-\gamma_j^2} & -\gamma_j & 0 \\ \gamma_j & \sqrt{1-\gamma_j^2} & 0 \end{array}\right], \quad (2.15)$$

where $-1 < \gamma_j < 1$. This realization will be used in Subsection 2.6.

2.5 Cascade Structures

We now have a powerful tool to construct simple networks of cascaded all-pass filters, for which the individual state transfer functions form orthogonal basis functions. Let us now consider some special cases.

Finite Impulse Response (FIR) Networks: Consider the tapped delay line network in Figure 2.2. Here $H(z) = z^{-m}$, and $H_i(z) = z^{-1}$. The corresponding orthogonal state-space model is obtained by taking

$$x(t) = [u(t-1)\, u(t-2) \ldots u(t-n)]^T.$$

Consequently

$$\begin{aligned} x(t+1) &= Ax(t) + Bu(t) \\ y(t) &= Cx(t) + Du(t), \end{aligned}$$

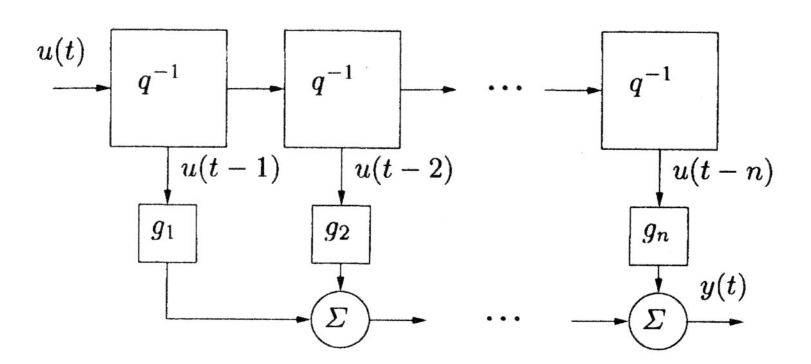

Fig. 2.2. FIR network.

where

$$A = \begin{pmatrix} 0\,0 \ldots 0\,0 \\ 1\,0 \ldots 0\,0 \\ 0\,1 \ldots 0\,0 \\ \vdots \quad \ddots \quad \vdots \\ 0\,0 \ldots 1\,0 \end{pmatrix}, \quad B = \begin{pmatrix} 1 \\ 0 \\ 0 \\ \vdots \\ 0 \end{pmatrix},$$

$$C = \begin{pmatrix} 0\,0 \ldots 0\,1 \end{pmatrix}, \quad D = 0. \tag{2.16}$$

Laguerre Networks: By taking n identical first-order all-pass filters

$$H_i(z) = \frac{1-az}{z-a}, \quad -1 < a < 1,$$

using the orthogonal state-space realizations (2.13), the input-to-state transfer functions equal the so-called Laguerre basis functions

$$F_k(z) = \frac{\sqrt{1-a^2}}{z-a}\left[\frac{1-az}{z-a}\right]^{k-1}. \tag{2.17}$$

The corresponding Laguerre model has a pole at a with multiplicity n.

Two-parameter Kautz Networks: By using identical second-order all-pass filters,

$$H_i(z) = \frac{-cz^2 + b(c-1)z + 1}{z^2 + b(c-1)z - c}, \quad -1 < b < 1, \; -1 < c < 1. \tag{2.18}$$

with orthogonal state-space realizations (2.14), the so-called two-parameter Kautz basis functions

$$\begin{aligned} F_{2k-1}(z) &= \frac{\sqrt{1-c^2}(z-b)}{z^2+b(c-1)z-c}\left[\frac{-cz^2+b(c-1)z+1}{z^2+b(c-1)z-c}\right]^{k-1} \\ F_{2k}(z) &= \frac{\sqrt{1-c^2}\sqrt{1-b^2}}{z^2+b(c-1)z-c}\left[\frac{-cz^2+b(c-1)z+1}{z^2+b(c-1)z-c}\right]^{k-1}. \\ &\quad -1<b<1, \quad -1<c<1, \quad k=1,2\ldots, \end{aligned}$$

are obtained. The corresponding two-parameter Kautz model will have the pole structure $\{\xi_1, \xi_2, \xi_1, \xi_2, \ldots \xi_1, \xi_2\}$, where ξ_1 and ξ_2 can be either real or complex conjugated.

Generalized Orthonormal Basis Functions: The class of generalized orthonormal basis functions (GOBF) is obtained by cascading identical n_b^{th} order all-pass filters,

$$H_i(z) = G_b(z).$$

The corresponding state-space forms of $G_b(z)$ is just restricted to be orthogonal. The corresponding GOBF basis functions can then be written in vector form

$$V_k(z) = (zI - A)^{-1}B[G_b(z)]^{k-1}, \tag{2.19}$$

and thus

$$F_{j+(k-1)nb}(z) = e_j^T(zI - A)^{-1}B[G_b(z)]^{k-1}, \quad j = 1, \ldots nb.$$

Observe that further structure can be imposed by factorizing $G_b(z)$ into first and second order all-pass filters. The vector function

$$V_1(z) = (zI - A)^{-1}B$$

will play an important role in the GOBF analysis to follow, and actually determines the specific structure.
The corresponding GOBF model, illustrated in Figure 2.3 will have the pole structure $\{\xi_1, \ldots \xi_{n_b}, \xi_1 \ldots \xi_{n_b}, \ldots \ldots \xi_1, \ldots \xi_{n_b}\}$.

Takenaka-Malmquist Functions: The general case corresponds to cascading different all-pass blocks of order one or two. These basis functions are sometimes called the Takenaka-Malmquist functions, *cf.* (2.1). The corresponding basis functions for the case with possible complex poles are

$$F_k(z) = \frac{\sqrt{1 - |\xi|_k^2}}{z - \xi_k} \prod_{i=1}^{k-1} \left[\frac{1 - \xi_i^* z}{z - \xi_i} \right]$$

Poles with non-zero imaginary part can be combined in complex conjugated pairs into the corresponding real Kautz form discussed above.

Observe that the ordering of the all-pass filters will influence the basis functions. This freedom can be used to incorporate other consideration, such as compactness. See [26] for a detailed discussion on optimal basis selection.

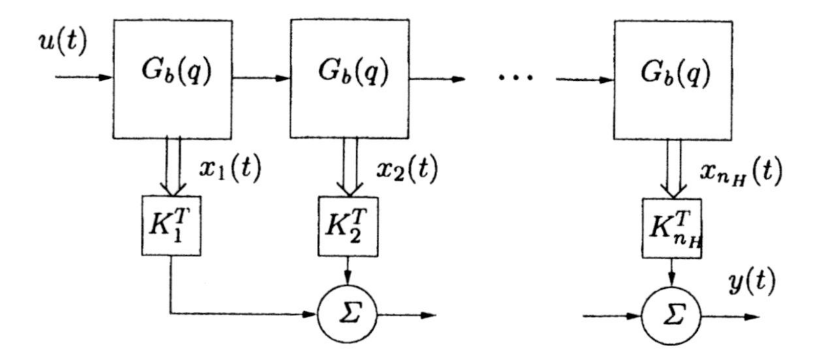

Fig. 2.3. GOBF model. Notice that the gains K_j are $1 \times n_b$ row-vectors.

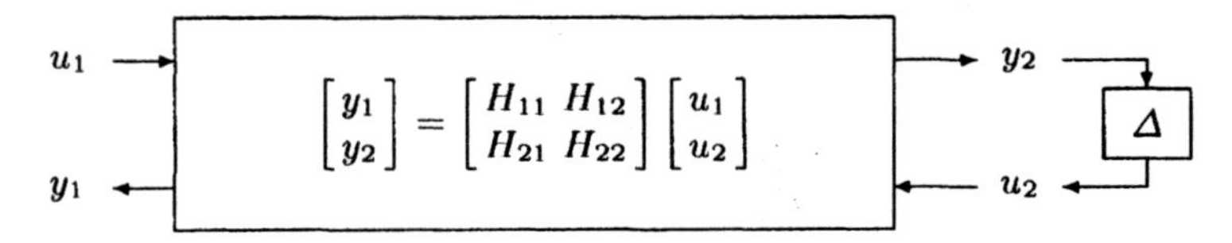

Fig. 2.4. A general feedback connection.

2.6 Ladder Structures

In addition to the cascaded connection of balanced all-pass systems described in the previous subsection, feedback connections of balanced all-pass systems are also important. Let us begin with the feedback connection presented in Figure 2.4. It is quite simple to verify that the transfer function from $u_1(t)$ to $y_1(t)$ is

$$H_{11}(z) + H_{12}(z)\Delta(z)(I - H_{22}(z)\Delta(z))^{-1}H_{21}(z). \tag{2.20}$$

Note that this is the linear fractional transform widely used in robust control theory to describe uncertainties in a system [340]. According to Theorem 2.2, if both

$$H(z) = \begin{bmatrix} H_{11}(z) & H_{12}(z) \\ H_{21}(z) & H_{22}(z) \end{bmatrix}$$

and $\Delta(z)$ have orthogonal realizations, then so has the system of Figure 2.4. Consider, in particular, the case $H(z) = H(z,\gamma)$, where

$$H(z,\gamma) = \begin{bmatrix} H_{11}(z) & H_{12}(z) \\ H_{21}(z) & H_{22}(z) \end{bmatrix} = \begin{bmatrix} \sqrt{1-\gamma^2} & \gamma \\ -\gamma & \sqrt{1-\gamma^2} \end{bmatrix} \begin{bmatrix} 1 & 0 \\ 0 & z^{-1} \end{bmatrix}, \qquad |\gamma| < 1.$$

Note that one possible orthogonal realization of this system is given by (2.15). The block diagram of this all-pass transfer function is depicted in Figure 2.5.

2.6.1 Schur Algorithm

For this $H(z,\gamma)$ the linear fractional transform (2.20) takes the form

$$H_n(z) = -\gamma + \frac{(1-\gamma^2)\Delta z^{-1}}{1-\gamma\Delta z^{-1}} = \frac{z^{-1}\Delta - \gamma}{1-\gamma z^{-1}\Delta}, \tag{2.21}$$

which is a bilinear transformation of $z^{-1}\Delta(z)$. Suppose that $H_n(z)$ in (2.21) is a rational all-pass transfer function of degree n. Using the previous formula we may write, with H_{n-1} instead of Δ,

$$H_n(z) = \frac{H_{n-1}(z) - \gamma_n z}{z - \gamma_n H_{n-1}(z)} \qquad \text{with} \qquad H_{n-1}(z) = z\,\frac{H_n(z) + \gamma_n}{1 + \gamma_n H_n(z)}.$$

Represent $H_n(z)$ in the form

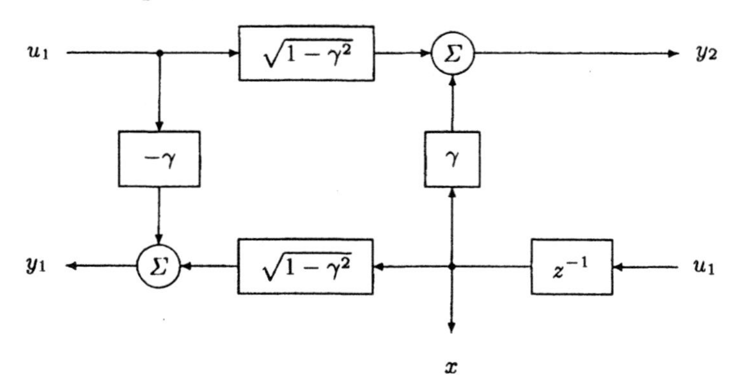

Fig. 2.5. The block diagram of the 2-input 2-output all-pass transfer function $H(z,\gamma)$.

$$H_n(z) = \epsilon D_n(z)/N_n(z),$$

where $N_n(z)$ is the (monic) denominator of $H_n(z)$, $D_n(z) = z^n N_n(1/z)$ is the reciprocal polynomial of $N_n(z)$, and where $|\epsilon| = 1$. It is then possible to verify that $H_{n-1}(z)$ becomes a rational all-pass function of degree less than n if and only if

$$\gamma_n = -\lim_{z\to\infty} H_n(z) = -1/\lim_{z\to 0} H_n(z).$$

For all other values of γ_n, the degree of $H_{n-1}(z)$ remains equal to n. In fact, if $|\gamma_n| \neq 1$, then the degree of $H_{n-1}(z)$ will be exactly $n-1$. This degree reduction step constitutes the first iteration of the famous Schur algorithm [140,272] applied to $H_n(z)$ and is also the first iteration of the well-known Schur-Cohn stability test applied to $N_n(z)$. If the same degree reduction formulas are applied to $H_{n-1}(z)$, $H_{n-2}(z)$, and so on, eventually a constant $H_k(z)$ will be reached. If $H_n(z)$ is asymptotically stable, which, according to Section 3.3, implies that $|H_n(z)| < 1$ for $|z| > 1$, then $|\gamma_n| < 1$. Because for these values of γ_n the bilinear transform (2.21) maps the exterior of the unit circle into the exterior of the unit circle, *cf.* the Laguerre case of Subsection 3.2, it follows than when $H_{n-1}(z)$ is not a constant, then $|H_{n-1}(z)| < 1$ for $|z| > 1$; if $H_{n-1}(z)$ is a constant, then $|H_{n-1}(z)| = 1$ for all z, and no more degree reduction steps can be performed.

Thus, when $H_n(z)$ is asymptotically stable (and all-pass), Schur's iterations can be applied exactly n times, giving rise to asymptotically stable all-pass transfer functions $H_{n-1}(z), \ldots, H_0(z)$, the last of which is constant. Moreover, it also follows that $|\gamma_k| < 1$ for $k = 1, \ldots, n$.
This constitutes the Schur-Cohn stability test. It can be shown that if one of the γ_k has a modulus not strictly less than 1 then $H_n(z)$ is not asymptotically stable.

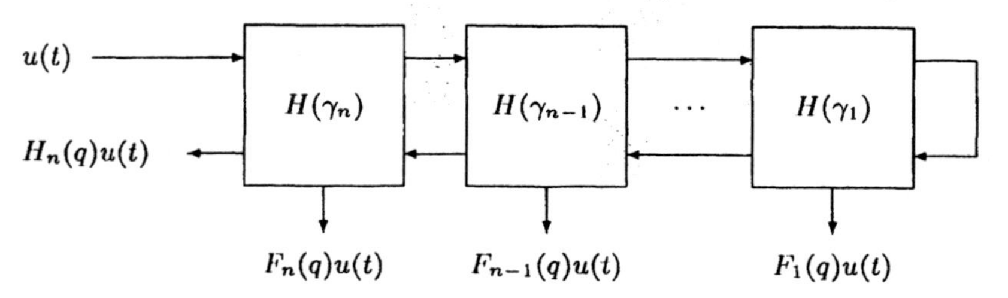

Fig. 2.6. Orthogonal realization of an all-pass scalar transfer function. The $H(\gamma)$ blocks are presented in Figure 2.5. The signals $F_k(q)u(t)$ are the state variables.

2.6.2 Recursion

The degree reduction steps give rise to orthogonal feedback-based realizations for the $H_k(z)$ all-pass transfer functions, as shown in Figure 2.6 for $H_n(z)$. These orthogonal filter structures are precisely the classical Gray-Markel normalized ladder filters [107]. Similar, but more sophisticated orthogonal structures can be found in [76].
The recursion can be applied to calculate the polynomials

$$\begin{bmatrix} N_k(z) \\ D_k(z) \end{bmatrix} = \begin{bmatrix} 1 & -\gamma_k \\ -\gamma_k & 1 \end{bmatrix} \begin{bmatrix} N_{k-1}(z) \\ z\, D_{k-1}(z) \end{bmatrix},$$

with $N_0(k) = D_0(k) = 1$. As mentioned above, the functions

$$F_k(z) = \left[\prod_{l=k}^{n_b} \sqrt{1-\gamma_l^2} \right] \frac{N_{k-1}(z)}{D_{n_b}(z)}$$

are orthogonal and

$$G_b(z) = \frac{N_{n_b}(z)}{D_{n_b}(z)}.$$

Recall that $N_k(z)$ and $D_k(z)$ are reciprocal polynomials, *i.e.* $N_k(z) = z^{-k} D_k(1/z)$. It is worthwhile to note that the all-pass filter used in Laguerre networks corresponds to the case $n = 1$ and $\gamma_1 = a$ and that the all-pass filter used in two-parameter Kautz networks corresponds to the case $n = 2$ and $\gamma_1 = b$ and $\gamma_2 = c$,

$$G_b(z) = \frac{1-\gamma_1 z}{z-\gamma_1},$$

$$G_b(z) = \frac{1-\gamma_1(1-\gamma_2)z-\gamma_2 z^2}{z^2-\gamma_1(1-\gamma_2)z-\gamma_2}.$$

2.6.3 Gray-Markel Normalized Ladder Form

Another way to look at the classical Gray-Markel normalized ladder realization of an all-pass function corresponds to using Givens rotations as elementary factors of the orthogonal system matrix

$$\begin{bmatrix} A & B \\ C & D \end{bmatrix},$$

which should have a Hessenberg structure (zeros below the first sub-diagonal). The book [254] discusses how to use this structure as a IIR model structure for system identification and adaptive filtering. It seems that the shift structure obtained by cascading all-pass filters allows a simpler asymptotic theory. Also notice that all ladder orthonormal basis functions are of the same order, while the orders of the basis functions obtained by cascading all-pass filters increases with higher index.

Finally, we would like to mentioned lattice/ladder structures for colored input signals. The corresponding basis function and networks are introduced and used in Chapter 11.

2.7 Reproducing Kernels

The theory of orthogonal functions are closely related to reproducing kernel Hilbert spaces. This section, partly based on [208], gives an overview of how to use reproducing kernels to analyze problems related to orthogonal rational functions.

2.7.1 Derivation

Let X_n denote the space spanned by the functions $\{\bar{F}_1(z)\ldots\bar{F}_n(z)\}$, and assume $G(z) \in Y$ where $X_n \subset Y$. Consider the problem of finding the best approximation of $G(z)$ in X_n by solving the optimization problem

$$\min_{\bar{G}\in X_n} ||G - \bar{G}||^2.$$

Because $\bar{G}(z) \in X_n$, we can use the parameterization

$$\bar{G}(z) = \bar{\theta}^T \bar{V}(z), \quad \bar{V}(z) = \left[\bar{F}_1(z)\ldots\bar{F}_n(z)\right]^T, \, \bar{\theta} \in \mathbb{R}^n,$$

to obtain

$$||G - \bar{G}||^2 = \langle G, G\rangle + [\![G, \bar{V}]\!]\,\bar{\theta} + \bar{\theta}^T\,[\![\bar{V}, G]\!] + \bar{\theta}^T\,[\![\bar{V}, \bar{V}]\!]\,\bar{\theta}. \tag{2.22}$$

This is a quadratic function of $\bar{\theta}$ and it is easy to show that

$$\bar{\theta}_n = [\![\bar{V}, \bar{V}]\!]^{-1}\,[\![G, \bar{V}]\!]^T$$

minimizes the cost function (2.22). The corresponding optimal approximation of G in X_n is thus given by

$$G_n(\mu) = [\![G, \bar{V}]\!] \, [\![\bar{V}, \bar{V}]\!]^{-1} \, \bar{V}(\mu)$$

Here we have used the notation subindex n to denote that this is the optimal approximation of $G(z)$ in the n-dimensional subspace X_n. We have also used the argument $z = \mu$ for reasons that soon will become clear.

By defining the so-called reproducing kernel for X_n,

$$K(z, \mu, X_n) = \bar{V}^*(\mu) \, [\![\bar{V}, \bar{V}]\!]^{-1} \, \bar{V}(z),$$

the optimal approximation can be written as

$$G_n(\mu) = \langle G, K(., \mu, X_n) \rangle.$$

Notice that $\langle G, K(., \mu, X_n) \rangle$ is just the orthogonal projection of G on X_n. This means that

$$F(\mu) = \langle F, K(., \mu, X_n) \rangle, \quad \forall F \in X_n,$$

and in particular

$$G_n(\mu) = \langle G_n, K(., \mu, X_n) \rangle = \langle G, K(., \mu, X_n) \rangle.$$

The reproducing kernel for a space is unique and independent of basis. A change of basis $V(z) = T\bar{V}(z)$, for a non-singular square matrix T, gives

$$\begin{aligned} K(z, \mu, X_n) &= \bar{V}^*(\mu) \, [\![\bar{V}, \bar{V}]\!]^{-1} \, \bar{V}(z) \\ &= V^*(\mu)[T^T]^{-1} \left(T^{-1} [\![V, V]\!] \, [T^T]^{-1}\right)^{-1} T^{-1} V(z) \\ &= V^*(\mu) \, [\![V, V]\!]^{-1} \, V(z). \end{aligned}$$

2.7.2 Orthonormal Basis

For an orthonormal basis $V(z) = [F_1(z) \dots F_n(z)]^T$ we have $[\![V, V]\!] = I$, and the expression for the reproducing kernel simplifies to

$$K(z, \mu, X_n) = \sum_{k=1}^{n} F_k^*(\mu) F_k(z).$$

The function

$$K(z, z, X_n) = \sum_{k=1}^{n} |F_k(z)|^2 \tag{2.23}$$

plays an important role in calculating certain Toeplitz forms and parameter covariance matrices. First, notice that the normalization gives

$$\frac{1}{2\pi}\int_{\pi}^{\pi} K(e^{i\omega}, e^{i\omega}, X_n)d\omega = n.$$

The matrix

$$\bar{P} = [\![\bar{V}, \bar{V}]\!]$$

can be viewed as a state covariance matrix, see (2.5). We then have the relation

$$\bar{V}^*(z)\bar{P}^{-1}\bar{V}(z) = K(z, z, X_n),$$

where the right-hand side can be calculated using (2.23). For the Takenaka-Malmquist functions (of which FIR, Laguerre and Kautz are special cases), there is an explicit expression of $K(z, z, X_n)$:

$$F_k(z) = \frac{\sqrt{1-|\xi|_k^2}}{z-\xi_k}\prod_{i=1}^{k-1}\left[\frac{1-\xi_i^* z}{z-\xi_i}\right], \quad |\xi_i| < 1, \quad \Rightarrow$$

$$K(z, z, X_n) = \sum_{i=1}^{n}\frac{1-|\xi_i|^2}{|z-\xi_i|^2}. \tag{2.24}$$

2.7.3 Weighted Spaces and Gramians

Weighting can be incorporated in this framework by studying the space X_n^H spanned by

$$\{F_1(z)H(z)\ldots F_n(z)H(z)\}.$$

By defining the Gramian

$$P_H = [\![VH, VH]\!] = \frac{1}{2\pi i}\oint_{\mathbb{T}} V(z)V^T(1/z)\Phi(z)\frac{dz}{z}, \quad \Phi(z) = H(z)H(1/z),$$

the reproducing kernel for X_n^H can be written

$$K_n(z, \mu, X_n^H) = H^*(\mu)V^*(\mu)P_H^{-1}V(z)H(z). \tag{2.25}$$

The right-hand side of this expression has a nice interpretation in terms of Toeplitz covariance matrices. Let $V(z) = [1\ldots z^{-(n-1)}]^T$. The Gramian P_H is then the covariance matrix corresponding to the spectral density $\Phi(z)$, and from (2.25) we have

$$\frac{V^*(z)P_H^{-1}V(z)}{K(z, z, X_n^H)} = \frac{1}{\Phi(z)}, \tag{2.26}$$

which is closely related to (3.20). As mentioned earlier, it is easy to calculate $K(z, z, X_n^H)$ for certain spaces. For example, if $V(z) = [1\ldots z^{-(n-1)}]^T$ and $\Phi(z)$ corresponds to an autoregressive process $H(z) = z^n/A(z)$, where $A(z) = \prod_{i=1}^{n}(z-\xi_i)$, it follows from (2.24) that

$$K(z, z, X_n^H) = \sum_{i=1}^{n}\frac{1-|\xi_i|^2}{|z-\xi_i|^2}. \tag{2.27}$$

This result can be used to give explicit variance expression for estimated models, see [208,332] and Chapter 5.

Another interesting expression is

$$\begin{aligned}||HK(.,\mu,X_n)||^2 &= \langle HK(.,\mu,X_n), HK(.,\mu,X_n)\rangle \\ &= \frac{1}{2\pi i}\oint_{\mathbb{T}} H(z)\bar{V}^*(\mu)\bar{P}^{-1}\bar{V}(z)\bar{V}^T(1/z)\bar{P}^{-1}\bar{V}(\mu)H(1/z)\frac{dz}{z} \\ &= \bar{V}^*(\mu)\bar{P}^{-1}\bar{P}_H\bar{P}^{-1}\bar{V}(\mu), \end{aligned} \tag{2.28}$$

where

$$\bar{P} = [\![\bar{V},\bar{V}]\!] = \frac{1}{2\pi i}\oint_{\mathbb{T}} \bar{V}(z)\bar{V}^T(1/z)\frac{dz}{z},$$
$$\bar{P}_H = [\![\bar{V}H,\bar{V}H]\!] = \frac{1}{2\pi i}\oint_{\mathbb{T}} \bar{V}(z)\bar{V}^T(1/z)\Phi(z)\frac{dz}{z}.$$

Now

$$\langle HK(.,\mu,X_n), HK_n(.,\mu,X_n)\rangle = \langle \Phi K(.,\mu,X_n), K_n(.,\mu,X_n)\rangle.$$

2.7.4 Convergence

If $\Phi(z)$ would have been an $\mathcal{H}_2(\mathbb{E})$-function and for a complete basis $\{F_k(z)\}_{k=1\ldots\infty}$, we would have that

$$\langle \Phi K_n(.,\mu,X), K_n(.,\mu,X)\rangle \approx \Phi(\mu)K_n(\mu,\mu,X),$$

for large n. Now, $\Phi(z) = H(z)H(1/z)$ has poles outside the unit circle and is not in $\mathcal{H}_2(\mathbb{E})$. However, the result still holds! Define the positive function

$$P(z,\mu,X_n) = \frac{|K(z,\mu,X_n)|^2}{K(\mu,\mu,X_n)},$$

which will behave as a real-valued Dirac function for large n. Also $\Phi(e^{i\omega})$ will be a real valued positive function of ω, and

$$\langle HK(.,\mu,X_n), HK(.,\mu,X_n)\rangle = K(\mu,\mu,X_n)\langle \Phi, P(.,\mu,X_n)\rangle.$$

Now a rather involved analysis, given in [208], proves that

$$\langle \Phi, P(.,\mu,X_n)\rangle \to \Phi(\mu)$$

as the dimension n tends to infinity. This implies

$$\frac{\bar{V}^*(z)\bar{P}^{-1}\bar{P}_H\bar{P}^{-1}\bar{V}(z)}{K(z,z,X_n)} \approx \Phi(z), \quad \text{for large } n.$$

This expression is very useful in calculating the asymptotic variance of model estimates, see Chapter 5.

3

Transformation Analysis

Bo Wahlberg

KTH, Stockholm, Sweden

3.1 Introduction

This chapter describes a framework to use certain transformations and Laurent series expansion to derive orthonormal basis function representations. The presentation is partly based on [317]. The idea is to first reduce the time constants of the transfer function $G(z)$ to be approximated. This is done by a transformation approach, similar to techniques to modify the bandwidth in classical filter design. The transformed system, with faster dynamics, can then be efficiently approximated by a low-order finite impulse response (FIR) model. The corresponding 'FIR' basis functions are next re-mapped to the original domain to obtain infinite impulse response (IIR) orthogonal basis functions, which finally are used to approximate the original transfer function $G(z)$.

3.1.1 Outline

We will study the following approach: Map the transfer function $G(z)$ from the z-domain to what will be called the λ-domain in such a way that the poles of the transformed systems are faster (closer to the origin) than for the original system $G(z)$. The key difficulties are how to preserve stability and orthogonality. This will be done using an all-pass transfer function $G_b(z)$: $z \mapsto 1/\lambda$, that, via the inverse $N(\lambda) : \lambda \mapsto 1/z$, forms an isomorphism that preserves metric and orthogonality. The transfer function $G(z)$ is mapped to transfer function

$$\breve{G}(\lambda) = \sum_{k=1}^{\infty} \breve{g}_k \lambda^{-k},$$

with faster dynamics then $G(z)$, *i.e.* the impulse response $\breve{g}_k$ tends faster to zero, as k goes to infinity, than the impulse response of the original system. This means that the truncated transfer function

$$\breve{G}_n(\lambda) = \sum_{k=1}^{n} \breve{g}_k \lambda^{-k}$$

gives an accurate finite impulse response (FIR) approximation of $\breve{G}(\lambda)$ for a reasonable low value of n. The orthonormal 'delay' basis functions $\{\lambda^{-k}\}$ are then transformed back to the z-domain to give the basis functions $\{F_k(z)\}$. This leads to the approximation

$$G(z) \approx G_n(z) = \sum_{k=1}^{n} \breve{g}_k F_k(z).$$

Preservation of orthogonality implies that $\{F_k(z)\}$ are orthonormal and that the size of the approximation error is the same in both domains. The poles of the set of orthonormal basis function $\{F_k(z)\}$ depend on the poles of all-pass transfer function $G_b(z)$, which hence will be the design variable to generate of a variety of rational orthonormal basis functions.

The transformation approach will now be used to re-derive the classical Laguerre functions (2.17). This analysis will then be extended to cover more general functions.

3.1.2 Notation

Let $\mathbb{E}$ be the exterior of the unit disc including infinity, and $\mathbb{T}$ the unit circle. By $\mathcal{H}_2(\mathbb{E})$ we denote the Hardy space of square integrable functions on $\mathbb{T}$, analytic in the region $\mathbb{E}$. We will in this chapter only study transfer functions with real-valued impulse responses, $G^*(z) = G(z^*)$ where * denotes complex conjugation. For rational functions, this means transfer functions with real valued coefficients. We denote the inner product of $X(z), Y(z) \in \mathcal{H}_2(\mathbb{E})$ by

$$\langle X, Y \rangle := \frac{1}{2\pi} \int_{-\pi}^{\pi} X(e^{i\omega}) Y^*(e^{i\omega}) d\omega = \frac{1}{2\pi i} \oint_{\mathbb{T}} X(z) Y(1/z) \frac{dz}{z},$$

where we have used $Y^*(1/z^*) = Y(1/z)$ for transfer functions with real valued impulse response. For matrix-valued functions $X(z) \in \mathcal{H}_2^{n \times p}(\mathbb{E})$, $Y(z) \in \mathcal{H}_2^{m \times p}(\mathbb{E})$, we will use the cross-Gramian

$$[\![X, Y]\!] := \frac{1}{2\pi i} \oint_{\mathbb{T}} X(z) Y^T(1/z) \frac{dz}{z}.$$

The cross-Gramian is sometimes called the matrix 'outer product'.

3.2 The Laguerre Functions

We will first illustrate how to derive the classical Laguerre functions by the use of a bilinear transformation.

3.2.1 Bilinear Transformation

Consider the bilinear transformation

$$\lambda^{-1} = G_b(z), \; G_b(z) = \frac{1 - az}{z - a}, \; -1 < a < 1,$$

which corresponds to a first-order all-pass transfer function. The inverse transformation equals

$$z^{-1} = N(\lambda), \; N(\lambda) = \frac{1 + a\lambda}{\lambda + a}, \; -1 < a < 1,$$

and is also related to a first-order all-pass transfer function, $N(\lambda)N(1/\lambda) = 1$. The relation between the measures of integration is

$$\frac{d\lambda}{\lambda} = V_1(z)V_1(1/z)\frac{dz}{z}, \quad V_1(z) = \frac{\sqrt{1 - a^2}}{z - a}.$$

The bilinear transformation, illustrated in Figure 3.1, has several important properties. In particular,

$$|z| = 1 \Rightarrow \left| \frac{z - a}{1 - az} \right| = 1,$$

$$|z| > 1 \Rightarrow \left| \frac{z - a}{1 - az} \right| > 1,$$

$$|z| < 1 \Rightarrow \left| \frac{z - a}{1 - az} \right| < 1,$$

which imply that it maps the unit circle into the unit circle. Furthermore the bilinear mapping is one-to-one and the value $z = a$ is mapped to $\lambda = 0$.

3.2.2 Isomorphism

An isomorphism is a map/transformation that preserves sets and relations among elements. Here we are interested in preserving the metric and in particular orthonormality. The problem is then to map the two strictly proper

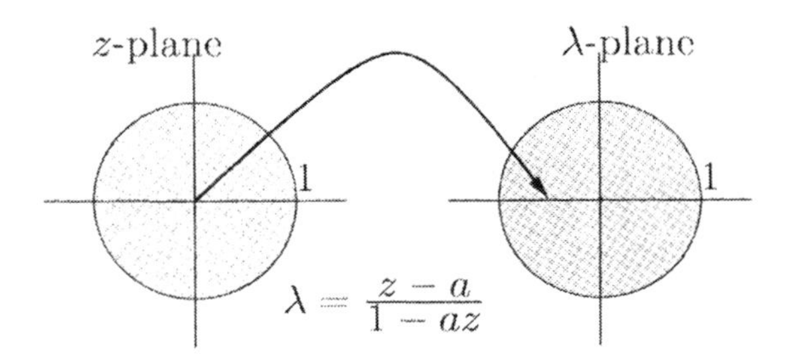

Fig. 3.1. Bilinear transformation.

stable transfer functions $G_1(z)$ and $G_2(z)$ to the λ-domain transfer functions $\breve{G}_1(\lambda)$ and $\breve{G}_2(\lambda)$, $z^{-1} = N(\lambda)$, so that

$$\langle G_1, G_2 \rangle = \langle \breve{G}_1, \breve{G}_2 \rangle.$$

This means that orthogonal functions in the z-domain are transformed into orthogonal functions in the λ-domain, and *vice versa.*
Now change the measure of integration in

$$\begin{aligned}\langle G_1, G_2 \rangle &= \frac{1}{2\pi i} \oint_{\mathbb{T}} G_1(z) G_2(1/z) \frac{dz}{z} \\ &= \frac{1}{2\pi i} \oint_{\mathbb{T}} \left[\frac{G_1(z) G_2(1/z)}{V_1(z) V_1(1/z)} \right]_{z^{-1}=N(\lambda)} \frac{d\lambda}{\lambda}.\end{aligned}$$

Hence, the transformation

$$G(z) \mapsto \breve{G}(\lambda) = \lambda^{-1} \left[\frac{G(z)}{V_1(z)} \right]_{z^{-1}=N(\lambda)} \tag{3.1}$$

gives $\langle G_1, G_2 \rangle = \langle \breve{G}_1, \breve{G}_2 \rangle$ directly. The extra shift λ^{-1} in $\breve{G}(\lambda)$ is introduced to synchronize the impulse responses. The corresponding inverse mapping equals

$$\breve{G}(\lambda) \mapsto G(z) = V_1(z) \left[\lambda \breve{G}(\lambda) \right]_{\lambda^{-1}=G_b(z)}.$$

Introduce the functions

$$\begin{aligned}\tilde{G}(\lambda) &= [G(z)]_{z^{-1}=N(\lambda)} \\ W_0(\lambda) &= \lambda^{-1} \left[\frac{1}{V_1(z)} \right]_{z^{-1}=N(\lambda)} = \frac{\sqrt{1-a^2}}{1+a\lambda},\end{aligned} \tag{3.2}$$

to write

$$\breve{G}(\lambda) = \tilde{G}(\lambda) W_0(\lambda).$$

Due to the factor $W_0(\lambda)$ the transfer function $\breve{G}(\lambda)$ appears to have an unstable pole at $\lambda = -1/a$. This pole is, however, cancelled, since we have assumed that $[G(z)]_{z^{-1}=N(-1/a)} = G(\infty) = 0$, *i.e.* that the relative degree of $G(z)$ is larger than or equal to one. The bilinear mapping preserves stability, since it maps the unit circle into the unit circle.

Example 3.1. Study the first-order transfer function

$$G(z) = \frac{1}{z - \bar{a}} \quad \mapsto \breve{G}(\lambda) = \frac{\sqrt{1-a^2}}{1 - a\bar{a}} \left[\frac{1}{\lambda - (\bar{a} - a)/(1 - a\bar{a})} \right].$$

Hence, the pole is mapped from $\bar{a}$ to

$$G_b(1/\bar{a}) = \frac{\bar{a} - a}{1 - a\bar{a}}.$$

The choice $a = \bar{a}$ makes $\breve{G}(\lambda)$ equal to a pure delay.

3.2.3 The Laguerre Functions

The Laguerre expansion of a transfer function $G(z)$ can now be obtained from the impulse response (Laurent series) expansion of $\breve{G}(\lambda)$,

$$\breve{G}(\lambda) = \sum_{k=1}^{\infty} \breve{g}_k \lambda^{-k} \mapsto G(z) = \sum_{k=1}^{\infty} \breve{g}_k F_k(z),$$

$$\lambda^{-k} \mapsto F_k(z) = \frac{\sqrt{1-a^2}}{z-a} \left[\frac{1-az}{z-a} \right]^{k-1}.$$

We have now derived the classical Laguerre functions , $\{F_k(z)\}$, which by the orthonormal property of $\{\lambda^{-k}\}$ and the isomorphism are orthonormal. This implies that almost all results for FIR functions can be mapped to a similar result for Laguerre functions.

3.2.4 Basic Properties

To conclude this section, we will summarize some basic properties of the Laguerre orthonormal basis functions $F_k(z)$:

- The frequency responses of Laguerre transfer functions

$$F_k(z) = \frac{\sqrt{1-a^2}}{z-a} \left[\frac{1-az}{z-a} \right]^{k-1}$$

  are uniformly bounded:

$$|F_k(e^{i\omega})| \leq \sqrt{\frac{1+a}{1-a}}.$$

  This leads to the error bound:

$$|G(e^{i\omega}) - \sum_{k=1}^{n} \breve{g}_k F_k(e^{i\omega})| \leq \sqrt{\frac{1+a}{1-a}} \sum_{k=n+1}^{\infty} |\breve{g}_k|.$$

- The frequency response of $F_k(z)$ for low frequencies ω can be approximated as

$$F_k(e^{i\omega}) = \frac{\sqrt{1-a^2}}{e^{i\omega}-a} \left[\frac{1-ae^{i\omega}}{e^{i\omega}-a} \right]^{k-1} \approx \frac{\sqrt{1-a^2}}{e^{i\omega}-a} e^{-Ti\omega k}, \quad T = \frac{1+a}{1-a}.$$

  This follows from that the first derivative (the slope) of $(1-ae^{i\omega})/(e^{i\omega}-a)$ for $\omega \to 0$ is $-i(1+a)/(1-a)$, which gives

$$\frac{1-ae^{i\omega}}{e^{i\omega}-a} \approx e^{-i\omega(1+a)/(1-a)}, \quad \text{for small } \omega.$$

For low frequencies, the Laguerre filters act as first-order systems with a time delay approximately equal to $k(1+a)/(1-a)$. The time-delay is increasing by larger a, which means that the Laguerre shift gives more phase-shift for lower frequencies that the ordinary delay shift. The impulse responses of some Laguerre transfer functions are given in Figure 3.2, and the corresponding step responses in Figure 3.3 illustrate this property. Notice that the delay also increases with larger k. Hence, the memory of the filters increases with increasing orders k, and one has to be careful to take the effects of initial conditions into account when using these filters.

- The Laguerre expansion coefficients can be calculated as

$$\breve{g}_k = \langle G, F_k \rangle = \frac{1}{2\pi i} \oint_{\mathbb{T}} G(z) F_k(1/z) \frac{dz}{z}.$$

This follows from the orthonormality of the Laguerre functions. Observing that the integral is the inverse Z-transform of a convolution gives the following interpretation. Let the input to the Laguerre filter $F_k(q)$ be $\{g_{-t}\}$, $-\infty < t < 0$, where g_k is the impulse response of $G(z)$. The corresponding output at $t = 0$ then equals $\breve{g}_k$.

- Let X_n be the space spanned by the Laguerre functions $\{F_k(z),\ k = 1 \dots n\}$, and study the optimal approximation

$$\begin{aligned} G_n(e^{i\omega_0}) &= \sum_{k=1}^{n} \breve{g}_k F_k(e^{i\omega_0}) = \frac{1}{2\pi} \int_{-\pi}^{\pi} G(e^{i\omega}) \left[\sum_{k=1}^{n} F_k(e^{-i\omega}) F_k(e^{i\omega_0}) \right] d\omega \\ &= \frac{1}{2\pi} \int_{-\pi}^{\pi} G(e^{i\omega}) K^*(e^{i\omega}, e^{i\omega_0}, X_n) d\omega = \langle G, K(., e^{i\omega_0}, X_n) \rangle. \end{aligned}$$

Here we have defined the Laguerre frequency window function (reproducing kernel)

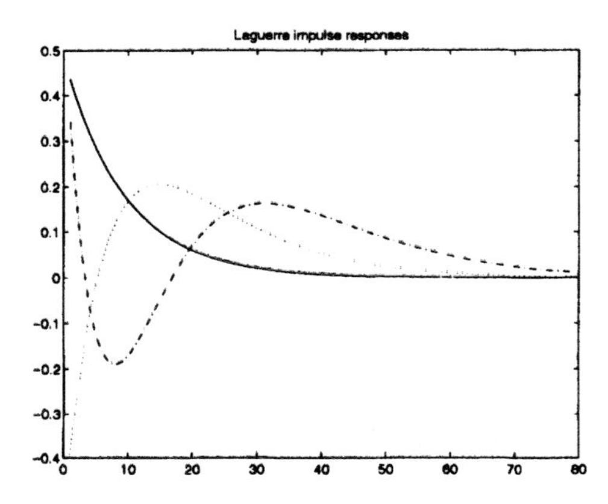

Fig. 3.2. Impulse response for $F_1(z)$ (solid line), $F_2(z)$ (dotted line), $F_3(z)$ (dash-dotted line) for $a = 0.9$.

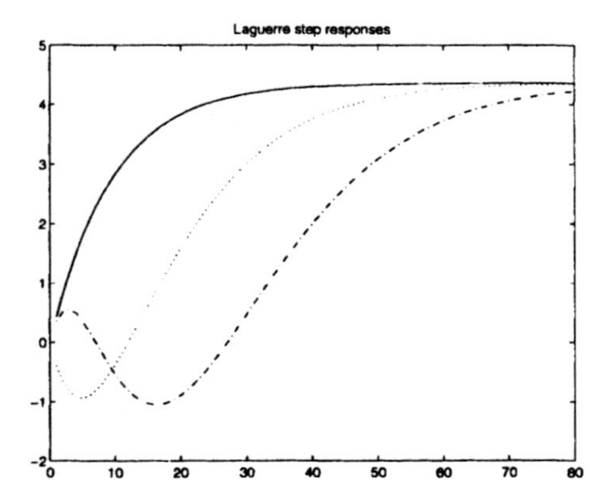

Fig. 3.3. Step response for $F_1(z)$ (solid line), $F_2(z)$ (dotted line), $F_3(z)$ (dash-dotted line) for $a = 0.9$.

$$K(z, \mu, X_n) = \sum_{k=1}^{n} F_k(z) F_k^*(\mu).$$

Hence, the truncated expansion approximation can be viewed as smoothed frequency response, compare Figure 3.4. We refer to Section 2.7, or to [58] for further details.

- For further information on classical results related to Laguerre orthogonal functions and polynomials, such as recursion formulas *etc.*, see *e.g.* the classical book by Szegö, [294], or [61, 96, 139, 201, 204, 261, 265].

3.3 Generalized Orthonormal Basis Functions

We will now extend the Laguerre transformation analysis by more or less repeating the steps in the previous section to cover more general function classes. Though this extension is in principle straightforward, it leads to some quite advanced results on all-pass systems.

3.3.1 Multi-linear Transformation

The way to extend the bilinear transformation to multiple poles is to use the (multi-linear) transformation

$$\lambda^{-1} = G_b(z), \quad G_b(z) = \prod_{j=1}^{m} \frac{1 - \xi_j^* z}{z - \xi_j},$$

where $G_b(z)$ is a mth-order stable all-pass transfer function with a real valued impulse response,

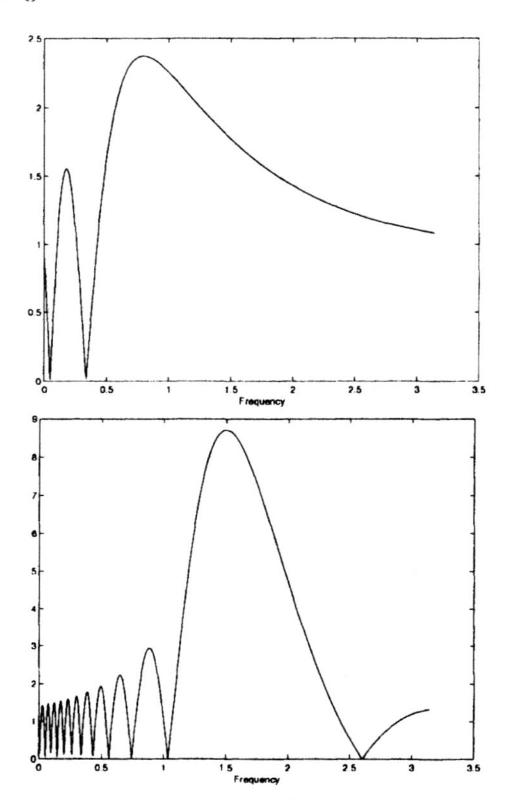

Fig. 3.4. The Laguerre frequency window function $|K(.,e^{i\omega_0},X_n)|$, with $a = 0.7$ and $\omega_0 = \pi/2$, for $n = 5$ (left figure) and $n = 25$ (right figure).

$$G_b(z)G_b(1/z) = 1,$$

specified (up to a sign) by its poles $\{\xi_j \in \mathbb{C};\ |\xi_j| < 1,\ j = 1\ldots m\}$. The mapping $\lambda^{-1} = G_b(z)$ or equally $\lambda = G_b(1/z)$ maps the unit disc onto the unit disc, because

$$\begin{aligned} |z| = 1 &\Rightarrow |G_b(1/z)| = 1, \\ |z| > 1 &\Rightarrow |G_b(1/z)| > 1, \\ |z| < 1 &\Rightarrow |G_b(1/z)| < 1. \end{aligned}$$

The all-pass property gives $G_b(z)G_b^*(1/z^*) = 1$, and if $|z| = 1$ we have $z = 1/z^*$, it follows that $|G_b(z)|^2 = 1$. The second inequality follows from the equality and from the maximum modulus theorem (as $G_b(1/z)$ does not have any poles in $|z| < 1$). The first inequality follows from the second, as $|G_b(z)| = 1/|G_b(1/z)|$, and if $|z| > 1$ then $|1/z| < 1$. Also notice that $\lambda^{-1} = G_b(z)$ maps $z = \xi_j$, $j = 1\ldots m$ to $\lambda = 0$.

3.3.2 Outline

The idea is now to mimic the transformation analysis of Laguerre models.

- First we would like to find a isomorphic transformation $G(z) \mapsto \breve{G}(\lambda)$ such that the poles of $G(z)$ are mapped as $\lambda^{-1} = G_b(z)$. The new difficulty is that the inverse mapping $\lambda \mapsto z$ is 1-to-m. It turns out that the function $\breve{G}(\lambda)$ will be vector-valued.
- The next challenge is to characterize the inverse transformation $\breve{G}(\lambda) \mapsto G(z)$ to be able to re-map the basis function λ^{-k} to give the new orthogonal basis functions $F_k(z)$.
- A by-product of this analysis will be results on how to calculate correlations between signals of the form $x_k(t) = F_k(q)u(t)$, where $u(t)$ is a stochastic process with spectral density $\Phi_u(z)$.

3.3.3 A Root-locus Approach

As mentioned above, the main problem is to describe the inverse mapping $\lambda \mapsto z$, which for the general case is 1-to-m. By writing $\lambda^{-1} = G_b(z)$ as

$$1 - \lambda G_b(z) = 0$$

we can see that the inverse mapping problem equals a standard root locus problem for stability analysis in feedback control. For a given (positive) feedback gain λ we would like to find the roots of $1 - \lambda G_b(z) = 0$. The roots are the poles of the closed loop system

$$\begin{aligned} y(t) &= G_b(q)u(t) \\ u(t) &= \lambda y(t), \end{aligned}$$

where q denotes the shift operator. Let (A, B, C, D) be an orthogonal state-space realization of $G_b(z)$, *i.e.* a state-space realization

$$\begin{bmatrix} x(t+1) \\ y(t) \end{bmatrix} = \begin{bmatrix} A & B \\ C & D \end{bmatrix} \begin{bmatrix} x(t) \\ u(t) \end{bmatrix},$$

which satisfies

$$\begin{bmatrix} A & B \\ C & D \end{bmatrix} \begin{bmatrix} A & B \\ C & D \end{bmatrix}^T = I. \tag{3.3}$$

Using

$$u(t) = \lambda y(t) = \lambda Cx(t) + \lambda Du(t) \;\Rightarrow\; u(t) = [1/\lambda - D]^{-1} Cx(t),$$

in $x(t+1) = Ax(t) + Bu(t)$, the closed loop system in state-space form equals

$$\begin{aligned} x(t+1) &= N(1/\lambda)x(t), \\ N(\lambda) &= A + B(\lambda - D)^{-1}C. \end{aligned} \tag{3.4}$$

The poles of the closed loop system is given by the eigenvalues of state-space matrix $N(1/\lambda)$. Consequently, the eigenvalues $\{z_j,\ j = 1 \dots m\}$ of $N(1/\lambda)$ satisfy the equation $\lambda^{-1} = G_b(z_j)$. We have now characterized the inverse mapping.

3.3.4 Eigenvalues and Eigenvectors

Taking the Z-transform of (3.4) gives

$$z[X(z) - x(0)] = N(1/\lambda)X(z), \quad X(z) = V_1(z)U(z) + z(zI - A)^{-1}x(0).$$

Evaluate these expressions for $x(0) = 0$,

$$[zI - N(1/\lambda)]X(z) = 0, \quad X(z) = V_1(z)U(z),$$

which only has a non-trivial solution if $z = z_j$, the eigenvalues of $N(1/\lambda)$. Hence, $\{V_1(z_j),\ j = 1 \dots m\}$ are eigenvectors of $N(1/\lambda)$. We have now used properties of the closed loop system to derive the Hambo-domain shift function $N(1/\lambda)$ introduced using more advanced arguments by Van den Hof *et al.* [306].

3.3.5 The Christoffel-Darboux Formula

The so-called (generalized) Christoffel-Darboux (C-D) formula

$$V_1^T(z_1)V_1(z_2) = \frac{G_b(z_1)G_b(z_2) - 1}{1 - z_1 z_2}. \tag{3.5}$$

will be very useful in the forthcoming analysis. The following simple direct proof is due to [131]: Using (3.3), $C^T D = -A^T B$, $C^T C = I - A^T A$, $D^T D = I - B^T B$ together with $B = (zI - A)V(z)$ directly gives $G_b(z_1)G_b(z_2) - 1 = V_1^T(z_1)[1 - z_1 z_2]V_1(z_2)$.

If $G_b(1/z_i) = G_b(1/z_j) = \lambda$ and $z_i \neq z_j$, the C-D formula gives

$$V_1^T(z_i)V_1(1/z_j) = \frac{G_b(z_i)G_b(1/z_j) - 1}{1 - z_1 z_2} = 0. \tag{3.6}$$

3.3.6 Eigenvalue Decomposition

From the orthogonality property (3.6), we directly obtain the following eigenvalue decomposition

$$N(\lambda) = \sum_{k=1}^{m} \frac{1}{z_k} \frac{1}{V_1^T(z_k)V_1(1/z_k)} V_1(1/z_k)V_1^T(z_k). \tag{3.7}$$

If $|\lambda| = 1$, $N(\lambda)$ is a unitary matrix and thus has orthogonal eigenvectors. Furthermore, the corresponding eigenvalues will be distinct. The C-D formula gives a generalization of this result as long as the eigenvalues are distinct and not equal to zero. Also notice that $N(\lambda)$ is a MIMO all-pass transfer function, $N(\lambda)N^T(1/\lambda) = I$, with a orthogonal state-space realization (D, C, B, A).

3.3.7 Measure of Integration

We also need to find $d\lambda/\lambda$. This can also be done using the C-D formula. Let

$$z_1 = \frac{1}{z+\Delta}, \; z_2 = z \quad \Rightarrow 1 - z_1 z_2 = \Delta/(z+\Delta).$$

By substituting this expression into (3.5) and letting $\Delta \to 0$, we obtain

$$\begin{aligned} V_1^T(1/z)V_1(z) &= zG_b(z)\frac{d}{dz}[G_b(1/z)] \\ \Rightarrow \quad \frac{d\lambda}{\lambda} &= V_1^T(1/z)V_1(z)\frac{dz}{z}. \end{aligned} \tag{3.8}$$

This shows how the measure of integration changes under the change of variable, *i.e.*

$$\lambda^{-1} = G_b(z), \quad \Rightarrow \quad \frac{d\lambda}{\lambda} = V_1^T(1/z)V_1(z)\frac{dz}{z}. \tag{3.9}$$

Remark 3.1. The inverse multi-valued mapping can be interpreted as single-valued using the theory of branches and Riemann surfaces. The difficult λ points are the so-called branch points, where the inverse mapping collapses from being 1-to-m to taking fewer values. The branch points satisfy $d/dz\, G_b(1/z) = 0$, and thus using (3.8), $G_b(1/z)\, V_1^T(1/z)V_1(z)/z = 0$. Hence, $\lambda = G_b(1/z) = 0$ may be a branch point if $G_b(z)$ has multiple poles. The other branch points are given by $V_1^T(1/z)V_1(z)/z = 0$. Because $V_1^T(1/z)V_1(z) = V_1^*(z)V_1(z) > 0$ for $|z| = 1$, the inverse function is always 1-to-m on the unit circle, *i.e.*, if $G_b(1/z_j) = \lambda$, $j = 1 \ldots m$, with $|\lambda| = 1$, then $z_j \neq z_k$ for $j \neq k$. This result can also be obtained by studying the phase function $\beta(\omega) : [0 : 2\pi] \mapsto [0 : 2m\pi]$ from $\lambda = e^{i\beta(\omega)} = G_b(e^{-i\omega})$. As noted in [238] and [270], $\beta(\omega)$ is a strictly increasing one-to one function and thus has a well defined inverse. See also Chapter 8.
For the special case $G_b(z) = z^{-m}$ we have $\beta(\omega) = m\omega$, and the inverse function can be taken to $\beta^{-1}(\bar{\omega}) = \bar{\omega}/m$. However, there are m values that are mapped to the same λ, namely $\bar{\omega}_j = \omega + 2\pi(j-1)$, $\quad j = 1 \ldots m$, $\omega \in [0, 2\pi]$, for which $z_j = e^{i\beta^{-1}(\bar{\omega}_j)}$, $\quad j = 1 \ldots m$. This is the standard Riemann surfaces interpretation of the function $z = \lambda^{1/m}$.

Assume that z_j, $j = 1 \ldots m$, are ordered and distinct, *cf.* the discussion above. Hence, given λ we can uniquely find $\{z_j\}$. In the sequel we will use a mixed notation with λ and $\{z_j\}$.

3.3.8 The Function $W_0(\lambda)$

Finally, we need to generalize the function $W_0(\lambda)$ defined by Equation (3.2), *i.e.* we want to find $W_0(\lambda)$ such that $V_1^T(z_j)W_0(\lambda) = \lambda^{-1}$. A solution is obtained by studying the inputs-to-state transfer function for the orthogonal state-space realization (D, C, B, A) of $N(\lambda)$,

$$W_0(\lambda)^T = \lambda^{-1}(\lambda^{-1}I - D)^{-1}C,$$

for which

$$\begin{aligned} V_1^T(z_j)W_0(\lambda) = W_0^T(\lambda)V_1(z_j) &= \lambda^{-1}(\lambda^{-1}I - D)^{-1}C[(z_jI - A)^{-1}B] \\ &= \lambda^{-1}(\lambda^{-1} - D)^{-1}(G_b(z_j) - D) = \lambda^{-1}. \end{aligned} \tag{3.10}$$

A more insightful interpretation of $W_0(\lambda)$ is given in Chapter 12.

3.3.9 Summary

All results and definitions have now been obtained in order to generalize the Laguerre analysis: We will use the transformation $\lambda^{-1} = G_b(z)$, where $G_b(z)$ is all-pass transfer function of order m with an orthogonal state-space realization (A, B, C, D). We have showed that

$$\frac{d\lambda}{\lambda} = V_1^T(1/z)V_1(z)\frac{dz}{z}, \quad V_1(z) = (zI - A)^{-1}B.$$

The transformation $z = N(1/\lambda)$ can be used as an 'inverse' of $\lambda^{-1} = G_b(z)$. We will, however, use the transformation $z^{-1} = N(\lambda) \Rightarrow z = N^T(1/\lambda)$ instead of $z = N(1/\lambda)$. The reason for using this extra transpose is to follow the standard Hambo transform definitions given in Chapter 12.

3.3.10 The Hambo Transform

The Hambo (system) transform of a transfer function $G(z)$ is defined as

$$G(z) \quad \mapsto \quad \tilde{G}(\lambda) = [G(z)]_{z^{-1}=N(\lambda)}.$$

The substitution $z^{-1} = N(\lambda)$ must by interpreted in terms of a series expansion. By using the orthogonality property of the eigenvalue-decomposition (3.7) of $N(\lambda)$, we obtain

$$\begin{aligned} \tilde{G}(\lambda) &= [\sum_{j=1}^{\infty} g_j z^{-j}]_{z^{-1}=N(\lambda)} = \sum_{j=1}^{\infty} g_j [N(\lambda)]^j \\ &= \sum_{k=1}^{m} \frac{\sum_{j=1}^{\infty} g_j z_k^{-j}}{V_1^T(z_k)V_1(1/z_k)} V_1(1/z_k)V_1^T(z_k) \\ &= \sum_{k=1}^{m} \frac{G(z_k)}{V_1^T(z_k)V_1(1/z_k)} V_1(1/z_k)V_1^T(z_k). \end{aligned}$$

Because $\tilde{G}(\lambda)V_1(1/z_j) = G(z_j)V_1(1/z_j)$, $G(z_j)$ is an eigenvalue of $\tilde{G}(\lambda)$.
It is easy to compute $\tilde{G}(\lambda)$ for a rational $G(z)$ using a matrix fraction description:

$$G(z) = \frac{B(z)}{A(z)} \mapsto \tilde{G}(\lambda) = \left([A(z)]_{z^{-1}=N(\lambda)}\right)^{-1} [B(z)]_{z^{-1}=N(\lambda)}.$$

The transformed system $\tilde{G}(\lambda)$ has many common properties with $G(z)$, for example, the same McMillan degree and the same Hankel singular values, see Chapter 12 or [131].

3.3.11 Transformation of Poles and Zeros

The relation between the poles and zeros of $G(z)$ and $\tilde{G}(\lambda)$ follows from

$$\det \tilde{G}(\lambda) = \prod_{k=1}^{m} G(z_k).$$

Hence, if $G(z)$ has a pole or a zero at $z = \alpha$, then $\tilde{G}(\lambda)$ has a pole or a zero at $\lambda = G_b(1/\alpha)$. This is exactly the feature we were aiming for!

3.3.12 Transformation of All-pass Transfer Function

Another important result is that the all-pass transfer function $G_b(z)$ is mapped to

$$\begin{aligned} \tilde{G}_b(\lambda) &= \sum_{k=1}^{m} \frac{G_b(z_k)}{V_1^T(z_k)V_1(1/z_k)} V_1(1/z_k)V_1^T(z_k) \\ &= \sum_{k=1}^{m} \frac{\lambda^{-1}}{V_1^T(z_k)V_1(1/z_k)} V_1(1/z_k)V_1^T(z_k) = \lambda^{-1} I. \end{aligned} \tag{3.11}$$

3.3.13 The Inverse Hambo Transform

The inverse transform can now easily be obtained using (3.10). Introduce the notation $z = z_1$, and note that $z_k = z_k(z)$, $k = 2 \ldots m$. Then

$$G(z) = V_1^T(z)[\lambda \tilde{G}(\lambda) W_0(\lambda)]_{\lambda^{-1}=G_b(z)}.$$

3.3.14 The Hambo Signal Transform

The next step is to scale $\tilde{G}(\lambda)$ with the measure of integration as in the Laguerre case (3.1) to prepare for the isomorphism. This leads to the to the vector valued Hambo signal transform:

$$\begin{aligned} G(z) &\mapsto \breve{G}(\lambda) = [G(z)]_{z^{-1}=N(\lambda)} W_0(\lambda), \\ \breve{G}(\lambda) &\mapsto G(z) = V_1^T(z) \left[\lambda \breve{G}(\lambda)\right]_{\lambda^{-1}=G_b(z)}. \end{aligned} \tag{3.12}$$

The poles of $\breve{G}(\lambda)$ are the same as of $\tilde{G}(\lambda)$. Notice that $W_0(\lambda)$ will introduce one unstable pole at $\lambda = 1/D$. However, by assuming that $G(z)$ is strictly proper, it will have one zero at infinity, which maps to a zero at $\lambda = 1/D$. Hence, this unstable pole is cancelled.

We have outlined a quite abstract theory for which we now will explicitly describe for the following example.

Example 3.2. Let us exemplify the theory by going back to the FIR case, $G_b(z) = z^{-m}$, which has the simple orthonormal state-space realization

$$\begin{bmatrix} A & B \\ C & D \end{bmatrix} = \begin{bmatrix} 0 \dots 0 & 1 \\ 1 \dots 0 & 0 \\ \vdots \ddots \vdots & \vdots \\ 0 \dots 1 & 0 \end{bmatrix}.$$

Here

$$V(z) = \begin{bmatrix} z^{-1} \\ \vdots \\ z^{-m} \end{bmatrix}, \qquad N(\lambda) = \begin{bmatrix} 0 \dots 0 & \lambda^{-1} \\ 1 \dots 0 & 0 \\ \vdots \ddots \vdots & \vdots \\ 0 \dots 1 & 0 \end{bmatrix}.$$

Furthermore, if $z^m = \lambda$ then

$$N(\lambda)V(1/z) = \begin{bmatrix} \lambda^{-1} z^m \\ z \\ \vdots \\ z^{m-1} \end{bmatrix} = \frac{1}{z} V(1/z).$$

The special case $\lambda = 1$ leads to the discrete Fourier transform frequency gridding, $z_j = e^{2\pi i j/m}$, for which the orthogonality of $V(1/z_j)$ is well-known.

3.3.15 Isomorphism

We now have exactly the same setup as for the Laguerre case. It only remains to establish the isomorphism.

Theorem 3.1. *Assume that $G_k(z) \in \mathcal{H}_2(\mathbb{E})$ and that $G_k(z) \mapsto \breve{G}_k(\lambda)$, $k = 1, 2$, according to (3.12). Then*

$$\langle G_1, G_2 \rangle = \left[\!\left[\breve{G}_1^T, \breve{G}_2^T \right]\!\right] \tag{3.13}$$

Proof. To prove this, use

$$\breve{G}_1(\lambda) = \lambda^{-1} \sum_{k=1}^{m} \frac{G_1(z_k)}{V_1^T(z_k) V_1(1/z_k)} V_1(1/z_k),$$

and the same for $\breve{G}_2(\lambda)$, and

$$\breve{G}_1^T(\lambda)\breve{G}_2(1/\lambda) = \sum_{k=1}^{m} \frac{G_1(z_k)G_2(1/z_k)}{V^T(z_k)V(1/z_k)}.$$

Now use the (one-to-one branch) change of variable of integration

$$\lambda^{-1} = G_b(z_k), \quad \Rightarrow \quad \frac{d\lambda}{\lambda} = V^T(z_k)V(1/z_k)\frac{dz_k}{z_k}$$

to obtain

$$\begin{aligned}\frac{1}{2\pi i}\oint_{\mathbb{T}} \breve{G}_1^T(\lambda)\breve{G}_2(1/\lambda)\frac{d\lambda}{\lambda} &= \sum_{k=1}^{m} \frac{1}{2\pi i}\oint_{\mathbb{T}_k} G_1(z)G_2(1/z)\frac{dz}{z} \\ &= \frac{1}{2\pi i}\oint_{\mathbb{T}} G_1(z)G_2(1/z)\frac{dz}{z},\end{aligned}$$

where $\mathbb{T}_k$ is the part of the unit circle corresponding to the one-to-one mapping $\lambda \mapsto z_k$. The last equality follows from $\cup_{k=1}^{m}\mathbb{T}_k = \mathbb{T}$.

■

Extension. The same calculation can be used to show that

$$[\![V_1G_1, V_1G_2]\!] = \left[\!\left[\tilde{G}_1^T, \tilde{G}_2^T\right]\!\right] \tag{3.14}$$

by noting that

$$\tilde{G}_1^T(\lambda)\tilde{G}_2(1/\lambda) = \sum_{k=1}^{m} \frac{G_1(z_k)G_2(1/z_k)}{V_1^T(z_k)V_1(1/z_k)}V_1(z_k)V_1^T(1/z_k).$$

For $G_1 = G_2 = 1$ and thus $\tilde{G}_1 = \tilde{G}_2 = I$, the result (3.14) confirms the orthogonality of $V_1(z)$.

3.3.16 Generalized Orthonormal Basis Functions

We have now completed the extension of the Laguerre transformation analysis, and it only remains to use this setup to derive the corresponding orthogonal basis functions. The transformed function $\breve{G}(\lambda)$ can be used to derive the so-called generalized orthonormal basis function (GOBF) expansion of a strictly proper stable transfer function $G(z)$. Study the impulse response (Laurent series) expansion

$$\breve{G}(\lambda) = \sum_{k=1}^{\infty} K_k\lambda^{-k},$$

where the impulse response $\{K_k\}$ of $\breve{G}(\lambda)$ is vector-valued. Using the inverse transformation (3.9) then gives

$$G(z) = \sum_{k=1}^{\infty} V_1^T(z) K_k [G_b(z)]^{k-1},$$

which is the GOBF expansion of $G(z)$. The corresponding vector valued basis functions, denoted by

$$V_k(z) = V_1(z)[G_b(z)]^{k-1}, \tag{3.15}$$

are by construction orthonormal.

3.3.17 Rate of Convergence

Because the poles of $\breve{G}(\lambda)$ are a subset of $\{G_b(1/p_j)\}$, where $\{p_j\}$ are the poles of $G(z)$,

$$K_k \sim \mathcal{O}(\max_j |G_b(1/p_j)|^{k-1}).$$

This result is closely related to the error expression (2.2). Hence, by proper choice of $G_b(z)$ related to *a priori* information about the poles of $G(z)$, we can approximate $G(z)$ by a low-order GOBF model

$$G_{OBF}(z) = \sum_{k=1}^{n} V_1^T(z) K_k [G_b(z)]^{k-1}. \tag{3.16}$$

An interesting special case is actually $n = 1$, *i.e.* just the first term of the expansion. This means that we only use the components of $V_1(z)$ as basis functions.

We have now re-derived the GOBF basis functions (2.19). At the same time, we have developed a framework for analyzing the corresponding GOBF approximations. This is the topic of the next section.

3.4 State Covariance Matrix Results

Consider a coloured input signal $u(t)$, with spectral density

$$\Phi_u(z) = H_u(z) H_u(1/z).$$

Here $H_u(z)$ is a stable spectral factor of $\Phi_u(z)$. We are interested in the stochastic properties of the 'orthogonal states' of $G_b(z)$ for this input signal, *i.e.* $X(z) = V_1(z)U(z)$. In case of a white input signal, the covariance matrix of $x(t)$ equals the identity matrix. This is, however, not the case for a colored input, but it will still have a very specific structure. This is the topic of this section.

3.4.1 Transformation of Spectra

The spectral density of the states $X(z) = V_1(z)U(z)$ of the all-pass filter equals

$$\Phi_x(z) = V_1(z)V_1^T(1/z)\Phi_u(z) = V_1(z)H_u(z)V_1^T(1/z)H_u(1/z),$$

and the corresponding state covariance matrix is

$$R_x = \frac{1}{2\pi i}\oint_{\mathbb{T}} V_1(z)V_1^T(1/z)\Phi_u(z)\frac{dz}{z} = [\![V_1 H_u, V_1 H_u]\!]\,. \tag{3.17}$$

We have here used a notation R_x, which is related to linear regression estimation instead of P, which will be exclusively used for the white noise case.

3.4.2 Key Results

By using the Hambo domain spectral density

$$\tilde{\Phi}_u(\lambda) = \tilde{H}_u(\lambda)\tilde{H}_u^T(1/\lambda),$$

and the results (3.14) and (3.17), we directly obtain the expression:

$$R_x = \frac{1}{2\pi i}\oint_{\mathbb{T}} \tilde{\Phi}_u(1/\lambda)\frac{d\lambda}{\lambda}. \tag{3.18}$$

This means that the R_x is the variance of a vector valued stochastic process with spectral density $\tilde{\Phi}_u(1/\lambda)$. This result can be generalized to determine the cross-correlations between the states $X_i(z) = V_i(z)U(z)$ and $X_j(z) = V_j(z)U(z)$, where $\{V_j(z)\}$ are the GOBF vector basis functions defined by (3.15). The result, which can be proved using (3.11), is

$$\frac{1}{2\pi i}\oint_{\mathbb{T}} V_i(z)V_j^T(1/z)\Phi_u(z)\frac{dz}{z} = \frac{1}{2\pi i}\oint_{\mathbb{T}} \tilde{\Phi}_u(1/\lambda)\lambda^{i-j}\frac{d\lambda}{\lambda}.$$

Remark 3.2. These results, first derived in [306], show that the covariance matrix of the (extended) state vector has a special structure directly related to the transformed spectra $\tilde{\Phi}_u(1/\lambda)$. Furthermore, the covariance matrix corresponding to the total input to states transfer function

$$V_{GOBF}(z) = [V_1^T(z)\dots V_{n_H}^T(z)]^T$$

has a block Toeplitz structure corresponding to the Hambo domain spectral density $\tilde{\Phi}_u(1/\lambda)$. This can then be used to generalize results from identification of FIR models, *cf.* Chapters 4 and 5.

3.4.3 Approximative Factorizations

Next we will derive approximate factorizations of R_x, which are valid for large dimensions (large m). Such results are well-known for Toeplitz matrices and are related to Fourier transforms and spectra, see [108].

The state covariance matrix R_x can for large values of m be approximated by

$$R_x \approx \tilde{\Phi}_u(1/\lambda).$$

We will not formalize this as a theorem nor describe in which sense (norm) the approximation holds. Instead, we will outline the key ideas with a certain abuse of stringency. The rigorous conditions for the approximations to hold together with the quite involved formal proofs of these results are given in [222], but in a different and more advanced framework.

Idea of Proof. The proof is based on the approximation

$$\begin{aligned}
&\frac{1}{2\pi i}\oint_{\mathbb{T}_k} V_1(z)V_1^T(1/z)\Phi_u(z)\frac{V_1^T(z)V_1(1/z)}{V_1^T(z)V_1(1/z)}\frac{dz}{z} \approx \\
&\quad \frac{\Phi_u(z_k)}{V_1^T(z_k)V_1(1/z_k)}V_1(z_k)V_1^T(1/z_k)\frac{1}{2\pi i}\oint_{\mathbb{T}_j} V_1^T(z)V_1(1/z)\frac{dz}{z} \\
&\quad = \frac{\Phi_u(z_k)}{V_1^T(z_k)V_1(1/z_k)}V_1(z_k)V_1^T(1/z_k)\frac{1}{2\pi i}\oint_{\mathbb{T}}\frac{d\lambda}{\lambda} \\
&\quad = \frac{\Phi_u(z_k)}{V_1^T(z_k)V_1(1/z_k)}V_1(z_k)V_1^T(1/z_k).
\end{aligned}$$

This directly gives

$$\begin{aligned}
R_x &= \frac{1}{2\pi i}\oint_{\mathbb{T}} V_1(z)V_1^T(1/z)\Phi_u(z)\frac{dz}{z} = \sum_{j=k}^{m}\frac{1}{2\pi i}\oint_{\mathbb{T}_k} V_1(z)V_1^T(1/z)\Phi_u(z)\frac{dz}{z} \\
&\approx \sum_{k=1}^{m}\frac{\Phi_u(z_k)}{V_1^T(z_k)V_1(1/z_k)}V_1(z_k)V_1^T(1/z_k) = \tilde{\Phi}_u(1/\lambda).
\end{aligned}$$

The main steps in the approximation are that $\mathbb{T}_j$ tends to z_j as $m \to \infty$, that the length of $\mathbb{T}_j$ approximately is $1/V_1^T(z_j)V_1(1/z_j)$, and that $V_1^T(z_j)V_1(1/z_j)$ behaves as a Dirac function as $m \to \infty$. Hence, z_j, $j = 1 \dots m$ form a dense grid of the unit circle. However, we must also be careful with end effects, meaning that the approximation may not be good in the corners of the matrix. Notice that the dimension of R_x tends to infinity as $m \to \infty$.

Remark 3.3. To start with, notice that the approximation indicates that $\tilde{\Phi}_u(\lambda)$ is approximately constant for large m and $|\lambda| = 1$. The natural approximation is then

$$R_x \approx \tilde{\Phi}_u(1) = \sum_{k=1}^{m} \frac{\Phi_u(z_k)}{V_1^T(z_k)V_1(1/z_k)} V_1(z_k)V_1^T(1/z_k), \tag{3.19}$$

where $G_b(1/z_k) = 1$, $k = 1 \ldots m$.

The FIR transformation $G_b(z) = z^{-m}$ and taking $\lambda = 1$ leads to the discrete Fourier transform, *cf.* Example 3.2. The corresponding covariance matrix R_x will then be Toeplitz, and it is well-known that it can be approximately diagonalized using the DFT transform. In the same way for a general all-pass filter $G_b(z)$, the solutions to $G_b(1/z_k) = 1$, and the corresponding vectors $V_1(z_k)$ result in a discrete orthonormal basis functions transform, and (3.19) is the corresponding approximate factorization of R_x.

3.4.4 Approximative Spectral Density Results

Let us use the approximations to derive some results that are instrumental in deriving high-order transfer function variance expressions (see Chapters 4 and 5). Study

$$\begin{aligned}
&\frac{V_1^T(1/z_i)R_xV_1(z_j)}{V_1^T(1/z_i)V_1(z_i)} \\
&\approx \sum_{k=1}^{m} \frac{\Phi_u(z_k)}{V_1^T(z_k)V_1(1/z_k)} \frac{1}{V_1^T(1/z_i)V_1(z_i)} V_1^T(1/z_i)V_1(z_k)V_1^T(1/z_k)V_1(z_j) \\
&= \frac{\Phi_u(z_i)}{V_1^T(z_i)V_1(1/z_i)} \frac{1}{V_1^T(1/z_i)V_1(z_i)} V_1^T(1/z_i)V_1(z_i)V_1^T(1/z_i)V_1(z_j) \\
&= \begin{cases} \Phi_u(z_i), & i = j \\ 0, & i \neq j \end{cases}.
\end{aligned}$$

We have extensively used the orthogonality condition $V_1^T(z_j)V_1(1/z_i) = 0$ for $i \neq j$. The approximation is only valid for large m. Similarly

$$\begin{aligned}
&\frac{V_1^T(1/z_i)R_x^{-1}V_1(z_j)}{V_1^T(1/z_i)V_1(z_i)} \\
&\approx \sum_{k=1}^{m} \frac{1}{\Phi_u(z_k)} \frac{1}{V_1^T(z_k)V_1(1/z_k)} \frac{1}{V_1^T(1/z_i)V_1(z_i)} V_1^T(1/z_i)V_1(z_k)V_1^T(1/z_k)V_1(z_j) \\
&= \frac{1}{\Phi_u(z_i)} \frac{1}{V_1^T(z_i)V_1(1/z_i)} \frac{1}{V_1^T(1/z_i)V_1(z_i)} V_1^T(1/z_i)V_1(z_i)V_1^T(1/z_i)V_1(z_l) \\
&= \begin{cases} \frac{1}{\Phi_u(z_i)}, & i = j \\ 0, & i \neq j \end{cases}.
\end{aligned} \tag{3.20}$$

Because the $\{z_j\}$ form a dense grid of the unit circle for $\lambda = 1$ when $m \to \infty$, we have now derived the results in [222]. As mentioned above, formal proofs of the above derivation require a much more rigorous treatment. However, the objective of our analysis is to give more intuitive insights. The application of these results in system identification is presented in Chapters 4 and 5.

4

System Identification with Generalized Orthonormal Basis Functions

Paul Van den Hof[1] and Brett Ninness[2]

[1] Delft University of Technology, Delft, The Netherlands
[2] University of Newcastle, NSW, Australia

4.1 Introduction

When identifying a dynamical system on the basis of experimentally measured data records, one of the most important issues to address carefully is the choice of an appropriate model structure or model set. The model set reflects the collection of models among which a best model is sought on the basis of the given data records. The choice of model set directly influences the maximum achievable accuracy of the identified model. On the one hand, the model set should be as large and flexible as possible in order to contain as many candidate models as possible. This reduces the structural or bias error in the model. On the other hand, when parameterizing the model set with (real-valued) parameters, the number of parameters should be kept as small as possible because of the principle of parsimony [34,283]. This principle states that the variability of identified models increases with increasing number of parameters. The conflict between large flexible model sets and parsimoniously parameterized model sets is directly related to the well-known bias/variance trade-off that is present in estimation problems.
It is easy to show that (partial) knowledge of the system to be identified can be used to shape the choice of model structure. This is indicated in a simple example.

Example 4.1. Consider a system $G(q) = \sum_{k=1}^{\infty} g_k^{(o)} q^{-k}$ with pulse response $\{g_k^{(o)}\}$. For simplicity, we will represent a system model by its pulse response. If one would know that the shape of the system's pulse response would be like the shape of $\{g_k^{(o)}\}$, then a very simple parameterization of an appropriate model set could be

$$g_k(\alpha) = \alpha \cdot g_k^{(o)} \qquad \alpha \in \mathbb{R},\ k \in \mathbb{Z}.$$

In this – trivial – situation of a model set with one parameter, the value $\alpha = 1$ leads to an exact presentation of the system.

If one would know the order (*e.g.* equal to 3) and the (distinct) real-valued poles $\{a_1, a_2, a_3\}$ of the system, then an appropriate model set could be

$$g_k(\theta) = \sum_{i=1}^{3} \theta_i a_i^{k-1} \quad k \geq 1 \qquad \theta = [\theta_1 \; \theta_2 \; \theta_3]$$

where now three parameters are used to represent a model set that captures the real system.
If one would specify only the order of the system (equal to 3) then a proper parameterization would be

$$G(q, \theta) = \frac{\theta_1 q^{-1} + \theta_2 q^{-2} + \theta_3 q^{-3}}{1 + \theta_4 q^{-1} + \theta_5 q^{-2} + \theta_6 q^{-3}}$$

and six parameters are required to represent all models in such a way that there is a guarantee that the real system is present in the set.

This example shows that the complexity of the model set that is required for identifying a system is not necessarily dependent on the complexity (order) of the system itself; it is dependent on the prior information on the system that one is able and/or willing to incorporate in the model set. Note also that it may not be realistic to require that we have an exact representation of our system present in the model set; it can be sufficient to know that – within the model set chosen – there exists a model that approximates the system dynamics very accurately.
Model structures induced by (orthonormal) basis functions have attractive properties in view of the bias/variance trade-off. This is caused by the fact that – when appropriately chosen – they require only a limited number of parameters to represent models that can accurately describe the dynamics of the underlying system. The choice of basis functions then becomes a principle design issue in the construction of model sets.
In this chapter, it will be shown how basis function model structures can be used in prediction error identification problems, both as an implementation tool and as an analysis tool.
First, a brief state-of-the-art will be presented of prediction error system identification, following closely the framework of [164]. Next in Section 4.3, attention will be focused on identification methods for basis functions model structures, while the linearity-in-the-parameters property of the structures directly leads to linear regression algorithms. Asymptotic bias and variance results will be discussed, generalizing the well-known classical results as present in the standard identification framework of Ljung, see [164]. In terms of asymptotic bias, the main result is that the bias-induced model error can be shown to explicitly become smaller when the choice of poles in the basis functions used is well adapted to the poles that are present in the system to be modelled. In terms of asymptotic variance, the influence of the choice of basis functions

on the asymptotic variability of the estimates is specified, both in terms of estimated parameters and in terms of related frequency responses.
The question whether orthonormality of the basis functions is essential or not is discussed in Section 4.4, where the relation is investigated with fixed denominator model structures. Here it is shown that basis function model structures not only play their role as an implementation tool but also as an analysis tool. This topic is also the subject of Chapter 5, where variance issues in identification are discussed in more detail.

4.2 Prediction Error Identification

4.2.1 Identification Setting and Prediction

As a framework for identification, the black-box identification setting according to Ljung [164] will be adopted. For analysis purposes, it is assumed that an available data sequence of input and output signals $\{u(t), y(t)\}_{t=1,\cdots N}$ is generated by a linear, time-invariant, discrete-time *data generating system*:

$$y(t) = G_0(q)u(t) + v(t) \tag{4.1}$$

with $G_0 \in \mathcal{H}_2$, u a quasi-stationary signal [164] and v a stationary stochastic process with rational spectral density, modelled by $v(t) = H_0(q)e(t)$, with e a zero-mean white noise process with variance σ_e^2 and H_0 a monic transfer function satisfying $H_0, H_0^{-1} \in \mathcal{H}_2$, where monicity implies that $\lim_{|z|\to\infty} H_0(z) = 1$. This data generating system is basically determined by the pair (G_0, H_0), which for future use will be denoted by $\mathcal{S}$.
Under the given assumptions on the data, the one-step-ahead predictor of the output signal $y(t)$ based on information on $\{y(t-1), y(t-2), \cdots\}$ and on $\{u(t), u(t-1), \cdots\}$ is known to be given by

$$\hat{y}(t|t-1) = [1 - H_0(q)]^{-1}y(t) + H_0(q)^{-1}G_0(q)u(t).$$

In prediction error identification, a parameterized model $(G(q,\theta), H(q,\theta))$ is hypothesized where $\theta \in \Theta \subset \mathbb{R}^d$ represents a parameter vector, composed of real-valued coefficients. This model representation leads to a one-step-ahead predictor

$$\hat{y}(t|t-1;\theta) = [1 - H(q,\theta)]^{-1}y(t) + H(q,\theta)^{-1}G(q,\theta)u(t) \tag{4.2}$$

that is used as a basis for identifying the unknown parameters, which is generally performed by minimizing a scalar-valued (least-squares) identification criterion:

$$V_N(\theta) := \frac{1}{N}\sum_{t=1}^{N} \varepsilon_f(t,\theta)^2 \tag{4.3}$$

$$\widehat{\theta}_N = \arg\min_{\theta} V_N(\theta) \tag{4.4}$$

with

$$\varepsilon_f(t,\theta) = L(q)\varepsilon(t,\theta)$$

a prefiltered version of the one-step-ahead prediction error

$$\varepsilon(t,\theta) := y(t) - \hat{y}(t|t-1;\theta). \tag{4.5}$$

The prefiltering of the prediction error can be employed to extend flexibility; it is one of the design variables that is available to the user. If the noise term e is Gaussian distributed with constant variance, the prediction error method is actually a maximum likelihood method.
The identification setup is schematically depicted in Figure 4.1.
Optimization of the identification criterion (4.3) will generally be a non-convex optimization problem for which iterative search (gradient) algorithms have to be applied. This also implies that convergence to the global optimum cannot easily be guaranteed. In some particular cases, the identification problem reduces to a convex optimization problem with an analytical solution.
Predictor models provide a (one-step-ahead) prediction of the output $y(t)$, given data from the past. The prediction error that is made can serve as a signal that indicates how well a model is able to describe the dynamics that underlies a measured data sequence. In this respect, the predictor model can be considered as a mapping from data u, y to a prediction error ε. If the data generating system G_0, H_0 is equal to the model G, H, then $\varepsilon = e$, being a

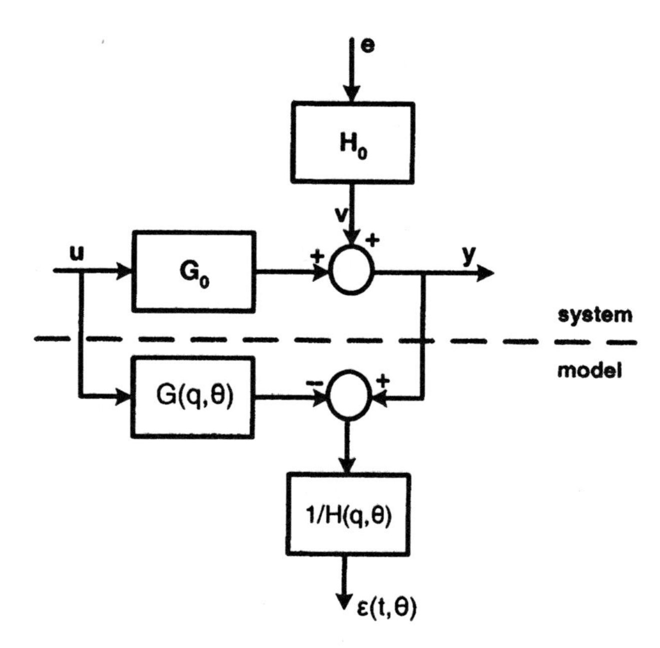

Fig. 4.1. Data-generating system G_0, H_0 and predictor model (G, H).

white noise process. However, also in the situation that the model differs from the data generating system, the model pertains its role as a predictor and provides a predicted output signal and a prediction error.

For dealing with quasi-stationary signals, use will be made of a generalized expectation operator, denoted by $\bar{\mathsf{E}}\{\}$ defined by

$$\bar{\mathsf{E}}\{y(t)\} = \lim_{N\to\infty} \frac{1}{N}\sum_{t=1}^{N} \mathsf{E}\{y(t)\},$$

with related (cross-)covariance functions

$$R_y(\tau) := \bar{\mathsf{E}}\{y(t)y(t-\tau)\}, \quad R_{yu}(\tau) := \bar{\mathsf{E}}\{y(t)u(t-\tau)\}$$

and (cross-)power spectral densities

$$\Phi_y(e^{i\omega}) := \sum_{\tau=-\infty}^{\infty} R_y(\tau)e^{-i\omega\tau}, \quad \Phi_{yu}(e^{i\omega}) := \sum_{\tau=-\infty}^{\infty} R_{yu}(\tau)e^{-i\omega\tau}.$$

4.2.2 Model Structures and Model Sets

There are several different block-box parameterizations (or model structures) available for the transfer functions $G(q,\theta)$ and $H(q,\theta)$. Most of them parameterize the two transfer functions in terms of ratio's of polynomials, where the coefficients in the polynomials act as parameters. The most common ones are collected in Table 4.1, where all of the characters A, B, C, D, F indicate polynomials in the delay operator q^{-1}. The several structures are known under the acronyms given in the table.
The most popular model structure is the auto-regressive exogenous (ARX) structure, determined by

$$A(q^{-1},\theta) = 1 + a_1 q^{-1} + a_2 q^{-2} + \cdots + a_{n_a} q^{-n_a} \tag{4.6}$$
$$B(q^{-1},\theta) = b_0 + b_1 q^{-1} + b_2 q^{-2} + \cdots + b_{n_b} q^{-n_b} \tag{4.7}$$

with

$$\theta := [a_1\ a_2\ \cdots\ a_{n_a}\ b_0\ b_1\ \cdots\ b_{n_b}]^T \in \Theta \subset \mathbb{R}^{n_b+n_a+1}$$

such that

$$G(q,\theta) = \frac{B(q^{-1},\theta)}{A(q^{-1},\theta)}, \tag{4.8}$$
$$H(q,\theta) = \frac{1}{A(q^{-1},\theta)}. \tag{4.9}$$

For this model structure, the predictor model satisfies the linear-in-θ equality

$$A(q^{-1},\theta)y(t) = B(q^{-1},\theta)u(t) + \varepsilon(t).$$

Table 4.1. Black-box model structures

Model structure	$G(q,\theta)$	$H(q,\theta)$
ARX	$\frac{B(q^{-1},\theta)}{A(q^{-1},\theta)}$	$\frac{1}{A(q^{-1},\theta)}$
ARMAX	$\frac{B(q^{-1},\theta)}{A(q^{-1},\theta)}$	$\frac{C(q^{-1},\theta)}{A(q^{-1},\theta)}$
OE	$\frac{B(q^{-1},\theta)}{F(q^{-1},\theta)}$	1
FIR	$B(q^{-1},\theta)$	1
BJ	$\frac{B(q^{-1},\theta)}{F(q^{-1},\theta)}$	$\frac{C(q^{-1},\theta)}{D(q^{-1},\theta)}$

Every model structure or parameterization induces a set of predictor models. Formally this will be denoted by the *model set*

$$\mathcal{M} := \{(G(q,\theta), H(q,\theta)) \in \mathcal{H}_2 \times \mathcal{H}_2 \mid \theta \in \Theta \subset \mathbb{R}^d\}.$$

This representation allows us to distinguish the following situations:

- $\mathcal{S} \in \mathcal{M}$: the data generating system is in the model set; *i.e.* an exact representation of the data generating system can be found in $\mathcal{M}$ by choosing an appropriate value for θ;
- $\mathcal{S} \notin \mathcal{M}$: no exact representation of $\mathcal{S}$ exists within $\mathcal{M}$.

When only directing attention to the input/output dynamical properties of a system, it is attractive to deal with the set of i/o models:

$$\mathcal{G} := \{G(q,\theta) \in \mathcal{H}_2 \mid \theta \in \Theta \subset \mathbb{R}^d\}.$$

This leads to expressions $G_0 \in \mathcal{G}$ and $G_0 \notin \mathcal{G}$ indicating whether the plant i/o dynamics can be captured exactly within the model set chosen. These notions can be used fruitfully when discussing the statistical properties of related identification algorithms.

Two important properties of the model structures should be mentioned here:

- For some particular model structures, the expression for the output predictor (4.2) becomes linear in the unknown parameters. Note that in order for this to happen, it is required that $(1-H(q,\theta))^{-1}$ as well as $H(q,\theta)^{-1}G(q,\theta)$ are polynomial. Using the property that $H(q,\theta)$ is defined to be monic, this linear-in-the-parameters property can be shown to hold for the structures: ARX and FIR (finite impulse response). In the ARX-case with $A(q^{-1},\theta)$ and $B(q^{-1},\theta)$ according to (4.6)-(4.7) it follows that

$$\hat{y}(t|t-1;\theta) = [1 - H(q,\theta)]^{-1}y(t) + H(q,\theta)^{-1}G(q,\theta)u(t)$$
$$= -\sum_{k=1}^{n_a} a_k q^{-k} y(t) + \sum_{k=1}^{n_b} b_k q^{-k} u(t) \qquad (4.10)$$

  which clearly shows the linear relationship between the coefficients in θ and the predictor $\hat{y}(t|t-1;\theta)$. This linearity can be fruitfully used in linear regression techniques where the least squares identification criterion can be minimized by solving a set of linear equations.
- If G and H are independently parameterized, there is an opportunity to estimate the two transfer functions independently. This refers *e.g.* to a possible consistent estimation of G even in the situation that the noise model H is misspecified. A further discussion of this will follow in the next section. For the model structures considered in this chapter the property of independent parameterization refers to the situation that there is no common denominator in G and H. This property holds for the model structures FIR, OE (output error), and BJ (Box-Jenkins).

Returning to (4.10), it can be observed that new model structures can be constructed by replacing the delay operators q^{-k} with any sequence of (orthogonal) basis functions $F_k(q)$ as presented in Chapters 1 and 2, leading to

$$\hat{y}(t|t-1;\theta) = -\sum_{k=1}^{n_a} a_k \cdot F_k(q)y(t) + \sum_{k=1}^{n_b} b_k \cdot F_k(q)u(t),$$

while the basis functions used for the y and the u component may even be chosen different from each other. This shows that the basis functions actually have the role of prefilters that operate on input/output signals, and that consequently have a potential influence on the noise effects in the filtered data.

It is particularly attractive to consider an FIR model structure as a starting point, as this has both the properties discussed before: a linear-in-the-parameters predictor and an independent parameterization of $G(q,\theta)$ and $H(q,\theta)$. The corresponding generalized structure:

$$\hat{y}(t|t-1;\theta) = \sum_{k=1}^{n_b} b_k \cdot F_k(q)u(t)$$

will be further analysed in Section 4.3.

4.2.3 Convergence and Consistency Results

When applying a quadratic identification criterion V_N (4.3), a parameter estimate results for which asymptotic properties can be derived in the situation that the number of data point N tends to ∞. When the noise influence on the measured data is normally distributed, the least squares estimator can be shown to be equivalent to a maximum likelihood estimator. However, also with other noise distributions, attractive properties hold as listed next.
The first result, known as the *convergence result*, states that the parameter estimator converges with probability 1 to the minimizing argument of the expected value of the loss function.

Convergence Result

$\widehat{\theta}_N \rightarrow \theta^\star$ with probability 1 as $N \rightarrow \infty$, where

$$\theta^\star = \arg\min_\theta \bar{V}(\theta) \qquad \text{with } \bar{V}(\theta) = \bar{\mathsf{E}}\{\varepsilon_f^2(\theta)\}.$$

As a result, the asymptotic parameter estimate is independent of the particular noise realization that is present in the data sequence. This property is particularly helpful in characterizing the structural (or bias) error in identified models, see Section 4.2.4. The question whether a data generating system can be recovered exactly from a measured data sequence is considered in the consistency result.

Consistency Result

If the input signal u is persistently exciting of a sufficient high-order, then the asymptotic parameter estimate $\theta^\star$ has the following properties:

a) If $\mathcal{S} \in \mathcal{M}$, then $G(q,\theta^\star) = G_0(q)$ and $H(q,\theta^\star) = H_0(q)$.
b) If $G_0 \in \mathcal{G}$, then $G(q,\theta^\star) = G_0(q)$ provided that the model set is independently parameterized.

The persistent excitation of the input signal u is required in order to guarantee that sufficient information on the dynamics of G_0 is present in the data. In general it is sufficient to require that $\Phi_u(e^{i\omega})$ has a non-zero contribution in the frequency range $-\pi < \omega \leq \pi$ in at least as many points as there are parameters to be estimated in $G(q,\theta)$.

Result a) shows that consistency is obtained under conditions that are rather appealing: the model set should be able to describe the system exactly, and the input signal has to be sufficiently exciting in order to be able to extract all of the dynamics of the system from the external signals. The importance of result b) is that consistency for G can also be obtained in situations that H is misspecified, provided that G and H do not have any common parameters. As mentioned before, this latter condition refers to a property of the model structure, which can be fully influenced by the designer.

Example 4.2. Consider a data generating system, determined by

$$G_0(q) = \frac{b_1^{(0)}q^{-1} + b_2^{(0)}q^{-2}}{1 + f_1^{(0)}q^{-1} + f_2^{(0)}q^{-2}}, \qquad H_0(q) = \frac{1 + c_1^{(0)}q^{-1}}{1 + d_1^{(0)}q^{-1}},$$

generating a data set according to $y(t) = G_0(q)u(t) + H_0(q)e(t)$.
For consistent identification, one will need a Box-Jenkins (BJ) model structure in order to guarantee that $\mathcal{S} \in \mathcal{M}$.
When identifying the system with an ARX model structure, then irrespective of the order of the models, no consistency will be achieved, and biased estimates will result.
When identifying the system with an output error (OE) model structure, having the independent parameterization property, a second-order OE model will be able to consistently identify G_0. This latter result holds in spite of a misspecification of H_0, through $H(q) = 1$.

4.2.4 Asymptotic Bias and Variance Errors

In every estimation, one has to deal with estimation errors. This is due to the fact that the information on the object to be estimated is usually only partial: measured data is contaminated by noise and the number of data is finite. A common and well-accepted approach is to decompose the estimation error (e.g. for a plant model $G(q, \widehat{\theta}_N)$) as:

$$G_0(q) - G(q, \widehat{\theta}_N) = G_0(q) - G(q, \theta^\star) + G(q, \theta^\star) - G(q, \widehat{\theta}_N). \tag{4.11}$$

In this decomposition, the first part, $G_0(q) - G(q, \theta^\star)$, is the ***structural or bias error*** usually induced by the fact that the model set is not rich enough to exactly represent the plant. The second part, $G(q, \theta^\star) - G(q, \widehat{\theta}_N)$, is the ***noise-induced or variance error*** which is due to noise disturbances on the measured data.

Bias

The question of characterizing the asymptotic bias in model estimates comes down to characterizing the difference between G_0 (or H_0) and $G(q, \theta^\star)$ (or

$H(q,\theta^\star))$. It is generally not possible to bound this difference in an explicit way. In an implicit way, one can use the convergence result of Section 4.2.3 which states that

$$\theta^\star = \arg\min_{\theta} \bar{V}(\theta).$$

By applying Parseval's relation, it follows that

$$\bar{V}(\theta) = \bar{\mathsf{E}}\{\varepsilon_f(t,\theta)^2\} = \frac{1}{2\pi}\int_{-\pi}^{\pi} \Phi_{\varepsilon_f}(\omega,\theta)d\omega$$

and by using the expressions for ε_f and for the data generating system, it follows that

$$\Phi_{\varepsilon_f}(\omega,\theta) = \frac{|G_0(e^{i\omega}) - G(e^{i\omega},\theta)|^2\Phi_u(e^{i\omega}) + \Phi_v(e^{i\omega})}{|H(e^{i\omega},\theta)|^2}|L(e^{i\omega})|^2$$

where

$$\Phi_v = \sigma_e^2|H_0(e^{i\omega})|^2.$$

As a result,

$$\theta^\star = \arg\min_{\theta}\frac{1}{2\pi}\int_{-\pi}^{\pi}\frac{|G_0(e^{i\omega}) - G(e^{i\omega},\theta)|^2\Phi_u(e^{i\omega}) + \Phi_v(e^{i\omega})}{|H(e^{i\omega},\theta)|^2}|L(e^{i\omega})|^2 d\omega. \tag{4.12}$$

Because $\theta^\star$ is the value (or set of values) to which the parameter estimate $\widehat{\theta}_N$ converges with probability 1, we now have a characterization of this limit estimate in the frequency domain. Note that it is valid irrespective of the fact whether $\mathcal{S} \in \mathcal{M}$ or $G_0 \in \mathcal{G}$. So also in an approximate setting, it –implicitly– characterizes the models that are identified.

In the situation of a model structure with a fixed noise model, *i.e.* $H(q,\theta) = H_\star(q)$, expression (4.12) reduced to

$$\theta^\star = \arg\min_{\theta\in\Theta}\frac{1}{2\pi}\int_{-\pi}^{\pi}|G_0(e^{i\omega}) - G(e^{i\omega},\theta)|^2\frac{\Phi_u(e^{i\omega})|L(e^{i\omega})|^2}{|H_\star(e^{i\omega})|^2}d\omega \tag{4.13}$$

which follows by neglecting the terms in the integrand that are independent of θ. It is clear now that the limiting estimate is obtained as that value of θ that makes $G(e^{i\omega},\theta)$ the best mean square approximation of $G_0(e^{i\omega})$ with a frequency weighting $\frac{\Phi_u(e^{i\omega})|L(e^{i\omega})|^2}{|H_\star(e^{i\omega})|^2}$. This frequency weighting function determines how the errors in the different frequency regions are weighted with respect to each other. For those values where the weighting function is large, the relative importance of error terms $|G_0(e^{i\omega}) - G(e^{i\omega},\theta)|^2$ in the total misfit is large, and consequently the estimated parameter will strive for a small error contribution $|G_0(e^{i\omega}) - G(e^{i\omega},\theta)|^2$ at that frequency. By choosing the fixed noise model, the input spectrum and the data prefilter L, this weighting function can be influenced by the user. Note that in this situation the asymptotic model is not dependent on the –unknown– noise spectrum Φ_v. As a result, there will be no noise-induced bias in model estimates when an output error (OE) model structure is chosen.

Variance

The variability of asymptotic parameter estimates can be specified by using dedicated versions of the central limit theorem. In its most general form, the characterization follows from the basic result:

$$\sqrt{N}(\widehat{\theta}_N - \theta^\star) \rightarrow \mathcal{N}(0, P_\theta) \quad \text{as } N \rightarrow \infty \tag{4.14}$$

i.e. the random variable $\sqrt{N}(\widehat{\theta}_N - \theta^\star)$ converges in distribution to a Gaussian probability density function with zero mean and covariance matrix P_θ. This result follows from a Taylor expansion representation of $\widehat{\theta}_N$ around $\theta^\star$ and holds under weak regularity conditions on the data.
As an additional result it holds that $\text{Cov}\{\sqrt{N}\widehat{\theta}_N\} \rightarrow P_\theta$ as $N \rightarrow \infty$.
However, the covariance matrix P_θ can be calculated explicitly only in a limited number of situations. One of these situations is the case where $\mathcal{S} \in \mathcal{M}$ and the parameter estimate is consistent.

In the situation $\mathcal{S} \in \mathcal{M}$, leading to a consistent parameter estimate $\theta^\star = \theta_0$, the asymptotic covariance matrix becomes

$$P_\theta = \sigma_e^2[\bar{\mathsf{E}}\{\psi(t,\theta_0)(\psi(t,\theta_0))^T\}]^{-1} \tag{4.15}$$

with

$$\psi(t,\theta_0) := -\frac{\partial}{\partial\theta}\varepsilon(t,\theta)\bigg|_{\theta=\theta_0}.$$

In the less restrictive situation that $G_0 \in \mathcal{G}$, a related result can be derived, although the expressions become more complex. Using

$$\theta = \begin{pmatrix} \rho \\ \eta \end{pmatrix}$$

where ρ is the parameter of G and η is the parameter of H the result reads:

In the situation where $G_0 \in \mathcal{G}$ and ρ_0 is estimated consistently, $\rho^\star = \rho_0$, the asymptotic covariance matrix for $\hat{\rho}_N$ becomes

$$P_\rho = \sigma_e^2[\bar{\mathsf{E}}\{\psi_\rho(t)\psi_\rho^T(t)\}]^{-1}\bar{\mathsf{E}}\{\tilde{\psi}(t)\tilde{\psi}^T(t)\}[\bar{\mathsf{E}}\{\psi_\rho(t)\psi_\rho^T(t)\}]^{-1} \tag{4.16}$$

with

$$\psi_\rho(t) = H^{-1}(q,\eta^*)\,\frac{\partial}{\partial\rho}G(q,\rho)\bigg|_{\rho=\rho_0} u(t) \tag{4.17}$$

$$\tilde{\psi}(t) = \sum_{i=0}^{\infty} f_i\psi_\rho(t+i) \tag{4.18}$$

and f_i determined by $\dfrac{H_0(z)}{H(z,\eta^*)} = \sum_{i=0}^{\infty} f_i z^{-i}$.

This result provides an expression for the asymptotic variance of parameter estimates for the case that model sets are applied that are not able to describe the noise/output transfer function H_0 accurately. This typically happens in the case of a model set having an output error (OE) model structure.
Note that when the noise model is estimated consistently, $H(q,\eta^*) = H_0(q)$, then $f_i = \delta(i)$ which leads to $\tilde{\psi}(t) = \psi_\rho(t)$, and consequently the expression for P_ρ reduces to the same expression as in the situation $\mathcal{S} \in \mathcal{M}$ discussed above.

The parameter covariance can be transformed to a related covariance expression on the frequency response of the model estimates. In the situation $\mathcal{S} \in \mathcal{M}$, this leads to

$$\sqrt{N} \cdot \mathrm{Cov}\{\begin{bmatrix} G(e^{i\omega},\widehat{\theta}_N) \\ H(e^{i\omega},\widehat{\theta}_N) \end{bmatrix}\} \to P(\omega) \quad \text{as } N \to \infty$$

where

$$P(\omega) = T_\theta'(e^{i\omega},\theta^\star) P_\theta T_\theta'(e^{-i\omega},\theta^\star)^T$$

and $T_\theta'(e^{i\omega},\theta^\star) := \frac{\partial}{\partial\theta}[G(e^{i\omega},\theta) \ H(e^{i\omega},\theta)]\Big|_{\theta=\theta^\star}$.

A simple and appealing expression can be given for the covariance of a model's frequency response when not only the number of data but also the model order tends to infinity.

Let n indicate the order of a model $(G(q,\theta), H(q,\theta))$ within $\mathcal{M}$.

- If $G_0 \in \mathcal{G}$ then

$$\mathrm{Cov}\{G(e^{i\omega},\widehat{\theta}_N)\} \sim \frac{n}{N}\frac{\Phi_v(e^{i\omega})}{\Phi_u(e^{i\omega})} \quad \text{for } n, N \to \infty, n << N$$

- If $\mathcal{S} \in \mathcal{M}$, then additionally

$$\mathrm{Cov}\{H(e^{i\omega},\widehat{\theta}_N)\} \sim \frac{n}{N}|H_0(e^{i\omega})|^2 \quad \text{for } n, N \to \infty, n << N$$

and $G(e^{i\omega},\widehat{\theta}_N)$ and $H(e^{i\omega},\widehat{\theta}_N)$ are uncorrelated.

This appealing result states that the variance of the plant model in each frequency is determined by the noise-to-input-signal ratio at that particular frequency multiplied by the ratio between model order and number of data. The more noise power is present in a particular frequency band, the higher the variance of the model estimate. In the same way, increasing the input power at a particular frequency decreases the variance of the estimates at that frequency. This (double)-asymptotic result has been shown for FIR models in [168] and for the other model structures of Table 4.1 in [162]. It may give the impression that it is applicable to all possible model structures. This issue will be critically addressed in Section 4.3 and Chapter 5, where a generalization of this result will be shown to model structures that incorporate orthogonal basis function expansions, and where it is shown that the approximations can be further improved.

4.2.5 Linear Regression

If the model structure has the property of being linear-in-the-parameters (*e.g.* ARX or FIR), the least squares problem (4.4) becomes a convex optimization problem that can be solved analytically:

$$\widehat{\theta}_N = \left[\frac{1}{N}\sum_{t=1}^{N}\varphi(t)\varphi^T(t)\right]^{-1} \cdot \frac{1}{N}\sum_{t=1}^{N}\varphi(t)y(t) \tag{4.19}$$

where $\varphi(t)$ is the vector containing measured data samples constituting the one-step-ahead predictor according to $\hat{y}(t|t-1;\theta) = \varphi^T(t)\theta$. For the ARX model structure (4.8)-(4.9), the regression vector becomes

$$\varphi^T(t) = [-y(t-1) \ \ -y(t-2) \ \cdots -y(t-n_a) \ u(t) \ u(t-1) \ \cdots u(t-n_b)],$$

while for the FIR model structure

$$\varphi^T(t) = [u(t) \ u(t-1) \ \cdots u(t-n_b)].$$

Note that on the basis of numerical considerations it is generally very unattractive to calculate the least-squares solution directly by the closed-form expression (4.19) because of the matrix inverse. Commonly, the solution is obtained through a QR-algorithm on the corresponding set of (normal) equations

$$\frac{1}{N}\Phi_N^T\Phi_N\widehat{\theta}_N = \frac{1}{N}\Phi_N^T\mathsf{Y}_N$$

with

$$\begin{aligned} \mathsf{Y}_N^T &:= [y(1) \ y(2) \ \cdots \ y(N)] \\ \Phi_N^T &:= [\varphi(1) \ \varphi(2) \ \cdots \ \varphi(N)]. \end{aligned} \tag{4.20}$$

The statistical analysis of this estimator follows from considering the data generating system (4.1), written in the format

$$\mathsf{Y}_N = \Phi_N^T\theta_0 + \mathsf{W}_N \tag{4.21}$$

where the noise vector $\mathsf{W}_N^T = [w_0(1) \ w_0(2) \ \cdots \ w_0(N)]$ is composed of samples $w_0(t) = e(t)$ in the ARX case, and $w_0(t) = v(t)$ in the case where the regressor

$\varphi(t)$ is composed of (noise-free) input signals only, as *e.g.* in the FIR-case. The asymptotic bias and variance expressions in the previous subsection of course remain valid. However also non-asymptotic results can now be obtained.
In the ARX situation of a stochastic regressor, the bias and variance of the estimator follow from

$$\widehat{\theta}_N = \theta_0 + \mathsf{E}\{\frac{1}{N}\Phi_N^T\Phi_N\}^{-1}\frac{1}{N}\Phi_N^T \mathsf{W}_N \tag{4.22}$$

$$\mathrm{Cov}\{\widehat{\theta}_N\} = \mathsf{E}\{(\frac{1}{N}\Phi_N^T\Phi_N)^{-1}\frac{1}{N}\Phi_N^T \mathsf{W}_N \mathsf{W}_N^T \Phi_N \frac{1}{N}(\frac{1}{N}\Phi_N^T\Phi_N)^{-1}\}. \tag{4.23}$$

Both expressions for bias and variance hold under the assumption that $\mathcal{S} \in \mathcal{M}$ and for finite values values of N. The expressions can, under minor conditions, be simplified in the asymptotic case $N \to \infty$ to satisfy

$$\mathsf{E}\{\widehat{\theta}_N\} \to \theta_0 \tag{4.24}$$

$$\mathrm{Cov}\{\widehat{\theta}_N\} \to \sigma_e^2 \cdot R_* \quad \text{with } R_* = \operatorname*{plim}_{N\to\infty} \mathsf{E}\{\frac{1}{N}\Phi_N^T\Phi_N\}. \tag{4.25}$$

If $\mathcal{S} \notin \mathcal{M}$ or $G_0 \notin \mathcal{G}$, the analysis of bias and variance becomes much more complicated.
If the regressor $\varphi(t)$ is deterministic, as is the case with FIR models with a deterministic input signal, the statistical analysis of the estimator (4.19) shows some particular simplifications. As this situation will also be treated in the sequel of this chapter, it deserves further attention. Denoting $y_o(t) := G_0(q)u(t)$, *i.e.* the noise free output signal, it follows from (4.19) that

$$\mathsf{E}\{\widehat{\theta}_N\} = (\frac{1}{N}\Phi_N^T\Phi_N)^{-1}\frac{1}{N}\sum_{t=1}^{N}\varphi(t)\cdot y_o(t) \tag{4.26}$$

$$\mathrm{Cov}\{\widehat{\theta}_N\} = (\frac{1}{N}\Phi_N^T\Phi_N)^{-1}\frac{1}{N}\Phi_N^T \cdot \mathsf{E}\{\mathsf{V}_N\mathsf{V}_N^T\}\cdot\Phi_N\frac{1}{N}(\frac{1}{N}\Phi_N^T\Phi_N)^{-1} \tag{4.27}$$

with the noise signal vector $\mathsf{V}_N^T := [v(1)\ v(2)\ \cdots\ v(N)]$. Note that these expressions are valid for finite N and irrespective of conditions like $G_0 \in \mathcal{G}$ or $\mathcal{S} \in \mathcal{M}$; they even hold true in the situation $G_0 \notin \mathcal{G}$. However if $G_0 \in \mathcal{G}$, then there exists a θ_0 such that $y_o(t) = \varphi^T(t)\theta_0$, thus leading to $\mathsf{E}\{\widehat{\theta}_N\} = \theta_0$. As a result, in this case the estimate is unbiased also for finite values of N. If the noise signal v is a stationary white noise process with variance σ_v^2, the covariance expression reduces to

$$\mathrm{Cov}\{\widehat{\theta}_N\} = \sigma_v^2 \cdot (\Phi_N^T\Phi_N)^{-1}. \tag{4.28}$$

This latter expression is closely related to the asymptotic parameter covariance matrix (4.15), that was presented for the general case in the situation $\mathcal{S} \in \mathcal{M}$.

4.3 Identification with Basis Function Model Structures

4.3.1 Introduction

Considering the 'classical' identification results described in the previous section, it appears that there are two attractive properties of model structures:

- The linearity-in-the-parameters property, shared by ARX and FIR, that leads to simple linear regression type of parameter estimation schemes (convex optimization); and
- Independent parameterizations of $G(q,\theta)$ and $H(q,\theta)$ (process and noise models), which implies that there will be no bias on the process model estimate that is induced by the noise properties.

Out of the presented classical model structures, a combination of these two properties can only be found in the FIR model structure. However, this structure has the additional disadvantage that it generally requires a large number of parameters to capture all dynamics of a dynamical process. This latter situation particularly holds when the process dynamics is moderately damped, *i.e.* when the pulse response of the system is 'long'. When applying an orthonormal basis functions model structure, the attractive properties can be retained without the need for excessive numbers of parameters.
In this section, attention will be focused to the model structure:

$$G(q,\theta) = \sum_{k=1}^{n} c_k F_k(q) \qquad H(q) = 1$$

where the sequence of basis functions $\{F_k(z)\}_{k=1,\cdots n}$ is constructed with one of the procedures presented in Chapter 2 after having specified n pole locations $\{\xi_i\}_{i=1,\cdots n}$ according to the general case in Section 2.5. The series expansion coefficients are collected in the parameter vector

$$\theta = [c_1 \; c_2 \; \cdots \; c_n]^T \in \mathbb{R}^n.$$

4.3.2 Least Squares Identification

Given data $\{u(t), y(t)\}_{t=1,\cdots,N}$ that is taken from experiments on the system, the one-step-ahead prediction error related to the considered model structure is given by

$$\begin{aligned} \varepsilon(t,\theta) &= H(q)^{-1}[y(t) - G(q,\theta)u(t)] \\ &= y(t) - \varphi^T(t)\theta \end{aligned} \tag{4.29}$$

with

$$\varphi(t) := \Gamma_n(q)u(t), \qquad \Gamma_n(q) := [F_1(q) \; F_2(q) \; \cdots F_n(q)]^T. \tag{4.30}$$

Note that (4.29) shows the linear regression structure of the identification problem. The difference with the classical FIR or ARX situation is that the regression variables are particularly filtered versions of the input signal, rather than delayed versions of u (and y).
The orthonormality property of the basis functions $\{F_k\}$ implies that

$$< F_k, F_\ell > := \frac{1}{2\pi}\int_{-\pi}^{\pi} F_k(e^{i\omega})F_\ell^*(e^{i\omega})d\omega = \begin{cases} 1 & \text{for } k = \ell \\ 0 & \text{for } k \neq \ell. \end{cases}$$

The least-squares parameter estimate is determined by

$$\widehat{\theta}_N = \arg\min_{\theta\in\Theta} \frac{1}{N}\sum_{t=1}^{N} \varepsilon(t,\theta)^2,$$

and can be constructed as a simple linear regression estimate, by solving the normal equations

$$\left[\frac{1}{N}\sum_{t=1}^{N}\varphi(t)\varphi^T(t)\right]\cdot\widehat{\theta}_N = \frac{1}{N}\sum_{t=1}^{N}\varphi(t)y(t), \tag{4.31}$$

through

$$\widehat{\theta}_N = \left[\frac{1}{N}\sum_{t=1}^{N}\varphi(t)\varphi^T(t)\right]^{-1}\cdot\frac{1}{N}\sum_{t=1}^{N}\varphi(t)y(t), \tag{4.32}$$

provided that the input signal is persistently exciting of a sufficiently high-order for the indicated inverse to exist.
Using the results from Section 4.2.3, it follows that the parameter estimate $\widehat{\theta}_N$ will converge with probability 1 to the limiting estimate

$$\begin{aligned} \theta^* &= \arg\min_{\theta\in\Theta} \bar{\mathsf{E}}\{\varepsilon(t,\theta)^2\} \\ &= R_n^{-1}\cdot f_n \end{aligned} \tag{4.33}$$

with, in accordance with (4.32),

$$R_n := \bar{\mathsf{E}}\{\varphi(t)\varphi^T(t)\} \qquad \text{and} \quad f_n := \bar{\mathsf{E}}\{\varphi(t)y(t)\}. \tag{4.34}$$

In the analysis of this parameter estimate and in its numerical conditioning, an important role is being played by the inverted matrix R_n. For the simple choice $F_k(z) = z^{-k}$, corresponding to a FIR model, R_n is a Toeplitz matrix:

$$\begin{aligned} R_n &= \bar{\mathsf{E}}\{\begin{bmatrix} u(t-1) \\ u(t-2) \\ \vdots \\ u(t-n) \end{bmatrix}\cdot[u(t-1)\ u(t-2)\ \cdots\ u(t-n)]\} \\ &= \begin{bmatrix} R_u(0) & R_u(1) & \cdots & R_u(n-1) \\ R_u(1) & R_u(0) & \cdots & \vdots \\ \vdots & \vdots & \ddots & \vdots \\ R_u(n-1) & R_u(n-2) & \cdots & R_u(0) \end{bmatrix} \end{aligned}$$

in which the element $[R_n]_{jk}$ can be written as

$$[R_n]_{jk} = \frac{1}{2\pi}\int_{-\pi}^{\pi} \Phi_u(e^{i\omega})e^{i\omega(j-k)}d\omega.$$

This implies that the Toeplitz matrix R_n is the covariance matrix related to the spectral density $\Phi_u(e^{i\omega})$. As a result, the parameter estimate results from solving a Toeplitz set of equations, for which there exist attractive analytical results [108, 121].

If the input signal is white, *i.e.* $\Phi_u(e^{i\omega}) = c$ (constant), then $R_n = cI$ and the matrix inverse is trivial. Consequently, the conditioning of the normal equations (4.31) is optimal. For more general choices of $F_k(z)$, it follows with Parseval's relation that

$$R_n = \frac{1}{2\pi}\int_{-\pi}^{\pi} \Gamma_n(e^{i\omega})\Gamma_n^*(e^{i\omega})\Phi_u(e^{i\omega})d\omega.$$

Again, if the input is white, then by orthonormality of the basis it follows that $R_n = cI$, and so calculation of the (asymptotic) least squares model is trivial. In this case it is not straightforward that R_n has a Toeplitz structure. However, whenever $\{F_k(z)\}_{k=1,\cdots n}$ is a sequence of (generalized) orthogonal basis functions as handled in this book, properties of R_n will match the properties of a (block) Toeplitz matrix. For instance, if the basis functions are determined by a repeating sequence of n_b poles, *i.e.*

$$\{\xi_1, \xi_2, \cdots \xi_{n_b}, \xi_1, \cdots \xi_{n_b}, \cdots, \xi_1, \cdots \xi_{n_b}\}$$

with $n = rn_b$ and r the number of repetitions, then it appears that R_n has a (block) Toeplitz structure with block-size n_b. This block-Toeplitz property simplifies the statistical and numerical analysis of the parameter estimates. More attention will be given to this issue in Section 4.3.4.

Example 4.3. Consider the Laguerre case where a repeating sequence of one single real-valued pole ξ is used to generate a sequence of orthonormal basis functions as described in Chapter 2. Then the (j, k)th entry of R_n is given by

$$[R_n]_{jk} = \frac{1}{2\pi}\int_{-\pi}^{\pi} F_j(e^{i\omega})F_k^*(e^{i\omega})\Phi_u(e^{i\omega})d\omega \tag{4.35}$$

where

$$F_j(e^{i\omega}) = \sqrt{1-\xi^2}\frac{(1-\xi e^{i\omega})^j}{(e^{i\omega}-\xi)^{j+1}}.$$

Substituting this in (4.35) and using a change of integration variable according to

$$e^{i\tilde{\omega}} = \frac{e^{i\omega}-\xi}{1-\xi e^{i\omega}} \Leftrightarrow e^{i\omega} = \frac{e^{i\tilde{\omega}}+\xi}{1+\xi e^{i\tilde{\omega}}}$$

and additionally

$$d\omega = \frac{1-\xi^2}{|e^{i\bar{\omega}}+\xi|^2} d\bar{\omega}, \qquad \frac{1}{e^{i\omega}-\xi} = \frac{e^{-i\bar{\omega}}+\xi}{1-\xi^2}$$

it follows that

$$[R_n]_{jk} = \frac{1}{2\pi}\int_{-\pi}^{\pi} e^{i\bar{\omega}(k-j)}\Phi_u(\frac{e^{i\bar{\omega}}+\xi}{1+\xi e^{i\bar{\omega}}})d\bar{\omega}.$$

This shows that R_n has a Toeplitz structure, being the covariance matrix related to the spectral density $\Phi(\frac{e^{i\omega}+\xi}{1+\xi e^{i\omega}})$.

As mentioned above, conditioning of the least-squares (LS) estimate is optimal in case the input signal is white. A further motivation for the use of orthonormnal basis functions, also for non-white input signal, is given in Chapter 6.

For analysis of the consistency/bias properties of the identified models, one can rely on the standard results that are presented in Section 4.2.3. For the considered model structure, these can be summarized as follows:

1. If $G_0 \in \mathcal{G}$, then $G(q, \widehat{\theta}_N)$ will be an unbiased and consistent estimate of G_0, provided that the input signal is persistently exciting of a sufficient high-order. As $\mathcal{G}$ now is composed of models with a finite series expansion in $\{F_k(q)\}$, the required condition $G_0 \in \mathcal{G}$ refers to the situation that the plant G_0 also has a finite expansion.
2. In all other cases a bias will result.

4.3.3 Asymptotic Bias Results

In general, the dynamical system G_0 will not have an exact representation in the form of a finite series expansion in the basis functions $\{F_k(q)\}$. As a result, the estimated models will only be able to approximate the system, and the challenging question will be to find a basis expansion that allows a small bias while only requiring a limited number of expansion parameters to be estimated.
In the sequel, it will be assumed that a model structure is chosen induced by a prespecified set of basis functions $\{F_k(q)\}_{k=1,\cdots,n}$ following any of the basis construction methods of Chapter 2.

Bias in Parameter Estimates

For analyzing this approximation (or bias) error, the original system G_0 is written in terms of the chosen basis functions as:

$$G_0(z) = \sum_{k=1}^{\infty} c_k^{(0)} F_k(z)$$

and the corresponding expansion coefficients constitute:

$$\theta_0 := [c_1^{(0)}\ c_2^{(0)}\ \cdots\ c_n^{(0)}]^T \text{ and}$$
$$\theta_e := [c_{n+1}^{(0)}\ c_{n+2}^{(0)}\ \cdots]^T.$$

For employing this system representation, an extension of basis functions has to be chosen beyond the set of n functions that determine the model structure. Without imposing any additional assumptions on this extension, it will be chosen only to satisfy the basic requirement of completeness, as presented in Section 2.2.

By substituting the system relation

$$y(t) = \varphi^T(t)\theta_0 + \varphi_e^T(t)\theta_e + v(t)$$

with

$$\varphi_e(t) := \Gamma_{n+1,\infty}(q)u(t), \qquad \Gamma_{n+1,\infty} := [F_{n+1}(q)\ F_{n+2}(q)\ \cdots]^T \tag{4.36}$$

into the expressions (4.33) and (4.34), the asymptotic bias error follows as

$$\theta^\star - \theta_0 = R_n^{-1}\bar{\mathsf{E}}\{\varphi(t)\varphi_e^T(t)\}\theta_e. \tag{4.37}$$

The following observations can be made:

- In (4.37), θ_e refers to the extension of the finite series expansion that is required to represent the system G_0. It reflects the fact that the model set is not able to capture all system dynamics. When $\theta_e = 0$, *i.e.* when G_0 does have a finite expansion, the asymptotic bias error will be zero also.
- The second factor in (4.37) can be rewritten by using Parseval's relation into the form

$$\bar{\mathsf{E}}\{\varphi(t)\varphi_e^T(t)\} = \frac{1}{2\pi}\int_{-\pi}^{\pi} \Gamma_n(e^{i\omega})\Gamma_{n+1,\infty}^*(e^{i\omega})\Phi_u(e^{i\omega})d\omega.$$

When the input spectrum $\Phi_u(e^{i\omega})$ is constant, this expression reduces to an inner product between different basis functions, which by virtue of the orthonormality property is equal to 0. This leads to the next result:

If the input spectrum is constant (white noise) the expansion coefficients are identified consistently.

Note that this statement is about the expansion coefficients in θ_0 and not about the model $G_0(q)$, i.e the tail of the expansion remains unmodelled.

The expression for the bias (4.37) can be bounded in several ways. By using the submultiplicative property of 2-norms, one can write:

$$\|\theta^\star - \theta_0\|_2 \le \|R(n)^{-1}\|_2 \cdot \|\bar{\mathsf{E}}\{\varphi(t)\varphi_e^T(t)\}\|_2 \cdot \|\theta_e\|_2 \le \frac{\max_\omega \Phi_u(e^{i\omega})}{\min_\omega \Phi_u(e^{i\omega})} \cdot \|\theta_e\|_2. \tag{4.38}$$

In this bound, use is made of the fact that $\|R_n^{-1}\|_2$ can be overbounded by $(\min_\omega \Phi_u(e^{i\omega}))^{-1}$, and the fact that $\|\bar{\mathsf{E}}\{\varphi(t)\varphi_e^T(t)\}\|_2$ can be overbounded by $\|R_\infty\|_2$ which is bounded by $\max_\omega \Phi_u(e^{i\omega})$. As R_n is completely known whenever input signals are given and basis functions have been specified, a less conservative bound on the parameter error is achieved by

$$\|\theta^\star - \theta_0\|_2 \leq \|R_n^{-1}\|_2 \cdot \max_\omega \Phi_u(e^{i\omega}) \cdot \|\theta_e\|_2.$$

Bias in Frequency Responses

The error bounds on the parameter estimates can directly be used for specifying also an upper bound on the $\mathcal{H}_2$ error between $G_0(z)$ and $G(z,\theta^\star)$. Due to the fact that the $\mathcal{H}_2$-norm of a transfer function is equal to the squared sum of its series of expansion coefficients in any orthonormal basis, it follows that

$$\|G(z,\theta^\star) - G_0(z)\|_{\mathcal{H}_2} = \|\theta^\star - \theta_0\|_2 + \|\theta_e\|_2 \leq \left[\frac{\max_\omega \Phi_u(e^{i\omega})}{\min_\omega \Phi_u(e^{i\omega})} + 1\right] \|\theta_e\|_2.$$

However, this bound can be rather conservative, as can be verified that for white input signal $\Phi_u(e^{i\omega}) = c$ the actual $\mathcal{H}_2$-norm of the model error is $\|\theta_e\|_2$ rather than $2\|\theta_e\|_2$ which results from substitution of the appropriate spectrum in the equation above.
A more careful analysis shows that when focusing on the frequency response of the model a corresponding bound can be formulated. By using the notation

$$G(e^{i\omega},\theta^\star) - G_0(e^{i\omega}) = [\Gamma_n^T(e^{i\omega}) \;\; \Gamma_{n+1,\infty}^T(e^{i\omega})] \begin{bmatrix} \theta^\star - \theta_0 \\ \theta_e \end{bmatrix}.$$

an upper bound on the frequency response error can be formulated as:

$$\begin{aligned} |G(e^{i\omega_1},\theta^\star) - G_0(e^{i\omega_1})| &\leq \max_{1\leq k\leq n} |F_k(e^{i\omega_1})| \cdot (\|\theta^\star - \theta_0\|_1 + \|\theta_e\|_1) \\ &\leq \max_{1\leq k\leq n} |F_k(e^{i\omega_1})| \cdot \left[\sqrt{n}\frac{\max_\omega \Phi_u(e^{i\omega})}{\min_\omega \Phi_u(e^{i\omega})} + 1\right] \|\theta_e\|_1. \end{aligned}$$

In these expressions $\|\cdot\|_1$ refers to the ℓ_1-norm of a vector, being defined as $\|x\|_1 = \sum_i |x_i|$, while in the latter inequality use is made of the standard relation that for any n-dimensional vector x, $\|x\|_1 \leq \sqrt{n}\|x\|_2$.

The important observation is that the bound on the bias in both the frequency response and the $\mathcal{H}_2$ model error is dependent on the basis functions chosen. This dependency is reflected in the factors $\|\theta_e\|_2$ and $\|\theta_e\|_1$, which represent that part of the dynamical system G_0 that cannot be modelled by our model, as it cannot be fitted into the finite expansion model structure. As the choice of basis functions crucially affects the extension sequence θ_e, the principal problem now is to find an upper bound of (some norm on) this extension sequence as a function of the choice of basis functions.
For a more detailed treatment of specifying model uncertainty bounds, the reader is referred to Chapter 7.

System Approximation and the Role of Kernels

The best approximation of $G_0(z)$ in the model class $G(z,\theta)$, measured in an $\mathcal{H}_2$ sense, is naturally given by $\hat{G}_n(z) = \sum_{k=1}^{n} c_k^{(0)} F_k(z)$. This is the expansion sequence with the coefficients of the original system, however restricted to a finite sequence of length n. This is also the limiting model that results when using a least squares criterion with a white input signal.
By using a projection mechanism, this limiting model can be described as a simple operation on the original system G_0. By writing

$$G_0(q) = \sum_{k=1}^{\infty} c_k^{(0)} F_k(q)$$

and using the fact that

$$c_k^{(0)} = < G_0, F_k >$$

it follows that

$$\begin{aligned}\hat{G}_n(z) &= \sum_{k=1}^{n} c_k^{(0)} F_k(z)\\ &= \sum_{k=1}^{n} \frac{1}{2\pi}\int_{-\pi}^{\pi} G_0(e^{i\omega})F_k^*(e^{i\omega})F_k(z)d\omega\\ &= \frac{1}{2\pi}\int_{-\pi}^{\pi} G_0(e^{i\omega})K_n(e^{i\omega},z)^* d\omega\end{aligned}$$

where the term

$$K_n(\eta, z) = \sum_{k=1}^{n} F_k(\eta)F_k^*(z) \tag{4.39}$$

determines the way in which the approximation of G_0 is being made. Because of the obvious importance of this term K_n, both in this context and a wider one of mathematical approximation theory [62, 293], it is given a particular name of *reproducing kernel.* This name originates from the characterizing property that

$$\hat{G}_n(z) = \frac{1}{2\pi}\int_{-\pi}^{\pi} \hat{G}_n(e^{i\omega})K_n(e^{i\omega},z)^* d\omega. \tag{4.40}$$

which can be easily verified from the equations given above, by using $c_k^{(0)} = < \hat{G}_n, F_k >$ for $k = 1, \cdots n$. It shows that the transfer function for a particular value of z can be reproduced by a kernel representation (4.40) involving the full frequency response of the system.
The critically important issue is that for the model structures considered in this chapter this reproducing kernel can actually be computed. The orthonormality of the basis functions structures here is the important issue. Utilizing the expression (4.39) together with a closed-form expression for the basis functions that are being used, as *e.g.* $\{F_k\}_{k=1,\cdots n}$ with

$$F_k(q) = \frac{\sqrt{1-|\xi_k|^2}}{q-\xi_k}\prod_{j=1}^{k-1}\left(\frac{1-\xi_j^* q}{q-\xi_j}\right). \tag{4.41}$$

and the related all-pass function

$$G_b(q) = \prod_{j=1}^{n}\left(\frac{1-\xi_j^* q}{q-\xi_j}\right)$$

the following relation can be shown to hold:

$$\begin{aligned}\sum_{k=1}^{n} F_k(z_1)F_k^*(z_2) &= \frac{G_b(z_1)G_b^*(z_2)-1}{1-z_1z_2^*} &&\text{for } z_1z_2^* \neq 1\\ &= n &&\text{for } z_1z_2^* = 1.\end{aligned} \tag{4.42}$$

A closed-form expression for the reproducing kernel, as given here, is known as a Christoffel-Darboux formula. The given closed-form expression is most simply verified by using an orthogonal state space realization (A, B, C, D) of $G_b(z)$, and by writing $[F_1(z)\ F_2(z)\ \cdots\ F_n(z)]^T = (zI - A)^{-1}B$.

These ingredients can now be used to characterize the approximation error $G_0 - \hat{G}_n$. For ease of expression, we consider G_0 to be represented in the partial fraction expansion:

$$G_0(z) = \sum_{j=1}^{n_0}\frac{b_j}{z-p_j^0}$$

where all (single) poles satisfy $|p_j^0| < 1$. Note that this restricts attention to systems that have all poles distinct. The system approximation error now is

$$G_0(z) - \sum_{k=1}^{n} c_k^{(0)}F_k(z) = \sum_{k=n+1}^{\infty} c_k^{(0)}F_k(z),$$

and employing the fact that, using Cauchy's integral theorem,

$$c_k^{(0)} =< G_0(z), F_k(z) >= \sum_{j=1}^{n_0}\frac{b_j}{p_j^0}F_k(\frac{1}{p_j^0})$$

the approximation error reads:

$$G_0(z) - \hat{G}_n(z) = \sum_{j=1}^{n_0}\sum_{k=n+1}^{\infty}\frac{b_j}{p_j^0}F_k(\frac{1}{p_j^0})F_k(z). \tag{4.43}$$

If the first n basis functions $\{F_k(z)\}_{k=1,\cdots,n}$ are considered to be generated by an all-pass function G_b having n poles, then we know that the extension

$\{F_k(z)\}_{k=n+1,\cdots}$ can be done by periodic repetition of the poles, such that $F_{k+n}(z) = G_b(z)F_k(z)$. Using this in expression (4.43) above, it follows that

$$G_0(z) - \hat{G}_n(z) = \sum_{j=1}^{n_0} \frac{b_j}{p_j^0} \sum_{k=1}^{n} F_k(\frac{1}{p_j^0})F_k(z) \sum_{r=1}^{\infty} [G_b(\frac{1}{p_j^0})G_b(z)]^r \tag{4.44}$$

$$= \sum_{j=1}^{n_0} \frac{b_j}{p_j^0} \sum_{k=1}^{n} F_k(\frac{1}{p_j^0})F_k(z) \cdot \frac{G_b(\frac{1}{p_j^0})G_b(z)}{1 - G_b(\frac{1}{p_j^0})G_b(z)}. \tag{4.45}$$

Substituting (4.42) into (4.45) with $z_1 = 1/p_j^0$ and $z_2^* = z$ leads to

$$G_0(z) - \hat{G}_n(z) = \sum_{j=1}^{n_0} \frac{b_j}{z - p_j^0} G_b(\frac{1}{p_j^0})G_b(z),$$

which is a closed-form and relatively simple expression for the system approximation error. When evaluating this error on the unit circle, it follows from the all-pass property of G_b that

$$|G_0(e^{i\omega}) - \hat{G}_n(e^{i\omega})| \leq \sum_{j=1}^{n_0} \left| \frac{b_j}{e^{i\omega} - p_j^0} \right| \prod_{k=1}^{n} \left| \frac{p_j^0 - \xi_k}{1 - \xi_k^* p_j^0} \right|. \tag{4.46}$$

This expression shows a tight bound on the approximation error. If for each pole p_j^0 in the system there exists a matching pole $\xi_j = p_j^0$ in the basis, then the upper bound is zero, and G_0 indeed has turned into a *finite* series expansion in terms of the basis functions $\{F_k(z)\}$.

If basis functions $\{F_k\}_{k=1,\cdots n}$ are chosen with an extension to infinity that is induced by a periodic choice of pole locations, *i.e.* $\xi_{jn+k} = \xi_{(j-1)n+k}$ for all $j, k \in \mathbb{Z}_+$, then it can be shown that the dynamics that governs the coefficient sequence $\{c_1^{(0)}, c_2^{(0)}, \cdots\}$ is described by the n_0 eigenvalues $\prod_{k=1}^{n} \frac{p_j^0 - \xi_k}{1 - p_j^0 \xi_k}$, $(j = 1, \cdots n_0)$, while the magnitude of the slowest eigenvalue is given by

$$\rho := \max_j \prod_{k=1}^{n} \left| \frac{p_j^0 - \xi_k}{1 - p_j^0 \xi_k} \right| \tag{4.47}$$

This slowest eigenvalue will determine the convergence rate of the expansion coefficients in θ_e, and thus it directly determines the norms $\|\theta_e\|$ in the bias expressions of the previous section.

Both in terms of parameter error and frequency response error, it is thus attractive to choose basis functions for which ρ is small. This comes down to choosing a basis with poles that closely match the poles of the underlying system to be modelled. An appropriately chosen set of basis functions can then lead to an increased rate of convergence of the corresponding expansion coefficients.

Table 4.2. Maximum eigenvalue of expansion coefficients of a system with pole locations $p_{1,2}^0 = 0.985 \pm 0.16i; p_3^0 = 0.75$, expanded in terms of several basis functions; number of expansion coefficients, $n_{0.01}$, that it takes before their magnitude is decayed below a level of 1%.

	ρ	$n_{0.01}$
FIR	0.9979	2191
Laguerre with $\xi_1 = 0.96$	0.9938	740
Kautz with $\xi_{1,2} = 0.97 \pm 0.1i$	0.9728	167
GOBF with $\xi_{1,2} = 0.97 \pm 0.1i$, $\xi_3 = 0.7$	0.9632	123
GOBF with $\xi_{1,2} = 0.98 \pm 0.15i$, $\xi_3 = 0.72$	0.7893	20

Example 4.4. As an example, consider a third-order system with poles

$$p_{1,2}^0 = 0.985 \pm 0.16i; \quad p_3^0 = 0.75$$

The value of ρ for several choices of basis functions is collected in Table 4.2, where it is also indicated how many expansion coefficients have to be taken into account before their magnitude is decayed below a level of 1% of the first one ($n_{0.01}$). Starting with an FIR model, requiring 2191 coefficients in order to capture the dominant part of the system's dynamics, the number of coefficients can be reduced by a factor 3 when choosing a single pole Laguerre model using only a very rough estimate of the pole. A factor of 100 can be obtained if a more accurate indication of the plant's poles is used.

From the approximation results presented above, the following direct result follows for the asymptotic bias in the case of an identified model.

Consider an orthonormal basis $\{F_k\}_{k=1,\cdots n}$ induced by the poles $\{\xi_1, \cdots \xi_n\}$, and suppose $G_0(z)$ has partial fraction expansion

$$G_0(z) = \sum_{j=1}^{n_0-1} \frac{b_j}{z - p_j^0}, \qquad |p_j^0| < 1$$

If the input u is white ($\Phi_u(e^{i\omega}) =$ constant), then the limiting estimate $\theta^\star$ satisfies

$$|G_0(e^{i\omega}) - G(e^{i\omega}, \theta^\star)| \leq \sum_{j=1}^{n_0} \left| \frac{b_j}{e^{i\omega} - p_j^0} \right| \prod_{k=0}^{n-1} \left| \frac{p_j^0 - \xi_k}{1 - \xi_k^* p_j^0} \right|. \qquad (4.48)$$

Note that this result only addresses the bias error for white input. For the case of coloured input, things become more complicated. One way of analyzing this situation – at least partially – is to present a numerical tool for calculating the bias error:

The asymptotic frequency response estimate $G(e^{i\omega}, \theta^\star)$ at $\omega = \omega_\circ$ is related to the true frequency response estimate $G_0(e^{i\omega})$ via

$$G(e^{i\omega_\circ}, \theta^\star) = \frac{1}{2\pi}\int_{-\pi}^{\pi} G_0(e^{i\omega})K_n(e^{i\omega}, e^{i\omega_\circ})^* \Phi_u(e^{i\omega})\, d\omega \qquad (4.49)$$

where the reproducing kernel $K_n(e^{i\omega}, e^{i\omega_\circ})$ is given by

$$K_n(e^{i\omega}, e^{i\omega_\circ}) = \Gamma^\star(e^{i\omega_\circ})R_n^{-1}\Gamma(e^{i\omega}). \qquad (4.50)$$

In the above expression, the reproducing kernel $K_n(e^{i\omega}, e^{i\omega_\circ})$ is the reproducing kernel for the space $Sp\{F_1(z), \cdots, F_n(z)\}$ with respect to the weighted inner product $\langle f, g\rangle_u = \frac{1}{2\pi}\int_{-\pi}^{\pi} f(e^{i\lambda})g^*(e^{i\lambda})\Phi_u(e^{i\lambda})\, d\lambda$.
The correctness of the statement can be verified by examining that for all $k = 1, \cdots n$,

$$\left\langle F_k(e^{i\lambda}), \Gamma^\star(e^{i\omega})R_n^{-1}\Gamma(e^{i\lambda})\right\rangle_u = F_k(e^{i\omega}).$$

This result also makes explicit the relationship between $G_0(e^{i\omega})$, the choice of the basis functions $\{F_k(z)\}$, the input spectrum $\Phi_u(e^{i\omega})$, and the estimate $G(e^{i\omega}, \theta^\star)$ of $G_0(e^{i\omega})$. It shows that the estimated response at some frequency $\omega_\circ$ depends on the integral of the whole true frequency response weighted by a kernel $K_n(e^{i\omega}, e^{i\omega_\circ})$. Obviously, we would like the kernel to be as much like a Dirac delta function $\delta(\omega - \omega_\circ)$ as possible. We can check this for a particular choice of $\{F_k(z)\}$ by using (4.50).
For example, suppose we envisage an input spectral density as shown in the upper right corner of Figure 4.2, and we are trying to decide between using one of two possible model structures corresponding to the following choice of basis functions:

$$\text{FIR:} \qquad F_n(z) = \frac{1}{z^n} \qquad n = 1, 2, \cdots, 10$$

$$\text{Laguerre: } F_n(z) = \frac{\sqrt{1-\xi^2}}{z-\xi}\left(\frac{1-\xi z}{z-\xi}\right)^{n-1} \qquad n = 1, 2, \cdots, 10; \qquad \xi = 0.9.$$

We can then calculate the reproducing kernels $K_n(e^{i\omega}, e^{i\omega_\circ})$ for both these choices via (4.50) and plot them to see if we can expect good estimates at frequencies of interest to us. This is done for $\omega_\circ$ set to the normalized frequencies of 0.12 rad/s and 0.79 rad/s in the lower left and right of Figure 4.2. The Laguerre kernel is shown as a solid line and the FIR kernel is shown as a dashed

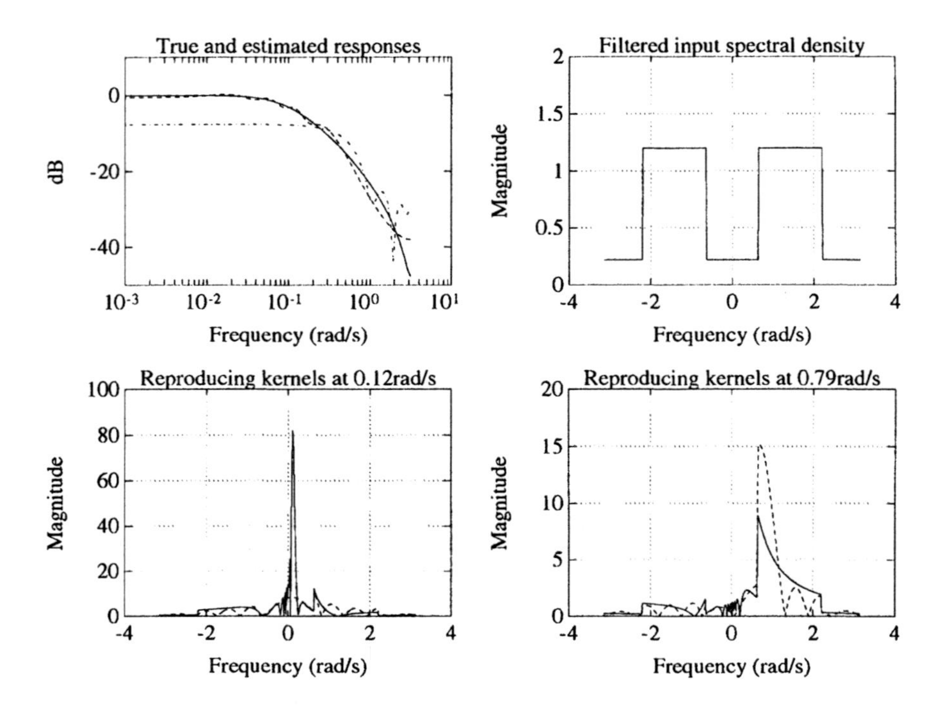

Fig. 4.2. Comparison of FIR and Laguerre basis function models for frequency response estimation by examination of their corresponding reproducing kernels. In top left plot, Laguerre estimate is dashed line, FIR estimate is dash-dot line, and true system is the solid line. In bottom plots, FIR kernel is shown as a dashed line and Laguerre kernel as solid line.

line. As can be seen, at 0.12 rad/s the Laguerre kernel is a good approximation to a delta function at that frequency, so we should expect accurate frequency response estimates from a Laguerre-based model at that frequency. In contrast, the FIR kernel is a very poor approximation to a delta function[1], so would not be a good basis function choice if accurate low frequency estimates are required. At the higher frequency of 0.79 rad/s, the reproducing kernels for the two basis function choices are shown in the lower right of Figure 4.2. Once the kernel $K_n(e^{i\omega}, e^{i\omega_\circ})$ has been calculated for a range of $\omega_\circ$ of interest, we can use (4.49) to calculate what the asymptotic estimate $G(e^{i\omega}, \theta^\star)$ will be for a hypothesized $G_0(e^{i\omega})$. For example, with a hypothetical $G_0(z)$ of

$$G_0(z) = \frac{0.4}{z - 0.6}$$

[1]The FIR kernel, shown as a dashed line, is hard to see in the lower left of Figure 4.2 because it is buried in the low-amplitude side-lobes of the Laguerre kernel.

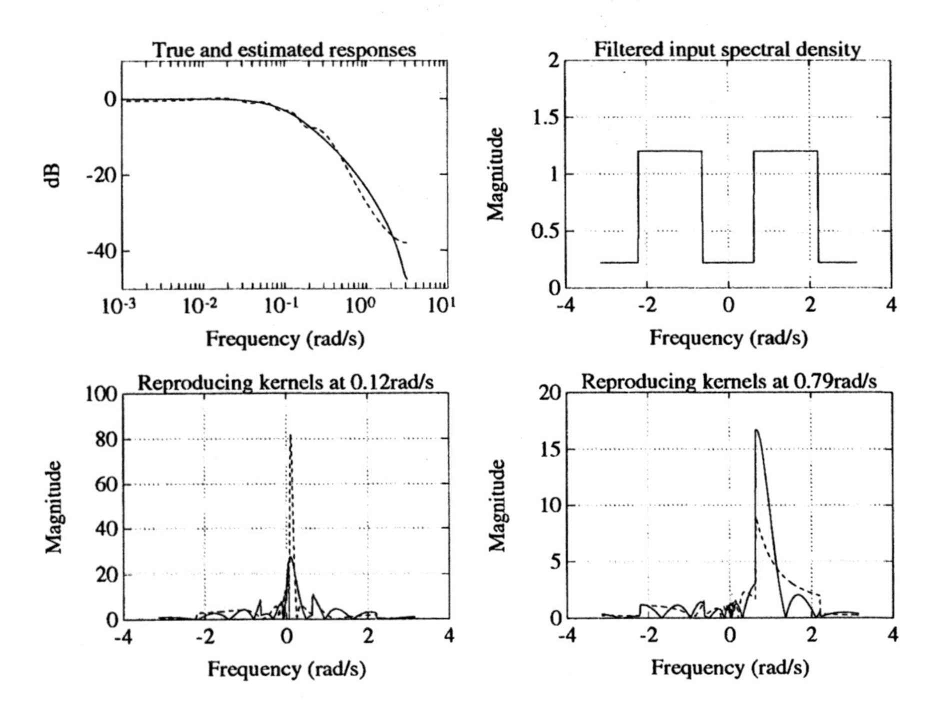

Fig. 4.3. Comparison of Legendre and Laguerre basis function models for frequency response estimation by examination of their corresponding reproducing kernels. In top left plot, Legendre estimate sits directly on top of true system, which appears as a solid line, Laguerre estimate is the dashed line. In bottom plots, Legendre kernel is shown as solid line and Laguerre kernel as dashed line.

whose frequency response is shown as a solid line in the upper left of Figure 4.2, we can predict via (4.49) what the asymptotic estimates will be for the FIR and Laguerre model structures. The Laguerre is shown as a dashed line and the FIR as a dash-dot line in the upper left of Figure 4.2. As can be seen, the Laguerre model is more accurate at low frequencies and we expected this because its reproducing kernel $K_n(e^{i\omega}, e^{i\omega_o})$ was very much like a Dirac delta at low frequencies. Note that these plots were not made from an estimation experiment, but were made by numerically evaluating (4.49).

Of course, the kernels for other basis function choices can also be calculated and integrated against hypothetical $G_0(e^{i\omega})$ in an effort to *a priori* choose the most suitable basis for frequency response estimation. For example, in the upper left of Figure 4.2 we note that against a hypothetical $G_0(e^{i\omega})$, a Laguerre basis estimate performs much better than an FIR estimate of the same order. Nevertheless, the Laguerre model is somewhat deficient at high frequencies. Motivated by this, we examine the use of the so-called Legendre basis, which is (4.41) with the choice of pole location as

$$\xi_k = \frac{2 - \xi(2k+1)}{2 + \xi(2k+1)}$$

for some fixed real ξ. Due to the progression in pole location in this basis it might be expected to perform better than a Laguerre basis at high frequencies. This indeed is the case as reference to Figure 4.3 shows. Note that in the upper left plot of Figure 4.3, the Legendre basis function estimate cannot be discriminated from the hypothetical $G_0(e^{i\omega})$ because it sits directly on top of it.

4.3.4 Asymptotic Variance Results

In Equation (4.11), the total estimation error $G_0(e^{i\omega}) - G(e^{i\omega}, \widehat{\theta}_N)$ was decomposed into a 'bias error' and a 'variance error' part as

$$G_0(e^{i\omega}) - G(e^{i\omega}, \widehat{\theta}_N) = [G_0(e^{i\omega}) - G(e^{i\omega}, \theta^\star)] + [G(e^{i\omega}, \theta^\star) - G(e^{i\omega}, \widehat{\theta}_N)].$$

The bias error term $G_0(e^{i\omega}) - G(e^{i\omega}, \theta^\star)$ has just been studied in the previous section. This section will consider the variance error term defined as being $G(e^{i\omega}, \theta^\star) - G(e^{i\omega}, \widehat{\theta}_N)$. It represents the error induced by the measurement noise process $v(t)$.
From classical analysis of prediction error identification methods, see *e.g.* [164] and Section 4.2, we know that, under fairly weak conditions,

$$\sqrt{N}(\widehat{\theta}_N - \theta^\star) \to \mathcal{N}(0, P_n) \quad \text{as } N \to \infty. \tag{4.51}$$

For output error identification schemes, as considered here, the asymptotic covariance matrix can be specified using the results of (4.16)-(4.18), leading to:

$$P_n = R_n^{-1} Q_n R_n^{-1}. \tag{4.52}$$

In this expression, R_n is as defined in (4.34) and Q_n is given by

$$Q_n := \sigma_e^2 \cdot \bar{\mathsf{E}}\{\tilde{\varphi}(t)\tilde{\varphi}^T(t)\}$$

where $\tilde{\varphi}(t) = \sum_{i=0}^{\infty} h_0(i)\varphi(t+i)$, and $h_0(i)$ is the pulse response of the transfer function H_0.
A full analysis of P_n for finite values of n is hardly tractable. However, when both N and n tend to infinity, nicely interpretable results can be achieved. For the FIR situation, it is known ([164]), as presented in Section 4.2.4, that under minor signal conditions on input spectrum and disturbance process,

$$\text{Var}\{G(e^{i\omega}, \widehat{\theta}_N\} \to \frac{n}{N} \frac{\Phi_v(e^{i\omega})}{\Phi_u(e^{i\omega})} \quad \text{as } n, N \to \infty, \; n << N. \tag{4.53}$$

This means that the frequency response variance of the estimated model tends to be proportional to the noise-to-signal power ratio at the considered frequency, which is a very appealing mechanism. This result can be generalized

to the situation of orthogonal basis functions generated by an arbitrary choice of pole-locations. First, the generalized result will be presented and discussed, and subsequently a sketch of its derivation will be given.
The generalized result for the variance of the model estimate is, that for a model structure in terms of a sequence of orthogonal basis functions $\{F_k(z)\}$:

$$\mathrm{Var}\{G(e^{i\omega},\widehat{\theta}_N\} \to \frac{1}{N}\frac{\Phi_v(e^{i\omega})}{\Phi_u(e^{i\omega})}\cdot\sum_{k=1}^{n}|F_k(e^{i\omega})|^2 \quad \text{as } n,N\to\infty,\ n<<N \tag{4.54}$$

Apparently, an additional weighting factor occurs in the variance expression that is completely determined by the dynamics in the basis functions. The weighting factor n in the FIR case (4.53) is replaced by $\sum_{k=1}^{n}|F_k(e^{i\omega})|^2$ in the generalized case (4.54). However, in both situations the weighting factor equals the reproducing kernel (4.39) related to the basis functions chosen, evaluated in a particular frequency:

$$K_n(e^{i\omega},e^{i\omega}) = \sum_{k=1}^{n}|F_k(e^{i\omega})|^2.$$

In the FIR-case, with $|F_k(e^{i\omega})| = |e^{-i\omega k}| = 1$, the expression reduces to $K_n(e^{i\omega},e^{i\omega}) = n$. This shows that the reproducing kernel not only plays a key role in quantifying the bias error but also in quantifying the variance error of the estimates.
Typically, as *e.g.* in the Laguerre case, F_k will contain low-pass dynamics and a roll-off at higher frequencies. This will imply that the frequency response variance is reduced at higher frequencies. See for example Figure 4.4 where the expression $K_n(e^{i\omega},e^{i\omega})$ is plotted for a variety of choices of $\{\xi_k\}$. In particular, note that for all poles fixed at the origin, $K_n(e^{i\omega},e^{i\omega}) = n$ so that in this special case of FIR modelling, (4.54) is identical to (4.53). However, the more poles that are not fixed at the origin, the more $K_n(e^{i\omega},e^{i\omega})$ will become frequency dependent and will differ from n, and hence the more influence the basis functions will have on the variance distribution of the model estimate.
Note that the additional weighting factor $K_n(e^{i\omega},e^{i\omega})$ in expression (4.54) has a 'waterbed-effect' on the frequency-dependent expression. By orthonormality of the basis functions it follows straightforwardly that

$$\frac{1}{2\pi}\int_{-\pi}^{\pi} K_n(e^{i\omega},e^{i\omega})d\omega = n.$$

Therefore, any 'peaks' in the weighting function at particular frequencies will be compensated for by reduced values in other frequency regions.

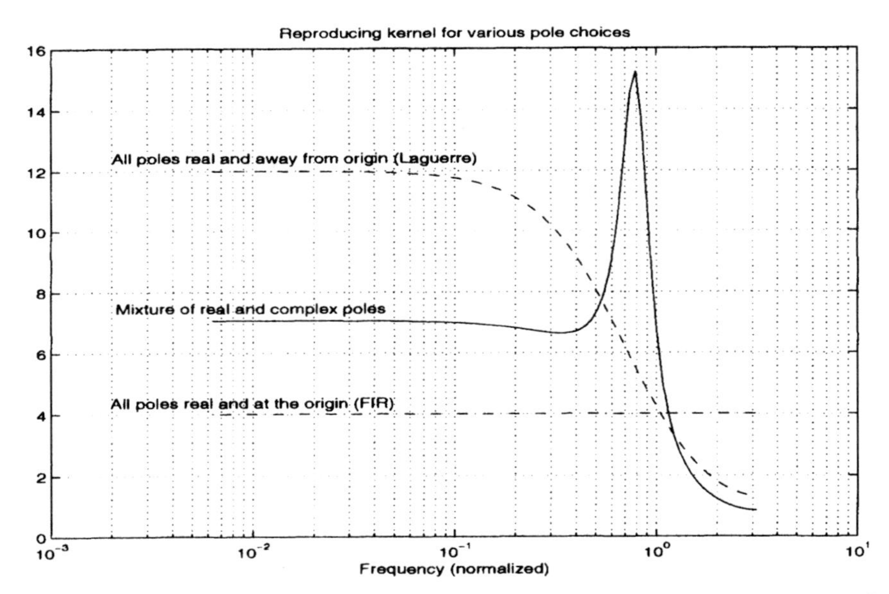

Fig. 4.4. Plot, for various choices of $\{\xi_k\}$, of term $K_n(e^{i\omega}, e^{i\omega}) = \sum_{k=1}^{n} |F_k(e^{i\omega})|^2$, which captures the effect of pole choice $\{\xi_k\}$ on transfer function estimate sensitivity to measurement noise. Here $n = 4$.

Background of the Generalized Variance Expression

When evaluating the roots of the generalized variance result (4.54), it appears that the theory of Toeplitz operators and their convergence properties play a crucial role. For the FIR situation, it can be observed that

$$R_n = \frac{1}{2\pi} \int_{-\pi}^{\pi} \Omega_n(e^{i\omega}) \Omega_n^*(e^{i\omega}) \Phi_u(e^{i\omega}) d\omega \tag{4.55}$$

with $\Omega_n(e^{i\omega}) = [e^{i\omega} \; e^{2i\omega} \; \cdots e^{ni\omega}]^T$. This means that R_n is the $n \times n$ Toeplitz matrix related to the symbol $\Phi_u(e^{i\omega})$, being denoted as $R_n = T_n(\Phi_u)$. Similarly it follows that

$$Q_n = T_n(\Phi_u \Phi_v). \tag{4.56}$$

Furthermore, employing the asymptotic-in-n behaviour of Toeplitz matrices [108, 121, 327] that for any continuous positive functions[2] f, g,

$$T_n(f)T_n(g) \approx T_n(fg)$$
$$T_n^{-1}(f) \approx T_n(f^{-1})$$

it follows from (4.52) that

[2]The precise meaning of the $\approx$ operator is that equality occurs in Hilbert–Schmidt weak matrix norm $|\cdot|$, defined by $|A_n|^2 = n^{-1}\text{Trace}\{A_n^T A_n\}$ as $n \to \infty$.

$$P_n = T_n(\frac{\Phi_v}{\Phi_u}).$$

Additionally, the result of [168] is used that

$$\lim_{n\to\infty} \frac{1}{n} \Omega_n^*(e^{i\omega_1}) T_n(f) \Omega_n(e^{i\omega_2}) = \begin{cases} f(\omega_1) & \omega_1 = \omega_2 \\ 0 & \omega_1 \neq \omega_2 \end{cases} \tag{4.57}$$

Then it follows that

$$\sqrt{N} \cdot \mathrm{Cov}\{G(e^{i\omega}, \theta^*)\} \to \Omega_n(e^{i\omega}) P_n \Omega_n^*(e^{i\omega}) \tag{4.58}$$

$$= n \cdot \frac{\Phi_v(e^{i\omega})}{\Phi_u(e^{i\omega})} \quad \text{as } n \to \infty. \tag{4.59}$$

For the generalization to the case of orthogonal basis functions $\{F_k\}_{k=1,\cdots n}$, two different paths have been followed.
The first path uses the fact that in the situation of basis functions induced by a periodic sequence of poles, the matrix P_n can be shown to be of a Toeplitz structure, with a symbol (related spectral density) that can be specified through a variable transformation on the original spectral density. This allows the use of the same theory as for the FIR situation, however now applied to transformed spectral densities. This approach has been pursued in [313] for Laguerre functions, in [315] for Kautz functions, and in [306] for the general case of repeating sequences of poles. In the latter situation, the Hambo transform is fruitfully used, see Section 3.3 and Chapter 12.
In the most general situation, which was pursued in [223], the required analysis is more involved, but very elegant in its results. First, the classical Toeplitz matrix formulation (4.55) is generalized to

$$M_n(f) := \frac{1}{2\pi} \int_{-\pi}^{\pi} \Gamma_n(e^{i\omega}) \Gamma_n^*(e^{i\omega}) f(\omega) \, d\omega \tag{4.60}$$

which, although formally identical to (4.55), is functionally quite different in that the underlying orthonormal basis is not fixed at the trigonometric one, but a generalization obtained by $\Gamma_n(z)$ defined in (4.30).
Matrices defined by (4.60) are called 'generalized Toeplitz' matrices, with the epithet deriving from the fact that if all the poles are chosen at the origin, then $M_n(f) = T_n(f)$ is a bona-fide Toeplitz matrix, but otherwise it is not.
These generalized Toeplitz matrices have the pleasant property that for large dimension n and any continuous positive definite f and g the same relations hold as for ordinary Toeplitz matrices:

$$M_n(f) M_n(g) \approx M_n(fg) \tag{4.61}$$

$$M_n^{-1}(f) \approx M_n(1/f). \tag{4.62}$$

Additionally, according to the definitions of R_n and Q_n, they will satisfy

$$R_n = M_n(\Phi_u), \qquad Q_n = M_n(\Phi_u \Phi_v)$$

so that as the matrix P_n governing the parameter space covariance according to (4.52) is given by $P_n = R_n^{-1} Q_n R_n^{-1}$ then using the approximations (4.61) and (4.62)

$$\begin{aligned} N\mathsf{E}\{(\widehat{\theta}_N - \theta^\star)(\widehat{\theta}_N - \theta^\star)^T\} &\approx P_n \\ &\approx M_n^{-1}(\Phi_u) M_n(\Phi_u \Phi_n) M_n^{-1}(\Phi_u) \\ &\approx M_n(\Phi_v / \Phi_u). \end{aligned}$$

Proceeding then as in the FIR situation, with $G(e^{i\omega}, \theta) = \Gamma_n^T(e^{i\omega})\theta$ the covariance of the frequency response estimate is governed by

$$N\mathsf{E}\{|G(e^{i\omega}, \theta^\star) - G(e^{i\omega}, \widehat{\theta}_N)|^2\} \approx \Gamma_n^\star(e^{i\omega}) M_n(\Phi_v/\Phi_u) \Gamma_n(e^{i\omega}). \tag{4.63}$$

The final step in the result then is a generalization of the convergence result (4.57). The generalization is that for every continuous function f

$$\lim_{n\to\infty} \frac{1}{K_n(e^{i\omega}, e^{i\omega})} \Gamma_n^\star(e^{i\omega_1}) M_n(f) \Gamma_n(e^{i\omega_2}) = \begin{cases} f(\omega_1)\ ; & \omega_1 = \omega_2, \\ 0 \ ; & \omega_1 \neq \omega_2. \end{cases} \tag{4.64}$$

Combining this convergence result with (4.63) then provides the result (4.54). For a more detailed analysis of variance issues related to orthonormal basis functions the reader is referred to Chapter 5.

4.4 Relations with Other Model Structures

4.4.1 Introduction

The material to this point has hopefully been a strong motivator for examining rational orthonormal bases as a tool for attacking various system theoretic problems – the emphasis being on those pertaining to system identification. The purpose of this section is to reinforce this motivation by revisiting certain key points raised in earlier chapters and examining them in greater detail. In particular, let us consider the devil's advocate position by posing the provocative challenge:

> *Surely 'orthonormal' is just an impressive sounding word. All that is really being done is that very simple model structures are being reparameterized in an equivalent form. Who cares whether this form is 'orthonormal' or not? Let's not confuse changes in model structure with fundamental changes in methods of estimation.*

For example, historically right from the 1950s, much of the motivation for using orthonormal bases in a system identification or system approximation setting has arisen from the desire to provide parsimonious representation of systems by a strategy of fixing poles near where the poles of the underlying dynamics are believed to lie.

However, once this idea is accepted, then one soon realizes that actually, the orthonormality property is not responsible for the advantage of achieving parsimony. It is achieved solely by the idea of fixing poles at arbitrary points. In this section, it will be discussed which role is played by orthonormal basis functions, taking into account that the corresponding model structure can be considered as a particular form of 'fixed-denominator model structure'. First, the relation between these two structures will be revealed, together with an analysis of the consequences for bias and variance issues. Additionally, extensions to relations with more general model structures will be discussed.

4.4.2 Fixed Denominator Models

Models with prespecified pole locations can be specified in a very simple way, by using the model structure

$$G(q,\beta) = \frac{b_1 q^{-1} + b_2 q^{-2} + \cdots + b_n q^{-n}}{D_n(q^{-1})} \tag{4.65}$$

where

$$D_n(q^{-1}) = \prod_{k=1}^{n} (1 - \xi_k q^{-1})$$

is a fixed denominator and $\beta^T = [b_1, b_2, \cdots, b_n]$ is a vector of parameters. This can be an appropriate alternative for the strategy of using an orthonormalized form

$$G(q,\theta) = \sum_{k=1}^{n} \theta_k F_k(q) \tag{4.66}$$

where the orthonormal functions $\{F_k(q)\}$ also have fixed poles given by the set $\{\xi_1, \cdots, \xi_n\}$.

Additionally, a parameter estimate for the simple structure (4.65) can be found by just using available software for estimation of FIR models (the numerator) after having pre-filtered the input $u(t)$ via the all-pole filter $1/D_n(q^{-1})$. If the estimate $\widehat{\theta}_N$ is to be found with respect to the orthonormal structure (4.66), then specialized software needs to be written. The relevant question now of course is: how equivalent are these model structures (4.65) and (4.66)?

If least squares estimation is performed relative to them, then both parameter estimates can be written as the result of a linear regression problem:

For (4.65):

$$\widehat{\beta}_N = \left(\frac{1}{N} \sum_{t=1}^{N} \gamma(t)\gamma(t)^T \right)^{-1} \cdot \frac{1}{N} \sum_{t=1}^{N} \gamma(t) y(t) \tag{4.67}$$

with

$$\gamma(t) = \Lambda_n(q)\, u(t), \qquad \Lambda_n(q) := [K_1(q)\ K_2(q)\ \cdots\ K_n(q)]^T \tag{4.68}$$

and $K_k(q) := \dfrac{q^{-k}}{D_n(q^{-1})}$.

Similarly for (4.66):

$$\widehat{\theta}_N = \left(\frac{1}{N}\sum_{t=1}^{N}\varphi(t)\varphi(t)^T\right)^{-1} \cdot \frac{1}{N}\sum_{t=1}^{N}\varphi(t)y(t) \tag{4.69}$$

with

$$\varphi(t) := \Gamma_n(q)u(t), \qquad \Gamma_n(q) = [F_1(q)\ F_2(q)\ \cdots\ F_n(q)]^T. \tag{4.70}$$

A key point is that under the condition that the poles in D_n and F_n are the same,

$$\mathrm{Span}\{K_1, K_1, \cdots, K_n\} = \mathrm{Span}\{F_1, F_1, \cdots, F_n\} \tag{4.71}$$

and as a result, there exists a linear relationship $\varphi(t) = J\gamma(t)$ for some non-singular J. Therefore, $\widehat{\beta}_N = J^T\widehat{\theta}_N$ and hence modulo numerical issues the least-squares frequency response estimate is invariant to the change in model structure between (4.65) and (4.66). Specifically:

$$\begin{aligned}
G(e^{i\omega}, \widehat{\beta}_N) &= \Lambda_n^T(e^{i\omega})\widehat{\beta}_N \\
&= \Lambda_n^T(e^{i\omega})\left(\sum_{t=1}^{N}\gamma(t)\gamma(t)^T\right)^{-1}\sum_{t=1}^{N}\gamma(t)y(t) \\
&= \Lambda_n^T(e^{i\omega})\left[J^{-1}\left(\sum_{t=1}^{n}\varphi(t)\varphi(t)^T\right)J^{-T}\right]^{-1} J^{-1}\sum_{t=1}^{N}\varphi(t)y(t) \\
&= [J\Lambda_n(e^{i\omega})]^T\left(\sum_{t=1}^{N}\varphi(t)\varphi(t)^T\right)^{-1}\sum_{t=1}^{N}\varphi(t)y(t) \\
&= \Gamma_n^T(e^{i\omega})\widehat{\theta}_N \\
&= G(e^{i\omega}, \widehat{\theta}_N).
\end{aligned}$$

Albeit that the implementation of the estimation algorithm can be different, the two model structures lead to the same estimated models, provided that we discard numerical considerations. In fact, it turns out that estimating the frequency response $G(e^{i\omega}, \widehat{\beta}_N)$ with respect to the model structure (4.65) is very often horribly ill conditioned to the point where it fails on modern high performance workstations. In these same cases, the estimation problem with respect to the orthonormalized form is well conditioned. Therefore, implementation with respect to the orthonormal form is very often the only way of implementing the idea of fixing the poles in a model structure. This issue is addressed in more detail in Chapter 6.
However, when disregarding numerical issues, we summarize the result:

In the prediction error identification context discussed in this chapter and considering orthogonal basis functions $\{F_k\}_{k=1,\cdots n}$ as defined before, the two model structures

$$G(q,\beta) = \frac{b_1 q^{-1} + b_2 q^{-2} + \cdots + b_n q^{-n}}{D_n(q^{-1})}$$

and

$$G(q,\theta) = \sum_{k=1}^{n} \theta_k F_k(q),$$

deliver estimated models that are identical if F_n and D_n share the same poles. In this situation, bias and variance properties of these models are identical.

This implies of course that all bias and variance results presented in Section 4.3 concerning the model's frequency response are valid also for the fixed-denominator model structures. So even in model structures that do not seem to be related to any orthogonal structures, the orthogonal basis functions underlying the fixed denominator play a crucial role in the asymptotic properties of the estimates.
Note *e.g.* that in the analysis of bias and variance properties in the previous section, orthonormality of the basis functions *did* play an important role. Both in the kernel representations leading to the bias results, and in the variance results based on the generalized Toeplitz forms, orthonormality of the functions was essentially used in the derivations. Apparently the orthonormal structure plays a crucial role, not (only) as an *implementation tool* for estimating models, but in particular as an *analysis tool* for specifying the bias and variance properties of identified models in a fixed denominator model structure.

This observation has another interesting consequence, that is related to the asymptotic variance expressions (4.53) and (4.54). For large n and N it has been derived that for FIR models ($\xi_k = 0$ for all k):

$$\mathsf{E}\{|G(e^{i\omega},\theta^\star) - G(e^{i\omega},\widehat{\theta}_N)|^2\} \approx \frac{n}{N}\frac{\Phi_v(e^{i\omega})}{\Phi_u(e^{i\omega})} \tag{4.72}$$

while for fixed denominator model structures this relation transforms to

$$\mathsf{E}\{|G(e^{i\omega},\theta^\star) - G(e^{i\omega},\widehat{\theta}_N)|^2\} \approx \frac{1}{N}\frac{\Phi_v(e^{i\omega})}{\Phi_u(e^{i\omega})}\sum_{k=1}^{n}|F_k(e^{i\omega})|^2. \tag{4.73}$$

Suppose that we choose a fixed denominator structure with denominator $D_n(q)$. Then this can be implemented by simply pre-filtering the input signal according to

$$u_f(t) = L(q)u(t) \qquad \text{with } L(q) = \frac{1}{D_n(q^{-1})}$$

and by estimating a FIR-numerator, according to

$$\hat{y}(t|t-1;\theta) = G_{fir}(q,\theta)u_f(t).$$

After estimating G_{fir}, the corresponding (fixed-pole) plant model is constructed as

$$\hat{G} = \hat{G}_{fir} \cdot \frac{1}{D_n(q^{-1})}.$$

When using the FIR-relation (4.72) for specifying the variance result, it follows that

$$\text{Var}\{\widehat{G}(e^{i\omega})\} = |L(e^{i\omega})|^2 \text{Var}\{\hat{G}_{fir}(e^{i\omega})\} \tag{4.74}$$

$$\approx \frac{n}{N}|L(e^{i\omega})|^2 \frac{\Phi_v(e^{i\omega})}{|L(e^{i\omega})|^2 \Phi_u(e^{i\omega})} \tag{4.75}$$

$$= \frac{n}{N}\frac{\Phi_v(e^{i\omega})}{\Phi_u(e^{i\omega})} \tag{4.76}$$

which is unchanged from the normal FIR case.
However, when using the fixed-denominator result (4.73), the variance expression becomes:

$$\text{Var}\{\widehat{G}(e^{i\omega})\} \approx \frac{1}{N}\frac{\Phi_v(e^{i\omega})}{\Phi_u(e^{i\omega})} \sum_{k=1}^{n} |F_k(e^{i\omega})|^2, \tag{4.77}$$

thus containing the additional term induced by the basis functions originating from the fixed pole locations. These two expressions are clearly different.
To illustrate these different variance expressions, we consider the following example.

Example 4.5. Consider a simple simulation example where an $n = 12$th order FIR model of the true system

$$G_0(q) = \frac{0.1548q^{-1} + 0.0939q^{-2}}{(1-0.6065q^{-1})(1-0.3679q^{-1})}$$

is obtained by observing 10,000 samples of the true system's input and output sequence when the former is a realization of a stationary Gaussian process with spectral density $\Phi_u(e^{i\omega}) = 0.25/(1.25-\cos\omega)$ and the latter is corrupted by zero mean Gaussian white noise of variance $\sigma^2 = 0.001$. In this case, because both n and N can reasonably be considered 'large', the approximation (4.72) could be expected to hold. This can be checked by Monte-Carlo simulation over, say, 500 input and noise realizations so as to estimate the variance $E\{|G(e^{i\omega},\theta^\star) - G(e^{i\omega},\widehat{\theta}_N)|^2\}$ by its sample average, which can then be compared to the approximation (4.72). The results for just such an experiment

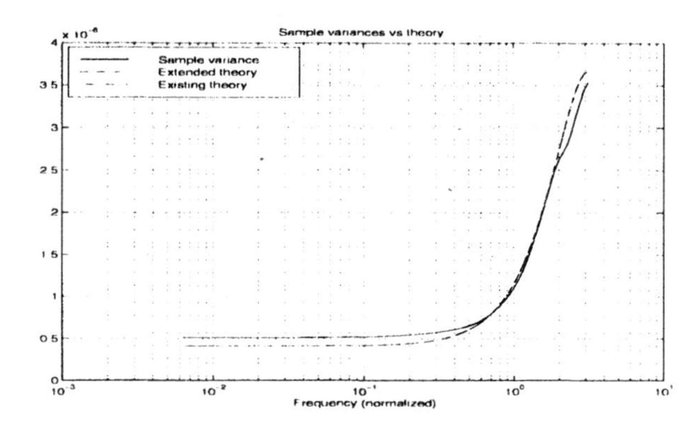

Fig. 4.5. Asymptotic variance expressions of FIR model ($n = 12$) with all poles at origin. Monte-Carlo estimate of sample variability (solid line), approximate expressions (4.72), and (4.73) (dash-dot). The latter two coincide, with (dashed-dot line) the approximate expression (4.72).

are shown in Figure 4.5 with the agreement between the expression (4.72) (dashed line) and the sample average (solid line) being excellent. Note that in this simulation (and in all the rest to follow in this section), the bias error is negligible, and hence the variance error represents the total error.
Continuing the simulation example, we now suppose that prior knowledge of the poles of $G_0(q)$ exists, so that in the interests of decreasing the bias error, it makes sense to try to incorporate this prior knowledge in the estimation process by fixing some poles in the model near where it is believed the true poles lie.

Suppose it is believed that the true pole is near $z = 0.75$, so that guesses of, say $z = 0.7, 0.72, 0.78, 0.8$ are to be incorporated into the model structure. This can be implemented by simply pre-filtering the input by
$L(q) = 1/(1 - 0.7q^{-1})(1 - 0.72q^{-1})(1 - 0.78q^{-1})(1 - 0.8q^{-1})$ before an FIR 'numerator' model is estimated of order 12.
Interestingly, when the expression (4.76) is compared to Monte-Carlo calculated sample variability as it is in Figure 4.6, then the agreement between the true variability (solid line) and approximation (4.76) (dash-dot line) is seen to be not nearly so good as is Figure 4.5. Nevertheless, the expression (4.76) still provides useful information on the qualitative 'high-pass' nature of how the true variability changes with frequency. The dashed line near the solid one in Figure 4.6 represents the expression (4.73), which in this case is shown to be a much more accurate approximation of the sample variance.

Now suppose even more guesses of system poles are made, say at the locations $z = \{0.7, 0.72, 0.78, 0.8, 0.75, 0.85, 0.82, 0.79\}$, with the sample variability again being compared to (4.76) and (4.73) in Figure 4.7. In this case, there

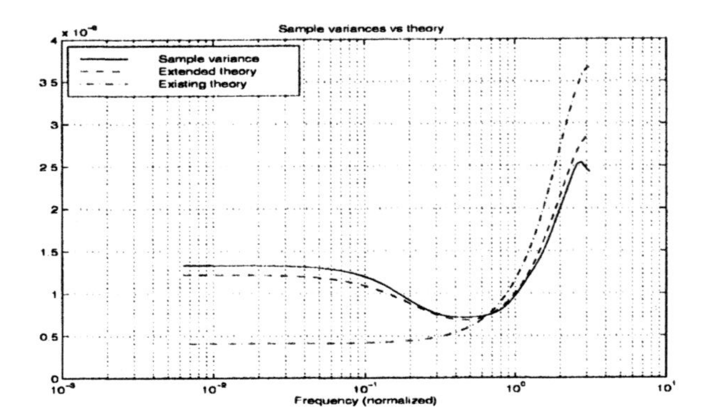

Fig. 4.6. Asymptotic variance expressions of FIR model ($n = 12$), with 4 poles away from origin. Monte-Carlo estimate of sample variability (solid line), approximate expression (4.72) (dash-dot line), and approximate expression (4.73) (dashed line).

is virtually no agreement (even qualitative) between true (sample) variability and variability predicted by (4.76), while the approximation by (4.73) is rather accurate.

The results of the example show that the asymptotic variance result as derived in the previous section, apparently is a better approximation of the variance than the known and standard approximation (4.72), at least in the case of prefiltered FIR model structures. This result, which will be made plausible in the sequel, can be considered an additional 'spin-off' of the orthogonal basis function identification results.

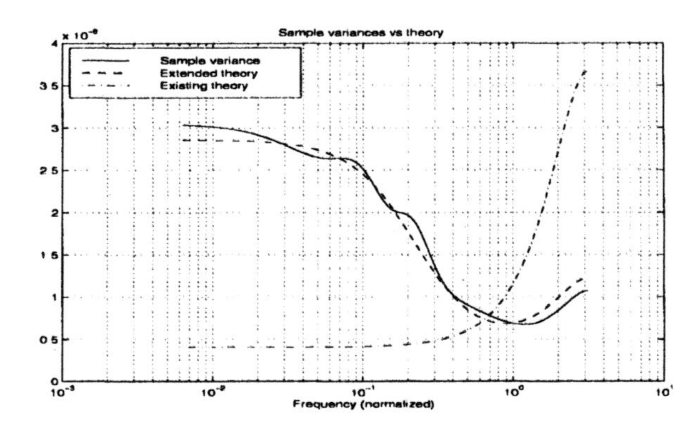

Fig. 4.7. Asymptotic variance expressions of FIR model ($n = 12$), with 8 poles away from origin. Monte-Carlo estimate of sample variability (solid line), approximate expression (4.72) (dash-dot line), and approximate expression (4.73) (dashed line).

Background of the Variance Result for Pre-filtered FIR Models

In order to understand why (4.72) is not a good approximation for arbitrary fixed pole location, it is necessary to understand the rudiments of how (4.72) is derived. There are two important principles underlying this derivation, namely, use of the asymptotic nature of Toeplitz matrices and the employment of the principles of Fourier series convergence.
Recalling the definition of R_n made in (4.55), note that via its Parseval-derived frequency domain characterization it may be written as

$$R_n = T_n(\Phi_u/|D_n|^2). \tag{4.78}$$

It is also possible, for the case of $H(q) = 1$, to establish via (4.56) that provided the measurement noise $v(t)$ is white, then $Q_n = \sigma^2 R_n$ so that $P_n = \sigma^2 R_n^{-1} = \sigma^2 T_n^{-1}(\Phi_u/|D_n|^2)$. Employing the asymptotic in n behaviour of Toeplitz matrices then delivers that

$$P_n = \sigma^2 T_n^{-1}(\Phi_u/|D_n|^2) \approx \sigma^2 T_n\left(\frac{|D_n|^2}{\Phi_u}\right).$$

Note that from the parameter space distributional result (4.51)

$$N\mathsf{E}\{(\widehat{\theta}_N - \theta^\star)(\widehat{\theta}_N - \theta^\star)^T\} \approx P_n \approx \sigma^2 R_n^{-1} \approx \sigma^2 T_n\left(\frac{|D_n|^2}{\Phi_u}\right). \tag{4.79}$$

In the case of using the model structure (4.65) with fixed denominator $D_n(z)$, which was considered in the previous simulation examples

$$G(e^{i\omega}, \theta) = \frac{1}{D_n(e^{i\omega})} \sum_{k=1}^{n} \theta_k e^{-i\omega k}$$

then in fact

$$\begin{aligned}
&\frac{N}{n}\mathsf{E}\{|G(e^{i\omega}, \theta^\star) - G(e^{i\omega}, \widehat{\theta}_N)|^2\} \approx \\
&\approx \frac{\sigma^2}{|D_n(e^{i\omega})|^2} \sum_{\ell=1}^{n}\sum_{m=1}^{n} e^{i\omega\ell} e^{-i\omega m} [T_n(|D_n|^2/\Phi_u)]_{m,\ell} \\
&= \frac{\sigma^2}{|D_n(e^{i\omega})|^2} \sum_{k=-n}^{n} \left(1 - \frac{|k|}{n}\right) c_k e^{i\omega k}
\end{aligned} \tag{4.80}$$

where $\{c_k\}$ are the Fourier coefficients of $|D_n|^2/\Phi_u$ defined by

$$c_k := \frac{1}{2\pi}\int_{-\pi}^{\pi} \frac{|D_n(e^{i\omega})|^2}{\Phi_u(e^{i\omega})} e^{-i\omega k}\, d\omega.$$

Actually, the left-hand side of (4.80) is in fact a Cesàro mean (triangularly windowed) Fourier reconstruction of $|D_n|^2/\Phi_u$, which is known to converge uniformly to $|D_n|^2/\Phi_u$ on its domain and with increasing n, provided $|D_n|^2/\Phi_u$ is continuous, see [344]. Therefore, it should approximately hold that

$$\sum_{k=-n}^{n}\left(1-\frac{|k|}{n}\right)c_k e^{i\omega k} \approx \frac{|D_n(e^{i\omega})|^2}{\Phi_u(e^{i\omega})} \tag{4.81}$$

so that combining this approximation with the ones leading to (4.80) provides

$$\frac{N}{n}\mathsf{E}\{|G(e^{i\omega},\theta^\star)-G(e^{i\omega},\widehat{\theta}_N)|^2\} \approx \frac{\sigma^2}{|D_n(e^{i\omega})|^2}\frac{|D_n(e^{i\omega})|^2}{\Phi_u(e^{i\omega})}$$

and hence

$$\mathsf{E}\{|G(e^{i\omega},\theta^\star)-G(e^{i\omega},\widehat{\theta}_N)|^2\} \approx \frac{n}{N}\frac{\sigma^2}{\Phi_u(e^{i\omega})} \tag{4.82}$$

which is (4.72) for $\Phi_v(e^{i\omega}) = \sigma_v^2$. Now, in the case where all the poles are fixed at the origin and hence $|D_n(e^{i\omega})|^2 = 1$, then as illustrated in Figure 4.5 these approximating steps work rather well and give an informative indication of the true variance error.

The point to be addressed now is why, as illustrated in Figures 4.6 and 4.7, the same approximating arguments work so poorly when poles are fixed away from the origin? The reason is that the analysis leading to the approximation in (4.81) depends on the Fourier series (which (4.81) is) to have approximately converged. The number of terms n for which this can be expected to have occurred is well-known to depend on the smoothness of the function being approximated [152]. But this function is $\Phi_u(e^{i\omega})/|D_n(e^{i\omega})|^2$, which clearly becomes less smooth as more poles are moved away from the origin. So the reason why (4.82) fails in the general fixed denominator case is that the underlying Fourier series has not approximately converged.
For the special FIR case of $|D_n(e^{i\omega})| = 1$, the smoothness is fixed as the smoothness of $\Phi_u(e^{i\omega})$ so that the approximation can be expected to monotonically improve with increasing n. However, the more poles that are chosen that are away from the origin, and hence the more dependent on n that $|D_n(e^{i\omega})|$ is, the less one should rely on (4.72) applying for finite n, because the less likely it is that the underlying Fourier series has come close to convergence. This is precisely the behaviour demonstrated in Figures 4.5, 4.6, and 4.7.

4.4.3 ARX and More General Structures

Although the fixed denominator model structure (4.65) and its generalization (4.66) have many practical advantages, for the estimation of accurate low-order models, they require accurate prior knowledge on the pole locations of the underlying system.
A common strategy for estimating the pole locations of $G_0(q)$ while still involving a predictor that is linear in θ, is to apply an ARX model structure:

$$G(q,\theta) = \frac{B(q^{-1},\theta)}{A(q^{-1},\theta)}, \quad H(q,\theta) = \frac{1}{A(q^{-1},\theta)} \tag{4.83}$$

where dynamics and noise model share parameters in θ. As is well-known and discussed in Section 4.2, this can lead to bias if the model structure is not rich enough. This motivates the inclusion of a fixed numerator term in $H(q,\theta)$, according to:

$$G(q,\theta) = \frac{B(q^{-1},\theta)}{A(q^{-1},\theta)}, \quad H(q,\theta) = \frac{D_n(q^{-1})}{A(q^{-1},\theta)}. \tag{4.84}$$

The $D_n(q^{-1})$ term in the noise model is to avoid bias by allowing $H(q,\theta^\star)$ to approximate $H_0(q)$ accurately for some $\theta^\star$. Simultaneously, pole zero cancellations in $H(q,\theta^\star)$ are possible to occur, in order to allow $G(q,\theta^\star)$ and $H(q,\theta^\star)$ to have distinct poles. In this way, the bias in $G(q,\theta^\star)$ can be substantially reduced through an appropriate selection of D_n.
Note that the predictor for the model structure (4.84) is given by:

$$\begin{aligned}
\hat{y}(t|t-1;\theta) &= H(q,\theta)^{-1}[y(t) - G(q,\theta)u(t) \\
&= \frac{1}{D_n(q^{-1})}[A(q^{-1},\theta)y(t) - B(q^{-1},\theta)u(t)]
\end{aligned}$$

which, for a fixed $D_n(q^{-1})$, remains linear in the parameters θ.
Most commonly the model structure (4.84) appears with the choice $\xi_k = 0$ in which case it is known as the 'equation error' or sometimes 'ARX' model structure and for which the variance analysis provides the same asymptotic result as for the FIR model:

$$\mathsf{E}\{|G(e^{i\omega},\theta^\star) - G(e^{i\omega},\widehat{\theta}_N)|^2\} \approx \frac{n}{N}\frac{\Phi_v(e^{i\omega})}{\Phi_u(e^{i\omega})}. \tag{4.85}$$

If in (4.84) the $\{\xi_k\}$ are not all chosen at the origin, then (similar to the fixed-denominator case) the approximation (4.85) can be quite inaccurate, even for large model order and data length. Based on similar arguments as for the FIR models, the asymptotic variance expression has to be replaced by the more accurate approximation (4.73).

4.5 Summary

In this chapter, it has been shown that orthonormal basis functions can be fruitfully used in problems of system identification in the classical prediction error framework. The combination of a flexible linearly parameterized model structure together with the possibility to decouple plant and noise dynamics has attractive properties both in terms of algorithm complexity (linear regression) and statistical properties (bias and variance) of the estimated models.

Analysis of the related results is supported by the fact that the least-squares problem in system identification remains to be governed by a (block) Toeplitz system of equations.
One of the basic features in identification with orthogonal basis functions is the fact that a relatively small of number of coefficients need only be estimated for arriving at highly accurate models. This situation is created when the user has been able to construct a basis of which the dynamics closely matches with the dynamics of the system to be modelled. The question how to optimally choose pole locations will be extensively discussed in Chapters 10 and 11. Note, however, that any choice of basis allows the parameterization of the same set of models. In other words: a choice for a particular basis does not limit the model class that can be taken into account. The more accurate the choice of basis is, the fewer parameters will be required to construct an appropriate model. This mechanism allows the incorporation of prior knowledge concerning the system poles, of which in many situations a rough indication will exist. Unlike the general situation of incorporating prior knowledge in identification algorithms, the required knowledge here does not necessarily have to be 100% correct.

4.6 Bibliography

The prediction error identification results presented in this chapter were published in a sequence of papers dealing with the Laguerre case ($n_b = 1$) [313], the Kautz case ($n_b = 2$) [315], and the generalized cases with repeated poles (n_b finite) [306] and any sequence of poles (n_b infinite) [214]. The fundamental role of orthogonal basis functions in identification is discussed in detail in [224]. Consequences for the variance expressions for estimated models are discussed in detail in the subsequent Chapter 5. Basis functions have also been used to perform prediction error type identification on the basis of frequency domain data [58, 71, 211] as well as in spectral modelling [318]. Frequency domain identification will be further treated in Chapter 8. The handling of multivariable models is considered in [209, 211] and will be briefly discussed in Section 10.5; a case study of multivariable prediction error identification is presented in [309]. Besides the use of these basis functions in nominal model identification, they also can be fruitfully employed in the problem of quantifying model uncertainty on the basis of measurement data [212]. In a prediction error framework, basis functions are used in this respect in [69, 71, 119]. This latter subject is further explored in Chapter 7.

5

Variance Error, Reproducing Kernels, and Orthonormal Bases

Brett Ninness[1] and Håkan Hjalmarsson[2]

[1] University of Newcastle, NSW, Australia
[2] KTH, Stockholm, Sweden

5.1 Introduction

The preceding chapters have developed a general theory pertaining to rational orthonormal bases. They have also presented a general framework for addressing system identification problems and have then provided a link between the two by illustrating the advantages of parameterizing models for system identification purposes using rational orthonormal bases.
This chapter continues on these themes of exposing the links between rational orthonormal bases and the system identification problem. The main feature of this chapter is that it will progress to consider more general model structures, especially those not specifically formulated with respect to an orthonormal basis. At the same time that we generalize this aspect, we will focus another by considering only the noise-induced error. We will not discuss bias error, as it will be analysed in detail in the following chapters, and was also addressed in the previous one.
Many of the underlying ideas necessary for the developments here have already been introduced, but in the interests of the reader who is seeking to 'dip into' chapters individually, a brief introduction to essential background ideas and notation will be made at the expense of some slight repetition.
In précis then, the sole focus of this chapter is to examine and quantify noise-induced estimation errors in the frequency domain, and there are four essential points to be made here in relation to this.

1. The quantification of noise-induced error (variance error) is equivalent to the quantification of the reproducing kernel for a particular function space X_n;
2. This function space X_n depends on the model structure, as well as on input and noise spectral densities. Hence, it is *not* independent of model structure;

3. The quantification of the reproducing kernel, and hence the quantification of variance error, depends crucially on the construction of a rational orthonormal basis for the afore-mentioned function space X_n;
4. The variance error quantifications that result are quite different to certain pre-existing ones, while at the same time are often much more accurate. This latter point will be established here empirically by simulation example.

5.2 Motivation

To make the context and purpose of this chapter more concrete, and to also motivate the presented material, we begin with a simulation example to illustrate the practical impact of what will follow.
Consider a linear and time invariant system that in continuous time has transfer function

$$G(s) = \frac{0.0012(1-3.33s)^3}{(s+0.9163)^2(s+0.3567)^2(s+0.2231)^3}.$$

Suppose that the response of this system is measured via the collection of input-output data sampled at 1 second intervals with zero-order-held inputs. This implies a discrete time representation

$$G(q) = \frac{\begin{array}{c}-0.0177(q^2-2.7192q+1.8489)(q-4.1377)\times \\ (q-1.3298)(q+0.4466)(q+0.0463)\end{array}}{(q-0.8)^3(q-0.7)^2(q-0.4)^2}.$$

Suppose further that we seek to estimate this transfer function on the basis of the observed input-output data via the use of the prediction error estimation methods described in the previous chapter. For this purpose, consider the case of a 7th order output-error model structure where we observe a length $N = 10,000$ sample record for which the output is corrupted by white Gaussian noise of variance $\sigma^2 = 0.01$, and with input that is a realization of a stationary Gaussian process with spectral density

$$\Phi_u(\omega) = \frac{1}{1.25 - \cos\omega}.$$

For this scenario, the sample mean square error in the estimation of the frequency response $G(e^{i\omega})$ and more then 10,000 estimation experiments with different input and noise realizations is used as an estimate of the variability $\text{Var}\{G(e^{i\omega}, \widehat{\theta}_N)\}$ of the estimated frequency response, and is plotted as a solid line in Figure 5.1.
In relation to this estimation scenario, a seminal result is that this variability of the frequency response estimate $G(e^{i\omega}, \widehat{\theta}_N)$ may be approximated as [162, 164, 167, 168]

$$\operatorname{Var}\{G(e^{i\omega},\widehat{\theta}_N)\} \approx \frac{m}{N}\frac{\Phi_\nu(\omega)}{\Phi_u(\omega)}. \tag{5.1}$$

Here, Φ_ν and Φ_u are, respectively, the measurement noise and input excitation spectral densities, and $\widehat{\theta}_N$ is the prediction error estimate based on the N observed data points of a vector $\theta \in \mathbb{R}^n$ that parameterizes a model structure $G(q,\theta)$ for which (essentially) the model order $m = \dim\theta/(2^d)$, where d is the number of denominator polynomials to be estimated in the model structure. Apart from its simplicity, a key factor underlying the importance and popularity of the approximation (5.1) is that, according to its derivation [162, 164, 167, 168], it applies for a very wide class of so-called shift invariant model structures. For example, all the well known FIR, ARX, ARMAX, output-error and Box-Jenkins structures are shift invariant [162]. Additionally, as shown in [164], it also applies when non-parametric (spectral based) estimation methods [36, 164] are employed provided that the m term in (5.1) is replaced by one dependent on the number of data points (and the windowing function) used.

Therefore, the only influence that the chosen model structure has on the right-hand side of (5.1) is in terms of its order, and because of this the belief that $\operatorname{Var}\{G(e^{i\omega},\widehat{\theta}_N)\}$ is invariant to the particular choice of mth order model structure has become a fundamental tenet of system identification.

However, when the 'classical' approximation (5.1), shown as a dash-dot line in Figure 5.1, is compared to the true variability shown as the solid line, it is clearly a poor approximation to the true variability. For example, apart from quantitative errors, the approximation (5.1) is also qualitatively misleading by being of a 'high-pass' nature when the true variability appears to be 'low pass'.

By way of contrast, this chapter will derive a new variance error quantification for this output error case, which may be expressed as follows:

$$\operatorname{Var}\{G(e^{i\omega},\widehat{\theta}_N)\} \approx \frac{2}{N}\frac{\Phi_\nu(\omega)}{\Phi_u(\omega)}\sum_{k=1}^{m}\frac{1-|\xi_k|^2}{|e^{i\omega}-\xi_k|^2}. \tag{5.2}$$

Here, the terms $\{\xi_1,\cdots,\xi_m\}$ are the estimated poles of $G(q,\widehat{\theta}_N)$. This new approximation (5.2) is shown as the dashed line in Figure 5.1, and in contrast to (5.1), it is seen to be an accurate approximation to the variability of the model estimate.

An important aspect of this new quantification (5.2) is that it indicates, in contrast to pre-existing thought, that the true variability is not invariant to either the model structure, or the true dynamics. Therefore, accurate approximation of the variance may need to take these into account, as the new expression (5.2) does. The remainder of this chapter is now devoted to establishing a theoretical framework for understanding these issues and illustrating how this framework implies (5.2) and further new results.

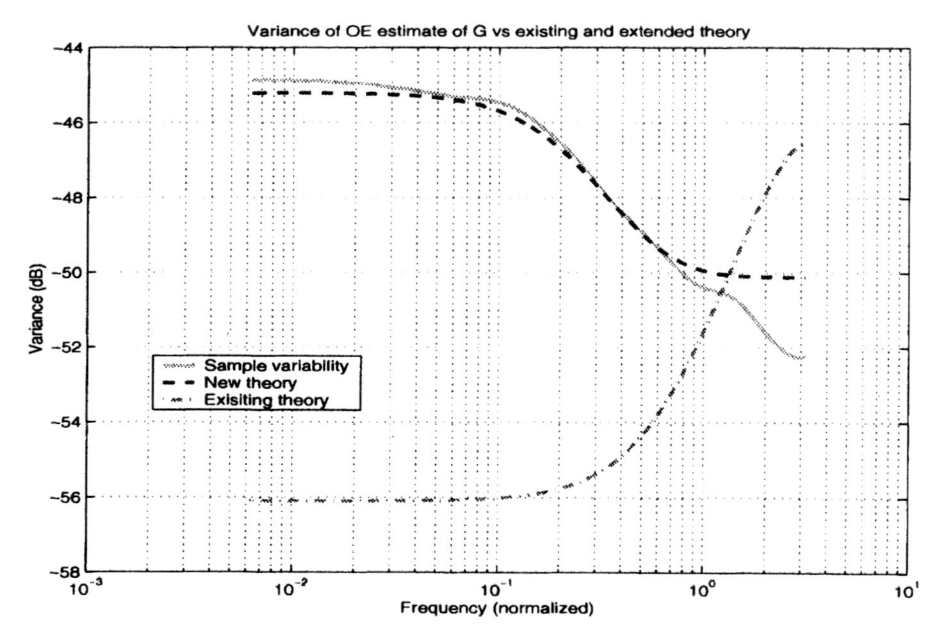

Fig. 5.1. Variability of output-error estimate: True variability vs. theoretically derived approximations. Solid line is Monte-Carlo estimate of true variability, dash-dot line is the pre-existing approximation (5.203), which does not account for system poles or model structure. The dashed line is the new approximation presented in (5.2) whereby estimated system pole positions $\{\xi_1, \cdots, \xi_m\}$ and the fact that an output-error structure is employed are both accounted for.

5.3 Problem Formulation

In order to analyse the phenomenon illustrated in Figure 5.1, it is necessary to precisely define the problem setting, which is one in which a model structure is used to describe the relationship between an observed input data record $\{u_t\}$ and output data record $\{y_t\}$ as

$$y_t = G(q,\theta)u_t + \nu_t, \qquad \nu_t = H(q,\theta)e_t. \tag{5.3}$$

Here, $\{e_t\}$ is a zero-mean white noise sequence that satisfies $\mathbf{E}\{e_t^2\} = \sigma^2$, $\mathbf{E}\{e_t^8\} < \infty$, and $G(q,\theta)$, $H(q,\theta)$ are transfer functions, rational in the backward shift operator q^{-1}, and parameterized by a vector $\theta \in \mathbb{R}^n$ in such as way that $H(q,\theta)$ is monic; *i.e.* $\lim_{|q|\to\infty} H(q,\theta) = 1$. In particular, $G(q,\theta)$ and $H(q,\theta)$ are of the following form

$$G(q,\theta) = \frac{B(q,\theta)}{A(q,\theta)}, \qquad H(q,\theta) = \frac{C(q,\theta)}{D(q,\theta)} \tag{5.4}$$

where

$$A(q,\theta) = 1 + a_1 q^{-1} + a_2 q^{-1} + \cdots + a_{m_a} q^{-m_a}, \tag{5.5}$$
$$B(q,\theta) = b_1 q^{-1} + b_2 q^{-1} + \cdots + b_{m_b} q^{-m_b}, \tag{5.6}$$
$$D(q,\theta) = 1 + d_1 q^{-1} + d_2 q^{-1} + \cdots + d_{m_d} q^{-m_d}, \tag{5.7}$$
$$C(q,\theta) = 1 + c_1 q^{-1} + c_2 q^{-1} + \cdots + c_{m_c} q^{-m_c}, \tag{5.8}$$

for some integers m_a, m_b, m_c, m_d. Depending on the choices for these polynomials and their integer orders, the above model structure encompasses all of the common finite impulse-response (FIR), fixed denominator, auto-regressive exogenous (ARX), auto-regressive moving-average exogenous (ARMAX), output-error and Box-Jenkins model structures [44, 164, 278].
With regard to all these cases encompassed by (5.3), the mean-square optimal one-step ahead predictor $\widehat{y}_t(\theta)$ based on the model structure (5.3) is [164]

$$\widehat{y}_t(\theta) = H^{-1}(q,\theta)G(q,\theta)u_t + \left[1 - H^{-1}(q,\theta)\right] y_t$$

with associated prediction error

$$\varepsilon_t(\theta) \triangleq y_t - \widehat{y}_t(\theta) = H^{-1}(q,\theta)\left[y_t - G(q,\theta_n)u_t\right]. \tag{5.9}$$

Using this, a quadratic estimation criterion may be defined as

$$V_N(\theta) = \frac{1}{2N}\sum_{t=1}^{N} \varepsilon_t^2(\theta)$$

and then used to construct the prediction error estimate $\widehat{\theta}_N$ of θ as

$$\widehat{\theta}_N \triangleq \arg\min_{\theta \in \mathbb{R}^n} V_N(\theta). \tag{5.10}$$

Forming system estimates via the techniques (5.3)–(5.10) has become quite standard, in large part due to the availability of sophisticated software tools implementing the method [165], but also because of extensive theoretical understanding of the properties of such an approach.
For example, as has been established in [161, 164], under certain mild assumptions on the nature of the input $\{u_t\}$ (which will be discussed in detail later), the estimate $\widehat{\theta}_N$ converges with increasing N according to

$$\lim_{N\to\infty} \widehat{\theta}_N = \theta_\circ \triangleq \arg\min_{\theta\in\mathbb{R}^n} \lim_{N\to\infty} \mathsf{E}\{V_N(\theta)\} \qquad \text{w.p.1.} \tag{5.11}$$

As well, it also holds that as N increases, the estimate $\widehat{\theta}_N$ converges in law to a Normally distributed random variable with mean value $\theta_\circ$ according to [44, 164, 166]

$$\sqrt{N}(\widehat{\theta}_N - \theta_\circ) \xrightarrow{\mathcal{D}} \mathcal{N}(0, P_n) \qquad \text{as } N \to \infty. \tag{5.12}$$

The $n \times n$ 'covariance matrix' P_n in (5.12) is defined in terms of two other matrices R_n and Q_n as

$$P_n \triangleq R_n^{-1} Q_n R_n^{-1} \tag{5.13}$$

which themselves are specified as

$$R_n \triangleq \lim_{N\to\infty} \frac{1}{N} \sum_{t=1}^{N} \left[\mathsf{E}\{\psi_t(\theta_\circ)\psi_t^T(\theta_\circ)\} - \mathsf{E}\{\varepsilon_t(\theta_\circ)\left(\frac{\mathrm{d}\psi_t(\theta)}{\mathrm{d}\theta}\right)^T\} \right] \tag{5.14}$$

and

$$Q_n \triangleq \lim_{N\to\infty} \frac{1}{N} \sum_{t=1}^{N}\sum_{\ell=1}^{N} \mathsf{E}\{\psi_t(\theta_\circ)\psi_\ell^T(\theta_\circ)\varepsilon_t(\theta_\circ)\varepsilon_\ell(\theta_\circ)\}. \tag{5.15}$$

The quantity $\psi_t(\theta)$ in the preceding expressions is the prediction error gradient given by

$$\psi_t(\theta) \triangleq \frac{\mathrm{d}\widehat{y}_t(\theta)}{\mathrm{d}\theta} = H^{-1}(q,\theta)\left[\frac{\mathrm{d}G(q,\theta)}{\mathrm{d}\theta}, \frac{\mathrm{d}H(q,\theta)}{\mathrm{d}\theta}\right]\zeta_t(\theta), \quad \zeta_t(\theta) \triangleq \begin{bmatrix} u_t \\ \varepsilon_t(\theta)\end{bmatrix}. \tag{5.16}$$

Although an asymptotic distributional result like (5.12) is very satisfying theoretically, for practical applications it is rather less appealing, mainly due to the (just presented) intricate definition of P_n via Q_n, R_n and $\psi_t(\theta)$.
In response to this, the chapter here extends the methods employed seminal works [120, 162, 164, 167, 168]. This involves an approach of investigating how (5.12) manifests itself in the variability of the frequency responses $G(e^{i\omega}, \widehat{\theta}_N)$ and $H(e^{i\omega}, \widehat{\theta}_N)$. The results will be the development of new quantifications such as (5.2).
The path toward achieving this involves noting that, with the definition

$$\Pi(q,\theta) \triangleq [G(q,\theta), H(q,\theta)] \tag{5.17}$$

then according to a first-order Taylor expansion, the relationship between frequency domain and parameter space estimation errors is given as

$$\Pi^T(e^{i\omega},\widehat{\theta}_N) - \Pi^T(e^{i\omega},\theta_\circ) = \left[\left.\frac{\mathrm{d}\Pi(e^{i\omega},\theta)}{\mathrm{d}\theta}\right|_{\theta=\theta_\circ}\right]^T (\widehat{\theta}_N - \theta_\circ) + o(\|\widehat{\theta}_N - \theta_\circ\|^2). \tag{5.18}$$

Therefore, a consequence of (5.12) is that

$$\sqrt{N}\begin{bmatrix} G(e^{i\omega},\widehat{\theta}_N) - G(e^{i\omega},\theta_\circ) \\ H(e^{i\omega},\widehat{\theta}_N) - H(e^{i\omega},\theta_\circ)\end{bmatrix} \xrightarrow{\mathcal{D}} \mathcal{N}(0, \Delta_n(\omega)), \qquad \text{as } N\to\infty \tag{5.19}$$

where

$$\Delta_n(\omega) \triangleq \left[\left.\frac{\mathrm{d}\Pi(e^{i\omega},\theta)}{\mathrm{d}\theta}\right|_{\theta=\theta_\circ}\right]^T P_n \left[\left.\frac{\mathrm{d}\Pi(e^{-i\omega},\theta)}{\mathrm{d}\theta}\right|_{\theta=\theta_\circ}\right]. \tag{5.20}$$

The remainder of this chapter is devoted to quantifying the above variance error $\Delta_n(\omega)$ for various model structures. Key to this process is the establishment of a link between the right-hand side of (5.20), and the reproducing kernel for a particular space. The role of the orthonormal bases introduced in the previous chapters will then be to quantify this reproducing kernel, and hence quantify the variance error.
In order to avoid possible confusion, it needs to be highlighted that there is a point to using the notation $\Delta_n(\omega)$ rather than just $\Delta(\omega)$ since, as will become clear later, the nature of quantifications we will establish for the right-hand side of (5.20) will, in some cases, only be approximations for a given finite parameter dimension n, while in other cases they will be exact for finite n.
An essential point is that the use of an orthonormal basis will prove to be fundamental, even when the underlying model structure *is not parameterized* with respect to this basis. That is, the results of this chapter will establish that the orthonormal bases studied in this book play a central role in the theory of system identification, that goes will beyond whether or not they happen to be chosen as an implementation tool when coding a particular algorithm.

5.4 Reproducing Kernels

As just mentioned, in this chapter the idea of what is called a 'reproducing kernel' for a space will prove to be a vital tool that allows for the direct simplification of complicated quantities via what is essentially a geometric principle. Although this idea has been presented earlier, this section will refresh the main points, and also extend them to the multi-dimensional setting that will be required in this chapter.
For this purpose, consider a subspace X_n of L_2 defined by a sequence $\{g_k\}$ of elements in L_2 as

$$X_n \triangleq \mathrm{Sp}\{g_1, \cdots, g_n\}. \tag{5.21}$$

The so-called reproducing kernel $\varphi_n(\lambda, \omega) : [-\pi, \pi] \times [-\pi, \pi] \to \mathbb{C}^{p \times p}$ for this space X_n of $\mathbb{C}^p$ valued functions is an entity such that for any $\alpha \in \mathbb{C}^p$ [17,86],

$$\varphi_n(\cdot, \omega)\alpha \in X_n \qquad \forall\omega \in [-\pi, \pi] \tag{5.22}$$

and for any $f \in X_n$

$$\langle f(\cdot), \varphi_n(\cdot, \omega)\alpha \rangle = \alpha^\star f(\omega). \tag{5.23}$$

The above inner product is defined for arbitrary functions $f, g : [-\pi, \pi] \to \mathbb{C}^p$ as follows

$$\langle f, g \rangle = \frac{1}{2\pi} \int_{-\pi}^{\pi} g^\star(\lambda) f(\lambda)\, \mathrm{d}\lambda. \tag{5.24}$$

Here and in the sequel, for inner product expressions like (5.23), the implied integration according to (5.24) will be over the common argument (in the above case indicated by $\cdot$).

Because the mapping $f \mapsto f(\omega)$ is a (bounded) linear functional $X_n \to \mathbb{C}^p$, then it is a consequence of the Riesz representation theorem [256] that a function $\varphi_n(\lambda, \omega)$ satisfying (5.22) and (5.23) exists. Furthermore, $\varphi_n(\lambda, \omega)$ is 'Hermitian symmetric' in that for any $\alpha, \beta \in \mathbb{C}^p$

$$\alpha^\star \varphi_n(\lambda, \omega)\beta = \langle \varphi_n(\rho, \omega)\beta, \varphi_n(\rho, \lambda)\alpha\rangle = \overline{\langle \varphi_n(\rho, \lambda)\alpha, \varphi_n(\rho, \omega)\beta\rangle} = \overline{\beta^\star \varphi_n(\omega, \lambda)\alpha}.$$

This implies that $\varphi_n(\lambda, \omega)$ is the *unique* element that has the property (5.23) *and also* satisfies $\varphi_m(\lambda, \omega)\alpha \in X_n$ for any $\alpha \in \mathbb{C}^p$, as if another function $\gamma_n(\lambda, \omega)$ also had these properties, then it would hold that for arbitrary $\alpha, \beta \in \mathbb{C}^p$

$$\begin{aligned} \alpha^\star \gamma_n(\mu, \omega)\beta &= \langle \gamma_n(\lambda, \omega)\beta, \varphi_n(\lambda, \mu)\alpha\rangle \\ &= \overline{\langle \varphi_n(\lambda, \mu)\alpha, \gamma_n(\lambda, \omega)\beta\rangle} \\ &= \overline{\beta^\star \varphi_n(\omega, \mu)\alpha} = \alpha^\star \varphi_n(\mu, \omega)\beta. \end{aligned} \quad (5.25)$$

This last uniqueness property will be particularly vital in later developments. Despite it, there may (of course) be several different ways of expressing the unique reproducing kernel, two of which are particularly important in later developments.

Lemma 5.1 (Expressions for the Reproducing Kernel). *Consider the subspace $X_n \subseteq L_2$ defined via (5.21) and define the matrix valued function $\Psi : [-\pi, \pi] \to \mathbb{C}^{n\times p}$ and the matrix $W_n \in \mathbb{R}^{n\times n}$ according to*

$$\Psi(\lambda) \triangleq [g_1(\lambda), \ldots, g_n(\lambda)]^T, \qquad W_n \triangleq \frac{1}{2\pi}\int_{-\pi}^{\pi} \overline{\Psi(\lambda)}\, \Psi^T(\lambda)\, \mathrm{d}\lambda. \quad (5.26)$$

Then the reproducing kernel $\varphi_n(\lambda, \omega)$ for X_n may be expressed as

$$\varphi_n(\lambda, \omega) = \Psi^T(\lambda)\, W_n^{-1}\, \overline{\Psi(\omega)}. \quad (5.27)$$

Furthermore, with $\{F_k\}$ being an orthonormal basis such that

$$X_n = \mathrm{Sp}\{F_1(\lambda), \cdots, F_n(\lambda)\}, \qquad \langle F_k, F_\ell\rangle = \delta(k-\ell) \quad (5.28)$$

then the reproducing kernel $\varphi_n(\lambda, \omega)$ on X_n may also be expressed as

$$\varphi_n(\lambda, \omega) = \sum_{k=1}^{n} F_k(\lambda) F_k^\star(\omega). \quad (5.29)$$

Proof. To prove (5.27), denote by $e_k \in \mathbb{C}^n$ the vector of all zeros save for a 1 in the kth position so that $g_k(\lambda) = \Psi^T(\lambda)e_k$. Then using the formulation (5.27) and for $\alpha \in \mathbb{C}^p$ arbitrary

$$\begin{aligned} \left\langle g_k(\lambda), \Psi^T(\lambda)W_n^{-1}\overline{\Psi(\omega)\alpha}\right\rangle &= \frac{1}{2\pi}\int_{-\pi}^{\pi} \alpha^\star \Psi^T(\omega)T_n^{-1}\overline{\Psi(\lambda)}\Psi^T(\lambda)e_k\, \mathrm{d}\lambda \\ &= \alpha^\star \Psi^T(\omega)T_n^{-1}\left[\frac{1}{2\pi}\int_{-\pi}^{\pi} \overline{\Psi(\lambda)}\Psi^T(\lambda)\mathrm{d}\lambda\right] e_k \\ &= \alpha^\star \Psi^T(\omega)e_k = \alpha^\star g_k(\omega)\ \forall k = 1, \cdots, n. \end{aligned} \quad (5.30)$$

The result (5.29) follows directly from (5.27) by noting that, first, the reproducing kernel is unique, and hence invariant to the basis used. Second, with the choice $g_k(z) = F_k(z)$ then $W_n = I$ due to the orthonormality of the set $\{\mathcal{B}_k\}$.

■

In what follows, we will mainly be concerned with functions $f(\omega) : [-\pi, \pi] \to \mathbb{C}^p$ that arise as the restriction of $f(z) : \mathbb{C} \to \mathbb{C}^p$ to the domain $z = e^{i\omega}$, $\omega \in [-\pi, \pi]$. As such, this chapter will alternate between notation $f(\omega), f(e^{i\omega})$ and $f(z)$ as convenient.
Furthermore, in the case that will be exclusively considered in this chapter where all the $g_k(z)$ in (5.26) have real valued Laurent expansion co-efficients (corresponding to dynamic systems with real valued impulse responses), then W_n defined by (5.26) has all real valued entries, and hence

$$W_n = \frac{1}{2\pi} \int_{-\pi}^{\pi} \Psi(\lambda)\, \Psi^\star(\lambda)\, \mathrm{d}\lambda \tag{5.31}$$

is an alternate formulation of (5.27) that will sometimes prove useful.
The relevance of these reproducing kernel ideas to the problem of quantification of variance error for frequency function estimates will be seen to stem from the formulation (5.27) and (5.31) because, as will be shown, when the prediction errors are white, then the associated variance error in the frequency domain can be expressed (modulo a known scalar factor) as the quadratic form in (5.27), (5.31) for some particular choice of elements of $\Psi(z)$ which depend on the model structure.
Therefore, since the preceding lemma also establishes that the reproducing kernel, which is unique, can be expressed as (5.29), then this provides a means for exact quantification of variance error provided that an explicit expression for the orthonormal basis $F_k(z)$ spanned by the elements of $\Psi(z)$ can be found. With this is in mind, in the scalar case $p = 1$, the following lemma details an important situation in which the necessary concrete orthonormal basis formulation is available.

Lemma 5.2 (Orthonormal Basis for Fixed Denominator Spaces). *Consider the space*

$$X_n \triangleq \mathrm{Sp}\left\{ \frac{z^{-1}}{L_n(z)}, \frac{z^{-2}}{L_n(z)}, \cdots, \frac{z^{-m}}{L_n(z)} \right\} \tag{5.32}$$

where

$$L_n(z) = \prod_{k=1}^{n} (1 - \xi_k z^{-1}), \qquad |\xi_k| < 1 \tag{5.33}$$

for some set of specified poles $\{\xi_1, \cdots, \xi_n\}$ and where $m \geq n$. Then it holds that

$$X_n = \mathrm{Sp}\{F_1(z), \cdots, F_n(z)\} \tag{5.34}$$

where for $k = 1, \ldots, m$,

$$F_k(z) \triangleq \frac{\sqrt{1-|\xi_k|^2}}{z-\xi_k} \cdot \phi_{k-1}(z), \quad \phi_k(z) \triangleq \prod_{\ell=1}^{k} \frac{1-\overline{\xi_\ell} z}{z-\xi_\ell}, \quad \phi_0(z) \triangleq 1. \tag{5.35}$$

For $k = m+1, \cdots, n$, the use of (5.35) is continued, but with ξ_k being used for k in this range. Furthermore, the functions $\{F_k(z)\}$ defined in (5.35) satisfy $\langle F_k, F_\ell \rangle = \delta(k-\ell)$ and hence form an orthonormal basis.

Proof. This was established previously in Chapter 2.

■

Finally, in the multidimensional situation, one example with $p = 2$ will be especially important for later applications.

Lemma 5.3. *Let X_n be the subspace*

$$X_n \triangleq \mathrm{Sp}\{f_1, \cdots, f_n, g_1, \ldots, g_m\} \tag{5.36}$$

where $f_k(z) = [\widetilde{f}_k(z),\ 0]^T$ and $g_k(z) = [0,\ \widetilde{g}_k(z)]^T$ with all the $\widetilde{g}_k, \widetilde{f}_k$ being scalar valued. Suppose that $\varphi_n^g(\lambda,\omega)$ and $\varphi_m^f(\lambda,\omega)$ are the reproducing kernels for

$$X_n^g \triangleq \mathrm{Sp}\{\widetilde{g}_1, \cdots, \widetilde{g}_n\}, \qquad X_n^f \triangleq \mathrm{Sp}\left\{\widetilde{f}_1, \cdots, \widetilde{f}_m\right\} \tag{5.37}$$

respectively. Then the reproducing kernel for X_n is

$$\varphi_n(\lambda,\omega) = \begin{bmatrix} \varphi_n^f(\lambda,\omega) & 0 \\ 0 & \varphi_m^g(\lambda,\omega) \end{bmatrix}. \tag{5.38}$$

Proof. Follows by inspection.

■

5.5 Variance Error and the Reproducing Kernel

This section presents the core result for this chapter, which is the principle that the problem of quantifying variance error is equivalent to that of quantifying a certain reproducing kernel.

To proceed, recall from the introductory Section 5.3 that the focus of this chapter is on quantifying the frequency domain variability $\Delta_n(\omega)$ given by (5.19), (5.20) as

$$\begin{aligned} \Delta_n(\omega) &\triangleq \lim_{N\to\infty} N \cdot \mathrm{Var}\left\{ \begin{bmatrix} G(e^{i\omega}, \widehat{\theta}_N) - G(e^{i\omega}, \theta_\circ) \\ H(e^{i\omega}, \widehat{\theta}_N) - H(e^{i\omega}, \theta_\circ) \end{bmatrix} \right\} \\ &= \left[\left. \frac{\mathrm{d}\Pi(e^{i\omega},\theta)}{\mathrm{d}\theta} \right|_{\theta=\theta_\circ} \right]^T P_n \left[\left. \frac{\mathrm{d}\Pi(e^{-i\omega},\theta)}{\mathrm{d}\theta} \right|_{\theta=\theta_\circ} \right] \end{aligned} \tag{5.39}$$

where $\Pi(q,\theta) \triangleq [G(q,\theta), H(q,\theta)]$ and P_n is the parameter space covariance matrix $\mathrm{Cov}\{\widehat{\theta}_N\}$ given by (5.13)- (5.15).
Now, consider the case $\mathcal{S} \in \mathcal{M}$ in which the true system $\mathcal{S}$ is contained within the chosen model structure $\mathcal{M}$. Then under mild assumptions (as discussed in the previous chapter), the prediction error $\varepsilon_t(\theta_\circ)$ evaluated at the limiting estimate $\theta_\circ$ satisfies $\varepsilon_t(\theta_\circ) = e_t$ where $\{e_t\}$ is a white noise process.
As such, $\varepsilon_t(\theta_\circ) = e_t$ is uncorrelated with any signal that is a function of $e_{t-1}, e_{t-2}, \cdots$ and hence with regard to (5.14)

$$\mathsf{E}\{\varepsilon_t(\theta_\circ)\left(\frac{\mathrm{d}\psi_t(\theta)}{\mathrm{d}\theta}\right)^T\} = \mathsf{E}\{e_t\}\cdot\mathsf{E}\{\left(\frac{\mathrm{d}\psi_t(\theta)}{\mathrm{d}\theta}\right)^T\} = 0 \tag{5.40}$$

while with regard to (5.15)

$$\begin{aligned}\mathsf{E}\{\psi_t(\theta_\circ)\psi_\ell^T(\theta_\circ)\varepsilon_t(\theta_\circ)\varepsilon_\ell(\theta_\circ)\} &= \mathsf{E}\{\psi_t(\theta_\circ)\psi_\ell^T(\theta_\circ)\}\cdot\mathsf{E}\{e_t e_\ell\}\\ &= \sigma^2\mathsf{E}\{\psi_t(\theta_\circ)\psi_\ell^T(\theta_\circ)\}\delta(t-\ell).\end{aligned} \tag{5.41}$$

Substituting (5.40) and (5.41) into (5.13)–(5.15) then leads to

$$Q_n = \sigma^2 R_n = \lim_{N\to\infty}\frac{1}{N}\sum_{t=1}^{N}\mathsf{E}\{\psi_t(\theta_\circ)\psi_t^T(\theta_\circ)\} \tag{5.42}$$

and hence $P_n = \sigma^2 R_n^{-1}$. Furthermore, according to (5.16)

$$\psi_t(\theta_\circ) = H^{-1}(q,\theta_\circ)\left.\frac{\mathrm{d}\Pi(q,\theta)}{\mathrm{d}\theta}\right|_{\theta=\theta_\circ}\zeta_t(\theta_\circ), \quad \zeta_t(\theta_\circ) \triangleq \begin{bmatrix} u_t \\ \varepsilon_t(\theta_\circ)\end{bmatrix}. \tag{5.43}$$

Therefore, if we denote by $\Phi_{\zeta_\circ}(\omega)$ the power spectral density of $\{\zeta_t(\theta_\circ)\}$, and by $S_{\zeta_\circ(z)}$ the (assumed rational) spectral factor of $\Phi_{\zeta_\circ}(\omega)$ in that

$$\Phi_{\zeta_\circ} = S_{\zeta_\circ}(e^{i\omega})S_{\zeta_\circ}(e^{i\omega})^\star, \tag{5.44}$$

then by employing the formulation

$$\Psi(z,\theta_\circ) \triangleq H^{-1}(z,\theta_\circ)\left.\frac{\mathrm{d}\Pi(z,\theta)}{\mathrm{d}\theta}\right|_{\theta=\theta_\circ} S_{\zeta_\circ}(z) \tag{5.45}$$

for the spectral factor of the predictor gradient $\{\psi_t(\theta_\circ)\}$, then the use of Parseval's theorem permits the covariance matrix P_n to be written as

$$P_n = \sigma^2 W_n^{-1}, \quad W_n \triangleq \frac{1}{2\pi}\int_{-\pi}^{\pi}\Psi(e^{i\lambda})\,\Psi^\star(e^{i\lambda})\,\mathrm{d}\lambda. \tag{5.46}$$

In addition, according to (5.45)

$$\left.\frac{\mathrm{d}\Pi(z,\theta)}{\mathrm{d}\theta}\right|_{\theta=\theta_\circ} = H(z,\theta_\circ)\Psi(z,\theta_\circ)S_{\zeta_\circ}^{-1}(z). \tag{5.47}$$

Substituting this and (5.46) into (5.39) then allows the variance error $\Delta_n(\omega)$ to be expressed as

$$\Delta_n(\omega) = |H(e^{i\omega}, \theta_\circ)|^2 S_{\zeta_\circ}^{-\star}(e^{i\omega}) \Psi^\star(e^{i\omega}) W_n^{-1} \Psi(e^{i\omega}) S_{\zeta_\circ}^{-1}(e^{i\omega}). \tag{5.48}$$

Now here is the key point. It was just established in Lemma 5.1, that according to (5.26) and (5.27), if we define a space X_n as the linear span of the rows of $\Psi(z)$ as follows

$$X_n \triangleq \mathrm{Sp}\left\{ [\Psi(z, \theta_\circ)]_1^T, \cdots, [\Psi(z, \theta_\circ)]_n^T \right\}, \tag{5.49}$$

then the middle factor in the right-hand side of (5.48) can be expressed as

$$\Psi^T(e^{i\lambda})\, W_n^{-1}\, \overline{\Psi(e^{i\omega})} = \varphi_n(\lambda, \omega) \tag{5.50}$$

where $\varphi_n(\lambda, \omega)$ is the reproducing kernel for X_n. This result is so fundamental to this chapter that it will be repeated for emphasis.

Theorem 5.1 (Frequency Domain Variability: $\mathcal{S} \in \mathcal{M}$). *Suppose that $\widehat{\theta}_N$ is calculated via (5.10) using the model structure (5.3) and that the following assumptions are satisfied*

1. *$\varepsilon_t(\theta_\circ) = e_t$ where $\{e_t\}$ is a zero mean i.i.d. process that satisfies $\mathsf{E}\{|e_t|^8\} < \infty$;*
2. *The relationship (5.43) holds for some $\Pi(q, \theta)$, and some quasi-stationary (possibly vector valued) signal $\{\zeta_t(\theta)\}$ and for which the power spectral density $\Phi_{\zeta_\circ}(\omega)$ of $\{\zeta_t(\theta_\circ)\}$ satisfies $\Phi_{\zeta_\circ}(\omega) > 0\, \forall \omega \in [-\pi, \pi]$;*
3. *Neither of $G(z, \theta_\circ)$ or $H(z, \theta_\circ)$ contain any pole-zero cancellations.*

Then denoting $S_{\zeta_\circ}(z)$ as a spectral factor of the power spectral density $\Phi_{\zeta_\circ}(\omega)$ of $\{\zeta_t(\theta_\circ)\}$

$$\lim_{N \to \infty} N \cdot Cov \left\{ \begin{bmatrix} G(e^{i\omega}, \widehat{\theta}_N) \\ H(e^{i\omega}, \widehat{\theta}_N) \end{bmatrix} \right\} = \Delta_n(\omega) \tag{5.51}$$

where

$$\Delta_n(\omega) = \Phi_\nu(\omega)\, S_{\zeta_\circ}^{-\star}(e^{i\omega})\, \varphi_n(\omega, \omega)\, S_{\zeta_\circ}^{-1}(e^{i\omega}) \tag{5.52}$$

with $\varphi_n(\lambda, \omega)$ being the reproducing kernel for the space X_n defined via (5.49).

The significance of this key result is this. It establishes that the problem of deriving an expression $\Delta_n(\omega)$ for the estimate covariance in the frequency domain is equivalent to the problem of quantifying the reproducing kernel for a certain space X_n which is that spanned by the rows of $\Psi(z, \theta_\circ)$.
This space, and hence the corresponding reproducing kernel $\varphi_m(\lambda, \omega)$, will depend on the model structure employed. Therefore, the covariance $\Delta_n(\omega)$ of the dynamic system estimate, will also depend on the model structure.

5.6 Function Spaces Implied by Model Structures

We have just established, via Theorem 5.1, the fundamental link between the variance error, and the function space X_n implied by the model structure via the reproducing kernel for X_n. With this in mind, the section here now turns to the explicit characterisation of this function space X_n for several important classes of model structure. In what follows in this section, it will be assumed that the input spectral density $\Phi_u(\omega)$ has the following rational spectral factorisation

$$\Phi_u(\omega) = |\mu(e^{i\omega})|^2. \tag{5.53}$$

5.6.1 Fixed Denominator (Generalized FIR) Structure

Consider the simplest model structure case, which was introduced in the previous chapter, and which can be represented as

$$y_t = \frac{B(q,\theta)}{L_{m_a}(q)} u_t + e_t \tag{5.54}$$

where $B(q,\theta)$ is given by (5.6) and

$$L_{m_a}(z) = (1 - \xi_1 z^{-1}) \cdots (1 - \xi_{m_a} z^{-1}) \tag{5.55}$$

implements a fixed denominator. In previous, and subsequent chapters, this model structure will be considered mainly in it orthonormalized form. However, in practice it can be implemented by forming the pre-filtered signal

$$u_t^f = L_{m_a}^{-1}(q) u_t \tag{5.56}$$

and then performing estimation with respect to the FIR model structure

$$y_t = B(q,\theta) u_t^f + e_t. \tag{5.57}$$

This highlights that the model structure corresponds to a particular case of the general form (5.3) in which $m_a = m_c = m_d = 0$ and hence

$$A(q,\theta) = C(q,\theta) = D(q,\theta) = 1, \quad \zeta_t(\theta_\circ) = u_t^f, \tag{5.58}$$

$$G(q,\theta) = B(q,\theta), \quad H(q,\theta) = 1, \quad S_{\zeta_\circ}(z) = \frac{\mu(z)}{L_{m_a}(z)}. \tag{5.59}$$

This implies that the spectral factor $\Psi(z)$ of the predictor gradient $\{\psi_t(\theta_\circ)\}$ is given by (5.45) as

$$\Psi(z,\theta_\circ) = H^{-1}(z,\theta_\circ) \left. \frac{\mathrm{d}\Pi(z,\theta)}{\mathrm{d}\theta} \right|_{\theta=\theta_\circ} S_{\zeta_\circ}(z) \tag{5.60}$$

$$= \left. \frac{\mathrm{d}B(z,\theta)}{\mathrm{d}\theta} \right|_{\theta=\theta_\circ} \frac{\mu(z)}{L_{m_a}(z)} \tag{5.61}$$

$$= \left[z^{-1}, z^{-2}, \cdots, z^{-m_b}\right]^T \frac{\mu(z)}{L_{m_a}(z)}. \tag{5.62}$$

Therefore:

For the case of the fixed denominator model structure (5.54), the space X_n determined as the linear span of the rows of the spectral factor $\Psi(z)$ of the prediction error gradient ψ_t may be expressed as

$$X_n = \mathrm{Sp}\left\{ \frac{z^{-1}\mu(z)}{L_{m_a}(z)}, \frac{z^{-2}\mu(z)}{L_{m_a}(z)}, \cdots, \frac{z^{-m_b}\mu(z)}{L_{m_a}(z)} \right\}. \tag{5.63}$$

5.6.2 Box-Jenkins Structure

The fixed denominator case is the simplest model structure considered in this chapter. Let us now turn to the opposite end of the scale and consider the most complicated model structure which will be addressed in this chapter; namely, the Box-Jenkins structure whereby all the elements of (5.4) are to be estimated in that $m_a > 1, m_b > 1, m_c > 1, m_d > 1$. In this case, under an assumption that data is collected in open loop so that $\Phi_{ue}(\omega) = 0$

$$\zeta_t(\theta_\circ) = \begin{bmatrix} u_t, \\ e_t \end{bmatrix}, \qquad S_{\zeta_\circ}(z) = \begin{bmatrix} \mu(z) & 0 \\ 0 & \sigma \end{bmatrix}. \tag{5.64}$$

Furthermore, for any k in the appropriate range

$$\frac{\mathrm{d}}{\mathrm{d}a_k} G(z,\theta) = -\frac{G(z,\theta)z^{-k}}{A(z,\theta)}, \qquad \frac{\mathrm{d}}{\mathrm{d}b_k} G(z,\theta) = \frac{z^{-k}}{A(z,\theta)}, \tag{5.65}$$

$$\frac{\mathrm{d}H(z,\theta)}{\mathrm{d}d_k} = -\frac{H(z,\theta)z^{-k}}{D(z,\theta)}, \qquad \frac{\mathrm{d}H(z,\theta)}{\mathrm{d}c_k} = \frac{z^{-k}}{D(z,\theta)}. \tag{5.66}$$

and

$$\frac{\mathrm{d}}{\mathrm{d}c_k} G(z,\theta) = \frac{\mathrm{d}}{\mathrm{d}d_k} G(z,\theta) = \frac{\mathrm{d}}{\mathrm{d}a_k} H(z,\theta) = \frac{\mathrm{d}}{\mathrm{d}b_k} H(z,\theta) = 0. \tag{5.67}$$

Substituting all these expressions into (5.45) then indicates that the space X_n spanned by the rows of $\Psi(z,\theta_\circ)$ is given by

$$\begin{aligned} X_n = \mathrm{Sp}&\left\{ \left[\frac{G(z,\theta_\circ)\mu(z)z^{-1}}{H(z,\theta_\circ)A(z,\theta_\circ)}, 0\right]^T, \cdots, \left[\frac{G(z,\theta_\circ)\mu(z)z^{-m_a}}{H(z,\theta_\circ)A(z,\theta_\circ)}, 0\right]^T \right\} \oplus \\ \mathrm{Sp}&\left\{ \left[\frac{\mu(z)z^{-1}}{H(z,\theta_\circ)A(z,\theta_\circ)}, 0\right]^T, \cdots, \left[\frac{\mu(z)z^{-m_b}}{H(z,\theta_\circ)A(z,\theta_\circ)}, 0\right]^T \right\} \oplus \\ \mathrm{Sp}&\left\{ \left[0, \frac{z^{-1}}{D(z,\theta_\circ)}\right]^T, \cdots, \left[0, \frac{z^{-m_d}}{D(z,\theta_\circ)}\right]^T, \right. \\ &\left. \left[0, \frac{z^{-1}}{C(z,\theta_\circ)}\right]^T, \cdots, \left[0, \frac{z^{-m_c}}{C(z,\theta_\circ)}\right]^T \right\}. \end{aligned}$$

When the pairs $(A(q,\theta_\circ, B(q,\theta_\circ)$ and $(C(q,\theta_\circ, D(q,\theta_\circ)$ are co-prime, this may be summarized as follows.

For the case of a Box-Jenkins model structure and $\Phi_{ue}(\omega) = 0$, the space X_n determined as the linear span of the rows of the associated spectral factor $\Psi(z)$ of the prediction error gradient ψ_t may be expressed as

$$X_n = \mathrm{Sp}\left\{f_1(z), \cdots, f_{m_a+m_b}(z), g_1(z), \cdots, g_{m_c+m_d}(z)\right\} \tag{5.68}$$

where

$$f_k(z) \triangleq \left[\frac{z^{-k}\mu(z)}{A^2(z,\theta_\circ)H(z,\theta_\circ)}, 0\right]^T, g_k(z) \triangleq \left[0, \frac{z^{-k}}{C(z,\theta_\circ)D(z,\theta_\circ)}\right]^T. \tag{5.69}$$

5.6.3 Output-error Structure

Having considered both ends of the complexity spectrum, we will now consider two important cases in between. The first of these is the output-error structure for which (5.3) is used with the choice $m_c = m_d = 0$ and hence

$$C(q,\theta) = D(q,\theta) = H(q,\theta) = 1, \quad \zeta_t(\theta_\circ) = u_t, \quad S_{\zeta_\circ}(z) = \mu(z). \tag{5.70}$$

Because $H(q,\theta) = 1$, then $\mathrm{d}H(q,\theta)/\mathrm{d}\theta = 0$. Furthermore, the relationship (5.65) holds again in this case. Substituting all these expressions into (5.45) then indicates that the space X_n spanned by the rows of $\Psi(z,\theta_\circ)$ is given by

$$X_n = \mathrm{Sp}\left\{\frac{G(z,\theta_\circ)\mu(z)z^{-1}}{A(z,\theta_\circ)}, \cdots, \frac{G(z,\theta_\circ)\mu(z)z^{-m_a}}{A(z,\theta_\circ)}, \frac{\mu(z)z^{-1}}{A(z,\theta_\circ)}, \cdots, \frac{\mu(z)z^{-m_b}}{A(z,\theta_\circ)}\right\}. \tag{5.71}$$

Again, by way of emphasis:

For the case of an output-error model structure, the space X_n determined as the linear span of the rows of the associated spectral factor $\Psi(z)$ of the prediction error gradient ψ_t may be expressed as

$$X_n = \mathrm{Sp}\left\{\frac{z^{-1}\mu(z)}{A^2(z,\theta_\circ)}, \frac{z^{-2}\mu(z)}{A^2(z,\theta_\circ)}, \cdots, \frac{z^{-(m_a+m_b)}\mu(z)}{A^2(z,\theta_\circ)}\right\}. \tag{5.72}$$

5.6.4 ARMAX Structure

The final model structure to be considered is the ARMAX case in which the general structure (5.3) is used with the choices $m_a > 1, m_b > 1, m_c > 1, m_d = m_a$ and

$$D(q,\theta) = A(q,\theta). \tag{5.73}$$

In this case, under an assumption that data is collected in open loop so that $\Phi_{ue}(\omega) = 0$

$$\zeta_t(\theta_\circ) = \begin{bmatrix} u_t, \\ e_t \end{bmatrix}, \qquad S_{\zeta_\circ}(z) = \begin{bmatrix} \mu(z) & 0 \\ 0 & \sigma \end{bmatrix}. \tag{5.74}$$

Again, the relationship (5.65) holds. Furthermore, for any k

$$\frac{\mathrm{d}}{\mathrm{d}c_k} G(q,\theta) = \frac{\mathrm{d}}{\mathrm{d}b_k} H(q,\theta) = 0, \quad \frac{\mathrm{d}}{\mathrm{d}a_k} H(q,\theta) = -\frac{H(q,\theta)}{A(q,\theta)} \cdot q^{-k}, \tag{5.75}$$

$$\frac{\mathrm{d}}{\mathrm{d}c_k} H(q,\theta) = \frac{q^{-k}}{A(q,\theta)} \tag{5.76}$$

Therefore, because (5.122) also holds, then according to (5.45) the space X_n spanned by the rows of $\Psi(z,\theta_\circ)$ is given by

$$\begin{aligned} X_n = \mathrm{Sp}\Bigg\{ & \left[\frac{G(z,\theta_\circ)\mu(z)z^{-1}}{C(z,\theta_\circ)}, \frac{z^{-1}}{A(z,\theta_\circ)}\right]^T, \cdots, \\ & \left[\frac{G(z,\theta_\circ)\mu(z)z^{-m_a}}{C(z,\theta_\circ)}, \frac{z^{-m_a}}{A(z,\theta_\circ)}\right]^T \Bigg\} \oplus \\ & \mathrm{Sp}\left\{ \left[\frac{\mu(z)z^{-1}}{C(z,\theta_\circ)}, 0\right]^T, \cdots, \left[\frac{\mu(z)z^{-m_b}}{C(z,\theta_\circ)}, 0\right]^T \right\} \oplus \\ & \mathrm{Sp}\left\{ \left[0, \frac{z^{-1}}{C(z,\theta_\circ)}\right]^T, \cdots, \left[0, \frac{z^{-m_c}}{C(z,\theta_\circ)}\right]^T \right\}. \end{aligned} \tag{5.77}$$

That is:

For the case of an ARMAX model structure, the space X_n determined as the linear span of the rows of the associated spectral factor $\Psi(z)$ of the prediction error gradient ψ_t may be expressed as

$$X_n = \mathrm{Sp}\left\{f_1(z), \cdots, f_{m_a}(z), g_1(z), \cdots, g_{m_b}(z), h_1(z), \cdots, h_{m_c}(z)\right\} \tag{5.78}$$

where

$$f_k(z) \triangleq \left[\frac{G(z,\theta_\circ)\mu(z)z^{-k}}{C(z,\theta_\circ)}, \frac{z^{-k}}{A(z,\theta_\circ)}\right]^T, \tag{5.79}$$

$$g_k(z) \triangleq \left[\frac{\mu(z)z^{-k}}{C(z,\theta_\circ)}, 0\right]^T \qquad h_k(z) \triangleq \left[0, \frac{z^{-k}}{C(z,\theta_\circ)}\right]^T \tag{5.80}$$

5.7 Exact Variance Quantifications

Section 5.5 has established the link between variance error $\Delta_n(\omega)$ and the reproducing kernel associated with a particular function space X_n. The previous section then made explicit the various formulations for X_n for various common model structures. With this established, we are now finally in a position to establish explicit variance error quantifications by combining these ingredients.

In performing this, note that a fundamental step in the previous development was to recognise that on the one hand $\Delta_n(\omega)$ was, via (5.52), a known and simple function of a quantity

$$\Psi^\star(e^{i\omega})W_n^{-1}\Psi(e^{i\lambda}), \qquad W_n \triangleq \frac{1}{2\pi}\int_{-\pi}^{\pi} \Psi(e^{i\lambda})\,\Psi^\star(e^{i\lambda})\,\mathrm{d}\lambda. \tag{5.81}$$

Furthermore, via equation (5.27) of Lemma 5.1 this is the reproducing kernel $\varphi_n(\lambda,\omega)$ for the space X_n given as the linear span of the rows of $\Psi(z)$, the latter being the spectral factor of the prediction error gradient $\{\psi_t\}$.

This section will now use a further result of Lemma 5.1; namely, that according to (5.29) this reproducing kernel $\varphi_n(\lambda,\omega)$, can be quantified as

$$\varphi_n(\lambda,\omega) = \sum_{k=1}^{n} F_k(\lambda)F_k^\star(\omega) \tag{5.82}$$

where $\{F_1(z),\cdots,F_n(z)\}$ is an orthonormal basis for X_n. Hence according to (5.52), the variance error (5.52) is given as

$$\Delta_n(\omega) = \Phi_\nu(\omega)\,S_{\zeta_\circ}^{-\,\star}(e^{i\omega})\,\kappa_n(\omega)S_{\zeta_\circ}^{-1}(e^{i\omega}), \quad \kappa_n(\omega) \triangleq \sum_{k=1}^{n} F_k(\omega)F_k^\star(\omega). \tag{5.83}$$

This is significant. It has reduced the variance quantification question to one of constructing an orthonormal basis for a function space. As we shall see, in many important cases it is simple to formulate this orthonormal basis, and hence to quantify the variance error, in closed form. These cases will be ones where certain restrictions are placed on the nature of the input spectral density.

In other more general situations, it will not be possible to compute the relevant basis in closed form, and then attention will be turned to finding approximations for the reproducing kernel. This case will be dealt with later, but for the moment we consider the simpler scenario where the reproducing kernel can be found in closed form, and hence an 'exact' quantification of the variance error $\Delta_n(\omega)$ results.

5.7.1 Variability of Fixed Denominator Model Structures

This model structure was defined in (5.54), (5.55), and the relevant space X_n spanned by the rows of the spectral factor of the associated predictor gradient was found in (5.63) as

$$X_n = \mathrm{Sp}\left\{\frac{z^{-1}\mu(z)}{L_{m_a}(z)}, \frac{z^{-2}\mu(z)}{L_{m_a}(z)}, \cdots, \frac{z^{-m_b}\mu(z)}{L_{m_a}(z)}\right\}. \tag{5.84}$$

Recall that $\mu(z)$ is the (assumed rational) spectral factor of the input spectral density $\Phi_u(\omega)$. Consider first the simplest case of this input being white, so that $\mu(z) = 1$. Then, provided that the numerator order m_b is greater than or equal to the order $n = m_a$ of the fixed denominator $L_{m_a}(z)$ we can directly apply Lemma 5.2 to conclude that the associated reproducing kernel $\varphi_n(\lambda, \omega)$ evaluates at $\lambda = \omega$ to

$$\varphi_n(\omega, \omega) = (m_b - m_a) + \sum_{k=1}^{n} \frac{1 - |\xi_k|^2}{|e^{i\omega} - \xi_k|^2}. \tag{5.85}$$

As a consequence, substitution of (5.85) into (5.83) allows the variance error to be directly expressed as follows.

Consider the situation of using the fixed denominator model structure

$$y_t = \frac{B(q,\theta)}{L(q)} u_t + e_t \tag{5.86}$$

$$L_{m_a}(q) = (1 - \xi_1 q^{-1}) \cdots (1 - \xi_{m_a} q^{-1}) \tag{5.87}$$

$$B(q,\theta) = b_1 q^{-1} + b_2 q^{-1} + \cdots + b_{m_b} q^{-m_b}. \tag{5.88}$$

Suppose that $\mathcal{S} \in \mathcal{M}$, $m_b \geq m_a$ and that $\Phi_u(\omega)$ is white, and hence equal to a constant. Then

$$\Delta_n(\omega) = \lim_{N\to\infty} N \cdot \mathrm{Var}\{G(e^{i\omega}, \widehat{\theta}_N)\} = \frac{\sigma^2}{\Phi_u(\omega)} \left[(m_b - m_a) + \sum_{k=1}^{m_a} \frac{1 - |\xi_k|^2}{|e^{i\omega} - \xi_k|^2}\right] \tag{5.89}$$

This quantification for the normalized and asymptotic in N variability $\Delta_n(\omega)$ then leads to the following obvious approximation for the variance of the frequency response estimate:

$$\mathrm{Var}\{G(e^{i\omega}, \widehat{\theta}_N)\} \approx \frac{1}{N} \cdot \frac{\sigma^2}{\Phi_u(\omega)} \left[(m_b - m_a) + \sum_{k=1}^{n} \frac{1 - |\xi_k|^2}{|e^{i\omega} - \xi_k|^2}\right]. \tag{5.90}$$

Clearly, the accuracy of this approximation depends on the data length N being 'large'. However, it is the experience of the authors that, in fact, even for what might be considered quite small values (say, one hundred), the above approximation is very informative. This will be illustrated in later sections of this chapter.
For the sake of brevity, we will not further remind the reader of this passage from the asymptotic value $\Delta_n(\omega)$ to the approximate quantification of $\mathrm{Var}\{G(e^{i\omega}, \widehat{\theta}_N)\}$, although it is always implicit in what follows.

We congratulate readers who have persevered in their reading to reach this point! Hopefully, the dividend of this effort is becoming clear, namely, variance error quantification, while being *prima facie* difficult, in fact can be performed quite simply in cases where an orthonormal basis for the function space implied by the spectral factor of the prediction error spectral density (and hence implied by the model structure and input excitation spectral density) can be computed in closed form.

Furthermore, there is an important geometric principle that is exposed; namely, that all the model structure dependent parts of the variance error arise via the reproducing kernel, which (see the previous chapters for clarification) is the bilinear function that characterizes the orthogonal projection of an arbitrary function, onto the subspace of possible prediction error gradients that are implied by the model structure.

To illustrate further the utility of these principles, note that the above analysis can be very simply extended to the situation where the input spectral factor $\mu(z)$ is of an all-pole form

$$\mu(z) = \frac{1}{M_\ell(z)}, \quad M_\ell(z) = (1 - \eta_1 z^{-1}) \cdots (1 - \eta_\ell z^{-1}). \tag{5.91}$$

In this case, the underlying space X_n becomes

$$X_n = \mathrm{Sp}\left\{ \frac{z^{-1}}{L_{m_a}(z) M_\ell(z)}, \frac{z^{-2}}{L_{m_a}(z) M_\ell(z)}, \cdots, \frac{z^{-m_b}\mu(z)}{L_{m_a}(z) M_\ell(z)} \right\}. \tag{5.92}$$

Therefore, provided that the numerator order satisfies $m_b \geq m_a + \ell$, Lemma 5.2 may still be applied to find an orthonormal basis for X_n, and hence the previous result may be simply extended as follows.

Consider the situation of using the fixed denominator model structure (5.86)-(5.88). Suppose that $\mathcal{S} \in \mathcal{M}$ and that $\Phi_u(\omega) = 1/|M_\ell(e^{i}\omega)|^2$ where $M_\ell(z)$ is given by (5.91) and that $m_b \geq n + \ell$. Then

$$\Delta_n(\omega) = \lim_{N\to\infty} N \cdot \mathrm{Var}\{G(e^{i\omega}, \widehat{\theta}_N)\} =$$

$$\frac{\sigma^2}{\Phi_u(\omega)} \left[(m_b - m_a - \ell) + \sum_{k=1}^{m_a} \frac{1 - |\xi_k|^2}{|e^{i\omega} - \xi_k|^2} + \sum_{k=1}^{\ell} \frac{1 - |\eta_k|^2}{|e^{i\omega} - \eta_k|^2} \right]. \tag{5.93}$$

5.7.2 Variability of Output-error Model Structures

The basic mechanism of variance quantification has now been established, and this subsection together with the next one will simply repeat the methods already used in order to develop a catalogue of relevant expressions.

For this purpose, recall that in the output-error model structure case, the space X_n spanned by the rows of the spectral factor of the prediction error gradient spectral density were established in (5.72) to be

$$X_n = \mathrm{Sp}\left\{ \frac{z^{-1}\mu(z)}{A^2(z,\theta_\circ)}, \frac{z^{-2}\mu(z)}{A^2(z,\theta_\circ)}, \ldots, \frac{z^{-(m_a+m_b)}\mu(z)}{A^2(z,\theta_\circ)} \right\}. \tag{5.94}$$

Again, let us begin by supposing that the input spectral density is white so that $\mu(z) = 1$. Then again, Lemma 5.2 may be applied directly for the purposes of computing an orthonormal basis for X_n, and hence via (5.82), (5.83) for also computing an expression for the variance error.

Consider the situation of using the output-error model structure

$$y_t = \frac{B(q,\theta)}{A(q,\theta)} u_t + e_t, \tag{5.95}$$

$$A(q,\theta) = 1 + a_1 q^{-1} + a_2 q^{-1} + \cdots + a_{m_a} q^{-m_a}, \tag{5.96}$$

$$B(q,\theta) = b_1 q^{-1} + b_2 q^{-1} + \cdots + b_{m_b} q^{-m_b}. \tag{5.97}$$

Suppose that $\mathcal{S} \in \mathcal{M}$, $m_b \geq m_a$ and that $\Phi_u(\omega)$ is white, and hence equal to a constant. Then

$$\Delta_n(\omega) = \lim_{N\to\infty} N \cdot \mathrm{Var}\{G(e^{i\omega}, \widehat{\theta}_N)\} = \frac{\sigma^2}{\Phi_u(\omega)} \left[(m_b - m_a) + 2 \cdot \sum_{k=1}^{m_a} \frac{1 - |\xi_k|^2}{|e^{i\omega} - \xi_k|^2} \right] \tag{5.98}$$

where the $\{\xi_k\}$ above are defined as the zeros of the asymptotic in N denominator estimate

$$A(z,\theta_\circ) = (1 - \xi_1 z^{-1}) \cdots (1 - \xi_{m_a} z^{-1}). \tag{5.99}$$

It is worth commenting that the factor of two that appears in the last term of (5.98) is completely due to the fact that the function space X_n relevant to this case and given in (5.94) involves elements with $A^2(z)$ in the denominator, so that the zeros of $A(z)$ appear twice.

This is very important, because if we compare the quantification (5.98) to that of (5.93) which applies for the fixed denominator case, we see that in the case of $m_b = m_a$, then the variability involved with estimating the system poles (output-error case) is exactly twice the variability involved with fixing the system poles at correct values. Because twice as much information is estimated (a denominator as well as a numerator), this makes intuitive sense. However, as will be profiled later in this chapter, it was not implied by variance quantifications that pre-existed the results here.

Finally, in relation to this output-error case, note that it is again possible to extend variance quantification to the situation of the input spectral factor $\mu(z)$ being of the all-pole form (5.91) by noting that in this instance the associated space X_n becomes

$$X_n = \mathrm{Sp}\left\{ \frac{z^{-1}}{A^2(z,\theta_\circ)M_\ell(z)}, \frac{z^{-2}}{A^2(z,\theta_\circ)M_\ell(z)}, \ldots, \frac{z^{-(m_a+m_b)}}{A^2(z,\theta_\circ)M_\ell(z)} \right\}. \quad (5.100)$$

Therefore, provided that the numerator order satisfies $m_b \geq m_a + \ell$, Lemma 5.2 may still be applied to find an orthonormal basis for X_n, and hence the previous result may be simply extended as follows.

Consider the situation of using the output-error model structure (5.95)-(5.97). Suppose that $\mathcal{S} \in \mathcal{M}$, that $\Phi_u(\omega) = |M_\ell(e^{i}\omega)|^2$ where $M_\ell(z)$ is given by (5.91), and that $m_b \geq m_a + \ell$. Then

$$\Delta_n(\omega) = \lim_{N\to\infty} N \cdot \mathrm{Var}\{G(e^{i\omega}, \widehat{\theta}_N)\} =$$

$$\frac{\sigma^2}{\Phi_u(\omega)} \left[(m_b - m_a - \ell) + 2 \cdot \sum_{k=1}^{m_a} \frac{1 - |\xi_k|^2}{|e^{i\omega} - \xi_k|^2} + \sum_{k=1}^{\ell} \frac{1 - |\eta_k|^2}{|e^{i\omega} - \eta_k|^2} \right] \quad (5.101)$$

where the $\{\xi_k\}$ above are defined as the zeros of the asymptotic denominator:

$$A(z, \theta_\circ) = (1 - \xi_1 z^{-1}) \cdots (1 - \xi_{m_a} z^{-1}). \quad (5.102)$$

5.7.3 Variability of Box-Jenkins Model Structures

We conclude this section by addressing the most comprehensive model structure: the Box-Jenkins sort in which an independently parameterized dynamics model and noise model are both estimated. To address this situation, we will suppose that data is collected in open loop so that $\Phi_{ue}(\omega) = 0$ and hence

$$S_{\zeta_\circ}(\omega) = \begin{bmatrix} \Phi_u(\omega) & 0 \\ 0 & \sigma^2 \end{bmatrix}. \quad (5.103)$$

In this case, recall from Section 5.6.2 that the space X_n spanned by the rows of the prediction error gradient spectral factor is given by (5.68), (5.69) as

$$X_n = \mathrm{Sp}\left\{ f_1(z), \cdots, f_{m_a+m_b}(z), g_1(z), \cdots, g_{m_c+m_d}(z) \right\} \quad (5.104)$$

where

$$f_k(z) \triangleq \left[\widetilde{f}_k(z), 0 \right]^T, \quad g_k(z) \triangleq \left[0, \widetilde{g}_k(z) \right]^T, \quad (5.105)$$

$$\widetilde{f}_k(z) = \frac{z^{-k}\mu(z)}{A^2(z,\theta_\circ)H(z,\theta_\circ)}, \quad \widetilde{g}_k(z) = \frac{z^{-k}}{C(z,\theta_\circ)D(z,\theta_\circ)}. \quad (5.106)$$

In this situation, according to Lemma 5.2, the reproducing kernel for X_n is

$$\varphi_n(\lambda,\omega) = \begin{bmatrix} \varphi_n^f(\lambda,\omega) & 0 \\ 0 & \varphi_m^g(\lambda,\omega) \end{bmatrix} \tag{5.107}$$

where $\varphi_n^f(\lambda,\omega)$ and $\varphi_m^g(\lambda,\omega)$ are the reproducing kernels for the scalar function spaces

$$X_n^f \triangleq \mathrm{Sp}\left\{\widetilde{f}_1,\cdots,\widetilde{f}_m\right\}, \qquad X_n^g \triangleq \mathrm{Sp}\left\{\widetilde{g}_1,\cdots,\widetilde{g}_n\right\} \tag{5.108}$$

respectively. In order to use Lemma 5.2 to find a closed form expression for an orthonormal basis for X_n^f, and hence also to find an explicit expression for $\varphi_n(\lambda,\omega)$, it is necessary to impose an assumption that the quantity $A_\dagger(z)$ defined as

$$A_\dagger(z) = A^2(z,\theta_\circ)\frac{H(z,\theta_\circ)}{\mu(z)} \tag{5.109}$$

is a polynomial in z^{-1} of degree at most $m_a + m_b$. Under this assumption, the space X_n^f may be expressed as

$$X_n^f = \mathrm{Sp}\left\{\frac{z^{-1}}{A_\dagger(z,\theta_\circ)},\frac{z^{-2}}{A_\dagger(z,\theta_\circ)},\ldots,\frac{z^{-(m_a+m_b)}}{A_\dagger(z,\theta_\circ)}\right\} \tag{5.110}$$

which then permits Lemma 5.2 to be used to compute

$$\varphi_n^f(\omega,\omega) = \sum_{k=1}^{m_a+m_b} \frac{1-|\xi_k|^2}{|e^{i\omega}-\xi_k|^2} \tag{5.111}$$

where in the above, the $\{\xi_k\}$ are defined according to

$$z^{m_a+m_b}A_\dagger(z) = \prod_{k=1}^{m_a+m_b}(z-\xi_k). \tag{5.112}$$

It is worth emphasizing, that a point of including the $z^{m_a+m_b}$ term on the left of (5.112) is that it implies that $m_b - m_a$ of the zeros ξ_k on the right of (5.112) are equal to zero.
Similarly, the space X_n^g is, according to (5.108), given as

$$X_n^g = \mathrm{Sp}\left\{\frac{z^{-1}}{C(z,\theta_\circ)D(z,\theta_\circ)},\frac{z^{-2}}{C(z,\theta_\circ)D(z,\theta_\circ)},\ldots,\frac{z^{-(m_c+m_d)}}{C(z,\theta_\circ)D(z,\theta_\circ)}\right\} \tag{5.113}$$

and hence, with the definition of the zeros $\{\eta_k\}$ according to

$$z^{m_c+m_d}C(z,\theta_\circ)D(z,\theta_\circ) = \prod_{k=1}^{m_c+m_d}(z-\eta_k) \tag{5.114}$$

the use of Lemma 5.2 provides

$$\varphi_n^g(\omega,\omega) = \sum_{k=1}^{m_c+m_d} \frac{1-|\eta_k|^2}{|e^{i\omega}-\eta_k|^2}. \tag{5.115}$$

Substituting these expressions for φ_n^f, φ_n^g into the expression (5.107) for φ_n, and then substituting that into the general variance quantification result (5.83) with $S_{\zeta_\circ}$ given by (5.103) then implies the following variance quantification.

Consider the case of using a Box-Jenkins model structure and suppose that $\Phi_u(\omega) = |\mu(e^{i\omega})|^2$ is such that the polynomial condition (5.109) is satisfied and that $\Phi_{ue}(\omega) = 0$. Then provided $\mathcal{S} \in \mathcal{M}$, and with the zeros $\{\xi_k\}$ being defined by (5.112):

$$\lim_{N\to\infty} N \cdot \mathrm{Var}\{G(e^{i\omega},\widehat{\theta}_N)\} = \frac{\Phi_\nu(\omega)}{\Phi_u(\omega)} \sum_{k=1}^{m_a+m_b} \frac{1-|\xi_k|^2}{|e^{i\omega}-\xi_k|^2}. \tag{5.116}$$

Furthermore, with the zeros $\{\eta_k\}$ being defined by (5.114) and regardless of whether the polynomial condition (5.109) is satisfied

$$\lim_{N\to\infty} N \cdot \mathrm{Var}\{H(e^{i\omega},\widehat{\theta}_N)\} = \frac{\Phi_\nu(\omega)}{\sigma^2} \sum_{k=1}^{m_c+m_d} \frac{1-|\eta_k|^2}{|e^{i\omega}-\eta_k|^2}. \tag{5.117}$$

Clearly, the polynomial condition (5.109) that is necessary for this result is not a desirable one. It obviously holds in the case of white input $\mu(z) = 1$ and the true system being of output-error type in that $H(z) = 1$ (and hence $H(z,\theta_\circ) = 1$). However, there are several other, perhaps unexpected and non-trivial situations in which it can also apply.
For example, consider the case where the system to be identified is operating in closed loop as shown in Figure 5.2 and for which the relationship between the observed inputs, outputs and measurement noise are described by

$$y_t = T(q)r_t + S(q)H(q)e_t. \tag{5.118}$$

Here, $S(q)$ and $T(q)$ are the sensitivity and complementary sensitivity functions given (respectively) as

$$S(q) = \frac{1}{1+G(q)K(q)} = \frac{A(q)L(q)}{A_c(q)}, \qquad T(q) = \frac{G(q)K(q)}{1+G(q)K(q)} = \frac{B(q)P(q)}{A_c(q)} \tag{5.119}$$

where

$$K(q) = \frac{P(q)}{L(q)}, \qquad A_c(q) = A(q)L(q) + B(q)P(q). \tag{5.120}$$

A widely accepted strategy for estimating the dynamics $G(q)$ in this case is the so-called indirect method in which one first derives an estimate $T(q,\widehat{\theta}_N)$ of $T(q)$ on the basis of the observations $\{r_t\}$ and $\{y_t\}$, and then recovers an estimate $G(q,\widehat{\theta}_N)$ via the relationship

$$G(q,\widehat{\theta}_N) = \frac{T(q,\widehat{\theta}_N)}{K(q)[1-T(q,\widehat{\theta}_N)]}. \tag{5.121}$$

In this case, if the variability of the initial estimate $T(e^{i\omega},\widehat{\theta}_N)$ is to be quantified, then if the reference signal $\{r_t\}$ is white, the polynomial condition (5.109) becomes

$$A_\dagger(z) = A_c^2(z)\frac{A(z)L(z)}{A_c(z)} = A_c(z)A(z)L(z) \tag{5.122}$$

which is a polynomial, even though the estimated noise model $S(q,\theta_\circ)$,$H(q,\theta_\circ)$ will be very far from a constant white one.
The point is, that although the polynomial condition (5.109) can, at first glance, seem too restrictive tó useful, there are important situations when it still allows informative variance quantifications. For example, in this indirect estimation example, it exposes (and, in an obvious manner quantifies) via (5.122) the fact that the variability of the initial estimate $T(e^{i\omega},\widehat{\theta}_N)$ will depend on the closed loop poles, the open loop poles and the controller poles. Readers interested in a more detailed discussion of this issue should consult [221].

5.7.4 Further Simulation Example

A key point, raised at the end of Section 5.7.1, was that although the reproducing kernel methods profiled here can (as just illustrated) yield an exact expression for the asymptotic in data length N variance $\Delta_n(\omega)$, they still only provide an approximation for quantities like $\mathrm{Var}\{G(e^{i\omega},\widehat{\theta}_N)\}$; namely

$$\mathrm{Var}\{G(e^{i\omega},\widehat{\theta}_N)\} \approx \frac{1}{N}\Delta_n(\omega) \quad \text{where} \quad \Delta_n(\omega) = \lim_{N\to\infty} N\cdot \mathrm{Var}\{G(e^{i\omega},\widehat{\theta}_N)\}. \tag{5.123}$$

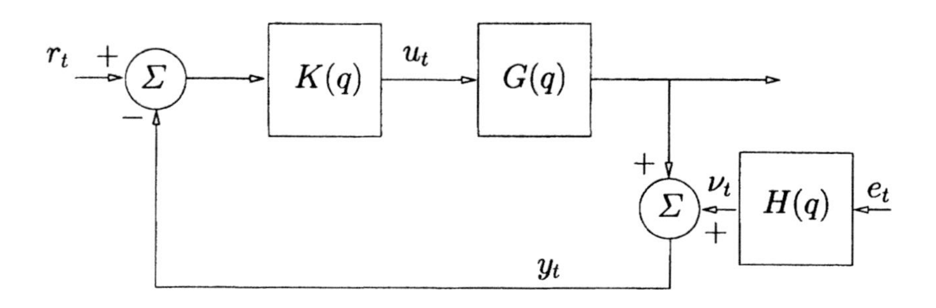

Fig. 5.2. Closed loop data collection.

In the authors' experience, this approximation is usually rather accurate, as we will now illustrate with a brief simulation example that complements the earlier motivational one of Section 5.2. For this purpose, consider the following very simple system

$$G(q) = \frac{0.05}{q - 0.95} \tag{5.124}$$

which is used to generate an $N = 10000$ sample input-output data record where the output $\{y_t\}$ is corrupted by white Gaussian noise of variance $\sigma^2 = 10$ and where the input $\{u_t\}$ is a realization of a stationary, zero mean, unit variance white Gaussian process. Note that this is a very heavy noise corruption, as it corresponds to an SNR of only -10dB.

On the basis of this observed data, a first-order output-error model structure $G(q, \widehat{\theta}_N)$ is estimated, and the sample mean square error in this estimate over 1000 estimation experiments with different input and noise realizations is used as an estimate of $\mathrm{Var}\{G(e^{i\omega}, \widehat{\theta}_N)\}$ which is plotted as a solid line in Figure 5.3(a). Furthermore, as established in (5.98), the asymptotic in data length variability in this case of $m = m_b = m_a$ is

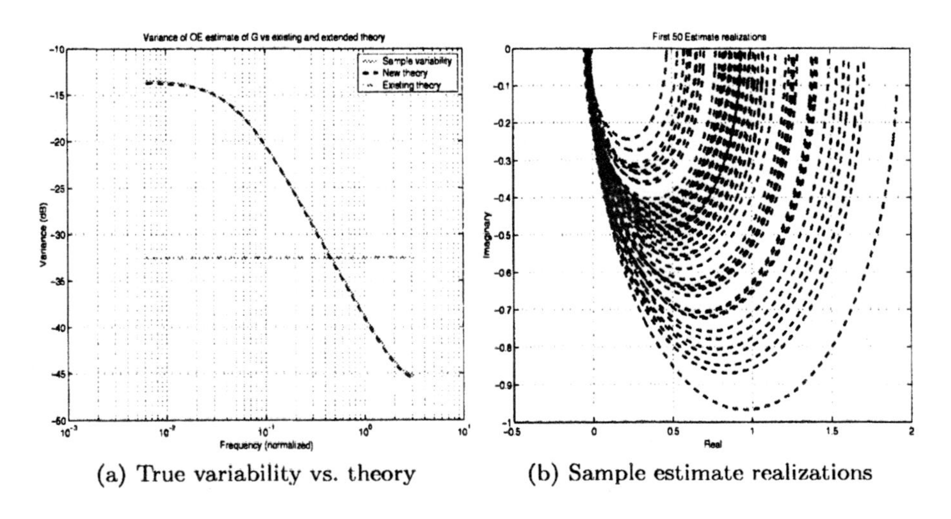

(a) True variability vs. theory

(b) Sample estimate realizations

Fig. 5.3. Figures illustrating variability of output-error Estimates: Part (a) shows the true variability vs. theoretically derived approximations. There the solid line is the Monte-Carlo estimate of the true variability, the dash-dot line is the pre-existing approximation (5.1), which does not account for system poles or model structure. The dashed line is the new approximation presented in (5.125) whereby estimated system pole positions $\{\xi_1, \cdots, \xi_m\}$ and the fact that an output-error structure is employed are both accounted for. Part (b) shows the first 50 (of 1000) estimate realizations to give a sense of the scale of the variability being quantified in Part (a).

$$\Delta_n(\omega) = \lim_{N\to\infty} N \cdot \mathrm{Var}\{G(e^{i\omega}, \widehat{\theta}_N)\} = 2 \cdot \frac{\sigma^2}{\Phi_u(\omega)} \sum_{k=1}^{m} \frac{1-|\xi_k|^2}{|e^{i\omega}-\xi_k|^2} \tag{5.125}$$

where the $\{\xi_k\}$ are the poles of the underlying true system (*i.e.* the input-output dynamics), which according to (5.123) implies the approximation

$$\mathrm{Var}\{G(e^{i\omega}, \widehat{\theta}_N)\} \approx \frac{2}{N} \cdot \frac{\sigma^2}{\Phi_u(\omega)} \sum_{k=1}^{m} \frac{1-|\xi_k|^2}{|e^{i\omega}-\xi_k|^2} \tag{5.126}$$

This approximation is shown as the dashed line in Figure 5.3. It is virtually indistinguishable from the solid line, which the reader will recall is the Monte-Carlo estimate of the true variability. That is, the approximation (5.125) is illustrated to be a highly accurate quantification!
To give the reader a sense of the scale of estimation error that is being quantified here, Figure 5.3(b) illustrates the first fifty (of one thousand) estimate realizations (represented via the corresponding Nyquist plot of $G(e^{i\omega}, \widehat{\theta}_N)$) that are averaged to produce an estimate of the true variability shown as the solid line in Figure 5.3(a). Clearly, these are large-scale errors that are being accurately quantified in Figure 5.3(a), which establishes that the results here are not restricted to the evaluation of minor effects.
Nevertheless, despite the very low signal to noise ration in this example, the data length $N = 10000$ could be considered quite large. Let us therefore turn to an other simulation example which involves only $N = 100$ data samples being collected from the resonant system

$$G(q) = \frac{0.0342q + 0.0330}{(q - 0.95e^{i*\pi/12})(q - 0.95e^{-i*\pi/12})} \tag{5.127}$$

with the same white input excitation as before, but now with white measurement noise corruption of variance $\sigma^2 = 0.01$ so that the SNR is raised to 20dB. For this scenario, the Monte-Carlo estimated true variability (solid line), and the approximation (5.125) (dashed line) are shown in Figure 5.4.
Clearly, the accuracy of the approximation (5.125), (5.123) is now degraded relative to the previous case. However, we would argue that the approximate quantification (5.125) is still a highly informative measure in both a qualitative and quantitative sense. In particular, with regard to qualitative aspects, as predicted by (5.126), there is a peak in the estimate variability due to the resonant poles of the true (and hence estimated) system.
In relation to these two simulation examples, there is a very obvious omission in our discussion. What is the significance of the flat dash-dot lines in Figures 5.3(a) and 5.4(a)? In fact, these lines represent an alternate quantification of variance error which was introduced back in Equation (5.1) of Section 5.2 (recall m is the system model order):

$$\mathrm{Var}\{G(e^{i\omega}, \widehat{\theta}_N)\} \approx \frac{m}{N} \frac{\Phi_\nu(\omega)}{\Phi_u(\omega)} \tag{5.128}$$

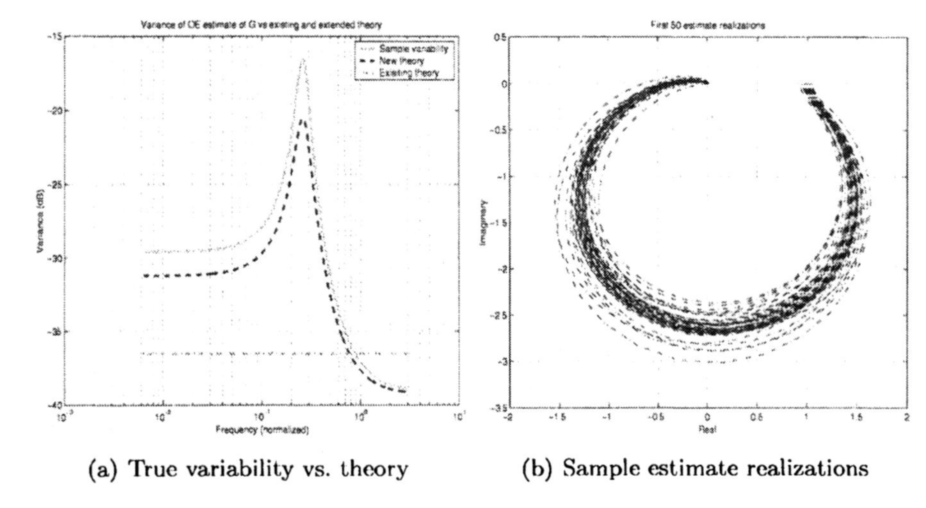

(a) True variability vs. theory (b) Sample estimate realizations

Fig. 5.4. Figures illustrating variability of output-error estimates: Part (a) shows the true variability vs. theoretically derived approximations. There the solid line is the Monte-Carlo estimate of the true variability, the dash-dot line is the pre-existing approximation (5.1), which does not account for system poles or model structure. The dashed line is the new approximation presented in (5.125) whereby estimated system pole positions $\{\xi_1, \cdots, \xi_m\}$ and the fact that an output-error structure is employed are both accounted for. Part (b) shows the first 50 (of 1000) estimate realizations to give a sense of the scale of the variability being quantified in Part (a).

In the examples just given, both $\Phi_u(\omega)$ and $\Phi_\nu(\omega)$ are constant due to whiteness assumptions, and this is why the left-hand side of (5.128) shown in Figures 5.3(a) and 5.4(a) is also a constant.

This discrepancy between the true variability, the pre-existing approximation (5.128), and the new one (5.125) is important. Considering Figure 5.3(a), if the approximation (5.128) were used to perform the very common procedure of judging the radius of error bounds on estimated Nyquist plots, then those radii would be approximately 100 times too small near zero frequency, and more than 10 times too large at the folding frequency.

Nevertheless, the quantification (5.128) also has a 20-year history of successful application since its seminal development in [162,168], and this is in large part due to the fact that it applies for a much wider class of scenarios than have yet been considered in this chapter.

That is, the quantification (5.128) applies for arbitrary input spectral density $\Phi_u(\omega)$, while that of (5.126) is one specially tailored to the case of $\Phi_u(\omega)$ being a constant. In this sense, any comparison between (5.126) and (5.128) is not fair, as the former trades off accuracy for generality, and the latter trades off the opposite: generality for accuracy.

However, especially with this in mind, we will devote the remainder of the chapter to arguing the following point. There are alternate trade-offs that can be made between accuracy and generality. In particular, note that in Figure 5.1 the input was not white, but (5.126) was still applied, and was significantly more accurate than (5.128).

5.8 Variance Quantification in the General Case

The remainder of this chapter will deal with the problem of quantifying estimate variability in the case where there are as few restrictions as possible placed on the input spectral density $\Phi_u(\omega)$. In relation to this, it is important to stress that under the very mild restriction that the spectral density has any rational spectral factorisation according to $\Phi_u(\omega) = |\mu(e^{i\omega})|^2$, then the fundamental result introduced in Section 5.5 applies; namely, that according to Theorem 5.1

$$\Delta_n(\omega) = \lim_{N\to\infty} N \cdot \mathrm{Cov}\left\{\begin{bmatrix} G(e^{i\omega}, \widehat{\theta}_N) \\ H(e^{i\omega}, \widehat{\theta}_N) \end{bmatrix}\right\} = \Phi_\nu(\omega)\, S_{\zeta_\circ}^{-\star}(e^{i\omega})\, \varphi_n(\omega,\omega)\, S_{\zeta_\circ}^{-1}(e^{i\omega}) \tag{5.129}$$

where $\varphi_n(\lambda,\omega)$ is the reproducing kernel for the space X_n spanned by the rows of the spectral factor of the prediction error gradient spectral density. In the previous sections, by placing restrictions on the input spectral factor $\mu(z)$, it was possible to explicitly formulate an orthonormal basis for X_n, and hence find a closed form expression for $\varphi_n(\lambda,\omega)$ and therefore also for the asymptotic in N variability $\Delta_n(\omega)$, $S_\zeta(z)$ is the spectral factor of $[u_t, \varepsilon_t(\theta_\circ)]^T$, and $\Phi_\nu(\omega) = \sigma^2|H(e^{i\omega}, \theta_\circ)|^2$.

However, in this section, we relax the restrictions on the spectral factor $\mu(z)$ of $\Phi_u(\omega)$. The price paid for this extra generality is that we lose the ability to form an appropriate orthonormal basis for the implied space X_n. This implies that a closed form exact expression for $\varphi_n(\lambda,\omega)$ also cannot be formulated. This will force us to pursue an alternate strategy of approximating $\varphi_n(\lambda,\omega)$ by a process of allowing the subspace dimension n to tend to infinity, and then using the ensuing infinite n quantification as if it applied for finite n. This, of course, is analogous to how the infinite in data length N expression is used as an approximation for finite N.

In order to briefly outline the essential details which follow, consider for the moment the situation of output-error modelling with $m_a = m_b \triangleq m$. Then according to the fundamental result (5.129) combined with the results of Section 5.6.3, we seek to quantify the reproducing kernel $\varphi_n(\lambda,\omega)$ for the space X_n defined back in (5.72) as

$$X_n = \mathrm{Span}\left\{\frac{z^{-1}\mu(z)}{A^2(z,\theta_\circ)}, \frac{z^{-2}\mu(z)}{A^2(z,\theta_\circ)}, \cdots, \frac{z^{-2m}\mu(z)}{A^2(z,\theta_\circ)}\right\} \tag{5.130}$$

where now the space dimension $n = 2m$. In the white input case of $\mu(z) = 1$, then an orthonormal basis $\{F_k(z)\}$ for this space may be explicitly constructed according to the results of Lemma 5.2, and hence the reproducing kernel $\varphi_n(\lambda, \omega)$ may be explicitly constructed according to Lemma 5.1.
However, when $\mu(z) \neq 1$, it is no longer possible to construct the required orthonormal basis in closed form. Nevertheless, as the model order grows, it is intuitively clear that the difference between the space (5.130) and the corresponding one involving $\mu(z) = 1$ becomes smaller. This suggests that it would be reasonable to approximate

$$\varphi_m(\omega, \omega) = 2\kappa_m(\omega) \triangleq 2\sum_{k=1}^{m} \frac{1 - |\xi_k|^2}{|e^{i\omega} - \xi_k|^2} \tag{5.131}$$

where the latter $\kappa_m(\omega)$ value was previously developed in Section 5.7.2. This then suggests, according to the fundamental result (5.129), the approximation for $\mu(z) \neq 1$ in the output-error case of

$$\Delta_{2m}(\omega) \approx \frac{2\sigma^2}{\Phi_u(\omega)} \sum_{k=1}^{m} \frac{1 - |\xi_k|^2}{|e^{i\omega} - \xi_k|^2}. \tag{5.132}$$

The accuracy of this approximation (recall, it was an equality when $\mu(z) = 1$) will be seen in what follows to depend on the smoothness of $\Phi_u(\omega) = |\mu(e^{i\omega})|^2$ as well as on the size of m (the larger the better).
In this work to come, the essential aspect of making the above intuitive developments rigorous will be to note that the reproducing kernel of interest $\varphi_n(\lambda, \omega)$ is related to a further reproducing $\phi_n(\lambda, \omega)$. This kernel applies for a space Y_n, which is X_n with $\mu(z) = 1$ and with respect to an inner produce in which the integrand is *weighted* with $\Phi_u(\omega)$. These two kernels will be seen to be related according to

$$\varphi_n(\lambda, \omega) = \mu(e^{i\lambda})\overline{\mu(e^{i\omega})}\, \phi_n(\lambda, \omega). \tag{5.133}$$

The value of this arizes due to $\phi_n(\lambda, \omega)$ being amenable to rigorous analysis of its properties as m grows. This then delivers, via (5.133) and (5.129), an argument for approximations such as (5.132), which make the above intuitive developments precise.
These ideas could be dealt with in the most direct manner possible, but in fact we elect a more tortuous route that detours via an initial overview of the analysis underlying the well-known approximation (5.128).
The first purpose of this strategy is to elucidate a primary reason behind the improved accuracy of quantifications derived here (such as (5.126)) relative to the older one (5.128); namely, that this pre-existing one implicitly requires Fourier reconstruction of a function dependent on $\Phi_u(\omega)$, with this same function scaled in such as way as to (perhaps dramatically) increase the total variation of that function, and hence to also upset the required Fourier

convergence. Further analysis in this section then highlights a core principle exposed in all chapters of this book – that the use of an orthonormal basis *adapted* to the function that is to be approximated can yield very significant benefits in accuracy.
The second purpose of the non-direct presentation chosen here is that it highlights a core principle of this chapter; Namely, recognizing the fundamental role of the reproducing kernel in the variance quantification problem is essential if maximum accuracy of quantification is to be achieved. The above-mentioned technique of adapting a basis will be seen to improve variance quantification accuracy. However, because it only implicitly addressed kernel approximation, it will also be seen to miss an essential quantification feature. This essential feature will be exposed only once the kernel quantification problem is addressed explicitly.
With this desiderata for the organization of this section in mind, we now take a further very brief diversion to introduce some required notation and background results.

5.8.1 Technical Preliminaries

This section establishes certain technical notation and results related to the properties of large Toeplitz matrices and generalizations of these matrices.

Generalized Toeplitz Matrices

Recall the definition (5.35) of a rational orthonormal basis made in Lemma 5.2 of Section 5.4. This will be used to define what is called a 'generalized' block Toeplitz matrix $M_n(F)$ as follows

Definition 5.1. *An $n \times n$ 'generalized' block-Toeplitz matrix is defined by a $p \times p$ positive definite matrix valued function $F(\omega)$ as*

$$M_n(F) \triangleq \frac{1}{2\pi} \int_{-\pi}^{\pi} [\Gamma_m(e^{i\omega}) \otimes I_p] F(\omega) [\Gamma_m^{\star}(e^{i\omega}) \otimes I_p] \, d\omega \tag{5.134}$$

where $mp = n$, I_p is a $p \times p$ identity matrix and

$$\Gamma_m(z) \triangleq [F_1(z), F_1(z), \cdots, F_m(z)]^T \tag{5.135}$$

with $\{F_k(z)\}$ given by (5.35).

Here, $\otimes$ is the Kronecker tensor product of matrices, which is defined when A is $m \times n$, and B is $\ell \times p$ according to the $m\ell \times np$ dimensional matrix $A \otimes B$ being given by

$$A \otimes B \triangleq \begin{pmatrix} a_{11}B & a_{12}B & \cdots & a_{1n}B \\ a_{21}B & a_{22}B & \cdots & a_{2n}B \\ \vdots & & & \vdots \\ a_{m1}B & a_{m2}B & \cdots & a_{mn}B \end{pmatrix}.$$

There are various useful identities applicable to this tensor product, but the only ones to be employed here are that $(A \otimes B)^T = A^T \otimes B^T$ and that for C and D of appropriate dimension $(A \otimes B)(C \otimes D) = AC \otimes BD$. For more detail on this topic of Kronecker product properties see [35].
The 'generalized' epithet in Definition 5.1 is derived from the fact that for the special case of $\xi_k = 0$ for all k, then $\{F_k(e^{i\omega})\}$ reverts to the trigonometric basis $\{e^{-i\omega k}\}$ and in this case, $M_n(F)$ possesses a more familiar 'block-banded' Toeplitz matrix structure. This chapter reserves the notation $T_n(F)$ for this special case of $M_n(F)$ as follows.

Definition 5.2. *An $n \times n$ block-Toeplitz matrix is defined by a $p \times p$ positive definite matrix valued function $F(\omega)$ as*

$$T_n(F) \triangleq \frac{1}{2\pi} \int_{-\pi}^{\pi} [\Lambda_m(e^{i\omega}) \otimes I_p] F(\omega) [\Lambda_m^\star(e^{i\omega}) \otimes I_p] \, d\omega \tag{5.136}$$

where $pm = n$ and

$$\Lambda_m(z) \triangleq [1, z, z^2, \cdots, z^{m-1}]^T. \tag{5.137}$$

Generalized Fourier Results

This section considers Lipschitz continuous functions of some positive order α: $f \in \text{Lip}(\alpha)$ meaning that that for some fixed constant $C < \infty$

$$|f(x) - f(y)| < C|x - y|^\alpha. \tag{5.138}$$

As will be profiled shortly, the preceding Toeplitz matrix definitions realize their fundamental role in deriving approximations like (5.1) and (5.2) by way of frequency dependent quadratic forms, and the approximations actually arize by virtue of these forms representing convergent Fourier reconstructions. In the 'traditional' block-banded Toeplitz matrix case, this principle appears as follows.

Theorem 5.2. *Provided $F(\omega)$ of dimension $p \times p$ is positive definite and (component-wise) Lipschitz continuous of order $\alpha > 0$ for all $\omega \in [-\pi, \pi]$, then*

$$\lim_{m \to \infty} \frac{1}{m} [\Lambda_m^\star(e^{i\omega}) \otimes I_p] T_{mp}^{-1}(F) [\Lambda_m(e^{i\lambda}) \otimes I_p] = \begin{cases} F^{-1}(\omega) & ; \omega = \lambda \\ 0 & ; \omega \neq \lambda \end{cases}$$

component-wise and uniformly on $\omega \in [-\pi, \pi]$. Here I_p is a $p \times p$ identity matrix.

Proof. See [121]. ∎

It is this result that underlies the approximation (5.1), and to derive the improved accuracy approximation (5.2), a generalized Toeplitz matrix form of it will be employed, which is given as follows.

Theorem 5.3. *Suppose that the poles $\{\xi_k\}$ in (5.35) are such that*

$$\sum_{k=1}^{\infty}(1-|\xi_k|)=+\infty.$$

Provided $F(\omega)$ of dimension $p \times p$ is positive definite and (component-wise) Lipschitz continuous of order $\alpha > 0$ for all $\omega \in [-\pi,\pi]$, then for any integer k (possibly negative)

$$\lim_{m\to\infty}\frac{1}{\kappa_m(\omega)}[\Gamma_m^\star(e^{i\omega})\otimes I_p]M_{mp}^{-k}(F)[\Gamma_m(e^{i\lambda})\otimes I_p]=\begin{cases}F^{-k}(\omega) & ;\omega=\lambda\\ 0 & ;\omega\neq\lambda\end{cases}$$

component-wise and uniformly on $\omega\in[-\pi,\pi]$ where

$$\kappa_m(\omega)\triangleq\sum_{k=1}^{m}\frac{1-|\xi_k|^2}{|e^{i\omega}-\xi_k|^2} \tag{5.139}$$

and with $\Gamma_m(z)$ being defined in (5.135).

Proof. See [224].

■

Note that the 'classical' result of Theorem 5.2 is a special case of Theorem 5.3 in which $\xi_k = 0$ for all k.

5.8.2 Kernel and Variance Approximation: A First Approach

With these technical preliminaries dispensed with, we now employ them for the purposes of approximating the reproducing kernel $\varphi_n(\lambda,\omega)$. To facilitate this, we will now turn to the representation (5.50), which is given as

$$\varphi_n(\lambda,\omega)=\Psi^T(e^{i\lambda})\,W_n^{-1}\,\overline{\Psi(e^{i\omega})} \tag{5.140}$$

where, according to (5.45)

$$\Psi(z,\theta_\circ)\triangleq H^{-1}(z,\theta_\circ)\left.\frac{\mathrm{d}\Pi(z,\theta)}{\mathrm{d}\theta}\right|_{\theta=\theta_\circ}S_{\zeta_\circ}(z). \tag{5.141}$$

Furthermore, note that by assuming $m_a = m_b = m_c = m_d = m$ and that θ is organized as

$$\theta=[a_0,b_0,d_0,c_0,a_1,b_1,d_1,c_1,\cdots,a_{m-1},b_{m-1},d_{m-1},c_{m-1}] \tag{5.142}$$

then

$$\frac{\mathrm{d}\Pi(z,\theta)}{\mathrm{d}\theta} = \left[\frac{\mathrm{d}G(q,\theta)}{\mathrm{d}\theta}, \frac{\mathrm{d}H(q,\theta)}{\mathrm{d}\theta}\right] = [\Lambda_m(q) \otimes I_4] Z(q,\theta) \qquad (5.143)$$

where $\Lambda_m(q)$ has been defined in (5.137) and

$$Z(q,\theta) \triangleq \begin{bmatrix} -A^{-1}(q,\theta)G(q,\theta) & 0 \\ A^{-1}(q,\theta) & 0 \\ 0 & -D^{-1}(q,\theta)H(q,\theta) \\ 0 & D^{-1}(q,\theta) \end{bmatrix}. \qquad (5.144)$$

Therefore, using the Toeplitz matrix notation introduced in the previous section, the reproducing kernel can be expressed as

$$\begin{aligned} \varphi_n(\lambda,\omega) = S_{\zeta_\circ}^T(e^{i\lambda}) Z_\circ^T(e^{i\lambda}) [\Lambda_m^T(e^{i\lambda}) \otimes I_4] \times \\ T_n^{-1}\left(\frac{Z_\circ \Phi_\zeta Z_\circ^\star}{|H_\circ^n|^2}\right) \overline{[\Lambda_m(e^{i\omega}) \otimes I_4] Z_\circ(e^{i\omega}) S_{\zeta_\circ}(e^{i\omega})}. \end{aligned} \qquad (5.145)$$

Now, the principle issue we address in this latter part of the chapter is that under relaxed assumptions on $\Phi_u(\omega)$, and as opposed to the situation in the previous sections, it is not possible for derive an exact closed form expression for this kernel.
Nevertheless, the fundamental relationship (5.129) still holds and provides the variance quantification

$$\Delta_n(\omega) = \Phi_\nu(\omega)\, S_{\zeta_\circ}^{-\star}(e^{i\omega})\, \varphi_n(\omega,\omega)\, S_{\zeta_\circ}^{-1}(e^{i\omega}) =$$

$$\Phi_\nu(\omega) Z_\circ^\star(e^{i\omega}) [\Lambda_m^\star(e^{i\omega}) \otimes I_4] T_n^{-1}\left(\frac{Z_\circ \Phi_\zeta Z_\circ^\star}{|H_\circ^n|^2}\right) [\Lambda_m(e^{i\omega}) \otimes I_4] Z_\circ(e^{i\omega}) \quad (5.146)$$

There is a potential difficulty in continuing the analysis of this expression in that the so-called symbol $Z_\circ \Phi_u Z_\circ^\star / |H_\circ^n|^2$ defining the above Toeplitz matrix is, by construction, singular even though the matrix $T_n(Z_\circ \Phi_u Z_\circ^\star / |H_\circ^n|^2)$ it implies is not[1]. On the other hand, as will soon be apparent, it will be necessary to work with an approximation that involves the inverse of this symbol. To address this, one idea (due to [162]) is to work with a perturbed matrix (note the new symbol of T_n^{-1})

$$\begin{aligned} \Delta_n(\omega,\delta) \triangleq \sigma^2 Z_\circ^\star(e^{i\omega}) [\Lambda_m^\star(e^{i\omega}) \otimes I_4] \times \\ T_n^{-1}\left(\frac{Z_\circ \Phi_\zeta Z_\circ^\star}{|H_\circ|^2} + \delta I_4\right) [\Lambda_m(e^{i\omega}) \otimes I_4] Z_\circ(e^{i\omega}). \end{aligned} \qquad (5.147)$$

It can then be argued that [162]

$$\Delta_n(\omega) = \lim_{\delta \to 0} \Delta_n(\omega,\delta). \qquad (5.148)$$

---

[1] Under an assumption of no pole-zero cancellations in $G(z,\theta_\circ)$ and $H(z,\theta_\circ)$.

In relation to this, note that [162] establishes that this procedure can be interpreted as one of modifying $V_N(\theta)$ to provide a 'regularized' estimate, in that $\Delta_n(\omega,\delta)$ measures the variability of $G(e^{i\omega},\widehat{\theta}_N(\delta))$, $H(e^{i\omega},\widehat{\theta}_N(\delta))$ where

$$\widehat{\theta}_N^n(\delta) \triangleq \arg\min_{\theta\in\mathbb{R}^n} \frac{1}{2N}\sum_{t=1}^{N}\varepsilon_t^2(\theta) + \frac{\delta}{2}\|\theta-\theta_\circ\|^2. \tag{5.149}$$

With this in mind, the next step in forming an approximate variance quantification is to note that the quadratic form defining $\Delta_n(\omega)$, by virtue of being formulated in terms of inverses of Toeplitz matrices parameterized by spectral densities, can be viewed as an mth-order Fourier reconstruction of the inverse of the spectral density. Specifically, by Theorem 5.2 and (5.147), and now making explicit that the subspace dimension is $n = 4m$

$$\lim_{m\to\infty}\frac{1}{m}\Delta_{4m}(\omega,\delta) \triangleq \Delta(\omega,\delta) = \sigma^2 Z_\circ^\star(e^{i\omega})\left(\frac{Z_\circ\Phi_\zeta Z_\circ^\star}{|H_\circ|^2} + \delta I_4\right)^{-1} Z_\circ(e^{i\omega}). \tag{5.150}$$

Therefore, using the matrix inversion lemma [106],

$$[A+BCD]^{-1} = A^{-1} - A^{-1}B[C^{-1}+DA^{-1}B]^{-1}DA^{-1} \tag{5.151}$$

and dropping the ω dependence for readability

$$\begin{aligned}\Delta(\omega,\delta) &= \sigma^2 Z_\circ^\star\left[\delta^{-1}I - \delta^{-1}Z_\circ\left[|H_\circ|^2\Phi_\zeta^{-1} + \delta^{-1}Z_\circ^\star Z_\circ\right]^{-1}\delta^{-1}Z_\circ^\star\right]Z_\circ \\ &= \sigma^2|H_\circ|^2 Z_\circ^\star Z_\circ\left[\Phi_\zeta Z_\circ^\star Z_\circ + \delta|H_\circ|^2 I\right]^{-1}.\end{aligned} \tag{5.152}$$

Therefore,

$$\lim_{\delta\to 0}\lim_{m\to\infty}\frac{1}{m}\Delta_{4m}(\omega,\delta) = \sigma^2|H(e^{i\omega},\theta_\circ)|\Phi_\zeta^{-1}(\omega). \tag{5.153}$$

By using an argument that the above convergence with increasing m has approximately occurred for the finite model order m of interest, an approximate quantification is provided as

$$\lim_{N\to\infty} N\cdot\mathrm{Cov}\{\Pi(e^{i\omega},\widehat{\theta}_N)\} = \Delta_{4m}(\omega) \approx m\Phi_\nu(\omega)|\Phi_\zeta^{-1}(\omega), \tag{5.154}$$

$$\Phi_\nu(\omega) = \sigma^2|H(e^{i\omega},\theta_\circ)|^2. \tag{5.155}$$

In a similar vein, assuming that convergence of the above has approximately occurred for finite N then leads to the further approximation

$$\mathrm{Cov}\{\Pi(e^{i\omega},\widehat{\theta}_N)\} \approx \frac{m}{N}\Phi_\nu(\omega)\Phi_\zeta^{-1}(\omega). \tag{5.156}$$

Furthermore, in situations where a noise model is not estimated, such as as when employing output-error of FIR-like fixed denominator models, the same analysis implies the slightly simpler expression

$$\mathrm{Var}\{G(e^{i\omega}, \widehat{\theta}_N)\} \approx \frac{m}{N} \frac{\Phi_\nu(\omega)}{\Phi_u(\omega)}. \tag{5.157}$$

These quantifications, via the thinking outlined in this section, were first established in the landmark works [162,168] and extended to the multi-variable case in [335].
However, we hope it has already occurred to the (highly attentive!) reader that (5.157) is precisely the same expression as (5.128), and (5.1) which has been earlier illustrated in Figures 5.1, 5.3, and 5.4 as being a somewhat poor approximation. Clearly, some explanation of this situation is necessary, which is the purpose of the following sections.

5.8.3 Genesis of Impaired Approximation

In fact, there are two main influences that act to degrade the fidelity of (5.156), (5.157) that is illustrated in the earlier Figures 5.1 and 5.3.
The first of these influences, explained in this section, arises because (5.156) and (5.157) are predicated on the convergence in (5.153) having approximately occurred for finite model order m so that (5.154) can be concluded. However, as we shall explain, whether this approximate convergence has occurred can be problematic.

To provide more detail, and in the interests of most clearly exposing the idea, consider the simplified output-error case in which only a dynamics model $G(q, \theta)$ is estimated. This implies that $H(q, \theta_\circ) = 1$ and

$$Z(q, \theta) = \begin{bmatrix} -G(q, \theta) \\ 1 \end{bmatrix}, \qquad \Phi_\zeta(\omega) = \Phi_u(\omega). \tag{5.158}$$

Note, that in the above, a slight redefinition of $Z(q, \theta)$ has been made which involved factoring out the term of $A^{-1}(q, \theta)$ from the original definition (5.144). By doing so, $Z(z, \theta_\circ)$ is now a function whose smoothness does not depend on model order. This will be important in a moment. It also implies that the limiting value of (5.153) arises from employing Theorem 5.2 in the following way

$$\lim_{m\to\infty} \frac{1}{m} [\Lambda_m^\star(e^{i\omega}) \otimes I_2] T_n^{-1} \left(\frac{Z_\circ \Phi_u Z_\circ^\star}{|A_\circ|^2} + \delta I_2 \right) [\Lambda_m(e^{i\omega}) \otimes I_2] = \left[\frac{Z_\circ \Phi_u Z_\circ^\star}{|A_\circ|^2} + \delta I_2 \right]^{-1} \tag{5.159}$$

That is, the previous asymptotic argument, in the output-error case, depends on a Fourier series reconstruction via (5.150) of a matrix valued function

$$\frac{Z_\circ(e^{i\omega})\Phi_u(\omega)Z_\circ^\star(e^{i\omega})}{|A_\circ(e^{i\omega})|^2} + \delta I_2. \tag{5.160}$$

As is well-known [84], the rate of Fourier series convergence is governed by the smoothness of the function being reconstructed. Furthermore, the variation (and hence non-smoothness) of (5.160) can be significantly *increased* due to division by the $|A_\circ(e^{i\omega})|^2$ term.
To see this, suppose for simplicity that all the zeroes of $A_\circ(z)$ are in the left half plane. Then (see Figure 5.5) $|A_\circ(e^{i0})| \leq \eta^m$ for some $\eta < 1$ and $|A_\circ(e^{i\pi})| \geq \gamma^m$ for some $\gamma > 1$ so that division of a function by $|A_\circ(e^{i\omega})|^2$ can magnify the maximum value of (5.160) by a factor of $\gamma^{2m} \gg 1$ and shrink the minimum values by a factor of $\eta^{2m} \ll 1$.

Therefore, as the model order m grows, the function (5.160) being implicitly Fourier reconstructed in (5.150) can develop greater variation, which necessitates more terms in its Fourier expansion before approximate convergence can be assumed in the step (5.150) leading to (5.156), (5.157). However, a key point is that the number of terms in the implicit Fourier reconstruction (5.150) is also given by the quantity m.

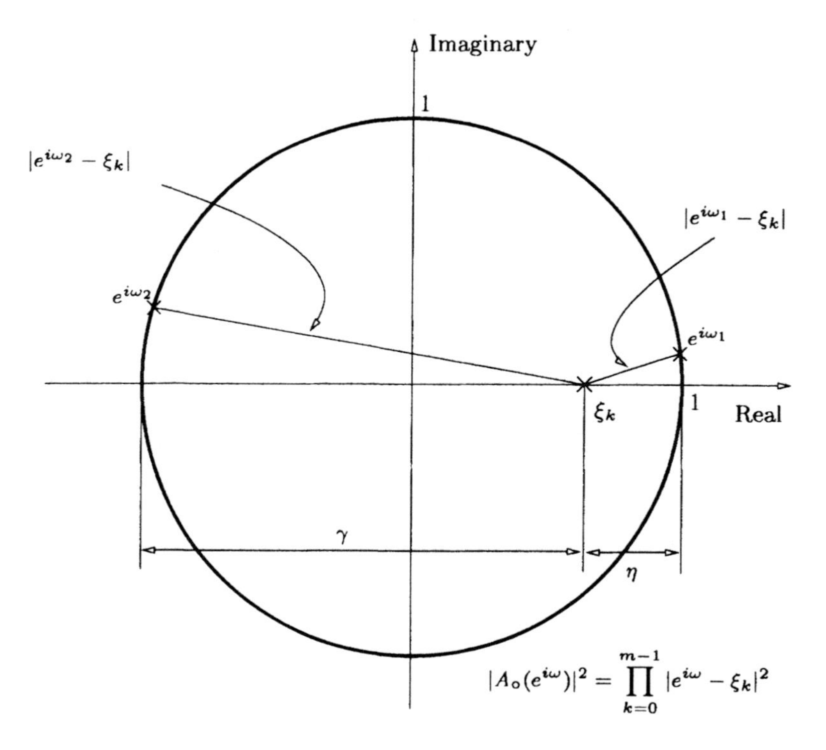

Fig. 5.5. Graphical illustration of how the magnitude variation of $A_\circ(e^{i\omega})$ depends on ω and hence increases for increasing order m.

That is, a circular situation arises in which just as the approximation order grows with m, the function being approximated develops greater variation with increasing m, and hence also requires a higher reconstruction order m in order to attain accurate approximation.
The net result is that, depending on the problem setting, it is problematic as to whether Fourier convergence can be assumed to hold in such a way that the approximation (5.156) via (5.154) can be concluded.

5.8.4 Improved Variance Approximation via Basis Adaptation

One key idea that will lead to an improved approximation, such as that shown in the original motivating Figure 5.1, is that we can change the orthonormal basis involved in the underlying Fourier series approximation to one that is adapted to the function (5.160) being reconstructed.
In particular, again focusing on the simplified output-error case for the sake of clarity, the problem of quantifying $\Delta_n(\omega)$ in (5.146) may be re-parameterized in terms of the orthonormal basis $\{F_k(z)\}$ introduced in Equation (5.35) of Lemma 5.2 after it is 'adapted' to the underlying denominator polynomial $A(z,\theta_\circ)$ in the sense that the zeros $\{\xi_1,\xi_2,\cdots,\xi_m\}$ of $A(z,\theta_\circ)$ are used as the poles of $\{F_k(z)\}$ according to the formulation (5.35).

The use of this basis to obtain variance quantifications then involves the analysis up until (5.146) being retained. However, because the poles $\{\xi_k\}$ of the bases (5.35) are chosen the same as the zeros of $A(z,\theta_\circ)$, then using the definition (5.135) it holds that for some non-singular $2m\times 2m$ matrix J_m, the matrix $\Lambda_m(z)$ appearing in (5.146), and defined in (5.137) in terms of a trigonometric basis $\{e^{i\omega n}\}$, is expressible as

$$A^{-1}(z,\theta_\circ)\Lambda_m(z) = J_m\Gamma_m(z).$$

Therefore

$$A^{-1}(z,\theta)(\Lambda_m(z)\otimes I_2) = J_m\Gamma_m(z)\otimes I_2 = (J_m\otimes I_2)(\Gamma_m(z)\otimes I_2). \quad (5.161)$$

Further developments then depend on using the generalized Toeplitz matrix $M_n(F)$ of definition 5.1 which via (5.161) is related to the conventional 'banded' Toeplitz matrix $T_n(F)$ of definition 5.2 according to

$$T_n\left(\frac{F}{|A_\circ^2|}\right) = (J_m\otimes I_2)M_n(F)(J_m^T\otimes I_2) \quad (5.162)$$

and hence the quantity $\Delta_n(\omega,\delta)$ in (5.147) previously analysed via Fourier theory with respect to the trigonometric basis becomes

$$\begin{aligned}
\Delta_{2m}(\omega,\delta) =& \frac{\sigma^2}{|A_\circ(e^{i\omega})|^2} Z_\circ^\star(e^{i\omega})[\Lambda_m^\star(e^{i\omega})\otimes I_2](J_m^{-T}\otimes I_2)\times \\
& M_n^{-1}\left(Z_\circ \Phi_u Z_\circ^\star + \delta|A_\circ|^2 I_2\right)(J_m^{-1}\otimes I_2)[\Lambda_m(e^{i\omega})\otimes I_2]Z_\circ(e^{i\omega}) \\
=& \sigma^2 Z_\circ^\star(e^{i\omega})[\Gamma_m^\star(e^{i\omega})\otimes I_2]M_n^{-1}\left(Z_\circ\Phi_u Z_\circ^\star + \delta|A_\circ|^2 I_2\right)\times \\
& [\Gamma_m(e^{i\omega})\otimes I_2]Z_\circ(e^{i\omega}).
\end{aligned} \tag{5.163}$$

In [223,224] a generalized Fourier theory involving the generalized basis (5.35) is developed, for which the most pertinent result in the current context is the convergence result of Theorem 5.3. Applying it together with the well-known matrix inversion lemma [106] provides the conclusion

$$\begin{aligned}
\lim_{m\to\infty}\frac{\Delta_{2m}(\omega,\delta)}{\kappa_m(\omega)} &= \sigma^2 Z_\circ^\star(e^{i\omega})\left[Z_\circ(e^{i\omega})\Phi_u(\omega)Z_\circ^\star(e^{i\omega})+\delta|A_\circ(e^{i\omega})|^2 I_2\right]^{-1} Z_\circ(e^{i\omega}) \\
&= \frac{\sigma^2 Z_\circ^\star(e^{i\omega})Z_\circ(e^{i\omega})}{\Phi_u(\omega)Z_\circ^\star(e^{i\omega})Z_\circ(e^{i\omega}) + \delta|A_\circ(e^{i\omega})|^2}
\end{aligned} \tag{5.164}$$

so that

$$\lim_{\delta\to 0}\lim_{m\to\infty}\frac{\Delta_{2m}(\omega,\delta)}{\kappa_m(\omega)} = \frac{\sigma^2}{\Phi_u(\omega)}. \tag{5.165}$$

Therefore, following the same line of argument (established in [164]) and leading to (5.156) then provides

$$\mathrm{Var}\{G(e^{i\omega},\widehat{\theta}_N)\} \approx \frac{\sigma^2}{N}\frac{\kappa_m(\omega)}{\Phi_u(\omega)} = \frac{1}{N}\frac{\Phi_\nu(\omega)}{\Phi_u(\omega)}\sum_{k=1}^{m}\frac{1-|\xi_k|^2}{|e^{i\omega}-\xi_k|^2} \tag{5.166}$$

which, aside from a numerator factor of 2, is the new approximation (5.125) which applies for (almost) arbitrary Φ_u and is illustrated in Figure 5.1.
A vital point here is that, by virtue of using the basis (5.35), the implicit Fourier reconstruction operating here (via (5.165)), is of a function $Z_\circ(e^{i\omega})\Phi_u(\omega)Z_\circ^\star(e^{i\omega})$ *whose Lipschitz smoothness is constant with respect to Fourier reconstruction length* m.

This property, which implies that the quality of the approximation improves monotonically with increasing m, is a key factor that imbues (5.166) with improved accuracy when compared with (5.157).
For example, as just shown, it is via recognizing the importance of monotonic convergence that the term $\kappa_m(\omega) = \sum_{k=1}^{m}(1-|\xi_k|^2)|e^{i\omega}-\xi_k|^{-2}$ is introduced in (5.2). By its frequency dependent nature, this is the key element in quantifying the variability shown in, for example, Figure 5.1 as 'low-pass' rather than 'high-pass'.

5.8.5 Further Improvement via a Focus on Kernel Approximations

This chapter began by exposing that variance quantification, via the fundamental expression (5.52) of Theorem 5.1, was equivalent to quantification of

a particular reproducing kernel. It then proceeded to derive closed forms for this kernel, and hence for the estimation variance, in cases where this was possible due to certain restrictions being imposed on $\Phi_u(\omega)$.
The subsequent sections 5.8.3 and 5.8.4 then provided, what might be considered a detour, in that they led to a new variance quantification (5.166) that is not, in fact, what we will finally propose. The latter is of the form (5.2), which differs from (5.166) by a constant factor of two.
Nevertheless, this diversion was included for the following two purposes. It allowed us to highlight some essential points that can lead to inaccuracy of the well-known approximation (5.1), (5.157), and it will now also emphasize the need to explicitly, rather than implicitly, deal with the need to quantify a reproducing kernel when evaluating variance quantifications.

In relation to this, it is important to recognize a fundamental point that underlies new quantifications (such as (5.126)) that are developed here; namely, when employing an asymptotic in model order result for a finite value of m, it is vital to distinguish between cases where this finite m is expected to be equal to, or higher than, the true model order.

This is necessary because a key step in existing analysis involves progressing from (5.147) to (5.150) by using the Fourier convergence result of Theorem 5.2 to argue that (again, for sake of clarity, we are considering the simpler output-error case whereby (5.158) applies)

$$\lim_{m\to\infty}\frac{1}{m}[\Lambda_m^\star(e^{i\omega})\otimes I_2]T_n^{-1}\left(\frac{Z_\circ\Phi_u Z_\circ^\star}{|A_\circ|^2}+\delta I_2\right)[\Lambda_m(e^{i\omega})\otimes I_2] =$$
$$\left[\frac{Z_\circ(e^{i\omega})\Phi_u(\omega)Z_\circ^\star(e^{i\omega})}{|A_\circ(e^{i\omega})|^2}+\delta I_2\right]^{-1}. \quad (5.167)$$

The pre-existing analysis profiled in Section 5.8.2 leading to the well-known quantifications (5.156), (5.157) then involves assuming that the above equality holds approximately for the finite model order m actually used.
However, if the chosen finite m is less than or equal to the true model order, then the Toeplitz matrix, and hence the whole matrix on the left-hand side of (5.167) will be full rank, even for $\delta = 0$. At the same time, when $\delta = 0$, the matrix on the right-hand side of (5.167) is only of rank one. It is therefore unlikely, as was hoped for in the analysis of Section 5.8.2, that the approximation

$$\frac{1}{m}[\Lambda_m^\star(e^{i\omega})\otimes I_2]T_n^{-1}\left(\frac{Z_\circ\Phi_u Z_\circ^\star}{|A_\circ|^2}+\delta I_2\right)[\Lambda_m(e^{i\omega})\otimes I_2] \approx$$
$$\left[\frac{Z_\circ(e^{i\omega})\Phi_u(\omega)Z_\circ^\star(e^{i\omega})}{|A_\circ(e^{i\omega})|^2}+\delta I_2\right]^{-1} \quad (5.168)$$

will be accurate, as the left-hand side is full rank, but the right-hand side is close to rank one!

For example, as was established in previous sections via the explicit formulae for reproducing kernels derived there, when $\Phi_u(\omega) = \gamma$ a constant and m is less than or equal to the true model order, then the following *exact* equality holds, even for finite m

$$Z_\circ^\star(e^{i\omega})[\Lambda_m^\star(e^{i\omega})\otimes I_2]T_n^{-1}\left(\frac{\gamma Z_\circ Z_\circ^\star}{|A_\circ|^2}\right)[\Lambda_m(e^{i\omega})\otimes I_2]Z_\circ(e^{i\omega}) = \frac{2}{\gamma}\sum_{k=1}^{m}\frac{1-|\xi_k|^2}{|e^{i\omega}-\xi_k|^2}. \tag{5.169}$$

As an aside, the authors consider that this simple expression for a quantity as seemingly complex as the left-hand side of (5.169) is a most surprising result! However, it is by recognizing it that the important factor of 2 arises in the new quantification (5.1). In particular, note that this frequency-dependent exact expression (5.169) is quite different to the approximate non-frequency-dependent value of 1, which is argued via the existing analysis [162] as summarized in (5.147)–(5.152).

Note further, that these difficulties associated with using the result (5.167) when $\delta = 0$ are not ameliorated via the strategy of the previous section in which the basis was adapted to the denominator term $|A(e^{i\omega}, \theta_\circ)|^2$, as this involves employing the convergence result

$$\lim_{m\to\infty}\frac{1}{\kappa_m(\omega)}[\Gamma_m^\star(e^{i\omega})\otimes I_2]M_n^{-1}\left(Z_\circ\Phi_u Z_\circ^\star + \delta|A_\circ|^2 I_2\right)[\Gamma_m(e^{i\omega})\otimes I_2]$$
$$= \left[Z_\circ(e^{i\omega})\Phi_u(\omega)Z_\circ^\star(e^{i\omega}) + \delta|A_\circ(e^{i\omega})|^2 I_2\right]^{-1} \tag{5.170}$$

which again, when $\delta = 0$, is full rank on the left-hand side but of rank one on the right-hand side, and hence unlikely to provide a good approximation for small δ.

Nevertheless, the previous section still established the fundamental ideas of recognizing underlying Fourier reconstructions and then adapting a basis to the function being reconstructed. Here, we couple this to the strategy of approximating the reproducing kernel that, according to (5.129), governs the variance quantification problem, and this will be seen to be key to establishing quantifications such as (5.1) for the case of general $\Phi_u(\omega)$.

For this purpose, in the interests of most clearly exposing ideas, we restrict our explanation to the output-error case with $m_a = m_b = m$ (later, results for the Box-Jenkins case will be presented), and then recognize that according to the fundamental result (5.129), we seek to quantify the reproducing kernel $\varphi_n(\lambda, \omega)$ for the space X_n defined back in (5.72) as

$$X_n = \mathrm{Sp}\left\{\frac{z^{-1}\mu(z)}{A^2(z,\theta_\circ)}, \frac{z^{-2}\mu(z)}{A^2(z,\theta_\circ)}, \cdots, \frac{z^{-2m}\mu(z)}{A^2(z,\theta_\circ)}\right\}. \tag{5.171}$$

(recall that in this section it is assumed that $m_a = m_b = m$ (5.142) so that $n = 2m$) and with respect to the inner product defined in (5.24) as

$$\langle f, g \rangle = \frac{1}{2\pi} \int_{-\pi}^{\pi} g^\star(\lambda) f(\lambda)\,\mathrm{d}\lambda. \tag{5.172}$$

However, it will be useful to also consider the related space

$$Y_n = \mathrm{Sp}\left\{ \frac{z^{-1}}{A^2(z,\theta_\circ)}, \frac{z^{-2}}{A^2(z,\theta_\circ)}, \cdots, \frac{z^{-2m}}{A^2(z,\theta_\circ)} \right\}. \tag{5.173}$$

Now, as established in Section 5.4, the reproducing kernel $\varphi_n(\lambda,\omega)$ is an element of X_n and hence according to (5.171), it must be of the form

$$\varphi_n(\lambda,\omega) = \phi_n(\lambda,\omega)\mu(e^{i\lambda}) \tag{5.174}$$

for some function $\phi_n(\lambda,\omega)$ that satisfies, for any fixed ω

$$\phi_n(\lambda,\omega) \in Y_n. \tag{5.175}$$

Furthermore, because any $f(z) \in Y_n$ can be expressed as $h(z)/\mu(z)$ for some $h(z) \in X_n$, then by the reproducing property of $\varphi_n(\lambda,\omega)$, and for any $f(z) \in Y_n$

$$f(e^{i\omega}) = \frac{h(e^{i\omega})}{\mu(e^{i\omega})} = \frac{1}{\mu(e^{i\omega})}\frac{1}{2\pi}\int_{-\pi}^{\pi} h(e^{i\lambda})\,\overline{\varphi_n(\lambda,\omega)}\,\mathrm{d}\lambda \tag{5.176}$$

$$= \frac{1}{2\pi}\int_{-\pi}^{\pi} \frac{h(e^{i\lambda})}{\mu(e^{i\lambda})}\overline{\left(\frac{\varphi_n(\lambda,\omega)}{\mu(e^{i\omega})\mu(e^{i\lambda})}\right)}|\mu(e^{i\lambda})|^2\,\mathrm{d}\lambda \tag{5.177}$$

$$= \frac{1}{2\pi}\int_{-\pi}^{\pi} f(e^{i\lambda})\overline{\left(\frac{\varphi_n(\lambda,\omega)}{\mu(e^{i\omega})\mu(e^{i\lambda})}\right)}|\mu(e^{i\lambda})|^2\,\mathrm{d}\lambda. \tag{5.178}$$

This establishes the following important principle.

If $\varphi_n(\lambda,\omega)$ is the reproducing kernel for the function space X_n defined in (5.171) and with respect to the 'unweighted' inner product (5.172), then

$$\phi_n(\lambda,\omega) = \frac{\varphi_n(\lambda,\omega)}{\mu(e^{i\omega})\mu(e^{i\lambda})} \tag{5.179}$$

is the reproducing kernel for the space Y_n defined in (5.173) and with respect to the 'weighted' (by $|\mu(e^{i\lambda})|^2$) inner product

$$\langle f, g \rangle = \frac{1}{2\pi}\int_{-\pi}^{\pi} g^\star(\lambda) f(\lambda)|\mu(e^{i\lambda})|^2\,\mathrm{d}\lambda. \tag{5.180}$$

The importance of this principle is that it allows us to reformulate the problem of quantifying output-error variance quantification $\Delta_n(\omega)$ via evaluation of $\varphi_n(\lambda,\omega)$ as follows. According to the fundamental principle (5.129)

$$\Delta_n(\omega) = \frac{\sigma^2}{\Phi_u(\omega)}\varphi_n(\omega,\omega) = \frac{\sigma^2}{\Phi_u(\omega)}|\mu(e^{i\omega})|^2\phi_n(\omega,\omega) = \sigma^2\phi_n(\omega,\omega) \quad (5.181)$$

where $\phi_n(\lambda,\omega)$ is the reproducing kernel for the space Y_n defined in (5.173) and with respect to the weighted inner product (5.180).

Additionally, according to Lemma 5.2, Y_n may be expressed as

$$Y_n = X_{2m} = \mathrm{Sp}\{F_1(z), F_2(z), \cdots, F_m(z), F_{m+1}(z), \cdots, F_{2m}(z)\}$$

where the poles of $\{F_1,\cdots,F_{2m}\}$ must must be set as the zeros of $A^2(z,\theta_\circ)$, and hence without loss of generality, we will assume that the poles of $\{F_{m+1},\cdots,F_{2m}\}$ are taken to be those same as the poles of $\{F_1,\cdots,F_m\}$ (ie. the zeros of $A(z,\theta_\circ)$) repeated.

Finally, via the definition (5.134) (5.135) with $p=1$,

$$\frac{1}{2\pi}\int_{-\pi}^{\pi} \Gamma_{2m}(e^{i\lambda})\left[\Gamma_{2m}^\star(e^{i\lambda})M_n^{-1}(\Phi_u)\Gamma_{2m}(e^{i\omega})\right]\Phi_u(\lambda)\,\mathrm{d}\lambda$$

$$= \left[\frac{1}{2\pi}\int_{-\pi}^{\pi}\Gamma_{2m}(e^{i\lambda})\Gamma_{2m}^\star(e^{i\lambda})\Phi_u(\lambda)\,\mathrm{d}\lambda\right]M_n^{-1}(\Phi_u)\Gamma_{2m}(e^{i\omega}) \quad (5.182)$$

$$= M_n(\Phi_u)M_n^{-1}(\Phi_u)\Gamma_{2m}(e^{i\omega}) = \Gamma_{2m}(e^{i\omega}). \quad (5.183)$$

Therefore, with respect to the weighted inner product (5.180) with $\Phi_u(\lambda) = |\mu(e^{i\lambda})|^2$, the function $\Gamma_{2m}^\star(\omega)M_n^{-1}(\Phi_u)\Gamma_{2m}(\lambda)$ reproduces all the elements of $\Gamma_{2m}(\lambda)$, and hence acts as a reproducing kernel $\phi_n(\lambda,\omega)$ for Y_n with respect to the weighted inner product (5.180). But the reproducing kernel for a space is unique, and hence the following is a formulation for $\phi_n(\lambda,\omega)$:

$$\phi_n(\lambda,\omega) = \Gamma_{2m}^\star(e^{i\omega})M_n^{-1}(\Phi_u)\Gamma_{2m}(e^{i\lambda}). \quad (5.184)$$

That is, we have re-expressed the reproducing kernel $\varphi_n(\lambda,\omega)$ via the relationship (5.179) as $\varphi_n(\lambda,\omega) = \phi_n(\lambda,\omega)\overline{F^{-1}(e^{i\omega})}F^{-1}(e^{i\lambda})$ and now, via (5.184) with respect to a basis $\{F_k(z)\}$, which is tailored (as explained in the previous section) for maximally accurate approximation.
Furthermore, by focusing directly on the reproducing kernel, we have arrived at an expression (5.184) that does not require a perturbation argument using δ factors in order to circumvent inversion of rank deficient quantities.
Instead, we simply draw on Theorem 5.3 to conclude that

$$\lim_{m\to\infty}\frac{1}{\kappa_{2m}(\omega)}\phi_n(\omega,\omega) = \lim_{m\to\infty}\frac{1}{\kappa_{2m}(\omega)}\Gamma_{2m}^\star(\omega)M_n^{-1}(\Phi_u)\Gamma_{2m}(\omega) = \frac{1}{\Phi_u(\omega)}. \quad (5.185)$$

We now recognize that $\kappa_{2m}(\omega) = 2\kappa_m(\omega)$ because the poles in the basis definition are repeated, and then further assume that for a finite model order of interest, approximate convergence in model order m has occurred. These conditions then establish the approximate reproducing kernel quantification

$$\phi_n(\omega, \omega) \approx \frac{\kappa_{2m}(\omega)}{\Phi_u(\omega)} = 2\frac{\kappa_m(\omega)}{\Phi_u(\omega)}. \tag{5.186}$$

When substituted into (5.181), this finally leads to our new approximate asymptotic in N variance quantification for output-error model structures and general $\Phi_u(\omega)$:

$$\Delta_{2m}(\omega) \approx 2\sigma^2 \frac{\kappa_m(\omega)}{\Phi_u(\omega)} = \frac{2\sigma^2}{\Phi_u(\omega)} \sum_{k=1}^{m} \frac{1 - |\xi_k|^2}{|e^{i\omega} - \xi_k|^2}.$$

According to the principles discussed immediately after (5.164), this then leads to the new approximation (5.2) of this chapter.
Now, as discussed in Section 5.8.4, the accuracy of the above approximation and hence that of the new quantification (5.2) depends on the length of the underlying Fourier reconstruction. Furthermore, the preceding argument has shown that the uniqueness of the reproducing kernel provides a means for re-expressing the complicated quantity (5.163) in a simpler form via the geometric principle of recognizing it as a subspace invariant, where the subspace dimension is exactly *twice* that of the model order m which is the reason for the factor of 2 appearing in the new approximation (5.2).

5.8.6 Variance Quantification for General Input Spectral Densities

For the sake of most clearly exposing underlying ideas, the previous few sections have concentrated on the output-error modelling situation and have been deliberately vague about assumptions on the input spectral density $\Phi_u(\omega)$.

However, with these new principles now fixed, this section moves on to a more formal treatment that precisely catalogues assumptions and consequent approximate variance quantifications for both the output-error case just profiled, as well as the more general Box-Jenkins setting. We begin by considering assumptions on the input spectral density.

Assumption Set 5.8.1 *Assumptions on the input excitation. The input sequence $\{u_t\}$ is uniformly bounded (with probability one) as $|u_t| \leq C_u < \infty$ for all t and is 'quasi-stationary' in the sense defined by Ljung such that the associated covariance function $R_u(\tau)$ is absolutely summable and the associated spectral density $\Phi_u(\omega) > 0$ and is Lipschitz continuous of some order $\alpha > 0$.*

□

It will also be necessary to define certain zeros in a manner that is an analogue of (5.109) but relaxed to account for the case of less restrictive input assumptions.

Definition 5.3. *Definition of Zeros $\{\xi_k\}$, $\{\eta_k\}$. Let $H_\dagger(z)$ be a rational and bi-proper function such that $A_\dagger(z)$ defined as*

$$A_\dagger(z) = \frac{A^2(z,\theta_\circ)H(z,\theta_\circ)}{H_\dagger(z)} \tag{5.187}$$

is a polynomial of order less than or equal to $2m = 2m_a$. Use this to define the zeros $\{\xi_k\}$ and $\{\eta_k\}$ according to

$$z^{2m}A_\dagger(z) = \prod_{k=1}^{2m}(z-\xi_k), \quad z^{2m}D(z,\theta_\circ)C(z,\theta_\circ) = \prod_{k=1}^{2m}(z-\eta_k) \tag{5.188}$$

with all these zeros $\{\xi_k\}$ and $\{\eta_k\}$ contained in the open unit disk $\mathbb{D}$ for any m. Finally, use these zeros to define the functions

$$\kappa_m(\omega) \triangleq \sum_{k=1}^{2m}\frac{1-|\xi_k|^2}{|e^{i\omega}-\xi_k|^2}, \qquad \widetilde{\kappa}_m(\omega) \triangleq \sum_{k=1}^{2m}\frac{1-|\eta_k|^2}{|e^{i\omega}-\eta_k|^2}. \tag{5.189}$$

The factorization (5.187) has appeared mysteriously and requires comment. It involves a refinement relative to what has been considered in the immediately preceding sections. There, it was illustrated that, in the output-error case, the reproducing kernel approximation problem is reduced to one of generalized Fourier reconstruction of $\Phi_u(\omega)$, and that hence the accuracy of approximation depends on the smoothness of $\Phi_u(\omega)$.

In the special situation of $\Phi_u(\omega)$ being a constant, then as shown in the previous Section 5.7.2, we can provide exact quantifications. In the case of $\Phi_u(\omega)$ non-constant, then we require the generalized Fourier convergence (5.185) to hold, and for this purpose we allow for a filtered version of Φ_u to be used, which is as smooth as possible in order to deliver maximal accuracy. This is the point of the factorization (5.187). It so happens that $H_\dagger$ specifies the filter applied to $\Phi_u(\omega)$ and hence implies an underlying Fourier reconstruction of

$$\frac{\Phi_u(\omega)}{|H_\dagger(e^{i\omega})|^2}. \tag{5.190}$$

In some cases, by judicious choice of a function $H_\dagger(z)$, the above can be made close to a constant. The consequence of this is that the kernel approximation implied by the ensuing $A_\dagger(z)$ specified by (5.187) is then of maximal accuracy.

This point will be explicitly illustrated later, but under the above assumptions, and using the ideas of the previous sections, the following result is central to addressing the problem of approximately quantifying the reproducing kernel in the general case.

Theorem 5.4. *Provided that input assumption 5.8.1 is satisfied and both of $G(q,\theta_\circ)$ or $H(q,\theta_\circ)$ contain no pole-zero cancellations for any model order m, then in the full Box-Jenkins modelling case, the associated reproducing kernel $\varphi_{4m}(\lambda,\omega)$ satisfies*

$$\lim_{m\to\infty} \varphi_{4m}(\omega,\omega)\cdot K_m^{-1}(\omega) = \Phi_\zeta^{-1}(\omega) \tag{5.191}$$

where, with relevant quantities prescribed by Definition 5.3

$$K_m(\omega) \triangleq \begin{bmatrix} \kappa_m(\omega) & 0 \\ 0 & \widetilde{\kappa}_m(\omega) \end{bmatrix}, \quad \Phi_\zeta(\omega) \triangleq \begin{bmatrix} \Phi_u(\omega) & \Phi_{ue}(\omega) \\ \overline{\Phi_{ue}(\omega)} & \sigma^2 \end{bmatrix}.$$

Furthermore, in the output-error modelling case

$$\lim_{m\to\infty} \varphi_{2m}(\omega,\omega)\cdot \kappa_m^{-1}(\omega) = \Phi_u^{-1}(\omega). \tag{5.192}$$

In order to employ this reproducing kernel quantification, it is also important to emphasize that a crucial aspect of the approach here is the recognition of the need to carefully consider the relationship between the model order m for which a variance error quantification is required and any underlying 'true' system order. Indeed, given the usual complexity of real-world dynamics, any assumption of the existence of a true model order could be quite inappropriate.

In relation to this issue, the work here takes the perspective that, while on the one hand it is reasonable to assume that under-modelling-induced error decreases with increasing model order m, it is also reasonable to assume that the model order of interest has not surpassed any underlying true order, and hence does not imply pole-zero cancellations in the (asymptotic in N) estimated system.

This last premise is considered to be a realistic way of avoiding the supposition of a true model order, while still considering that some sort of model validation procedure, which checks for the appropriateness of the model structure (5.3) (*e.g.* in terms of residual whiteness), and at the very least checks for pole-zero cancellation, is part of an overall estimation and error-quantification process [133, 275].
Incorporating these ideas implies that, unlike the previous sections where exact asymptotic variance quantifications were presented, it is now *not* presumed that the asymptotic (in N) estimation residual $\varepsilon_t(\theta_\circ) = e_t =$ white noise.

Instead, a more general situation is considered as follows.

Assumption Set 5.8.2 *Assumptions on Modelling Error.*

1. *Both $G(z,\theta_\circ)$ and $H(z,\theta_\circ)$ contain no pole-zero cancellation for any model order m;*
2. *It holds that $y_t = G(q)u_t + H(q)e_t$ for some asymptotically stable true system $G(q)$ and such that the bounds (5.193), (5.194) are satisfied;*
3. *It is assumed that $\varepsilon_t(\theta_\circ)$ can be decomposed as*

$$\varepsilon_t(\theta_\circ) = e_t + r_t^m \tag{5.193}$$

where r_t^m is independent of e_t (as in §5.3, $\{e_t\}$ is an i.i.d. zero mean white noise sequence for which $\mathsf{E}\{e_t^8\} < \infty$). Furthermore, it is also assumed that for some $\rho, \beta > 1$ and finite C

$$\frac{1}{N}\sum_{t=1}^{N} \mathsf{E}\{|r_t^m r_{t-\tau}^m|\} \le \frac{C}{m^\beta(1+|\tau|^\rho)}, \quad \frac{1}{N}\sum_{t=1}^{N} \mathsf{E}\{|r_t^m e_{t-\tau}|\} \le \frac{C}{m^\beta(1+|\tau|^\rho)}. \tag{5.194}$$

These requirements correspond to an assumption on the rate at which a finite dimensional model structure $G(q,\theta)$ is able to approximate an underlying true $G(q)$ that generates the observed input-output data via (5.3). More specifically, the bounds (5.194) imply an assumption on the worst-case frequency domain estimation error of

$$\|G(e^{i\omega})G(e^{i\omega},\theta_\circ)\|_\infty, \|H(e^{i\omega}) - H(e^{i\omega},\theta_\circ)\|_\infty = o(1/m^\beta) \quad \text{as}\, m \to \infty. \tag{5.195}$$

With these assumptions in mind, the following results provides the main variance quantification result of this section.

Theorem 5.5. *Under the assumptions sets 5.8.1 and 5.8.2 and in the full Box-Jenkins modelling case considered in this section where*

$$\Delta_{4m}(\omega) = \lim_{N\to\infty} N \cdot Cov\left\{\begin{bmatrix} G(e^{i\omega},\widehat{\theta}_N) \\ H(e^{i\omega},\widehat{\theta}_N)\end{bmatrix}\right\} \tag{5.196}$$

then with associated quantities being specified in Definition 5.3

$$\lim_{m\to\infty} \Delta_{4m}(\omega)K_m^{-1}(\omega) = \Phi_\nu(\omega)\Phi_\zeta^{-1}(\omega). \tag{5.197}$$

As per the previous arguments, this result clearly suggests the approximate quantifications

$$\mathrm{Var}\{G(e^{i\omega},\widehat{\theta}_N)\} \approx \frac{1}{N}\frac{\Phi_\nu(\omega)}{\Phi_u(\omega)}\sum_{k=1}^{2m}\frac{1-|\xi_k|^2}{|e^{i\omega}-\xi_k|^2}, \tag{5.198}$$

$$\mathrm{Var}\{H(e^{i\omega},\widehat{\theta}_N)\} \approx \frac{1}{N}\frac{\Phi_\nu(\omega)}{\sigma^2}\sum_{k=1}^{2m}\frac{1-|\eta_k|^2}{|e^{i\omega}-\eta_k|^2} \tag{5.199}$$

The output-error model structure case, which was developed in the immediately preceding section under slightly less general assumptions, can now be more formally handled as follows.

Corollary 5.1. *Under the assumptions sets 5.8.1 and 5.8.2, and in the output-error modelling case where*

$$\Delta_{2m}(\omega) = \lim_{N\to\infty} N\cdot Cov\left\{G(e^{i\omega},\widehat{\theta}_N)\right\} \tag{5.200}$$

then with associated quantities being specified in Definition 5.3

$$\lim_{m\to\infty}\Delta_{2m}(\omega)\kappa_m^{-1}(\omega) = \frac{\sigma^2}{\Phi_u(\omega)}. \tag{5.201}$$

This clearly suggests that the approximation (5.198) also holds for the output-error case. Furthermore, in the particular situation where $H_\dagger = 1$ is chosen in (5.187), then as $H(z,\theta_\circ) = 1$ in this case, all the zeros of $A_\dagger(z)$ are those of $A^2(z,\theta_\circ)$ and hence appear twice as $\xi_{m+k} = \xi_k$ so that

$$\kappa_m(\omega) = \sum_{k=1}^{2m}\frac{1-|\xi_k|^2}{|e^{i\omega}-\xi_k|^2} = 2\sum_{k=1}^{m}\frac{1-|\xi_k|^2}{|e^{i\omega}-\xi_k|^2} \tag{5.202}$$

which leads to the quantification

$$\mathrm{Var}\left\{G(e^{i\omega},\widehat{\theta}_N)\right\} \approx \frac{2}{N}\frac{\sigma^2}{\Phi_u(\omega)}\sum_{k=1}^{m}\frac{1-|\xi_k|^2}{|e^{i\omega}-\xi_k|^2}. \tag{5.203}$$

In fact, this case of fixed noise model warrants special mention because, coupled with an open loop measurement scenario, it is a situation in which variance error quantification can be achieved without requiring that the model class be able to asymptotically encapsulate the true underlying noise properties.

Corollary 5.2. *Under the same conditions as Theorem 5.5 save for the modifications that*

1. *In the model structure (5.3) the noise model is fixed at $H(q,\theta) = H_*(q)$ which is not necessarily equal to any true underlying one;*
2. *It holds that $y_t = G(q)u_t + H(q)e_t$ for some asymptotically stable true system $G(q)$ such that with $\theta_\circ$ defined by (5.11) and for some $C < \infty$, $\beta > 1$*
$$|G(e^{i\omega}) - G(e^{i\omega}, \theta_\circ)| \le Cm^{-\beta};$$
3. *The cross spectrum $\Phi_{ue}(\omega) = 0$.*

Then

$$\lim_{m\to\infty}\lim_{N\to\infty}\frac{N}{\kappa_m(\omega)}\, Var\left\{G(e^{i\omega},\widehat{\theta}_N)\right\} = 2\left|\frac{H(e^{i\omega})}{H_*(e^{i\omega})}\right|^2\frac{\sigma^2}{\Phi_u(\omega)}. \tag{5.204}$$

This result has application to data pre-filtering situations, which are equivalent to employing a fixed noise model [164].

5.9 Further Simulation Example

As has already been emphasized, the work in this chapter leading to the new quantification (5.2), while being derived via asymptotic analysis, is designed to be as accurate as possible for finite model orders m at which it is likely to be used.

Certainly, the example shown in Figure 5.1 indicates that this is the case, but it is still what might be considered a relatively high-order ($m = 7$) situation. In consideration of this and other aspects surrounding (5.2), (5.198), (5.203), this section presents several additional simulation examples designed to provide comprehensive empirical evidence substantiating the utility of the new approximations (5.198), (5.198), (5.203).

These studies are organized according to the type of system simulated, the colouring of the input spectra, the amount N of observed data, the model structure type, the experimental conditions, and the input-output dynamics that are one of the following.

System1: Low-Order

$$G(q) = \frac{0.1}{(q-0.9)}, \tag{5.205}$$

System2: Mid-Order

$$G(q) = \frac{0.06(q-0.8)(q-0.9)}{(q-0.99)(q-0.7)(q-0.6)}, \tag{5.206}$$

System3: Low-Order Resonant

$$G(q) = \frac{0.0342q + 0.0330}{(q - 0.95e^{i\pi/12})(q - 0.95e^{-i\pi/12})}, \tag{5.207}$$

System4: Mid-Order Resonant

$$G(q) = \frac{0.1176(q + 8.0722)(q + 0.8672)(q + 0.0948)}{(q - 0.75e^{i\pi/3})(q - 0.75e^{-i\pi/3})(q - 0.95e^{i\pi/12})(q - 0.95e^{-i\pi/12})}. \tag{5.208}$$

For each of these systems, two possible input spectra are considered

$$\Phi_u(\omega) = \frac{1}{1.25 - \cos\omega} \qquad \text{and} \qquad \Phi_u(\omega) = 1$$

as well as both long ($N = 10{,}000$) and short ($N = 200$) data lengths.
In order to most directly illustrate quantification accuracy, attention is initially restricted to the simplest case of white Gaussian measurement noise of variance $\sigma^2 = 0.0001$ and the employment of an output-error model of order equal to the true system.
This structure is fitted over 10,000 different input and measurement noise realizations to allow for the computation of the true estimate variability via sample average over these Monte-Carlo simulations, which is then compared to the new expression (5.2), (5.190) as well as the pre-existing one (5.1) in Figures 5.6–5.13 and according to the organization given in Table 5.1.

Table 5.1. Organization of simulation examples

System	Input spectrum			
	Coloured		White	
	$N = 10{,}000$	$N = 200$	$N = 10{,}000$	$N = 200$
1: Low-order	Fig 5.6(a)	Fig 5.6(b)	Fig 5.7(a)	Fig 5.7(b)
2: Mid-order	Fig 5.8(a)	Fig 5.8(b)	Fig 5.9(a)	Fig 5.9(b)
3: Low-order resonant	Fig 5.10(a)	Fig 5.10(b)	Fig 5.11(a)	Fig 5.11(b)
4: High-order resonant	Fig 5.12(a)	Fig 5.12(b)	Fig 5.13(a)	Fig 5.13(b)

In each of these figures, (the estimate of) the true variability is shown as a solid line, the new variance expression (5.2), (5.190) of this chapter is shown as a dashed line, and the pre-existing approximation (5.1) is illustrated via a dash-dot line. The consideration of all these examples reveals some important points.
First, the new approximation (5.2) is clearly quite robust. It provides an informative quantification across the full range of scenarios, even for the case of very low model order $m = 1$ and very low data length $N = 200$ as shown in Figure 5.6(b).

Second, as shown in the cases of white input, the new approximation (5.2) is essentially exact in these cases regardless of model order, save for small errors at very low data lengths. This, of course, is consistent with the results of Section 5.7.

Third, as illustrated in the case of resonant systems, even when the true variability has a quite complicated nature, the new approximation (5.2) is able to provide an informative and accurate quantification.

Finally, as suggested by examination of the dash-dot line representing (5.1) in each of Figures 5.1–5.13, the pre-existing and widely used quantification (5.1) can be unreliable. In these and similar cases, the new quantification (5.2) and its generalizations (5.198), (5.203) are perhaps more useful.

Turning now to the case of Box-Jenkins model structures, consider the situation of the input-output dynamics being that of the low-order system (5.205), with output measurements now subject to noise coloured as

$$H(q) = \frac{1 - 0.1q^{-1}}{1 - 0.9q^{-1}}.$$

In this case, the definition of $A_\dagger(z)$ given in (5.187) becomes (using $G(q, \theta_\circ) = G(q)$, $H(q, \theta_\circ) = H(q)$)

$$A_\dagger(z) = \frac{(1 - 0.9z^{-1})^2(1 - 0.1z^{-1})}{(1 - 0.9z^{-1})} H_\dagger^{-1}(z) = (1 - 0.9z^{-1})(1 - 0.1z^{-1})H_\dagger^{-1}(z) \tag{5.209}$$

If $H_\dagger(z) = 1$ is chosen, this implies $A_\dagger(z)$ is a polynomial of order $2m = 2$ as required, and in the case of white input, it also implies that $\Phi_u/|H_\dagger|^2$ is a constant. In this case, according to the results of Section 5.7 the asymptotic in N variance quantification implied by the ensuing $A_\dagger(z)$ will be exact. Reference to Figure 5.14(a) shows that indeed this is the case, with the approximation (5.198) shown as a dashed line being identical to the Monte-Carlo estimated variability shown as a solid line. Again, for comparison the dash-dot line there is the pre-existing quantification (5.1), which is clearly far less accurate.

However, now consider a modification of this simulation experiment in which the noise colouring is changed to

$$H(q) = \frac{1}{1 - 0.85q^{-1}}.$$

In this case, the factorization (5.209) becomes

$$A_\dagger(z) = \frac{(1 - 0.9z^{-1})^2}{(1 - 0.85z^{-1})} H_\dagger^{-1}(z). \tag{5.210}$$

There are then two clear choices for $H_\dagger(z)$ with the following implications for $A_\dagger(z)$

$$H_\dagger(z) = \frac{1}{1-0.85z^{-1}} \Rightarrow A_\dagger(z) = (1-0.9z^{-1})(1-0.9z^{-1})$$
$$\Rightarrow \xi_1, \xi_2 = 0.9, 0.9 \tag{5.211}$$
$$H_\dagger(z) = \frac{1-0.9z^{-1}}{1-0.85z^{-1}} \Rightarrow A_\dagger(z) = (1-0.9z^{-1})$$
$$\Rightarrow \xi_1, \xi_2 = 0.9, 0 \tag{5.212}$$

However, as discussed earlier, the most appropriate choice, in the sense of being the one that maximizes the variance quantification accuracy, is that which implies that $\Phi_u(\omega)/|H_\dagger(e^{i\omega})|^2$ is as close to flat as possible.
Because in this example, Φ_u is a constant, then the smoothness is clearly greatest for the second factorization choice of $H_\dagger = (1-0.9q^{-1})(1-0.85q^{-1})^{-1}$, and this leads to the approximation shown as the dashed line in Figure 5.14(b). The alternative choice of $H_\dagger = (1-0.85q^{-1})^{-1}$ is shown as the dash-dot line and while still a reasonably accurate quantification, especially considering the very low model orders involved, is inferior to the alternative. The variability $\mathrm{Var}\{H(e^{i\omega}, \widehat{\theta}_N)\}$ versus the quantification (5.203) is shown in Figure 5.15(a) and for which we can see exact agreement even though the true model order is only $m = 1$.
Finally, because this is an open-loop example, then Corollary 5.2 indicates that the approximation (5.203) with fixed noise model set to $H_* = 1$ should quantify the variability of an output-error structure fitted to this coloured noise case. The validity of this is illustrated in Figure 5.15(b) where the upper solid line of Monte-Carlo estimated output-error variability can be compared with (5.203) shown as a dash-dot line. These true and quantified variabilities can be contrasted with the Box-Jenkins estimates for the same data sets shown below these curves. Although the difference between these cases is not great, it is clearly substantial enough to indicate that, despite pre-existing thought, the variability $\mathrm{Var}\{G(e^{i\omega}, \widehat{\theta}_N)\}$ is not invariant to the model structure choice, and this was predicted by the variance quantifications (5.198) and (5.203) since the former involves a $\kappa_m(\omega)$ defined via $[\xi_1, \xi_0] = [0.9, 0.0]$ while the latter involves a $\kappa_m(\omega)$ defined via $[\xi_1, \xi_0] = [0.9, 0.9]$.

5.10 Conclusions and Further Reading

This chapter has concentrated on the problem of quantifying the noise-induced error in a parametric frequency response estimate. The most fundamental result established here was the equivalence between this problem, and that of quantifying the reproducing kernel for a certain function space.
The importance of this was seen to be that under certain assumptions, especially on the input spectral density, this reproducing kernel, and hence the estimate variability, could be exactly quantified in closed form by the use of the explicit orthonormal basis formulations derived in Chapter 2.

However, under less restrictive assumptions on the input, the kernel could only be quantified approximately. Previous work on this topic has also dealt with associated approximations in this situation of general input. There, two approximation steps were introduced. One, in which an underlying partial Fourier sum is recognized and is interpreted as if it is has converged. The second, in which a rank deficiency is handled via a perturbation argument.

This chapter has shown that, first, by recognizing the reproducing kernel inherent in the variance quantification, the rank deficiency problem can be completely circumvented, with no need for a perturbation argument. This removes one approximating step. What is left is still a Fourier reconstruction. However, as shown here, the accuracy in approximating this can also be improved by using a re-parameterization with respect to an orthonormal basis adapted to the function being reconstructed.

The focus of this chapter was to explain these principles, and the ideas underlying them. Necessarily, there is a fair amount of technical argument involved. However, in order to try to keep this to a minimum and expose the main ideas as clearly as possible, most results presented here have not been formally proven.

Readers who want to dig deeper in this area and avail themselves of the proofs could consider [208] for a discussion of exact variance quantification, the papers [216, 217, 224] for approximate variance quantification, the work [221] for a study of the application of the results here to the study of closed loop estimation methods, and finally the work [220] for a rapprochement between the new variance quantifications derived here and the pre-existing ones that did not take model structure into account.

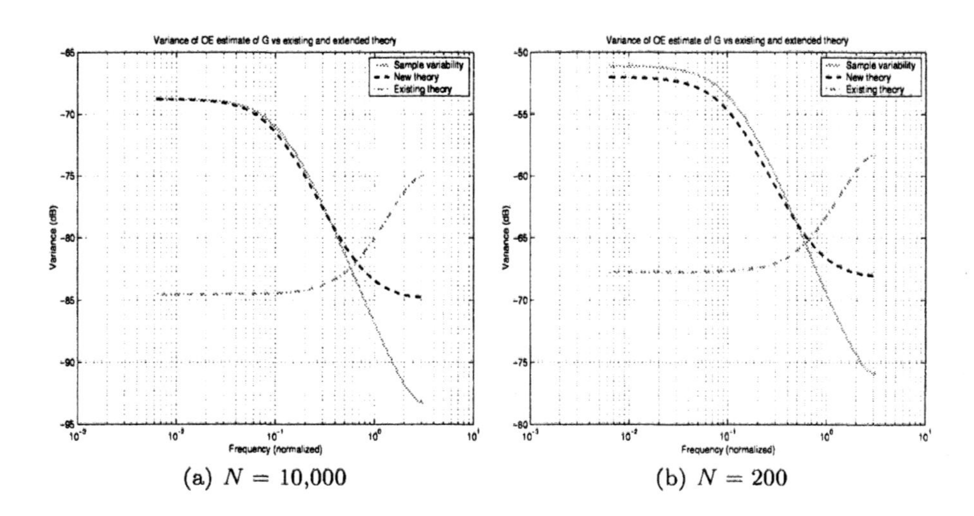

(a) $N = 10{,}000$ (b) $N = 200$

Fig. 5.6. System 1, very low-order, coloured Φ_u. True variability is solid line, new quantification (5.2) is the dashed line, and the existing quantification (5.1) is the dash-dot line.

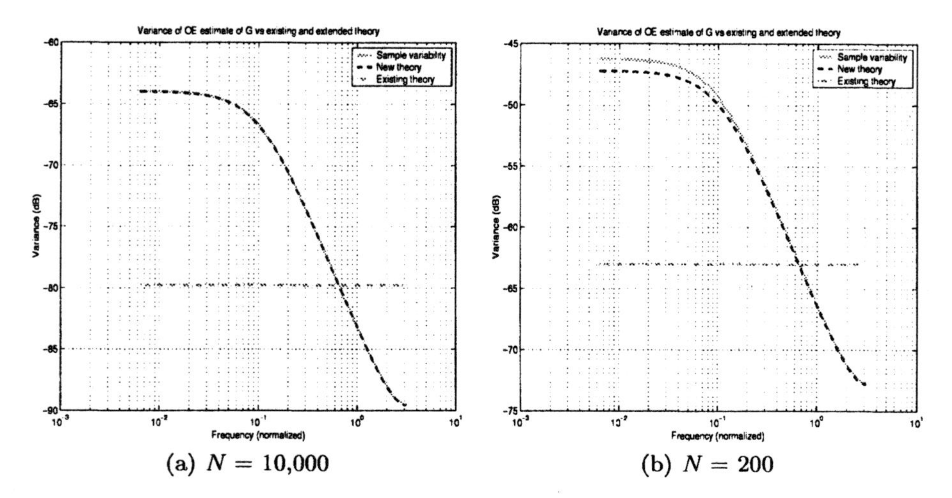

(a) $N = 10{,}000$ (b) $N = 200$

Fig. 5.7. System 1, very low-order, white Φ_u. True variability is solid line, new quantification (5.2) is the dashed line, and the existing quantification (5.1) is the dash-dot line.

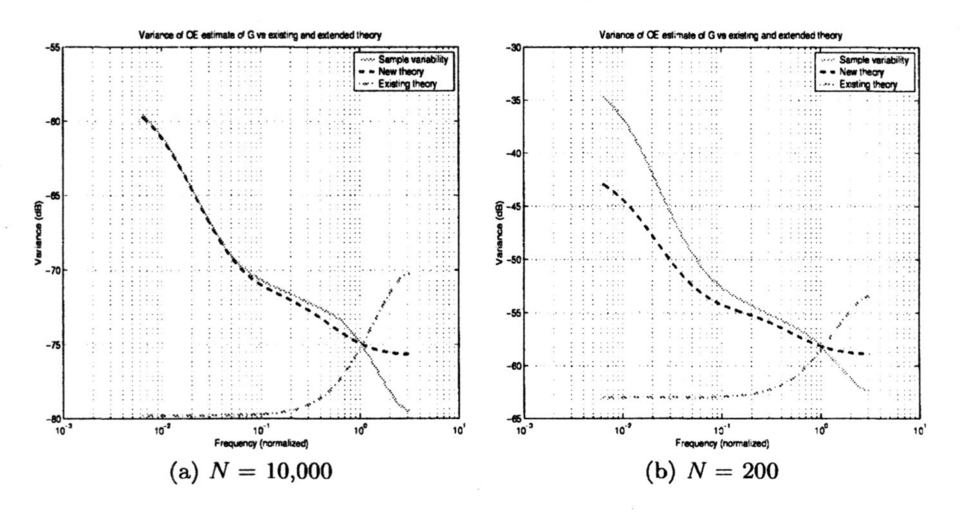

(a) $N = 10{,}000$ (b) $N = 200$

Fig. 5.8. System 2, mid-order, coloured Φ_u. True variability is solid line, new quantification (5.2) is the dashed line, and the existing quantification (5.1) is the dash-dot line.

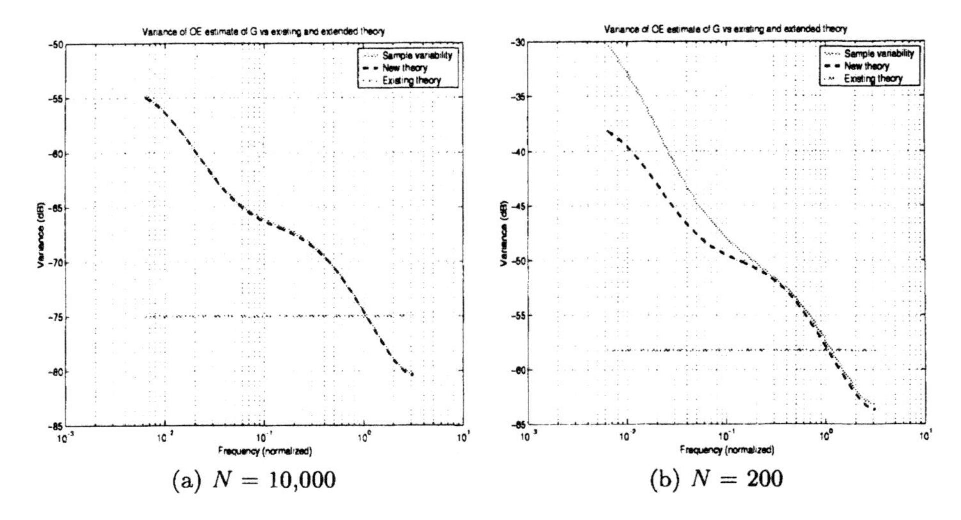

(a) $N = 10{,}000$ (b) $N = 200$

Fig. 5.9. System 2, mid-order, white Φ_u. True variability is solid line, new quantification (5.2) is the dashed line, and the existing quantification (5.1) is the dash-dot line.

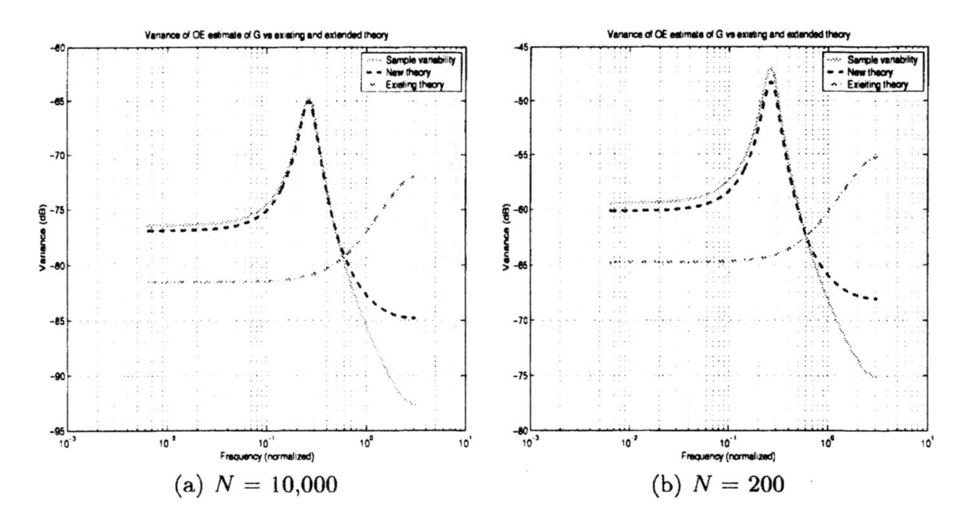

(a) $N = 10{,}000$ (b) $N = 200$

Fig. 5.10. System 3, low-order resonant, coloured Φ_u. True variability is solid line, new quantification (5.2) is the dashed line, and the existing quantification (5.1) is the dash-dot line.

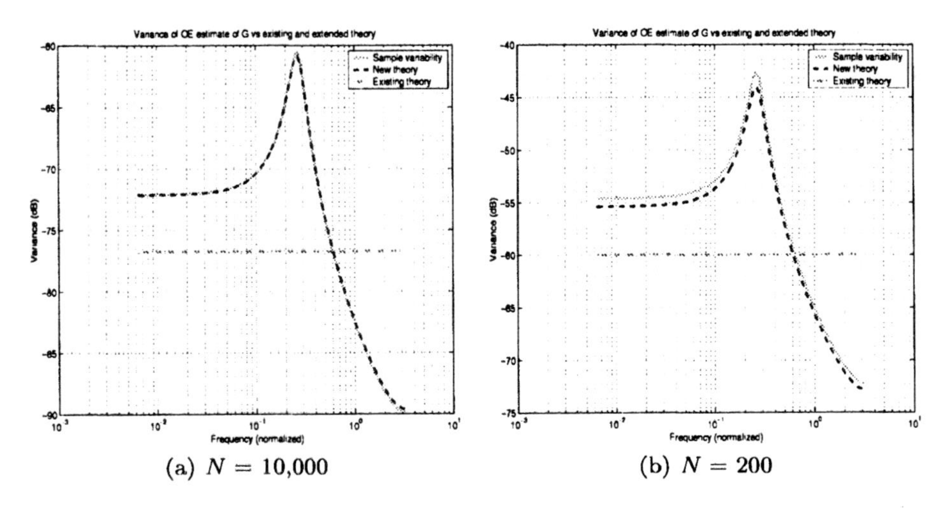

(a) $N = 10{,}000$ (b) $N = 200$

Fig. 5.11. System 3, low-order resonant, white Φ_u. True variability is solid line, new quantification (5.2) is the dashed line, and the existing quantification (5.1) is the dash-dot line.

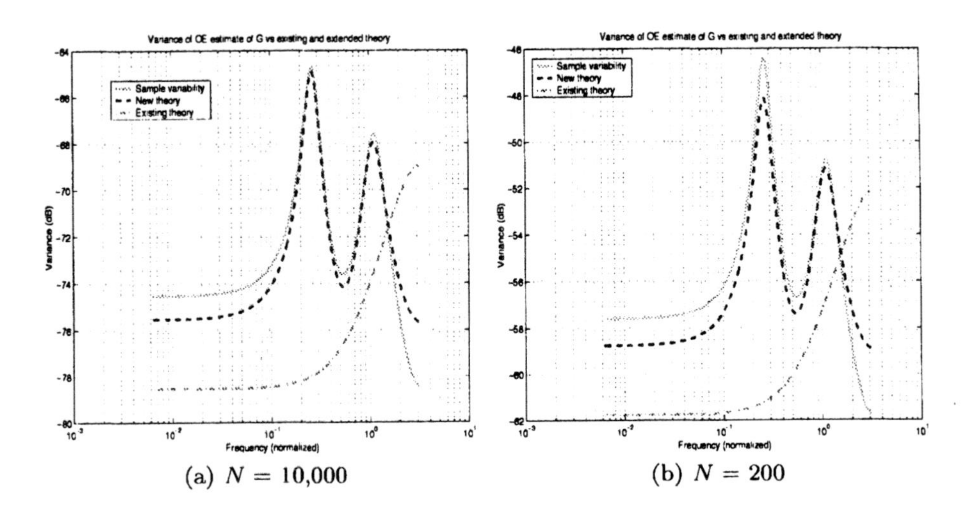

(a) $N = 10{,}000$ (b) $N = 200$

Fig. 5.12. System 4, mid-order resonant, coloured Φ_u. True variability is solid line, new quantification (5.2) is the dashed line, and the existing quantification (5.1) is the dash-dot line.

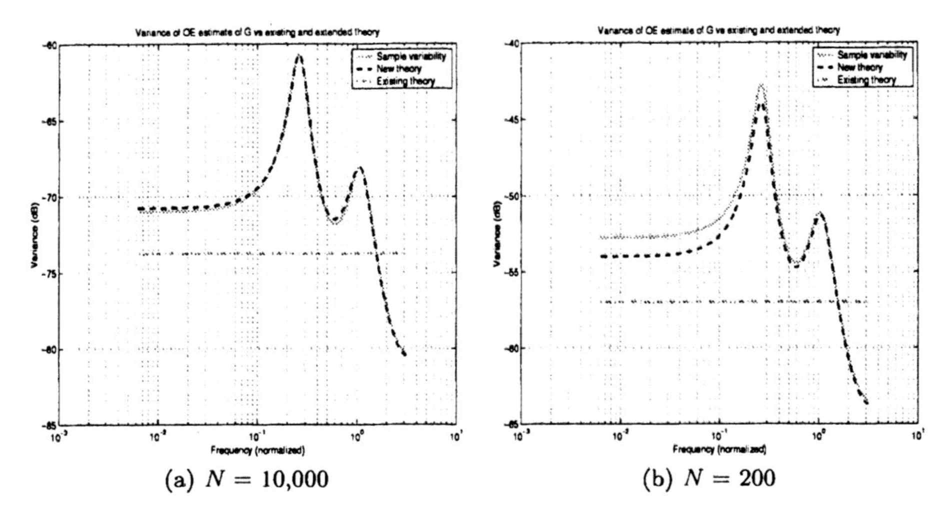

(a) $N = 10{,}000$ (b) $N = 200$

Fig. 5.13. System 4, mid-order resonant, white Φ_u. True variability is solid line, new quantification (5.2) is the dashed line, and the existing quantification (5.1) is the dash-dot line.

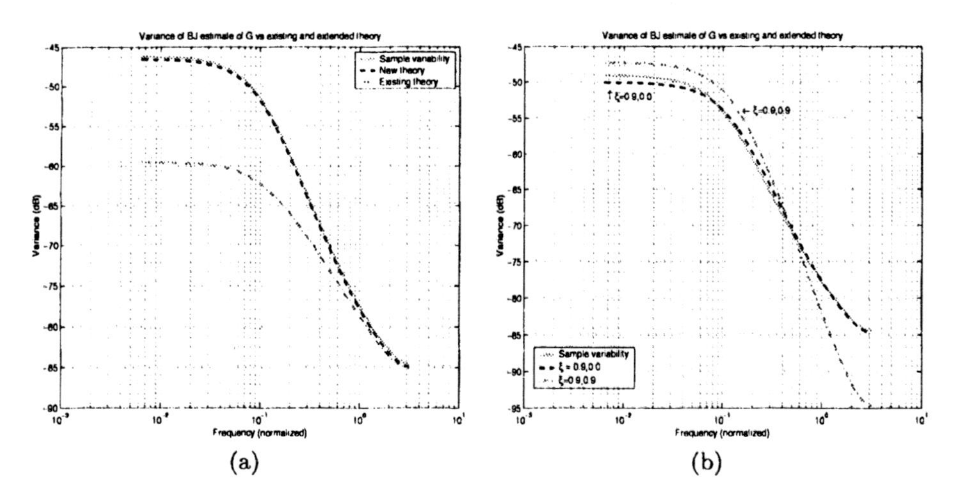

(a) (b)

Fig. 5.14. System 1, low-order, coloured Φ_u. Estimation using Box-Jenkins model structure. True variability is solid line, new quantification (5.198) is the dashed line. (a) Left figure shows, for $H = (1 - 0.1q^{-1})(1 - 0.9q^{-1})^{-1}$, the noise spectral factor $H(q)$ estimate variability, with (5.1) shown as the dash-dot line. (b) Right figure shows, for $H = (1 - 0.85q^{-1})^{-1}$, the dynamics $G(q)$ variability for two choices of $H_\dagger$.

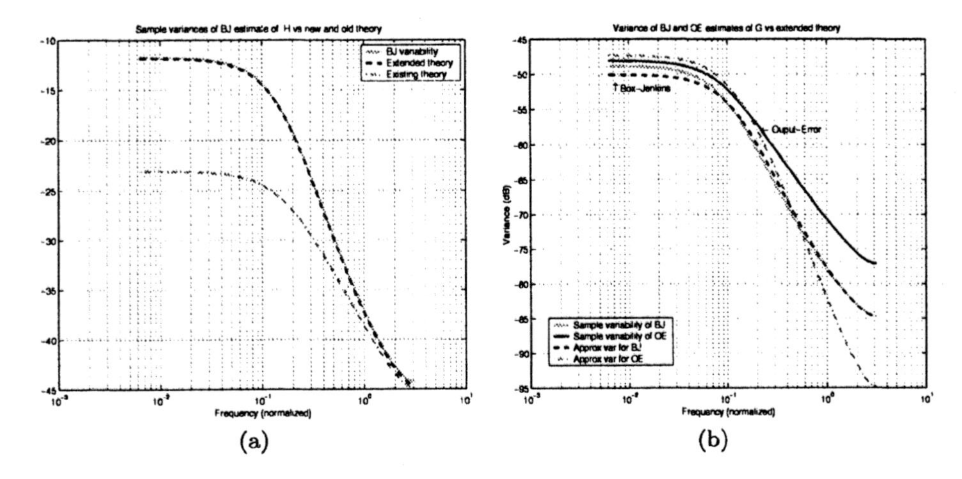

Fig. 5.15. System 1, low-order, coloured Φ_u. $H = (1 - 0.85q^{-1})^{-1}$. (a) Left plot shows noise spectral factor $H(q)$ variability versus new quantification (5.198) (dashed) and existing one (5.1) (dash-dot). (b) Right plot shows dynamics $G(q)$ variability for BJ and OE structures.

6

Numerical Conditioning

Brett Ninness[1] and Håkan Hjalmarsson[2]

[1] University of Newcastle, NSW, Australia
[2] KTH, Stockholm, Sweden

6.1 Introduction

A range of orthonormal basis formulations have been developed in previous chapters. Their utility as a theoretical tool for the quantification of both bias and variance error was also illustrated. The latter was shown to arise often in situations where the original model structure was not formulated in terms of an orthonormal basis.
This chapter now turns to the different issue of examining the practical (as opposed to theoretical) dividend of employing orthonormally parameterized model structures. In particular, this chapter recognizes the following. It is natural to question why, when writing computer code for system identification purposes, the model should be parameterized in an orthonormal form rather than a simpler, but mathematically equivalent 'fixed denominator' form.
In fact, as will be discussed here, a key motivating factor in the employment of these orthonormal forms is that of improved numerical properties; namely, for white input, perfect conditioning of the least-squares normal equations is achieved by design.
However, for the more usual case of coloured input spectrum, it is not clear what the numerical conditioning properties are in relation to simpler and perhaps more natural model structures. This chapter therefore poses the question: what is the benefit of forming model structures that are orthonormal with respect to white spectra, but not for the more common case of coloured spectra?
The answer found here via theoretical and empirical argument is that the orthonormal model structures are, in numerical conditioning terms, particularly *robust* to spectral colouring while simpler more natural forms are particularly fragile in this regard.
Of course, the significance of this improved conditioning is open to question. In response, this chapter concludes with an example of adaptive estimation, where there is a clear dividend of improved conditioning; namely, improved convergence rate.

6.2 Problem Setting and Motivation

To be more specific on these points, this chapter focuses on certain estimation problems already discussed in earlier chapters but which are briefly represented here in the interests of a self-contained presentation.
In particular, this chapter concentrates on the situation in which N point data records of an input sequence $\{u_t\}$ and output sequence $\{y_t\}$ of a linear and time-invariant system are available, and it is assumed that this data is generated as follows

$$y_t = G(q)u_t + \nu_t. \tag{6.1}$$

Here, $G(q)$ is a stable (unknown) transfer function describing the system dynamics that are to be identified by means of the observations $\{u_t\}$, $\{y_t\}$, and the sequence $\{\nu_t\}$ is some sort of possible noise corruption. The input sequence $\{u_t\}$ is assumed to be quasi-stationary in the sense used by Ljung [164] so that the limit

$$R_u(\tau) \triangleq \lim_{N\to\infty} \frac{1}{N} \sum_{t=0}^{N-1} \mathsf{E}\{u_t u_{t+\tau}\}$$

exists such that $\sum_\tau |R_u(\tau)| < \infty$. In this case, $\{u_t\}$ has an associated spectral density

$$\Phi_u(\omega) \triangleq \sum_{t=-\infty}^{\infty} R_u(\tau) e^{-i\omega\tau}$$

and it is assumed in this chapter that $\Phi_u(\omega) > 0$.
The method of estimating the dynamics $G(q)$ that is of interest here is one wherein the following 'fixed denominator' model structure is used:

$$G(q,\beta) = \sum_{k=1}^{p} \beta_k \mathcal{F}_k(q). \tag{6.2}$$

Here, the $\{\beta_k\}$ are real valued coefficients and the transfer functions $\{\mathcal{F}_k(q)\}$ may be chosen in various ways, but in every case the poles of the transfer functions $\{\mathcal{F}_k(q)\}$ are selected from the set $\{\xi_1, \xi_2, \cdots, \xi_p\} \subset \mathbb{D}$ where $\mathbb{D} \triangleq \{z \in \mathbb{C} : |z| < 1\}$ with $\mathbb{C}$ being the field of complex numbers. These fixed poles $\{\xi_k\}$ are chosen by the user to reflect prior knowledge of the nature of $G(q)$. That is, as detailed in the previous chapters, in the interests of improved estimation accuracy, they are chosen as close as possible to where it is believed the true poles lie.

An advantage of this simple model structure is that it is linearly parameterized in $\{\beta_k\}$, so that with $\beta \triangleq [\beta_1, \beta_2, \cdots, \beta_p]^T$ then the least-squares estimate

$$\widehat{\beta} = \arg\min_{\beta\in\mathbb{R}^p} \left\{ \frac{1}{N} \sum_{t=0}^{N-1} (y_t - G(q,\beta)u_t)^2 \right\} \tag{6.3}$$

is easily computed. Specifically, the solution $\widehat{\beta}$ to (6.3) can be written in closed form once the model structure (6.2) is cast in familiar linear regressor form notation as $G(q,\beta)u_t = \psi_t^T\beta$ where

$$\psi_t = \Lambda_p(q)\,u_t, \qquad \Lambda_p(q) \triangleq [\mathcal{F}_1(q), \mathcal{F}_2(q), \cdots, \mathcal{F}_p(q)]^T \tag{6.4}$$

so that (6.3) is solved as

$$\widehat{\beta} = \left(\sum_{t=0}^{N-1} \psi_t\psi_t^T\right)^{-1} \sum_{t=0}^{N-1} \psi_t y_t \tag{6.5}$$

provided that the input is persistently exciting enough for the indicated inverse to exist.
However, the discussion of Chapter 4 suggested that instead of using the model structure (6.2), the so-called orthonormal form of it could be employed. That is, the model structure (6.2) could be re-parameterized as

$$G(q,\theta) = \sum_{k=1}^{p} \theta_k F_k(q) \tag{6.6}$$

where again the coefficients $\{\theta_k\}$ are real valued, and the $\{F_k(q)\}$ are transfer functions such that

$$\mathrm{Sp}\{\mathcal{F}_1, \mathcal{F}_2, \cdots, \mathcal{F}_p\} = \mathrm{Sp}\{F_1, F_2, \cdots, F_p\} \tag{6.7}$$

with the further requirement that the $\{F_k(q)\}$ are also orthonormal:

$$\langle F_n, F_m\rangle = \frac{1}{2\pi i}\oint_{\mathbb{T}} F_n(z)\overline{F_m(z)}\,\frac{dz}{z} = \begin{cases} 1\,; m = n \\ 0\,; m \neq n. \end{cases} \tag{6.8}$$

Here $\mathbb{T} \triangleq \{z \in \mathbb{C} : |z| = 1\}$ is the complex unit circle. A wide range of orthonormal basis constructions were described in Chapter 2, but the current chapter focuses on the particular choice

$$F_n(q) = \begin{cases} \dfrac{\sqrt{1-|\xi_n|^2}}{q-\xi_n}\displaystyle\prod_{k=1}^{n-1}\left(\frac{1-\overline{\xi_k}q}{q-\xi_k}\right) & ; n > 1 \\[2ex] \dfrac{\sqrt{1-|\xi_1|^2}}{q-\xi_1} & ; n = 1. \end{cases} \tag{6.9}$$

In this case, defining in a manner analogous to the previous case

$$\phi_t = \Gamma_p(q)u_t, \qquad \Gamma_p(q) \triangleq [F_1(q), F_2(q), \cdots, F_p(q)]^T \tag{6.10}$$

then the least squares estimate with respect to the model structure (6.6) is given as

$$\widehat{\theta} = \left(\sum_{t=0}^{N-1} \phi_t \phi_t^T\right)^{-1} \sum_{t=0}^{N-1} \phi_t y_t. \tag{6.11}$$

A key point is that because there is a linear relationship $\phi_t = J\psi_t$ for some non-singular J, then $\widehat{\beta} = J^T\widehat{\theta}$ and hence modulo numerical issues, the least-squares frequency response estimate is invariant to the change in model structure between (6.2) and (6.6). Specifically:

$$\begin{aligned}
G(e^{i\omega}, \widehat{\beta}) &= \Lambda_p^T(e^{i\omega})\widehat{\beta} \\
&= \Lambda_p^T(e^{i\omega}) \left(\sum_{t=0}^{N-1} \psi_t \psi_t^T\right)^{-1} \sum_{t=0}^{N-1} \psi_t y_t \\
&= \Lambda_p^T(e^{i\omega}) \left[J^{-1} \left(\sum_{t=0}^{N-1} \phi_t \phi_t^T\right) J^{-T}\right]^{-1} J^{-1} \sum_{t=0}^{N-1} \phi_t y_t \\
&= [J\Lambda_p(e^{i\omega})]^T \left(\sum_{t=0}^{N-1} \phi_t \phi_t^T\right)^{-1} \sum_{t=0}^{N-1} \phi_t y_t \\
&= \Gamma_p^T(e^{i\omega})\widehat{\theta} \\
&= G(e^{i\omega}, \widehat{\theta}).
\end{aligned}$$

Given this exact equivalence of frequency response estimates, it is important to question the motivation for using the structure (6.9) (which is complicated by the precise definition of the orthonormal bases (6.9) or whichever other one is used [25, 132]) in place of some other one such as (6.2). In particular, depending on the choice of the $\{\mathcal{F}_k(q)\}$, the structure (6.2) may be more natural and/or be more straightforward to implement, so it is important to examine the rationale for employing the equivalent ortho-normalised version (6.6).

To elaborate further on this point, it is well-known [103] that the numerical properties of the solution of the normal equations arising in least squares estimation using the model structures (6.2) and (6.6) are governed by the condition numbers $\kappa(R_\psi(N))$ and $\kappa(R_\phi(N))$ of the matrices

$$R_\psi(N) \triangleq \frac{1}{N} \sum_{t=0}^{N-1} \psi_t \psi_t^T, \qquad R_\phi(N) \triangleq \frac{1}{N} \sum_{t=0}^{N-1} \phi_t \phi_t^T$$

where the vectors ψ_t and ϕ_t are defined in (6.4) and (6.10), respectively. However, by the quasi-stationarity assumption and by Parseval's theorem, the following limits exist:

$$R_\psi \triangleq \lim_{N\to\infty} R_\psi(N) = \frac{1}{2\pi} \int_{-\pi}^{\pi} \Lambda_p(e^{i\omega})\Lambda_p^\star(e^{i\omega})\Phi_u(\omega)\, d\omega \tag{6.12}$$

$$R_\phi \triangleq \lim_{N\to\infty} R_\phi(N) = \frac{1}{2\pi}\int_{-\pi}^{\pi} \Gamma_p(e^{i\omega})\Gamma_p^\star(e^{i\omega})\Phi_u(\omega)\,\mathrm{d}\omega, \tag{6.13}$$

so that the numerical properties of least squares estimation using the model structures (6.2) and (6.6) should be closely related to the condition numbers $\kappa(R_\psi)$ and $\kappa(R_\phi)$. These condition number quantities, are defined for an arbitrary non-singular matrix R as [103]

$$\kappa(R) \triangleq \|R\|\,\|R^{-1}\|$$

which is clearly dependent on the matrix norm chosen. Most commonly, the matrix 2-norm is used [103], which for positive definite symmetric R is the largest positive eigenvalue. In this case $\kappa(R)$ is the ratio of largest to smallest eigenvalue of R, and is a measure of the Euclidean norm sensitivity of the solution vector x of the equation $Rx = b$ to errors in the vector b.

Now, for white input $\{u_t\}$, its associated spectrum $\Phi_u(\omega)$ is a constant (say α) so that by orthonormality $R_\phi = \alpha I$ and hence the normal equation associated with the solution (6.11) is perfectly numerically conditioned. However, an obvious question concerns how the condition numbers of R_ψ and R_ϕ compare for the more commonly encountered coloured input case. A key result in this context is that purely by virtue of the orthonormality in the structure (6.6), an upper bound on the conditioning of R_ϕ may be guaranteed for any Φ_u by virtue of the fact that [222, 306] ($\lambda(R)$ denotes the set of eigenvalues of the matrix R)

$$\min_{\omega\in[-\pi,\pi]} \Phi_u(\omega) \le \lambda(R_\phi) \le \max_{\omega\in[-\pi,\pi]} \Phi_u(\omega). \tag{6.14}$$

No such bounds are available for the matrix R_ψ corresponding to the general (non-orthonormal) structure (6.2). This suggests that the numerical conditioning associated with (6.6) might be superior to that of (6.2) across a range of coloured Φ_u, and not just the white Φ_u that the structure (6.6) is designed to be perfectly conditioned for.

However, in consideration of this prospect, it would seem natural to also suspect that even though $R_\phi = I$ is designed to occur for unit variance white input, that $R_\psi = I$ might equally well occur for some particular coloured input. If so, then in this latter scenario the structure (6.6) would actually be inferior to the 'orthonormal basis form' (6.2) in numerical conditioning terms. Therefore, in spite of the guarantee (6.14), it is not clear when and why numerical considerations would lead to the structure (6.6) being preferred over the often-times simpler one (6.2).

The rest of this chapter is devoted to examining these questions. In addressing them, we begin in Section 6.3 by establishing a general framework for studying the question of the existence of a spectrum Φ_u for which perfect numerical

conditioning occurs. Using this framework, Section 6.4 and Section 6.5 establish first by a simple two-dimensional motivating example, and then for the case of arbitrary dimension, the following facts.
First, it may easily be the case that R_ψ is never a perfectly conditioned diagonal matrix for any Φ_u. Second, the manifolds (parameterized by Φ_u) of all possible R_ψ and R_ϕ are not the complete manifold of all possible symmetric $p \times p$ dimensional positive definite matrices. Instead, the respective manifolds of R_ψ and R_ϕ are of much smaller dimension. Therefore, because a perfectly conditioned matrix is, by construction, in the manifold of possible R_ϕ, and because the possible manifolds of R_ψ and R_ϕ are restricted, this provides further evidence that parameterization with respect to an orthonormal basis may provide improved numerical conditioning across a range of possible input spectra.

Further aspects of this conjecture are examined in greater detail in Section 6.6 and Section 6.7 by a strategy of deriving approximations for the eigenvalue locations of R_ϕ. These refine (6.14) in that they are expressed directly in terms of the Φ_u (actually, in terms of its positive real part) and the location of the fixed poles $\{\xi_k\}$ in such a way as to illustrate that the numerical conditioning of R_ϕ is (as is intuitively reasonable) closely related to the smoothness of Φ_u. In Section 6.8, for the specific case of $p = 2$, and for specific examples of $\mathcal{F}_1, \mathcal{F}_2$, a class of Φ_u are derived for which R_ϕ is guaranteed to have smaller condition number than R_ψ. In Section 6.9, this is generalized by analysis that is asymptotic in p, and is such as to establish that for model structures (6.4) with the $\{\mathcal{F}_k(q)\}$ chosen so that essentially a numerator is being estimated and a denominator $D_p(q)$ is being fixed, then this leads to poorer numerical conditioning than if the equivalent orthonormal structure (6.6) is used provided that the variation (across $\omega \in [-\pi, \pi]$) of $\Phi_u(\omega)$ is smaller than that of $\Phi_u(\omega)/|D_p(e^{i\omega})|^2$.

Finally, Section 6.10 will illustrate the importance of improved numerical conditioning in the context of achieving optimal convergence rates for least mean-square (LMS) adaptive filtering.
Note that as previously mentioned, although there are a number of possible alternatives discussed in Chapter 2 for the construction of orthonormal bases that satisfy the span condition (6.7), the particular choice (6.9) will be used here.
The reason for this is that the formulation (6.9) offers an explicit formulation for the orthonormal bases, and this will prove to be essential for the precise characterization of the spectral properties of R_ϕ. Note also, that under the span condition (6.7), all choices of orthonormal bases will lead to matrices R_ϕ that are unitarily congruent to one another, and which therefore possess precisely the same spectral properties. Therefore, any spectral conclusions made relative to the basis (6.9) will in fact apply to any orthonormal basis (with the same poles) that was discussed in Chapter 2.

6.3 Existence of Spectra

This section addresses the question of the existence of a particular coloured Φ_u for which the non-orthonormal model structure (6.2) leads to perfect conditioning ($R_\psi = I$) and would thus make it a superior choice on numerical grounds relative to the 'orthonormal' structure (6.6). This issue is subsumed by that of designing a $\Phi_u(\omega)$ parameterized via real valued coefficients $\{c_k\}$ as

$$\Phi_u(\omega) = \sum_{k=-\infty}^{\infty} c_k e^{i\omega k} \tag{6.15}$$

and so as to achieve an arbitrary symmetric, positive definite R_ψ. In turn, this question may be formulated as the search for the solution set $\{c_k\}$ such that

$$\sum_{k=-\infty}^{\infty} c_k \left[\frac{1}{2\pi i}\oint_{\mathbb{T}} \Lambda_p(z)\Lambda_p^\star(z) z^k \,\frac{\mathrm{d}z}{z}\right] = R_\psi$$

which (on recognizing that because Φ_u is necessarily real valued then $c_k = c_{-k}$) may be more conveniently expressed as the linear algebra problem

$$\Pi \begin{bmatrix} c_0 \\ c_1 \\ c_2 \\ \vdots \end{bmatrix} = \mathrm{vec}\{R_\psi\} \tag{6.16}$$

where the vec$\{\cdot\}$ operator is one which turns a matrix into a vector by stacking its columns on top of one another in a left-to-right sequence and the matrix Π, which will be referred to frequently in the sequel, is defined as

$$\Pi \triangleq \frac{1}{2\pi i}\oint_{\mathbb{T}} [\Lambda_p(z)\otimes I_p]\overline{\Lambda_p(z)}[1, z+z^{-1}, z^2+z^{-2}, \cdots]\,\frac{\mathrm{d}z}{z}. \tag{6.17}$$

Here, $\otimes$ denotes the Kronecker tensor product of matrices defined for an $m\times n$ matrix A and an $\ell \times p$ matrix B to provide the $n\ell \times mp$ matrix $A\otimes B$ as

$$A\otimes B \triangleq \begin{bmatrix} a_{11}B & a_{12}B & \cdots & a_{1n}B \\ a_{21}B & a_{22}B & \cdots & a_{2n}B \\ \vdots & & & \vdots \\ a_{m1}B & a_{m2}B & \cdots & a_{mn}B \end{bmatrix}.$$

The solution of (6.16) must be performed subject to the constraint that the Toeplitz matrix

$$\begin{bmatrix} c_0 & c_1 & c_2 & \cdots \\ c_1 & c_0 & c_1 & \\ c_2 & & \ddots & \\ \vdots & & & \ddots \end{bmatrix}$$

is positive definite, which is a necessary and sufficient condition [250] for $\Phi_u(\omega) > 0$.

Now it might be supposed that because (6.16) is an equation involving $p(p+1)/2$ constraints (defined by $\mathrm{vec}\{R_\psi\}$), but with an infinite number of degrees of freedom in the choice $c_0, c_1, \cdots$, then it should be possible to solve for an arbitrary symmetric positive definite R_ψ.

Perhaps surprisingly, this turns out not to be the case, the reason being that (as established in Theorem 6.1 following) the rank of Π in (6.16),(6.17) is always only p. In fact, therefore, the achievable R_ψ live only in a sub-manifold of the $p(p+1)/2$ dimensional manifold of $p \times p$ symmetric matrices, and this sub-manifold *may not contain a perfectly conditioned matrix.* Furthermore, as can be seen by (6.17), this sub-manifold that the possible R_ψ lie in will be completely determined by the choice of the functions $\mathcal{F}_k(z)$ in the model structure (6.2) and hence also in the definition for $\Lambda_p(z)$ in (6.4). These principles are most clearly exposed by considering some simple two dimensional examples.

6.4 Two-dimensional Example

Consider the simplest case of $p = 2$ wherein there are only 3 constraints inherent in (6.16), and one may as well neglect the third row of $[\Lambda_p(z) \otimes I_p]\overline{\Lambda_p(z)}$ (since it is equal, by symmetry, to the second row) and instead consider

$$\begin{bmatrix} \mathcal{F}_1(z)\mathcal{F}_1(1/z) \\ \mathcal{F}_1(z)\mathcal{F}_2(1/z) \\ \mathcal{F}_2(z)\mathcal{F}_2(1/z) \end{bmatrix} = \begin{bmatrix} \mathcal{F}_1(1/\xi_1)\mathcal{F}_1(z) + (1/z\xi_1)\mathcal{F}_1(1/\xi_1)\mathcal{F}_1(1/z) \\ \mathcal{F}_2(1/\xi_1)\mathcal{F}_1(z) + (1/z\xi_2)\mathcal{F}_1(1/\xi_2)\mathcal{F}_2(1/z) \\ \mathcal{F}_2(1/\xi_2)\mathcal{F}_2(z) + (1/z\xi_2)\mathcal{F}_2(1/\xi_2)\mathcal{F}_2(1/z) \end{bmatrix}. \quad (6.18)$$

Here, in forming the right-hand side of the above equation, it has been assumed that $\mathcal{F}_1(z)$ has a pole at $z = \xi_1$, $\mathcal{F}_2(z)$ has a pole at $z = \xi_2$, that $\mathcal{F}_1(0) \neq 0$, $\mathcal{F}_2(0) \neq 0$ and that $\xi_1, \xi_2 \in \mathbb{R}$. That is, $\mathcal{F}_1(z)$ and $\mathcal{F}_2(z)$ are of the simple form

$$\mathcal{F}_1(z) \triangleq \frac{1}{z - \xi_1}, \quad \mathcal{F}_2(z) \triangleq \frac{1}{z - \xi_2}, \quad \xi_1, \xi_2 \in \mathbb{R}. \quad (6.19)$$

The advantage of this re-parameterization into causal and anti-causal components in (6.18) is that it is then straightforward to calculate Π from (6.17) as

$$\Pi = \begin{bmatrix} \mathcal{F}_1(1/\xi_1)(1/\xi_1) & 2\mathcal{F}_1(1/\xi_1) & 2\mathcal{F}_1(1/\xi_1)\xi_1 & \cdots \\ \mathcal{F}_1(1/\xi_2)(1/\xi_2) & \mathcal{F}_2(1/\xi_1) + \mathcal{F}_1(1/\xi_2) & \mathcal{F}_2(1/\xi_1)\xi_1 + \mathcal{F}_1(1/\xi_2)\xi_2 & \cdots \\ \mathcal{F}_2(1/\xi_2)(1/\xi_2) & 2\mathcal{F}_2(1/\xi_2) & 2\mathcal{F}_2(1/\xi_2)\xi_2 & \cdots \end{bmatrix}. \quad (6.20)$$

Given this formulation, it is then clear that

$$\left[\frac{\mathcal{F}_2(1/\xi_1)}{2\mathcal{F}_1(1/\xi_1)}, -1, \frac{\mathcal{F}_1(1/\xi_2)}{2\mathcal{F}_2(1/\xi_2)},\right] \Pi = [0, 0, 0, \cdots] \quad (6.21)$$

provided that

$$\mathfrak{F}_1(1/\xi_2)\xi_1 = \mathfrak{F}_2(1/\xi_1)\xi_2 \tag{6.22}$$

which is certainly true for the first-order $\mathfrak{F}_1(z), \mathfrak{F}_2(z)$ given in (6.19). Therefore, Π is of row (and hence column) rank no more than two. Therefore, regardless of the choice of Φ_u, it is only possible to manipulate (via change of Φ_u) the corresponding R_ψ in a two-dimensional sub-manifold of the full three-dimensional manifold of symmetric two-by-two matrices.

Furthermore, the identity matrix is not part of the two-dimensional sub-manifold, because if it were to lie in the subspace spanned by the columns of Π, it would have to be orthogonal to the normal vector specifying the orientation of this subspace (the left-hand row vector in (6.21)). But it isn't, as

$$\left[\frac{\mathfrak{F}_2(1/\xi_1)}{2\mathfrak{F}_1(1/\xi_1)}, -1, \frac{\mathfrak{F}_1(1/\xi_2)}{2\mathfrak{F}_2(1/\xi_2)}\right] \begin{bmatrix} 1 \\ 0 \\ 1 \end{bmatrix} \neq 0$$

provided $\mathfrak{F}_1, \mathfrak{F}_2$ are of the form shown in (6.19). In fact, by the same argument, no diagonal matrix with positive valued entries is part of the manifold of achievable covariance matrices.

Therefore, even though Φ_u can be viewed as an infinite-dimensional quantity, its effect on R_ψ is not powerful enough to achieve an arbitrary positive definite symmetric matrix. In particular, there is no Φ_u for which the simple and natural fixed denominator basis (6.19) is perfectly conditioned.

However, if instead of (6.19) the alternative simple and natural choice

$$\mathfrak{F}_1(z) \triangleq \frac{1}{(z-\xi_1)(z-\xi_2)}, \quad \mathfrak{F}_2(z) \triangleq \frac{z}{(z-\xi_1)(z-\xi_2)} \tag{6.23}$$

for the fixed denominator basis functions are made, then again straightforward (but tedious) calculation provides that Π can be written as

$$C \begin{bmatrix} \frac{\xi_1}{1-\xi_1^2} - \frac{\xi_2}{1-\xi_2^2} & \frac{\xi_1^2}{1-\xi_1^2} - \frac{\xi_2^2}{1-\xi_2^2} & \frac{\xi_1^3}{1-\xi_1^2} - \frac{\xi_2^3}{1-\xi_2^2} & \cdots \\ \frac{1}{2}\left(\frac{1+\xi_1^2}{1-\xi_1^2}\right) - \frac{1}{2}\left(\frac{1+\xi_2^2}{1-\xi_2^2}\right) & \left(\frac{1+\xi_1^2}{1-\xi_1^2}\right)\xi_1 - \left(\frac{1+\xi_2^2}{1-\xi_2^2}\right)\xi_2 & \left(\frac{1+\xi_1^2}{1-\xi_1^2}\right)\xi_1^2 - \left(\frac{1+\xi_2^2}{1-\xi_2^2}\right)\xi_2^2 & \cdots \\ \frac{\xi_1}{1-\xi_1^2} - \frac{\xi_2}{1-\xi_2^2} & \frac{\xi_1^2}{1-\xi_1^2} - \frac{\xi_2^2}{1-\xi_2^2} & \frac{\xi_1^3}{1-\xi_1^2} - \frac{\xi_2^3}{1-\xi_2^2} & \cdots \end{bmatrix}$$

where $C \triangleq (\xi_1 - \xi_2)^{-1}(1 - \xi_1\xi_2)^{-1}$ so that again Π is only of rank two, this time since

$$[1, 0, -1]\,\Pi = [0, 0, 0, \cdots].$$

However, the important difference in this case is that since (as shown above) the vector $[1, 0, -1]^T$ is orthogonal to the space spanned by the columns of Π, and as $[1, 0, 1]^T$ is also orthogonal to this vector, then the identity matrix does lie in the manifold of possible R_ψ that can be generated by the manipulation of Φ_u.

6.5 Higher Dimensions

Given these motivating arguments specific to a two-dimensional case, it is of interest to consider the case of arbitrary dimension. As the algebra considered in the previous section illustrated, such a study will become very tedious as the dimension is increased. To circumvent this difficulty, the key idea of this section is to replace the study of the rank of Π associated with an arbitrary basis $\{\mathcal{F}_n(q)\}$ (such as those in (6.19) or (6.23)) by its rank with respect to the orthonormal basis $\{F_n(q)\}$ specified in (6.9). Fundamental to this strategy is that via the span equivalence condition (6.7) the rank is invariant to the change of basis, so that the most tractable one may as well be employed. The suitability of $\{F_n\}$ in this context is embodied in the following lemma.

Lemma 6.1. *For $\{F_n(z)\}$ defined by (6.9), the inner product defined by (6.8) and assuming all the $\{\xi_k\}$ are distinct*

$$\langle F_m(z), F_n(z)z^k\rangle = \begin{cases} \xi_n^k & ; m = n, k \geq 0, \\ \overline{\xi}_n^{|k|} & ; m = n, k < 0, \\ \sum_{\ell=m}^{n} A_{m,n}^{\ell} \overline{\xi}_\ell^{-k} & ; n > m, k < 0, \\ 0 & ; n > m, k \leq 0 \end{cases}$$

where

$$A_{m,n}^{\ell} \triangleq \frac{\sqrt{(1-|\xi_m|^2)(1-|\xi_n|^2)}(1-|\xi_\ell|^2)}{(1-\xi_m\overline{\xi}_\ell)(1-\xi_n\overline{\xi}_\ell)} \prod_{\substack{k=m \\ k\neq\ell}}^{n} \left(\frac{1-\xi_k\overline{\xi}_\ell}{\overline{\xi}_\ell - \xi_k}\right) \tag{6.24}$$

and

$$\sum_{\ell=m}^{n} A_{m,n}^{\ell} = 0.$$

Proof. Suppose that $m = n$. Then using the formulation (6.9) and in the case of $k \geq 0$

$$\begin{aligned} \langle F_m(z), F_n(z)z^k\rangle &= \frac{1}{2\pi i}\oint_{\mathbb{T}} \frac{(1-|\xi_n|^2)z}{(1-\xi_n z)(z-\overline{\xi}_n)} z^{-k}\,\frac{\mathrm{d}z}{z} \\ &= \frac{1}{2\pi i}\oint_{\mathbb{T}} \frac{(1-|\xi_n|^2)z}{(z-\xi_n)(1-\overline{\xi}_n z)} z^{k}\,\frac{\mathrm{d}z}{z} \\ &= \xi_n^k \end{aligned}$$

where the change of variable $z \mapsto 1/z$ was employed in progressing to the last line.

Similarly, for $k < 0$ the result

$$\langle F_m(z), F_n(z)z^k\rangle = \overline{\xi_n}^{|k|}$$

will clearly emerge. Now suppose (without loss of generality by symmetry) that $n > m$. In this case, for $k \leq 0$

$$\langle F_m(z), F_n(z)z^k\rangle = \frac{1}{2\pi i}\oint_{\mathbb{T}} \frac{\sqrt{(1-|\xi_m|^2)(1-|\xi_n|^2)}}{(z-\xi_m)(1-\overline{\xi}_n z)} \prod_{\ell=m}^{n-1}\left(\frac{z-\xi_\ell}{1-\overline{\xi}_\ell z}\right) z^{|k|}\,\mathrm{d}z = 0$$

with the result following by using Cauchy's integral formula after recognizing that the integrand is analytic on the interior of $\mathbb{T}$. Now suppose that $k \geq 0$. Then again employing the change of variable $z \mapsto 1/z$ and Cauchy's residue theorem

$$\begin{aligned}\langle F_m(z), F_n(z)z^k\rangle &= \frac{1}{2\pi i}\oint_{\mathbb{T}} \frac{\sqrt{(1-|\xi_m|^2)(1-|\xi_n|^2)}}{(1-\xi_m z)(1-\xi_n z)} \prod_{\ell=m}^{n}\left(\frac{1-\xi_\ell z}{z-\overline{\xi}_\ell}\right) z^k\,\mathrm{d}z \\ &= \sum_{\ell=m}^{n} A^\ell_{m,n}\overline{\xi}_\ell^k\end{aligned}$$

where under the assumption that the $\{\xi_k\}$ are distinct, the $\{A^\ell_{m,n}\}$ terms are as given in (6.24). Finally, by setting $k = 1$ and using the orthonormality of the $\{F_n\}$ and Cauchy's residue theorem again:

$$0 = \langle F_m, F_n\rangle = \sum_{\ell=m}^{n} A^\ell_{m,n}.$$

■

This lemma is the key to providing a more important result in Theorem 6.1 on the fundamental flexibility of manipulating R_ϕ or R_ψ by changing Φ_u. However, in order to develop this most clearly, it is expedient to split Φ_u into 'causal' and 'anti-causal' components as

$$\Phi_u(\omega) = \varphi(e^{i\omega}) + \varphi(e^{-i\omega}) \tag{6.25}$$

where $\varphi(z)$ is known as the 'positive real' part of Φ_u and is given by the so-called Herglotz-Riesz transform [250] as

$$\varphi(z) = \frac{c_0}{2} + \sum_{k=1}^{\infty} c_k z^k = \frac{1}{4\pi}\int_{-\pi}^{\pi}\left(\frac{1+ze^{i\omega}}{1-ze^{i\omega}}\right)\Phi_u(\omega)\,\mathrm{d}\omega. \tag{6.26}$$

With this definition in hand, the following lemma is available which builds on the previous one.

Lemma 6.2. *The matrix R_ϕ defined via (6.13), (6.9) and (6.10) has entries given by*

$$[R_\phi]_{m,n} = \begin{cases} \varphi(\xi_n) + \varphi(\overline{\xi_n}) & ; n = m, \\ \sum_{\ell=m}^{n} A_{m,n}^{\ell} \varphi(\overline{\xi_\ell}) & ; n > m \end{cases}$$

where $A_{m,n}^{\ell}$ is defined in (6.24) and it is understood that the array indexing of R_ϕ begins at $m, n = 1$.

Proof. By the formulation (6.13)

$$[R_\phi]_{m,n} = \langle F_m, \mathcal{B}_n \Phi_u \rangle = c_0 \langle F_m, F_n \rangle + \sum_{k=1}^{\infty} c_k \langle F_m, F_n z^k \rangle + c_k \langle F_m, F_n z^{-k} \rangle.$$

Therefore, if $n = m$, then by Lemma 6.1

$$[R_\phi]_{n,n} = c_0 + \sum_{k=1}^{\infty} c_k (\xi_n^k + \overline{\xi}_n^k) = \varphi(\xi_n) + \varphi(\overline{\xi}_n).$$

On the other hand, if $n > m$ and again using Lemma 6.1

$$[R_\phi]_{m,n} = \sum_{k=1}^{\infty} c_k \sum_{\ell=m}^{n} A_{m,n}^{\ell} \overline{\xi}_\ell^k = \sum_{\ell=m}^{n} A_{m,n}^{\ell} \left[\varphi(\overline{\xi}_\ell) - \frac{c_0}{2} \right] = \sum_{\ell=m}^{n} A_{m,n}^{\ell} \varphi(\overline{\xi}_\ell)$$

where in progressing to the last line the fact that

$$\sum_{\ell=m}^{n} A_{m,n}^{\ell} = 0$$

has been used.

■

Although this lemma will be used later for further developments, its main purpose here is to settle the question raised earlier in Sections 6.3 and 6.4 as to just how much flexibility there is in the assignment of R_ϕ, R_ψ by manipulation of the spectral density Φ_u.

Theorem 6.1. *With Π defined as in (6.17), and for all bases that maintain the same span as in condition (6.7), then the rank of Π is given as*

$$\operatorname{Rank} \Pi = p.$$

Proof. The main idea of the proof is to recognize that the rank of Π defined in (6.17) is invariant to a change of the basis function $\{\mathcal{F}_k\}$ making up Λ_p involved in the definition of Π, and itself defined in (6.4). This statement

must be made subject to the proviso that in making the change of basis, the underlying space being spanned remains the same, which is condition (6.7). This is because under this assumption, and denoting the two matrices resulting from two different bases as R_ψ, and R'_ψ, then a non-singular $p \times p$ matrix J will exist such that $R_\psi = JR'_\psi J^T$. Because the rank of Π is the number of degrees of freedom in the choice of the components of R_ψ by manipulation of the $\{c_k\}$ parameterizing Φ_u via (6.15), then provided J is non-singular, these degrees of freedom are invariant to congruence transformations by J.
With this idea in hand, the proof proceeds by electing to make the span-preserving change of basis

$$\{\mathcal{F}_1, \mathcal{F}_2, \cdots, \mathcal{F}_p\} \mapsto \{F_1, F_2, \cdots, F_p\}$$

with the $\{F_n\}$ being as defined in (6.9). In this case, the rank of Π is the number of effective degrees of freedom in the formation of the elements of R_ϕ by means of the choice of the $\{c_k\}$. But this is the same as the effective degrees of freedom in forming R_ϕ by the choice of $\varphi(z)$, and Lemma 6.2 makes it clear that because all the terms in the $p \times p$ matrix R_ϕ are linear combinations of $\{\varphi(\xi_1), \cdots, \varphi(\xi_p)\}$, then in fact there are only p degrees of freedom in the formation of R_ϕ by the choice of Φ_u.

■

This theorem exposes the key feature imbuing orthonormal parameterizations with numerical robustness beyond the white input case. Specifically, for white input, $R_\phi = I$ is perfectly numerically conditioned, while for this same white input $R_\psi \triangleq \Sigma \neq I$, which has inferior conditioning. As Φ_u is changed from the white case, both R_ϕ and R_ψ will change, but *but only in p-dimensional sub-manifolds.*
This feature of highly-restricted mobility raises the possibility that because (by construction) I is in the manifold of possible R_ϕ, but may not (as the previous section illustrated) be in the manifold of possible R_ψ, then the orthonormal model structure (6.9) may impart a numerical robustness to the associated normal equations across a range of coloured Φ_u. Examining this issue consumes the remainder of the chapter which is motivated, as previously, by a simple two-dimensional example.

6.6 Robustness in Two-dimensional Case

To examine further the issue of numerical conditioning being preserved robustly across a range of non-white input spectra, it is again expedient to return to the simple 2×2 case for illustrative purposes. In conjunction with this, assume that the simple fixed denominator basis (6.19) is again under consideration, and which has associated Π matrix given by (6.20).
It has just been established that the space of possible R_ψ depend on the column range-space of Π, and that this latter space is two-dimensional. In

fact, if Π is restricted to have only three columns, then it is straightforward to verify from (6.20) that

$$\Pi \begin{bmatrix} 2\xi_1\xi_2 \\ -(\xi_1+\xi_2) \\ 1 \end{bmatrix} = \begin{bmatrix} 0 \\ 0 \\ 0 \end{bmatrix}$$

provided again that (6.22) holds. In this case, the first two columns of Π in (6.20) completely determine the whole column range space of Π. Therefore, denoting by Σ the matrix R_ψ for white input ($\Phi_u = 1$), then by (6.20)

$$\Sigma \triangleq \begin{bmatrix} \mathcal{F}_1(1/\xi_1)(1/\xi_1) & \mathcal{F}_1(1/\xi_2)(1/\xi_2) \\ \mathcal{F}_1(1/\xi_2)(1/\xi_2) & \mathcal{F}_2(1/\xi_2)(1/\xi_2) \end{bmatrix} = \begin{bmatrix} \frac{1}{1-\xi_1^2} & \frac{1}{1-\xi_1\xi_2} \\ \frac{1}{1-\xi_1\xi_2} & \frac{1}{1-\xi_2^2} \end{bmatrix} \tag{6.27}$$

which means that all possible R_ψ are expressible as a perturbation away from Σ as

$$R_\psi = (\alpha_1 + (\xi_1+\xi_2)\alpha_2)\Sigma + \alpha_2(\xi_1-\xi_2)\begin{bmatrix} \mathcal{F}_1(1/\xi_1)(1/\xi_1) & 0 \\ 0 & -\mathcal{F}_2(1/\xi_2)(1/\xi_2) \end{bmatrix}. \tag{6.28}$$

Here, the choice of $\alpha_1, \alpha_2 \in \mathbb{R}$ embody the two-degrees of freedom in the manifold of possible R_ψ.

Using the same ideas, but instead employing the orthonormal basis (6.9), then using Lemma 6.1, it is straightforward to see using the reasoning just employed that all possible R_ϕ can be interpreted as a perturbation from the identity matrix

$$R_\phi = (\beta_1 + (\xi_1+\xi_2)\beta_2)I + \beta_2(\xi_1-\xi_2)\begin{bmatrix} 1 & K \\ K & -1 \end{bmatrix} \tag{6.29}$$

where

$$K \triangleq A^1_{1,2} = \frac{\sqrt{(1-\xi_1^2)(1-\xi_2^2)}}{\xi_1-\xi_2}. \tag{6.30}$$

Again, the choice of the real variables $\beta_1, \beta_2 \in \mathbb{R}$ provides the two degrees of freedom in the assignment of R_ϕ. Therefore, because by (6.29) the matrix R_ϕ starts, for white input, at a perfectly conditioned matrix and then (as Φ_u becomes coloured) moves in a sub-manifold of 2×2 symmetric matrices, while at the same time R_ψ starts at the imperfectly conditioned matrix Σ and also moves only in a sub-manifold, which by the argument in Section 6.4 does not contain a perfectly conditioned matrix, it seems reasonable to suspect that the matrix R_ϕ might be better conditioned than R_ψ for *any* coloured input. In order to investigate this further, it is necessary to be more precise as to how the eigenvalues of R_ϕ and R_ψ depend on the choice of Φ_u, and for this purpose the following result will prove useful.

Theorem 6.2. *Let $A = [a_{k\ell}]$ be an $n \times n$ real symmetric matrix. For a fixed index k let α and β be positive numbers satisfying*

$$\alpha\beta \geq \sum_{\substack{\ell=1 \\ \ell \neq k}}^{n} |a_{k\ell}|^2$$

Then the interval $[a_{kk} - \alpha, a_{kk} + \beta]$ contains at least one eigenvalue of A.

Proof. See [22].

■

This result is employed in this chapter instead of the similar and more widely known Geršgorin disc theorem [136], because the latter can only assert the existence of bounds lying in a region if that region is disjoint from certain others. Theorem 6.2 clearly avoids this restriction.
Application of Theorem 6.2 then allows the two eigenvalues λ_1 and λ_2 of R_ψ given by (6.28) and R_ϕ given by (6.29) to be bounded as

$$\lambda(R_\psi) \in \left(\frac{(\alpha_1 + 2\alpha_2\xi_1)}{1 - \xi_1^2} \pm \Delta_\psi\right) \cup \left(\frac{(\alpha_1 + 2\alpha_2\xi_2)}{1 - \xi_2^2} \pm \Delta_\psi\right), \tag{6.31}$$

$$\Delta_\psi \triangleq \frac{\alpha_1 + \alpha_2(\xi_1 + \xi_2)}{1 - \xi_1\xi_2}$$

and

$$\lambda(R_\phi) \in ((\beta_1 + 2\xi_1\beta_2) \pm \Delta_\phi) \cup ((\beta_1 + 2\xi_2\beta_2) \pm \Delta_\phi), \tag{6.32}$$

$$\Delta_\phi \triangleq \beta_2\sqrt{(1 - \xi_1^2)(1 - \xi_2^2)}$$

where the notation $(x \pm y)$ is meant to denote the open interval $(x - y, x + y)$. These bounds illustrate an inherent numerical robustness of the orthonormal form for *any* input spectral density. Specifically, (6.32) shows the eigenvalues of R_ϕ to be in regions centred at $\beta_1 + 2\xi_1\beta_2$ and $\beta_1 + 2\xi_2\beta_2$ and bounded from these centres by a distance Δ_ϕ. But these centres are of the same form as those pertaining to $\lambda(R_\psi)$ save that the centres pertaining to $\lambda(R_\psi)$ are divided by $1 - \xi_1^2$ and $1 - \xi_2^2$. This latter feature will, particularly if one of ξ_1 or ξ_2 are near 1 and the other isn't, tend to make the centres of the eigenvalue bound regions very different.
Furthermore, the bound $\Delta_\phi = \beta_2\sqrt{(1 - \xi_1^2)(1 - \xi_2^2)}$ is forced to be small (regardless of Φ_u) if any one of the poles ξ_1 or ξ_2 to be near 1, while the bound Δ_ψ cannot be forced (by choice of ξ_1 and ξ_2) to be small in a way that is insensitive to the Φ_u. Therefore, the numerical conditioning of R_ϕ shows an inherent robustness to the particular Φ_u defining it.

6.7 Higher Dimensions Again

Having argued for the specific $p = 2$ dimensional case that the superior numerical conditioning advantage of the orthonormal model structure (6.6) is a property that is robust across a range of coloured spectral densities Φ_u, this section extends the argument to arbitrary dimension. Central to this is the following result.

Theorem 6.3. *The eigenvalues* $\{\lambda_1, \lambda_2, \cdots, \lambda_p\}$ *of* R_ϕ *are contained in regions* $\Delta_1, \Delta_2, \cdots, \Delta_p$ *defined by*

$$\Delta_m \triangleq \{x \in \mathbb{R} : |x - 2\mathrm{Re}\,\varphi(\xi_m)| \leq \alpha_m\}$$

where

$$\alpha_m^2 \triangleq \sum_{\substack{n=1\\n\neq m}}^{p} \left(\sum_{\ell=m}^{n-1} |A_{m,n}^\ell| |\varphi(\overline{\xi}_\ell) - \varphi(\overline{\xi}_{\ell+1})| \right)^2 .$$

Proof. By Theorem 6.2 the regions $\{\Delta_m\}$ are provided as being

$$\Delta_m = \{x \in \mathbb{R} : |x - [R_\phi]_{m,m}| \leq \alpha_m\}$$

where

$$\alpha_m^2 \geq \sum_{\substack{n=1\\n\neq m}}^{p} |[R_\phi]_{m,n}|^2 .$$

But by Lemma 6.2

$$[R_\phi]_{m,m} = \varphi(\xi_m) + \varphi(\overline{\xi}_m) = 2\mathrm{Re}\,\varphi(\xi_m)$$

and also by the same lemma

$$\begin{aligned}
\sum_{\substack{n=1\\n\neq m}}^{p} |[R_\phi]_{m,n}|^2 &= \sum_{\substack{n=1\\n\neq m}}^{p} \left| \sum_{\ell=m}^{n} A_{m,n}^\ell \varphi(\overline{\xi}_\ell) \right|^2 \\
&\leq \sum_{\substack{n=1\\n\neq m}}^{p} \left(\sum_{\ell=m}^{n-1} |A_{m,n}^\ell| |\varphi(\overline{\xi}_\ell) - \varphi(\overline{\xi}_{\ell+1})| \right)^2
\end{aligned}$$

where in progressing to the last line, the fact that

$$\sum_{\ell=m}^{n} A_{m,n}^\ell = 0$$

was employed.

■

Note that this theorem provides a tight characterization in the sense that for white input, $\varphi(\xi_k) = c_0/2$ a constant, in which case the theorem provides the eigenvalues as being all at $\lambda_k = c_0$ with tolerance $\alpha_k = 0$.

However, more generally the theorem provides further indication of the general robustness of the condition number of R_ϕ. Specifically, if φ is smooth and the pole locations $\{\xi_k\}$ are chosen to be relatively 'clustered' around a common point, then this will imply that the terms $|\varphi(\overline{\xi}_i) - \varphi(\overline{\xi}_{i+1})|$ will be small. Hence via Theorem 6.3, the bounds α_m on the eigenvalue locations $\{2\mathrm{Re}\,\varphi(\xi_m)\}$ will be tight, and so the true eigenvalues should be very near to the locations $\{2\mathrm{Re}\varphi(\xi_m)\}$, which again if $\varphi(z)$ is smooth will be relatively tightly constrained.

6.8 Conditions for Numerical Superiority

The most desirable result that a study such as this could produce would be one that precisely formulated the necessary and sufficient conditions on Φ_u and $\{\xi_1, \cdots, \xi_p\}$ such that the numerical conditioning of R_ϕ was superior to that of R_ψ. Unfortunately, this appears to be an extremely difficult question, mainly due to the very complicated manner in which the condition number of a matrix depends on the elements of that matrix.

Nevertheless, the purpose of this section is to at least establish sufficient conditions for when superiority exists, but because of this involved nature of the question, it is only answered for the limited case of dimension $p = 2$.
In order to proceed with this analysis of the numerical superiority (or not) of one model structure over another, it turns out to be better to avoid consideration of the condition number $\kappa(R)$ of a matrix R directly, but instead to consider a new function $f(R)$ of a matrix R which is monotonic in condition number $\kappa(R)$ and which is defined as

$$f(R) \triangleq \left(\frac{\kappa(R)-1}{\kappa(R)+1}\right)^2 = \left(\frac{\lambda_{\mathsf{max}}(R)/\lambda_{\mathsf{min}}(R)-1}{\lambda_{\mathsf{max}}(R)/\lambda_{\mathsf{min}}(R)+1}\right)^2$$
$$= \left(\frac{\lambda_{\mathsf{max}}(R)-\lambda_{\mathsf{min}}(R)}{\lambda_{\mathsf{max}}(R)+\lambda_{\mathsf{min}}(R)}\right)^2.$$

Using this idea, it is possible to establish the following result on the general superiority of the orthonormal structure from a numerical conditioning perspective.

Theorem 6.4. *For the two-dimensional case of $p = 2$, consider R_ϕ defined by (6.13) and associated with the orthonormal model structure (6.6) and R_ψ defined by (6.12) with the $\{\mathcal{F}_k(q)\}$ defined by (6.23). Then for $\xi_1, \xi_2 \in \mathbb{R}^+$*

$$\kappa(R_\phi) \leq \kappa(R_\psi)$$

provided that Φ_u is such that the associated $\varphi(z)$ satisfies

$$\frac{\varphi(\xi_1)-\varphi(\xi_2)}{\xi_1-\xi_2} > 0, \qquad \frac{\xi_1\varphi(\xi_2)-\xi_2\varphi(\xi_1)}{\xi_1-\xi_2} > 0. \tag{6.33}$$

Proof. With the definition $C \triangleq (\xi_1-\xi_2)^{-1}(1-\xi_1\xi_2)^{-1}$, then straightforward (but tedious) algebra provides that

$$R_\psi = C\begin{bmatrix} \frac{2\xi_1\varphi(\xi_1)}{1-\xi_1^2} - \frac{2\xi_2\varphi(\xi_2)}{1-\xi_2^2} & \left(\frac{1+\xi_1^2}{1-\xi_1^2}\right)\varphi(\xi_1) - \left(\frac{1+\xi_2^2}{1-\xi_2^2}\right)\varphi(\xi_2) \\ \left(\frac{1+\xi_1^2}{1-\xi_1^2}\right)\varphi(\xi_1) - \left(\frac{1+\xi_2^2}{1-\xi_2^2}\right)\varphi(\xi_2) & \frac{2\xi_1\varphi(\xi_1)}{1-\xi_1^2} - \frac{2\xi_2\varphi(\xi_2)}{1-\xi_2^2} \end{bmatrix}.$$

Also, using Lemma 6.2, and with K given by (6.30) then R_ϕ may be expressed as

$$R_\phi = \begin{bmatrix} 2\varphi(\xi_1) & K[\varphi(\xi_1)-\varphi(\xi_2)] \\ K[\varphi(\xi_1)-\varphi(\xi_2)] & 2\varphi(\xi_2) \end{bmatrix}.$$

As well, note that for a 2×2 symmetric matrix A of the form

$$A = \begin{bmatrix} a & b \\ b & c \end{bmatrix}$$

then the function $f(A)$ may be calculated as

$$f(A) = \frac{(a-c)^2+4b^2}{(a+c)^2}.$$

In this case the calculation of $f(R_\psi)$ and $f(R_\phi)$ become

$$\begin{aligned} f(R_\psi) &= \frac{\left[\left(\frac{1+\xi_1^2}{1-\xi_1^2}\right)\varphi(\xi_1) - \left(\frac{1+\xi_2^2}{1-\xi_2^2}\right)\varphi(\xi_2)\right]^2}{\left[\frac{\xi_1\varphi(\xi_1)}{1-\xi_1^2} - \frac{\xi_2\varphi(\xi_2)}{1-\xi_2^2}\right]^2} \\ &= \frac{\left[(1-\xi_1^2\xi_2^2)[\varphi(\xi_1)-\varphi(\xi_2)] + (\xi_1^2-\xi_2^2)[\varphi(\xi_1)+\varphi(\xi_2)]\right]^2}{[\xi_1(1-\xi_2^2)\varphi(\xi_1) - \xi_2(1-\xi_1^2)\varphi(\xi_2)]^2} \\ &= \frac{(1+\xi_1\xi_2)^2\left[(1-\xi_1\xi_2)\left(\frac{\varphi(\xi_1)-\varphi(\xi_2)}{\xi_1-\xi_2}\right) + \left(\frac{\xi_1+\xi_2}{1+\xi_1\xi_2}\right)[\varphi(\xi_1)+\varphi(\xi_2)]\right]^2}{\left[[\varphi(\xi_1)+\varphi(\xi_2)] - \left(\frac{\xi_1\varphi(\xi_2)-\xi_2\varphi(\xi_1)}{\xi_1-\xi_2}\right)(1-\xi_1\xi_2)\right]^2} \end{aligned}$$

and

$$f(R_\phi) = \frac{[\varphi(\xi_1) - \varphi(\xi_2)]^2 + \dfrac{[\varphi(\xi_1) - \varphi(\xi_2)]^2(1-\xi_1^2)(1-\xi_2^2)}{(\xi_1 - \xi_2)^2}}{[\varphi(\xi_1) + \varphi(\xi_2)]^2}$$

$$= \left(\frac{\varphi(\xi_2) - \varphi(\xi_2)}{\xi_1 - \xi_2}\right)^2 \left(\frac{1 - \xi_1\xi_2}{\varphi(\xi_1) + \varphi(\xi_2)}\right)^2$$

$$= \frac{\left(\dfrac{\varphi(\xi_2) - \varphi(\xi_2)}{\xi_1 - \xi_2}\right)^2 (1-\xi_1\xi_2)^2}{[\varphi(\xi_1) + \varphi(\xi_2)]^2}.$$

Now, by assumption $\xi_1, \xi_2 \in \mathbb{R}^+$, so the numerator term of the $f(R_\psi)$ term is clearly greater than that of the $f(R_\phi)$ term if the first condition in (6.33) is satisfied, and the denominator term of the $f(R_\psi)$ term is clearly smaller than that of the $f(R_\phi)$ term if the second condition in (6.33) is satisfied, so that in this case, $\kappa(R_\phi) \le \kappa(R_\psi)$ is guaranteed provided the conditions (6.33) are met.

■

The most important question now is how large the class of possible Φ_u is that satisfy the sufficient conditions (6.33). For the purpose of analysing this, it is expedient to use the representation (6.26) in which case condition (6.33) becomes

$$0 < \frac{\varphi(\xi_1) - \varphi(\xi_2)}{\xi_1 - \xi_2} = \frac{1}{2\pi}\int_{-\pi}^{\pi} \frac{(1+\xi_1\xi_2)\cos\omega - (\xi_1+\xi_2)}{|1-\xi_1 e^{i\omega}|^2 |1-\xi_2 e^{i\omega}|^2} \Phi_u(\omega)\,d\omega \tag{6.34}$$

and similarly, after some algebra

$$0 < \frac{\xi_1\varphi(\xi_2) - \xi_2\varphi(\xi_1)}{\xi_1 - \xi_2}$$
$$= \frac{1}{4\pi}\int_{-\pi}^{\pi} \frac{(1-\xi_1^2\xi_2^2) + (\xi_1+\xi_2)[(\xi_1+\xi_2) - 2\cos\omega]}{|1-\xi_1 e^{i\omega}|^2 |1-\xi_2 e^{i\omega}|^2} \Phi_u(\omega)\,d\omega \tag{6.35}$$

The weight functions

$$\chi_1(\omega) \triangleq \frac{(1+\xi_1\xi_2)\cos\omega - (\xi_1+\xi_2)}{|1-\xi_1 e^{i\omega}|^2 |1-\xi_2 e^{i\omega}|^2} \tag{6.36}$$

$$\chi_2(\omega) \triangleq \frac{(1-\xi_1^2\xi_2^2) + (\xi_1+\xi_2)[(\xi_1+\xi_2) - 2\cos\omega]}{|1-\xi_1 e^{i\omega}|^2 |1-\xi_2 e^{i\omega}|^2} \tag{6.37}$$

appearing in these integral characterizations of (6.33) are plotted for the case of $\xi_1 = 0.5, \xi_2 = 0.6$ in the left diagram of Figure 6.1. The weight function χ_1 being the solid line, and the weight χ_2 being the dash-dot line. These weight functions χ_1 and χ_2 clearly concentrate attention around $\omega = 0$, and in such a way that any Φ_u of general low-pass nature will, when weighted by them

and integrated as in (6.34) and (6.35), generally produce a positive result, and hence satisfy the necessary conditions on Theorem 6.4.
To emphasize this further, it is at least clear from (6.34) and the plot of $\chi_1(\omega)$ that in general any $\Phi_u(\omega)$ that decays as ω tends to π will be such as to satisfy

$$\frac{\varphi(\xi_1)-\varphi(\xi_2)}{\xi_1-\xi_2}=\frac{1}{2\pi}\int_{-\pi}^{\pi}\chi_1(\omega)\Phi_u(\omega)\,\mathrm{d}\omega>0.$$

What may not be so clear at first inspection is whether this same class of 'low-pass' $\Phi_u(\omega)$ also lead to the second necessary condition of Theorem 6.4 being satisfied; namely $(\xi_1\varphi(\xi_2)-\xi_2\varphi(\xi_1))/(\xi_1-\xi_2)>0$. This can be clarified by examining the positive sign definiteness of the product

$$\left(\frac{\xi_1\varphi(\xi_2)-\xi_2\varphi(\xi_1)}{\xi_1-\xi_2}\right)\left(\frac{\varphi(\xi_1)-\varphi(\xi_2)}{\xi_1-\xi_2}\right)=$$

$$=\frac{1}{8\pi^2}\int_{-\pi}^{\pi}\int_{-\pi}^{\pi}\chi_1(\omega)\chi_2(\sigma)\Phi_u(\omega)\Phi_u(\sigma)\,\mathrm{d}\omega\,\mathrm{d}\sigma. \tag{6.38}$$

The two-dimensional 'kernel' $\chi_1(\omega)\chi_2(\sigma)$ is plotted, again for the case of $\xi_1=0.5,\xi_2=0.6$ in the right-hand diagram of Figure 6.1. Clearly, the bulk of it over all values of ω and σ is positive, and consideration of it indicates that the only way that the product (6.38) can be negative is if $\Phi_u(\omega)$ is very strongly concentrated around $\omega=0$. Specifically, the low-pass nature of Φ_u would need to imply a roll-off at around 5% of the sampling frequency or, put another way, the sampling rate would need to be around 10 times larger than the minimum Nyquist rate implied by the bandwidth of Φ_u.

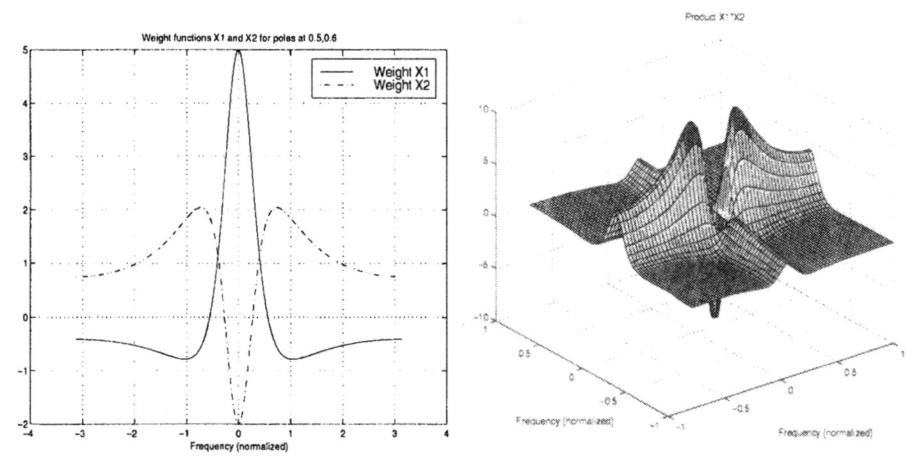

Fig. 6.1. The left figure shows the weight functions χ_1 and χ_2 defined in (6.36) and (6.37) for poles at $\xi_1=0.5,\xi_2=0.6$. The right figure shows the product of these weights (which is what is important in (6.38)) for poles at $\xi_1=0.8,\xi_2=0.95$.

The conclusion therefore is, at least in the specific $p = 2$ dimensional case, that R_ϕ has smaller condition number than R_ψ associated with $\mathfrak{F}_1, \mathfrak{F}_2$ given by the simple form (6.23) over a very wide range of input spectra Φ_u.
To examine this even more closely, specific classes of parameterized $\Phi_u(\omega)$ may be considered. For example, for the particularly simple class of Φ_u that have a spectral factorization $\Phi_u(z) = H(z)H(1/z)$ of the form

$$H(z) = 1 + az \tag{6.39}$$

then

$$\varphi(z) = \frac{c_0}{2} + c_1 z = \frac{1+a^2}{2} + az$$

so that the two conditions in (6.33) becomes $c_1 = a > 0$ $c_0 = 1 + a^2 > 0$, respectively. Therefore, $\kappa(R_\psi) > \kappa(R_\phi)$ for any input with spectral factor of the form (6.39) with $a > 0$. This is not a heavy restriction, because if $a < 0$, this implies that $\{u_t\}$ is differenced white noise, and hence $\Phi_u(e^{i\omega})$ is of the form shown in the left-hand diagram of Figure 6.2. This is necessarily increasing at the folding frequency $\omega = \pi$, which indicates the samples $\{u_t\}$ have been taken too slowly in that aliasing of the underlying continuous time signal is occurring. It is therefore not a reasonable situation.
Now consider the case of the spectral factor $H(z)$ being second order and of the form

$$H(z) = (1 + az)(1 + bz)$$

so that

$$\varphi(z) = \frac{c_0}{2} + c_1 z + c_2 z^2 = \frac{1 + (a+b)^2 + a^2 b^2}{2} + (a+b)(1+ab)z + abz^2.$$

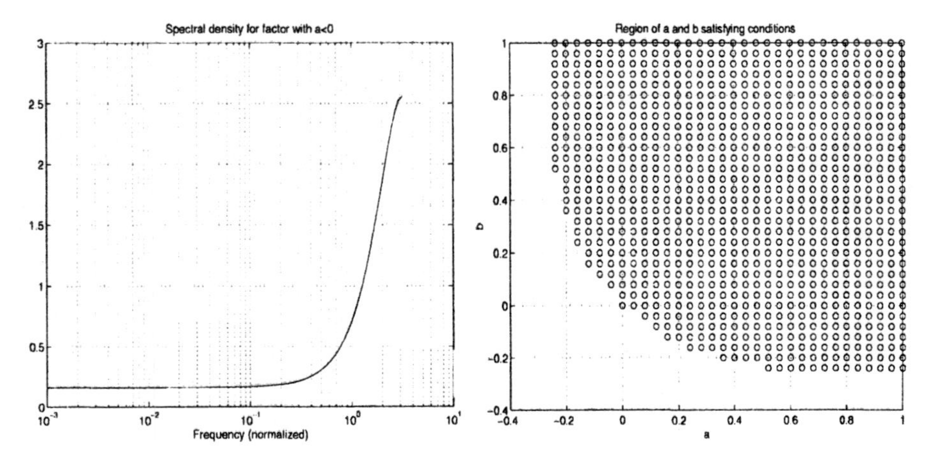

Fig. 6.2. (Left figure) Illustration of nature of spectral density $\Phi_u(\omega)$ for the case of its spectral factor being first order with a pole $a < 0$. (Right figure) Region of a and b in second-order $H(z)$ guaranteeing $\kappa(R_\psi) > \kappa(R_\phi)$ for all $\xi_1, \xi_2 \in \mathbb{R}^+$

It seems impossible to analytically derive conditions on $a, b \in \mathbb{R}^+$ such that the conditions (6.33) will be satisfied, but numerical experiment establishes that any a, b in the shaded region shown in the right-hand diagram of Figure 6.2 do the job. For example, $\kappa(R_\psi) \geq \kappa(R_\phi)$ for any $a \geq 0$ and $b \geq 0$.

6.9 Asymptotic Analysis

As mentioned in the introduction, a key feature of the orthonormal parameterization (6.6) is that associated with it is a covariance matrix with numerical conditioning guaranteed by the bounds [222, 306]

$$\min_{\omega \in [-\pi,\pi]} \Phi_u(\omega) \leq \lambda(R_\phi) \leq \max_{\omega \in [-\pi,\pi]} \Phi_u(\omega). \tag{6.40}$$

A natural question to consider is how tight these bounds are. In [222], this was addressed by a strategy of analysis that is asymptotic in the matrix size p. Specifically, define $M_\phi \triangleq \lim_{p\to\infty} R_\phi$. In this case, M_ϕ is an operator $\ell_2 \to \ell_2$, so that the eigenvalues of the finite dimensional matrix R_ϕ, generalize to the continuous spectrum $\lambda(M_\phi)$ which itself is defined as [33]

$$\lambda(M_\phi) = \{\lambda \in \mathbb{R} : \lambda I - M_\phi \text{ is not invertible}\}.$$

This spectrum can be characterized as follows.

Lemma 6.3. *Suppose that*

$$\sum_{k=1}^{\infty}(1 - |\xi_k|) = \infty.$$

Then

$$\lambda(M_\phi) = \text{Range}\{\Phi_u(\omega)\}.$$

Proof. Take any $\mu \in [\min_\omega \Phi_u(\omega), \max_\omega \Phi_u(\omega)]$ and suppose $\mu \notin \lambda(M)$. Then because by orthogonality $M(1)$ is the identity operator, then $M(\Phi_u - \mu)$ is an invertible operator from $\ell_2^+ \to \ell_2^+$ so that in particular $\exists x \in \ell_2^+$ such that for $e_0 = (1, 0, 0, \cdots)$

$$x^T M(\Phi_u - \mu) = e_0. \tag{6.41}$$

Therefore, defining $g(z) \in H_2(\mathbb{T})$ by $g(z) \triangleq (x_1 - 1)F_1(z) + \sum_{k=1}^{\infty} x_k B_k(z)$ gives that from (6.41) $(\Phi_u - \mu)g \perp \text{Sp}\{F_k\}$. But under the completeness condition $\sum(1 - |\xi_k|) = \infty$, $\text{Sp}\{F_k\} = H_2$ so that because $L_2 = H_2 \oplus H_2^\perp$ then $(\Phi_u - \mu)\overline{g} \in H_2$. But $g \in H_2$ by construction and the product of two H_2 functions is in H_1 [135].
Hence, $(\Phi_u - \mu)|g|^2$ is a real valued H_1 function, and the only such functions are constants [135]. However, because $\mu \in [\min_\omega \Phi_u(\omega), \max_\omega \Phi_u(\omega)]$, then

this function cannot be of constant sign, hence it cannot be a constant. This contradiction implies $[\min_\omega \Phi_u(\omega), \max_\omega \Phi_u(\omega)] \subset \lambda(M)$. Finally, according to the results provided in Chapter 2, $\lambda(M) \subset [\min_\omega \Phi_u(\omega), \max_\omega \Phi_u(\omega)]$.

■

This provides evidence that, at least for large p (when the issue of numerical conditioning is most important), the bounds (6.40) are in fact tight, and therefore

$$\kappa(R_\phi) \approx \frac{\max_\omega \Phi_u(\omega)}{\min_\omega \Phi_u(\omega)} \tag{6.42}$$

might be expected to be a reasonable approximation. Of course, what would also be desirable is a similar approximation for R_ψ, and naturally this will depend on the nature of the definition of the $\{\mathcal{F}_k(q)\}$. One obvious formulation is that of (6.23) extended to arbitrary dimension p as

$$\mathcal{F}_k(q) = \frac{z^k}{D_p(z)}, \qquad D_p(z) = \prod_{\ell=1}^{p}(z - \xi_\ell) \tag{6.43}$$

for $k = 1, 1, \cdots, p$ and $\{\xi_1, \cdots, \xi_p\} \in \mathbb{D}$ the fixed pole choices.
This case is considered important, as possibly the most straightforward way of realizing a fixed-pole estimate $G(q, \widehat{\beta})$ as originally defined in (6.3) of Section 6.1 would be to simply use pre-existing software for estimating FIR model structures, but after having pre-filtering the input sequence $\{u_t\}$ with the all-pole filter $1/D_p(q)$. This is identical to using the general model structure (6.2) with the $\{\mathcal{F}_k(q)\}$ choice of (6.43) above, with estimated FIR coefficients then simply being the numerator coefficient estimates $\{\beta_1, \cdots, \beta_p\}$.

Fortunately, for this common structure, it is also possible to develop an approximation of the condition number $\kappa(R_\psi)$ via the following asymptotic result which is a direct corollary of Theorem 6.3.

Corollary 6.1. *Consider the choice defined by (6.4) and (6.43) for $\{\mathcal{F}_k(q)\}$ defining R_ψ in (6.12). Suppose that only a finite number of the poles $\{\xi_k\}$ are chosen away from the origin so that*

$$D(\omega) \triangleq \lim_{p\to\infty} \prod_{\ell=1}^{p} |e^{i\omega} - \xi_\ell|^2 \tag{6.44}$$

exits. Define, in a manner analogous to that pertaining to Lemma 6.3, the operator $M_\psi : \ell_2 \to \ell_2$ as

$$M_\psi \triangleq \lim_{p\to\infty} R_\psi.$$

Then

$$\lambda(M_\psi) = \mathrm{Range}\left\{\frac{\Phi_u(\omega)}{D(\omega)}\right\}.$$

Proof. Note that for the choice of $\{\mathcal{F}_k(q)\}$ formulated in (6.43), then by the definition (6.4), (6.10), (6.12), and (6.13), R_ψ is the same as a matrix R_ϕ where the orthonormal basis involves the choice of all the poles $\{\xi_k\}$ taken at the origin, and the input spectrum $\Phi_u(\omega)$ is changed according to $\Phi_u(\omega) \mapsto \Phi_u(\omega)/|D_p(e^{i\omega})|^2$. By the assumptions on the pole locations, this latter quantity converges with increasing p to $\Phi_u(\omega)/|D(e^{i\omega})|^2$. Application of Lemma 6.3, which is invariant to the particular choice of the pole locations, then provides the result.

■

In analogy with the previous approximation, it is tempting to apply this asymptotic result for finite p to derive the approximation

$$\kappa(R_\psi) \approx \frac{\max_\omega \Phi_u(\omega)/|D_p(e^{i\omega})|^2}{\min_\omega \Phi_u(\omega)/|D_p(e^{i\omega})|^2}. \tag{6.45}$$

Now, considering that $|D_p(e^{i\omega})|^2 = \prod_{\ell=1}^{p} |e^{i\omega} - \xi_\ell|^2$ can take on both very small values (especially if some of the ξ_ℓ are close to the unit circle) and also very large values (especially if all the $\{\xi_\ell\}$ are chosen in the right-half plane so that aliasing is not being modelled), then the maxima and minima of $\Phi_u/|D_p|^2$ will be much more widely separated than those of Φ_u. The approximations (6.42) and (6.45) therefore indicate that estimation with respect to the orthonormal form (6.6) could be expected to be much better conditioned than that with respect to the model structure (6.4) with the simple choice (6.43) for a very large class of Φ_u – an obvious exception here would be $\Phi_u = |D_p|^2$ for which $R_\psi = I$.
However, this conclusion depends on the accuracy of applying the asymptotically derived approximations (6.42) and (6.45) for finite p. In the absence of theoretical analysis, which appears intractable, simulation study can be pursued. Consider p in the range 2-30 with all the $\{\xi_\ell\}$ chosen at $\xi_\ell = 0.5$, and $\Phi_u(\omega) = 0.36/(1.36 - \cos\omega)$ which is of a natural 'band-limited' sort. Then the maximum and minimum eigenvalues for R_ψ and R_ϕ are shown as solid lines in the left and (respectively) right diagrams in Figure (6.3). The dash-dot lines in these figures are the approximations (6.42) and (6.45). Clearly, in this case the approximations are quite accurate, even for what might be considered small p. Note that the minimum eigenvalue of R_ψ is shown only up until $p = 18$, as it was numerically impossible to calculate it for higher p. Again, this provides evidence that even though model structures (6.6) parameterized in terms of orthonormal $\{F_k(q)\}$ are only designed to provide superior numerical conditioning properties for white input, they seem to also provide improved conditioning over a very wide range of coloured inputs as well.

Finally, before leaving this topic of eigenvalue characterization in the asymptotic case, it should be mentioned that the recent work [39] has established

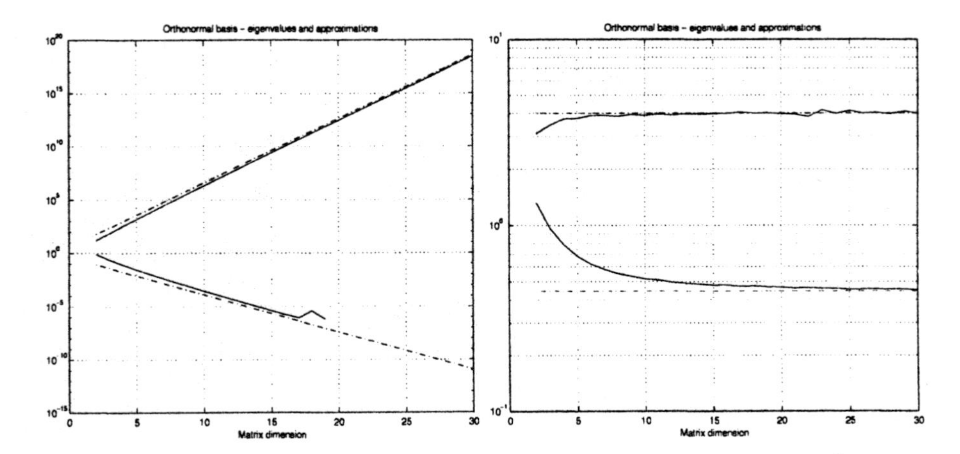

Fig. 6.3. Solid lines are maximum and minimum eigenvalues of (left figure) R_ψ and (right figure) R_ϕ for a range of dimensions p. The dash dot lines are the approximations (6.42) and (6.45).

important new results on the actual *distribution* of these eigenvalues. This contribution has generalized what was known since the 1950s [108] in relation to Toeplitz matrices; namely, that (using the notation of this chapter) if the eigenvalues $\lambda_1^{(p)}, \cdots, \lambda_p^{(p)}$ of $R_\psi \in \mathbb{R}^{p\times p}$ defined in (6.12) are assumed to ordered from smallest to largest, then the cumulative distribution of them is determined by $\Phi_u(\omega)/|D(e^{i\omega})|^2$ as follows

$$\lim_{p\to\infty} \frac{1}{p}\sum_{k=1}^{m} \lambda_k^{(p)} = \frac{1}{2\pi}\int_{\pi}^{\lambda_m^{(p)}} \frac{\Phi_u(\omega)}{|D(e^{i\omega})|^2}\, \mathrm{d}\omega. \tag{6.46}$$

This indicates that, for finite p, it can be expected that the eigenvalues of R_ψ will be clustered around regions where $\Phi_u/|D|^2$ is large. Because, as just explained, this same function can have quite large variation, this indicates a very uneven distribution of eigenvalues.
The work [39] (among other things) derives the corresponding eigenvalue characterization for the orthonormalized form R_ϕ defined in (6.13); namely, denoting the ordered eigenvalues of $R_\phi \in \mathbb{R}^{p\times p}$ as $\mu_1^{(p)}, \cdots, \mu_p^{(p)}$, then

$$\lim_{p\to\infty} \frac{1}{p}\sum_{k=1}^{m} \mu_k^{(p)} = \frac{1}{2\pi}\int_{\pi}^{\mu_m^{(p)}} \Phi_u(\chi^{-1}(\omega))\, \mathrm{d}\omega \tag{6.47}$$

where $\chi(\omega)$ is the asymptotic phase distribution of the Blaschke product component of $\{F_k\}$, which is defined in the following sense

$$\chi(\omega) = \lim_{p\to\infty} \frac{1}{p}\, \chi_p(\omega), \qquad e^{i\chi_p(\omega)} \triangleq \prod_{k=1}^{p} \left(\frac{1-\overline{\xi_k} q}{q-\xi_k}\right). \tag{6.48}$$

Because the distribution function (6.47) is now governed by Φ_u, which in general will have less variation than $\Phi_u/|D|^2$, then this new result establishes that the eigenvalue distribution of R_ϕ will be more uniform than that of R_ψ.

6.10 Convergence Rates and Numerical Conditioning

Having now argued that the numerical conditioning of normal equations can be improved, across a range of input spectral colourings, by employing an orthonormalized form, this final section provides an application example to illustrate the importance of this improved conditioning.
Of course, an obvious situation where improved conditioning will be an advantage is in the estimation of high-dimensional model structures whereby it is important to ameliorate the affects of finite precision computing.

However, this section will concentrate on a different scenario where, even in the presence of infinite precision computation, improved numerical conditioning still provides an important practical advantage. This situation is the widely employed one of adaptive filtering via the LMS algorithm. To discuss it, we recall the model structure (6.1), (6.6) in combination with the definition (6.10) for ϕ_t which may be expressed as the following 'linear regression' form

$$y_t = \phi_t^T \theta_\circ + \nu_t. \tag{6.49}$$

Here, the parameter vector $\theta_\circ \in \mathbb{R}^p$ is the true value that describes the underlying input-output dynamics $G(q)$ according to $G(q, \theta_\circ) = G(q)$.
An estimate of $G(q)$ may be obtained via the general recursive update method

$$\widehat{\theta}_{t+1} = \widehat{\theta}_\circ + L_t(y_t - \phi_t^T \widehat{\theta}_\circ), \tag{6.50}$$

where L_t is a gain vector that may be computed in various ways. A common choice for this gain vector is

$$L_t = \mu \phi_t, \quad \mu \in (0,1) \tag{6.51}$$

in which case (6.50) is known as the 'gradient' or 'least mean square' (LMS) algorithm, which is in extremely widespread use.
In order to study the performance of this estimation method, we define the estimation error $\widetilde{\theta}_\circ$ as

$$\widetilde{\theta}_\circ \triangleq \theta_\circ - \widehat{\theta}_t. \tag{6.52}$$

Substituting (6.52) into the general update equation (6.50) establishes that this error satisfies the following difference equation [113]

$$\begin{aligned}
\widetilde{\theta}_{t+1} &= \theta_\circ - \widehat{\theta}_t - L_t\left(\phi_t^T \theta_\circ + \nu_t - \phi_t \widehat{\theta}_t\right) \\
&= \left(\theta_\circ - \widehat{\theta}_t\right) - L_t \phi_t^T \left(\theta_\circ - \widehat{\theta}_t\right) - L_t \nu_t \\
&= \left(I - L_t \phi_t^T\right) \widetilde{\theta}_\circ - L_t \nu_t.
\end{aligned} \tag{6.53}$$

For the LMS algorithm where $L_t = \mu\phi_t$, then the average (over the ensemble of possible noise corruptions $\{\nu_t\}$) convergence properties are determined by the autonomous part of (6.53) as [113]

$$\mathsf{E}\{\widetilde{\theta}_{t+1}\} \approx (I - \mu\mathsf{E}\{\phi_t\phi_t^T\})\mathsf{E}\{\widetilde{\theta}_t\}.$$

This indicates [279] that the convergence properties are determined by the choice of μ combined with the eigenvalues of $R_\phi = \mathsf{E}\{\phi_t\phi_t^T\}$. Specifically, in order to guarantee stability it is necessary that the step-size μ be limited as

$$\lambda_{\max}(I - \mu R_\phi) < 1 \quad \Rightarrow \quad \mu < \frac{1}{\lambda_{\max}(R_\phi)}$$

As a consequence, the parameter space convergence in the direction of the eigenvector associated with $\lambda_{\min}(R_\phi)$ is limited to be no faster than the rate of decay of

$$\left[1 - \frac{\lambda_{\min}(R_\phi)}{\lambda_{\max}(R_\phi)}\right]^t = \left[1 - \frac{1}{\kappa(R_\phi)}\right]^t.$$

Therefore, in the interests of LMS convergence speed, it is desirable to choose a model structure $G(q,\theta)$ such that the ensuing regressor ϕ_t implies the minimum possible condition number for R_ϕ.

For the case of white input spectrum $\Phi_u(\omega)$ = constant, the orthonormal model structure (6.6) achieves this and is therefore optimal (in the white input convergence sense) among the class of all fixed denominator model structures, of which the more 'natural' or obvious choice (6.3) with $\mathcal{F}_k(q)$ given by (for example) (6.43) is also a member.

This optimality is illustrated via simulation in Figure 6.4. The top (slowly converging) plot is the observed mean square error (averaged over 500 simulations with different input and measurement noise realizations) obtained when using the LMS algorithm with the 'natural' fixed-denominator model structure (6.3), (6.43). The bottom (quickly converging) plot is the observed mean square error using the orthonormal model structure (6.6) and the same LMS algorithm, but with μ changed to keep the steady state error invariant to the change in model structure; the convergence rate improvement obtained by using the orthonormal structure is clear.

In both cases, the model structure was third order with the poles chosen at $\xi_1 = 0.4$, $\xi_2 = 0.85$, $\xi_2 = 0.6$, and the true system $G_t(q)$ was time invariant with true poles at $\gamma_1 = 0.9$ and $\gamma_2 = 0.37$. The output was corrupted by white Gaussian distributed noise of variance $\sigma_\nu^2 = 0.01$, and the input was white Gaussian distributed noise of variance $\Phi_u(\omega) = 10$.

6.11 Conclusions

A variety of arguments have been presented to indicate that the condition numbers $\kappa(R_\psi)$ and $\kappa(R_\phi)$, which govern the numerical properties of least squares estimation associated with (respectively) simple 'fixed-denominator' model structures and their orthonormalized forms, are such that $\kappa(R_\psi) \geq \kappa(R_\phi)$ for a very wide class of input spectra Φ_u. Although this might be considered somewhat surprising, as it is only designed to occur (by the construction of the 'orthonormal' model structure) for white Φ_u, it is also important because it provides a strong argument for why the extra programming effort should be expended to implement the various orthonormal model structures that have recently been examined in the literature.

This analysis is made in counter-argument to the charge (as posed in the introduction to this chapter), that a change of model structure is not the same as a change of estimation method – equivalent structures provide identical estimates but only modulo the numerical issues considered here.

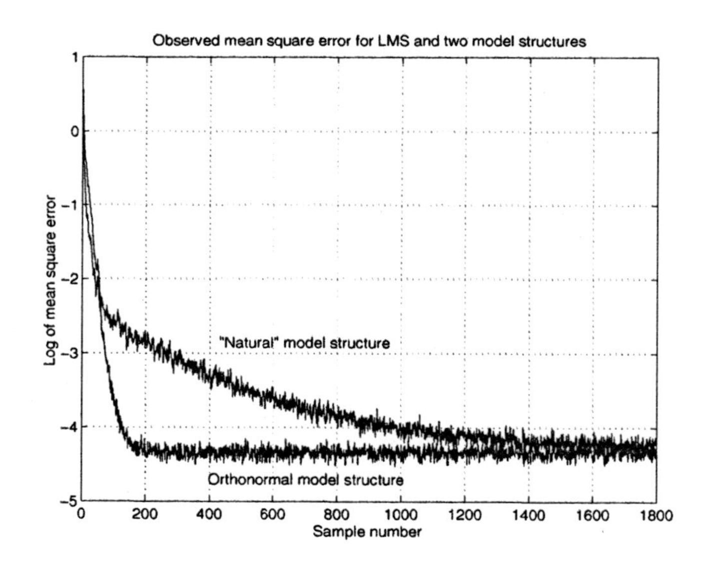

Fig. 6.4. Sample mean square error rates (averaged over 500 realizations) for LMS algorithm with (top plot) 'natural' fixed denominator model structure (6.3), (6.43) and (bottom plot) orthonormal model structure (6.6).

7

Model Uncertainty Bounding

Paul Van den Hof

Delft University of Technology, Delft, The Netherlands

7.1 Introduction

In system identification, every identified model needs to be accompanied by a quantification of its reliability. Probabilistic model error bounds are generally based on (asymptotic) variance expressions. For the orthonormal basis function model structures considered here, these variance expressions are analysed in detail in Chapter 5. In this chapter, it will be further explored how model uncertainty bounds can be specified when using either full-order or reduced-order model structures. In the latter situation not only do variance issues play a role, but also bias terms have to be incorporated. The interest in this analysis was initiated by the intention to use identified models (and their uncertainty bounds) as a basis for model-based robust control. In this approach, the model uncertainty bounds serve to guarantee robust stability and robust performance of the designed controller prior to its implementation on the real process [340].
For several reasons (*e.g.* computational complexity, analysis purposes), the orthonormal basis function model structures appear to be very attractive for model uncertainty quantification, in particular in those situations where also bias terms have to be incorporated due to under-modelling, as *e.g.* occurs when using so-called reduced-order model structures.

7.2 Prediction Error Identification with Full-order Model Structures

In prediction error identification ([164] and Chapter 4), the classical reasoning in model uncertainty bounding is to assume the availability of a full-order model structure, *i.e.* the situation 'system is in the model set' $\mathcal{S} \in \mathcal{M}$. In that situation, the central limit theorem provides the asymptotic distribution of the parameter estimate, *i.e.* for $N \to \infty$,

$$\sqrt{N}(\widehat{\theta}_N - \theta_0) \in \mathcal{N}(0, P_\theta)$$

and the variance error determines the (asymptotic) uncertainty in the parameter estimates according to

$$\mathrm{Cov}\{\widehat{\theta}_N\} = \frac{P_\theta}{N}$$

with N the number of data points, $\widehat{\theta}_N$ the estimated parameter, and P_θ the asymptotic covariance matrix.
The parameter uncertainty can then be represented by ellipsoidal areas, induced by the asymptotic covariance matrix, according to

$$\mathcal{D}_\theta = \{\theta \in \mathbb{R}^n \mid (\theta - \widehat{\theta}_N)^T P_\theta^{-1}(\theta - \widehat{\theta}_N) < \frac{c}{N}\}$$

where the probability level, indicated by c, is induced by the Gaussian distribution, as a consequence of applying the central limit theorem.
The parameter ellipsoids can be transformed to the frequency domain in order to construct asymptotic model uncertainty bounds for the frequency response $G(e^{i\omega}, \widehat{\theta}_N)$. If the model is linearly parameterized, as is the case for all orthogonal basis function model structures, this mapping is written as

$$\widehat{G}(e^{i\omega}) = G(e^{i\omega}, \widehat{\theta}_N) = \Gamma_n^T(e^{i\omega})\widehat{\theta}_N \tag{7.1}$$

with $\Gamma_n(z) = [F_1(z)\ F_2(z)\ \cdots\ F_n(z)]^T$, and $F_k(z)$ the kth basis function in any orthonormal basis within the class of basis functions considered in Chapter 2.
The ellipsoidal parameter uncertainty is then mapped to the frequency domain by using the two-dimensional representation

$$g(e^{i\omega}) = \begin{bmatrix} \mathrm{Re}(G(e^{i\omega})) \\ \mathrm{Im}(G(e^{i\omega})) \end{bmatrix}$$

and

$$g(e^{i\omega}, \widehat{\theta}_N) = T(e^{i\omega})\widehat{\theta}_N \text{ with } T(e^{i\omega}) = \begin{bmatrix} \mathrm{Re}(\Gamma_n^T(e^{i\omega})) \\ \mathrm{Im}(\Gamma_n^T(e^{i\omega})) \end{bmatrix}.$$

The corresponding ellipsoid in the frequency domain is then constructed according to

$$\mathcal{D}_\omega = \{g(e^{i\omega}) \in \mathbb{R}^{2\times 1} \mid$$
$$(g(e^{i\omega}) - g(e^{i\omega}, \widehat{\theta}_N))^T P(\omega)^{-1}(g(e^{i\omega}) - g(e^{i\omega}, \widehat{\theta}_N)) < \frac{c}{N}\}$$

with $P(\omega) = T(e^{i\omega})P_\theta T^T(e^{i\omega})$.
This ellipsoid can be calculated for every frequency ω, thus generating an – infinite – collection of ellipsoidal uncertainty sets representing the uncertainty in the frequency response $G_0(e^{i\omega})$. The ellipsoids are centred around the nominal model's frequency response, $G(e^{i\omega}, \widehat{\theta}_N)$. A relation between $\mathcal{D}_\theta$ and $\mathcal{D}_\omega$ can be formulated as follows.

Proposition 7.1. *([31])*

a. If $\theta \in \mathcal{D}_\theta$, then $G(e^{i\omega},\theta) \in \mathcal{D}_\omega$ for all ω.
b. If $G(e^{i\omega},\theta) \in \mathcal{D}_\omega$ for some ω, this does generally not imply that $\theta \in \mathcal{D}_\theta$.

The first result follows from the reasoning above, and the second part follows by observing that for $n > 2$, the mapping T is not bijective and so there exist $\theta \notin \mathcal{D}_\theta$ that lead to $G(e^{i\omega},\theta) \in \mathcal{D}_\omega$. Only if $n = 2$, can a one-to-one relation between the two sets be accomplished.

Note that the implication as formulated in the first part of the proposition is dependent on the fact that G is linearly parameterized. In the case of non-linear parameterizations, this result can be obtained only approximately, by considering first-order Taylor approximations of $\partial G(e^{i\omega},\theta)/\partial\theta$.
In classical prediction error identification, the uncertainty sets discussed above correspond to parameter confidence intervals:

$$\mathcal{D}_\theta = \{\theta \mid (\theta - \widehat{\theta}_N)^T P_\theta^{-1}(\theta - \widehat{\theta}_N) < \frac{c_\chi(\alpha,n)}{N}\}$$

where $c_\chi(\alpha,n)$ is the α-probability level for a χ^2_n-distributed random variable, having n degrees of freedom, *i.e.* $Pr\{\chi^2_n < c_\chi(\alpha,n)\} = \alpha$. It follows from asymptotic prediction error identification theory that

$$\theta_0 \in \mathcal{D}_\theta \text{ with probability } \alpha.$$

As a direct consequence of Proposition 7.1, it then follows that

$$G(e^{i\omega},\theta_0) \in \mathcal{D}_\omega \text{ with probability } \geq \alpha, \text{ where}$$

$$\mathcal{D}_\omega = \{g(e^{i\omega}) \in \mathbb{R}^{2\times 1} \mid (g(e^{i\omega}) - g(e^{i\omega},\widehat{\theta}_N))^T \cdot P(\omega)^{-1} \cdot \\ \cdot (g(e^{i\omega}) - g(e^{i\omega},\widehat{\theta}_N)) < \frac{c_\chi(\alpha,n)}{N}\}.$$

Note that the probability level in the frequency domain now is $\geq \alpha$, due to the rank properties of the mapping T as discussed above. Again, these results rely on the linearity of the parameterization $G(e^{i\omega},\theta)$. A more extensive discussion on probability levels related to ellipsoidal uncertainty regions in the Nyquist plane can be found in [31].
When given ellipsoids in the parameter or frequency domain, maximum amplitude and phase bounds can be calculated for the frequency response $G(e^{i\omega})$, as well as an upper bound for the $\mathcal{H}_\infty$-norm $\|G\|_\infty$ for all models G within $\mathcal{D}_\omega$.
For non-linearly parameterized model structures, the ellipsoidal uncertainty region in parameter space will generally not be mapped into an ellipsoidal uncertainty region in the frequency domain. However, procedures for calculating maximum amplitude and phase bounds directly on the basis of the parameter space uncertainty are available in terms of a sequence of simple algebraic tests

without requiring numerical optimization algorithms in [52]. Maximum $\mathcal{H}_\infty$-norm bounds and related robustness measures, as *e.g.* the Vinnicombe ν-gap metric, can also be calculated based on a constrained optimization technique involving linear matrix inequalities (LMIs) as shown in [32].
For linearly parameterized models, the ellipsoidal parameter uncertainty maps to an ellipsoidal uncertainty for each frequency ω, and maximum amplitude and phase bounds can directly be calculated in the frequency domain.

In this uncertainty bounding problem the importance of linearly parameterized output error model structures, one of the typical properties of orthogonal basis function models, is indicated by the following properties.

- The ellipsoidal parameter uncertainty bounds map directly to ellipsoidal frequency domain bounds.
- The parameter covariance matrix is analytically available also for finite number of data.
- Expressions for P_θ are available also in the situation $G_0 \in \mathcal{G}$ ($\mathcal{S} \notin \mathcal{M}$).
- Probability levels hold true not only asymptotically, but also for finite time if the noise signals are normally distributed.

Computationally simpler solutions for uncertainty bounding in phase and/or amplitude of the frequency response are obtained by using first-order approximations of the mappings from parameter to amplitude/phase. Consider the functions

$$f_{a,\omega}(\theta) = |G(e^{i\omega},\theta)|,$$
$$f_{p,\omega}(\theta) = \angle G(e^{i\omega},\theta)$$

then covariance information on $f_{a,\omega}$ and $f_{p,\omega}$ is obtained from the first-order approximations

$$\text{Cov}\{f_{a,\omega}(\widehat{\theta}_N)\} \approx \frac{1}{N} \left.\frac{\partial f_{a,\omega}(\theta)}{\partial\theta}^T\right|_{\theta=\widehat{\theta}_N} P_\theta \left.\frac{\partial f_{a,\omega}(\theta)}{\partial\theta}\right|_{\theta=\widehat{\theta}_N};$$

$$\text{Cov}\{f_{p,\omega}(\widehat{\theta}_N)\} \approx \frac{1}{N} \left.\frac{\partial f_{p,\omega}(\theta)}{\partial\theta}^T\right|_{\theta=\widehat{\theta}_N} P_\theta \left.\frac{\partial f_{p,\omega}(\theta)}{\partial\theta}\right|_{\theta=\widehat{\theta}_N}.$$

In Matlab's Identification Toolbox [165], this latter approximation is implemented for calculating amplitude and phase bounds of estimated frequency responses. A similar approach can be followed for bounding the uncertainty on pulse responses, step responses and pole/zero locations of estimated transfer functions.
Note that in consistent prediction error identification, reflected by the condition $\mathcal{S} \in \mathcal{M}$ or $G_0 \in \mathcal{G}$, model uncertainty is due to variance only, and there is no need to take account of bias issues, presuming that only the asymptotic case is considered ($N \to \infty$) and that issues of transient conditions can be neglected for sufficiently large number of data.

7.3 Prediction Error Identification with Reduced-order Model Structures

7.3.1 Introduction

If the classical assumption $\mathcal{S} \in \mathcal{M}$ is too strong and the situation occurs that the model structures used are not flexible enough to capture the real system, a different approach is required. Unmodelled dynamics will appear that cannot straightforwardly be incorporated in a variance error of the model. Incorporation of this unmodelled dynamics into a model uncertainty bound is rather complex if one pertains to a probabilistic framework for noise, as is common in the prediction error framework. Nevertheless, there exist a couple of approaches for incorporating this unmodelled dynamics in the model error bounds, as will be indicated in this section. The use of linearly parameterized model structures is instrumental in these approaches.
First, a computational approach will be presented due to Hakvoort [119] where bias errors are deterministically (worst-case) bounded, variance errors are probabilistically bounded, and frequency response uncertainty regions are calculated on a user-chosen frequency grid.
Second, we consider a related combined worst-case/probabilistic approach due to De Vries [69], which is particularly directed towards the use of periodic input signals.
Third, the stochastic embedding approach [105] is highlighted, in which also the under-modelling term is modelled in a stochastic framework, thus leading to a fully probabilistic approach.
In model error modelling [163], the under-modelling error is actually modelled by a high-order (linearly parameterized) auxiliary model, thus avoiding the problem of quantifying or bounding the bias error.

7.3.2 A Computational Approach to Probabilistic/Worst-case Error Bounding

In this approach to model uncertainty bounding, it is assumed that the measurement data has been generated by a dynamical system satisfying

$$y(t) = G_0(q)u(t) + v(t)$$

with v a stationary stochastic process, independent of u, and where G_0 allows a series expansion $G_0(z) = \sum_{k=1}^{\infty} g_0(k)F_k(z)$ with the expansion coefficients bounded by an *a priori* known bound

$$|g_0(k)| \leq \bar{g}(k), \quad k \geq 1, \tag{7.2}$$

and assuming an *a priori* known bound on the past input: $\max_{t \leq 0} |u(t)| \leq \bar{u}$. It is recognized that in an identification problem setting where the model structures are of limited complexity (reduced order), the difference between an estimated model $\widehat{G}(e^{i\omega})$ and the real system $G_0(e^{i\omega})$ will contain three terms:

- A variance error, due to noise and finite length of the data
- A bias (under-modelling) error
- An error due to non-zero initial conditions.

In the approach considered here, the first term is modelled in a probabilistic framework, whereas the second and third terms are modelled in a worst-case setting. In order to keep the technical exposition brief, the third term will be neglected here.
The type of result that follows is an expression of the type

$$\widehat{G}(e^{i\omega}) \in \mathcal{A}(\omega) \text{ with probability } \geq \alpha$$

with $\mathcal{A}(\omega)$ an (ellipsoidal) region in the frequency domain, and the expression holds true for all ω in a user-chosen frequency grid Ω.
The approach that is followed is:

- Use a linear model parameterization

$$\widehat{y}(t,\theta) = \sum_{k=1}^{n} g(k)F_k(q)u(t) = \varphi^T(t)\theta, \tag{7.3}$$

  with

$$\varphi^T(t) = [F_1(q)u(t) \ \cdots \ F_n(q)u(t)]$$

  and estimate the coefficients

$$\theta = [g(1) \ \cdots \ g(n)]^T$$

  in the series expansion by either a least squares (LS) or an instrumental variable (IV) prediction error technique.
- Overbound the effect of the neglected tail (worst-case setting) on the basis of the assumption in (7.2).
- Bound the noise-induced (variance) error in a probabilistic way.

With a least squares or IV parameter estimation technique, the parameter estimate is determined according to

$$\widehat{\theta}_N = \left[\frac{1}{N}\sum_{t=1}^{N}\zeta(t)\varphi^T(t)\right]^{-1}\frac{1}{N}\sum_{t=1}^{N}\zeta(t)y(t) \tag{7.4}$$

with $\zeta(t) = \varphi(t)$ in the LS situation, or $\zeta(t)$ a vector of instrumental signals, in the IV case.
The resulting estimated model is:

$$\widehat{G}(z) = \sum_{k=1}^{n}\widehat{g}(k)F_k(z) = \Gamma_n^T(z)\widehat{\theta}_N,$$

and when substituting expression (7.4), the frequency response of the model is given by

$$\widehat{G}(e^{i\omega}) = \Gamma_n^T(e^{i\omega}) \cdot \left[\frac{1}{N}\sum_{t=1}^{N}\zeta(t)\varphi^T(t)\right]^{-1} \frac{1}{N}\sum_{t=1}^{N}\zeta(t)y(t) \tag{7.5}$$

with $\Gamma_n^T(e^{i\omega}) = [F_1(e^{i\omega}) \ \cdots \ F_n(e^{i\omega})]$.
For each chosen value of ω, this can be written as

$$\widehat{G}(e^{i\omega}) = \frac{1}{N}\sum_{t=1}^{N}[r_1(t) + ir_2(t)]y(t)$$

with r_1, r_2 known real-valued deterministic signals being determined by Γ_n, ζ and φ to satisfy (7.5), and therefore being dependent on the particular choice of ω.
When denoting

$$G_0(q) = G(q, \theta_0) + G_a(q)$$

with

$$\begin{aligned} G(e^{i\omega}, \theta_0) &= \Gamma_n^T(e^{i\omega})\theta_0 \quad \text{ and } G(q,\theta_0)u(t) = \varphi^T(t)\theta_0 \\ G_a(q) &= \sum_{k=n+1}^{\infty} g_0(k)F_k(q), \end{aligned} \tag{7.6}$$

and substituting $y(t) = G_0(q)u(t) + v(t)$ into (7.5), it follows that

$$\widehat{G}(e^{i\omega}) - G(e^{i\omega}, \theta_0) = \frac{1}{N}\sum_{t=1}^{N}[r_1(t) + ir_2(t)][G_a(q)u(t) + v(t)]$$

and consequently

$$\widehat{G}(e^{i\omega}) - G_0(e^{i\omega}) = -G_a(e^{i\omega}) + \frac{1}{N}\sum_{t=1}^{N}[r_1(t) + ir_2(t)][G_a(q)u(t) + v(t)].$$

As a result, the estimated model $\widehat{G}(e^{i\omega})$ can be shown to satisfy, for a fixed frequency ω,

$$\widehat{G}(e^{i\omega}) - G_0(e^{i\omega}) = \delta(\omega) + \eta(\omega) + \mu(\omega)$$

where

- $\delta(\omega) := -G_a(e^{i\omega})$ reflects the neglected tail of the expansion and therefore refers to unmodelled dynamics; it is present due to the limited complexity of the model set.
- $\eta(\omega) = \frac{1}{N}\sum_{t=1}^{N}[r_1(t) + ir_2(t)]G_a(q)u(t)$ reflects a bias term on the estimated coefficients due to the neglected tail of the system's response;
- $\mu(\omega) = \frac{1}{N}\sum_{t=1}^{N}[r_1(t) + ir_2(t)]v(t)$ reflects a variance contribution due to the noise disturbance on the data.

Bounding the Bias Contributions

The unmodelled dynamics term $\delta(\omega) = -G_a(e^{i\omega})$ can be worst-case bounded by using the *a priori* known upper bound for $g_0(k), k > n$,

$$|\mathrm{Re}\{G_a(e^{i\omega})\}| \leq \bar{\delta}_R := \sum_{k=n+1}^{\infty} \bar{g}(k)|Re\{F_k(e^{i\omega})\}|,$$

$$|\mathrm{Im}\{G_a(e^{i\omega})\}| \leq \bar{\delta}_I := \sum_{k=n+1}^{\infty} \bar{g}(k)|Im\{F_k(e^{i\omega})\}|.$$

The bias term $\eta(\omega)$ can be worst-case bounded using the same prior information:

$$|\mathrm{Re}\{\eta(\omega)\}| \leq \bar{\eta}_R := \sum_{k=n+1}^{\infty} \bar{g}(k)\left|\frac{1}{N}\sum_{t=1}^{N} r_1(t)F_k(q)u(t)\right|,$$

$$|\mathrm{Im}\{\eta(\omega)\}| \leq \bar{\eta}_I := \sum_{k=n+1}^{\infty} \bar{g}(k)\left|\frac{1}{N}\sum_{t=1}^{N} r_2(t)F_k(q)u(t)\right|$$

Because all terms in the right-hand side expressions are known and the infinite sums converge, they can be calculated to within arbitrary accuracy. Note that effectively tighter bounds can also be calculated by bounding the two terms $\delta(\omega) + \eta(\omega)$ together, rather than constructing separate bounds for $\delta(\omega)$ and $\eta(\omega)$.

Bounding the Variance Contribution

For bounding the noise-induced term in the model uncertainty, use is made of the asymptotic prediction error theory, which can specify the asymptotic distribution of the random vector

$$\frac{1}{\sqrt{N}}\begin{bmatrix}\sum_{t=1}^{N} r_1(t)v(t) \\ \sum_{t=1}^{N} r_2(t)v(t)\end{bmatrix}. \tag{7.7}$$

The constructive result that can be employed is that

$$\frac{1}{N}\left(\sum_{t=1}^{N} r_1(t)v(t) \quad \sum_{t=1}^{N} r_2(t)v(t)\right)\left(\Lambda_{r_1r_2}^{N}\right)^{-1}\begin{pmatrix}\sum_{t=1}^{N} r_1(t)v(t) \\ \sum_{t=1}^{N} r_2(t)v(t)\end{pmatrix} \overset{N\to\infty}{\longrightarrow} \chi^2(2) \tag{7.8}$$

with the 2×2 matrix $\Lambda^N_{r_1 r_2}$ determined by

$$\Lambda^N_{r_1 r_2}(i,j) = \sum_{\tau=-N+1}^{N-1} \frac{N-|\tau|}{N} R^N_{r_i r_j}(\tau) R_v(\tau), \quad i,j = 1,2 \tag{7.9}$$

and

$$R^N_{r_i r_j}(\tau) := \frac{1}{N+\tau} \sum_{t=1}^{N+\tau} r_i(t) r_j(t-\tau), \ \tau = -N+1, \ldots, 0,$$

$$R^N_{r_i r_j}(\tau) := \frac{1}{N-\tau} \sum_{t=1}^{N-\tau} r_i(t+\tau) r_j(t), \ \tau = 1, \ldots, N-1, \quad i,j = 1,2.$$

For a fixed value of ω, the variance error is specified by an ellipsoidal region in the complex plane, determined by (7.8), which is valid at a pre-chosen level of probability determined by the χ^2 distribution. If the number of data N is finite, the result holds true in case the noise disturbance term $v(t)$ has a Gaussian distribution.

Bounding the Model Uncertainty

The worst-case bounded bias errors can be combined with the probabilistic variance errors. This is achieved by considering the random vector

$$\begin{bmatrix} \mathrm{Re}\left(\widehat{G}(e^{i\omega}) - G_0(e^{i\omega})\right) - \eta_R + \delta_R \\ \mathrm{Im}\left(\widehat{G}(e^{i\omega}) - G_0(e^{i\omega})\right) - \eta_I + \delta_I \end{bmatrix}. \tag{7.10}$$

For bounding this combined random variable, the following lemma can be employed:

Lemma 7.1. *Consider a real valued parameter estimate $\widehat{a} \in \mathbb{R}^n$ for a vector coefficient a_0, satisfying*

$$[\widehat{a} - a_0 - a_b]^T P^{-1} [\widehat{a} - a_0 - a_b] \leq c \quad w.p. \ \alpha$$

with a_b a fixed (bias) term and $P^{-1} = M^T M$. Then

$$[\widehat{a} - a_0]^T P^{-1} [\widehat{a} - a_0] \leq [\sqrt{c} + \|M a_b\|_2]^2 \quad w.p. \ \geq \alpha.$$

Proof. The lemma can simply be proved by applying the triangle inequality for norms, *i.e.* $\|M(\widehat{a} - a_0 - a_b)\|_2 \geq \|M(\widehat{a} - a_0)\|_2 - \|M a_b\|_2$.

Applying this lemma to the (two-dimensional) random variable of (7.10) leads to the result as formulated in the following proposition.

Proposition 7.2. *Consider the LS/IV-estimate $\widehat{\theta}_N$ and corresponding frequency response $\widehat{G}(e^{i\omega})$. Then for a fixed frequency ω, and $N \to \infty$,*

$$\begin{bmatrix} \mathrm{Re}\left(\widehat{G}(e^{i\omega}) - G_0(e^{i\omega})\right) \\ \mathrm{Im}\left(\widehat{G}(e^{i\omega}) - G_0(e^{i\omega})\right) \end{bmatrix}^T \left(\Lambda^N_{r_1 r_2}\right)^{-1} \begin{bmatrix} \mathrm{Re}\left(\widehat{G}(e^{i\omega}) - G_0(e^{i\omega})\right) \\ \mathrm{Im}\left(\widehat{G}(e^{i\omega}) - G_0(e^{i\omega})\right) \end{bmatrix} \leq$$

$$\left(\sqrt{\tfrac{c_{\chi,\alpha}}{N}} + \sqrt{\gamma_{11}^2 + \gamma_{21}^2}\,(\bar{\eta}_R + \bar{\delta}_R) + \sqrt{\gamma_{12}^2 + \gamma_{22}^2}\,(\bar{\eta}_I + \bar{\delta}_I)\right)^2, \quad w.p. \geq \alpha,$$

where the symmetric matrix $\Gamma = \begin{bmatrix} \gamma_{11} & \gamma_{12} \\ \gamma_{12} & \gamma_{22} \end{bmatrix}$ is determined by $\Gamma^T \Gamma = \left(\Lambda^N_{r_1 r_2}\right)^{-1}$, and $c_{\chi,\alpha}$ corresponds to a probability α in the χ^2-distribution with 2 degrees of freedom, such that $x \in \chi^2(2)$ implies $\mathrm{Prob}(x \leq c_{\chi,\alpha}) = \alpha$.

The proof of this proposition is found in [119].
Provided that the matrix $\Lambda^N_{r_1 r_2}$ is invertible, ellipsoidal model uncertainty regions are specified at a user-specified probability level. Invertibility of $\Lambda^N_{r_1 r_2}$ will generally be the case, except for frequencies $\omega = 0, \pi$. For these frequencies $r_2(t)$ will be zero, naturally implying that there is no imaginary model uncertainty for $\omega = 0, \pi$.

The different error sources can be traded off. In particular, the truncation value n can be used to make a trade-off between bias and variance. A larger value of n means a smaller bias, but a larger variance. An optimal value can be determined by varying n so as to minimize the size of the uncertainty.
An important step in the construction of uncertainty bounds is the choice of the *a priori* selected bound on the convergence rate of the expansion coefficients, reflected by $\bar{g}_0(k)$. Uncertainty bounds on identified parameters $\widehat{\theta}_N$ are used to adjust an exponential decay rate

$$\bar{g}_0(k) = M\rho^k$$

for real valued constants M, ρ, $|\rho| \leq 1$, so as to obtain tight upper bounds for the expansion coefficient uncertainty bounds. For this construction, an upper value k_m of k is selected in such a way that for $\ell > k_m$, zero is contained in the uncertainty set for $\hat{g}(\ell)$.

It is emphasized that the identification of the LS/IV model is not a goal as such, but serves only as a basis for the construction of system uncertainty regions. The analysis presented is essentially possible because of the linear regression type of estimation, induced by the use of generalized orthonormal basis functions. Through an appropriate choice of basis functions, low orders n of the identified model can result in both a small bias and a small variance contribution.

There is one unknown quantity left in the uncertainty description as formulated in Proposition 7.2, being the autocovariance function of the noise, $R_v(\tau)$, which appears in expression (7.9) for $\Lambda^N_{r_1 r_2}$. It can be shown that by using the residual signal

$$\hat{\varepsilon}(t) = y(t) - \varphi^T(t)\hat{\theta}_N$$

evaluated on a data set that is independent of the data used for identification, an adapted estimate:

$$\hat{\Lambda}^N_{r_i r_j} = \sum_{\tau=-w(N)}^{w(N)} c_w(\tau)\frac{N-|\tau|}{N} R^N_{r_i r_j} \hat{R}^N_{\hat{\varepsilon}}(\tau)$$

can be constructed with $R^N_{\hat{\varepsilon}}(\tau)$ the sample covariance of $\hat{\varepsilon}(t)$, that under weak assumptions on $c_w(\tau)$ satisfies $\lim_{N\to\infty} \hat{\Lambda}^N_{r_i r_j} \geq \Lambda_{r_i r_j}$. This implies that when using the new estimate $\hat{\Lambda}^N_{r_i r_j}$ in the uncertainty bounding procedure, the probabilistic expressions for the model uncertainty regions are still valid, albeit at the cost of introducing some conservatism. Note that as an alternative, also a parametric estimate can be used for estimating the covariance function $R^N_{\hat{\varepsilon}}(\tau)$.

The uncertainty bounds can be computed in any user-chosen frequency grid and lead to a Nyquist curve with uncertainty regions in user-specified frequencies, as illustrated in Figure 7.1. In all cases, the three different sources of uncertainty can be distinguished, which allows the user to determine which part is dominant and to adjust the experimental setup so as to reduce the overall uncertainty bound.

One point of difference between the method presented here and the classical statistical approach is that in the latter situation, there is a parametric uncertainty set, leading to a global confidence interval valid for a particular

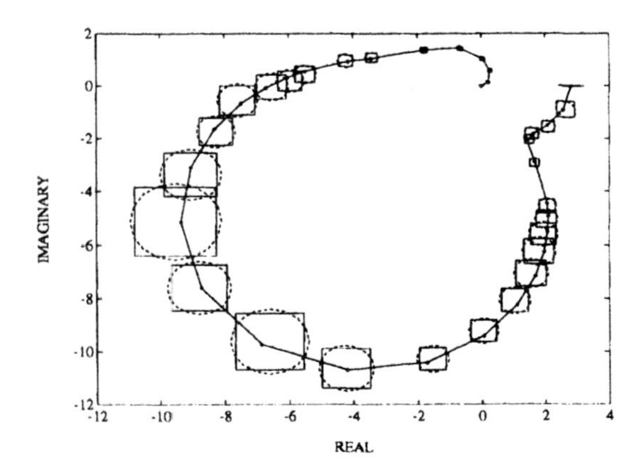

Fig. 7.1. Nyquist diagram with identified uncertainty bounds (ellipsoids) according to Proposition 7.2. The rectangular regions follow when bounding the real and imaginary parts separately.

probability level. When considering unmodelled dynamics, as considered here, the parametric structure seems to be lost, and relations between different frequencies are not simply taken into account. An alternative formulation of the method, indicated in the next subparagraph, shows that this problem can be circumvented by retaining the parametric structure of the problem.
The procedure sketched in this section is applicable to both open-loop and closed-loop identification problems, as well as to multi-variable systems [119]. Industrial application of the procedure can be found in [116] for an industrial glass tube manufacturing process and in [63] for the multi-variable servo control system in a wafer stepper.

A Formulation That Retains the Parametric Structure

The computational method described in this section can equivalently be formulated while retaining the parametric structure in the problem.
With the expression for the output data of the system

$$y(t) = \varphi^T(t)\theta_0 + \varphi_e^T(t)\theta_e + v(t)$$

with $\varphi_e^T(t)\theta_e = G_a(q)u(t)$ and $\theta_e = [g_0(n+1) \; g_0(n+2) \; \cdots]^T$ (see (7.6)) and using the notation

$$\begin{aligned} \Phi^T &= [\varphi(1) \; \cdots \; \varphi(N)] \\ {\Phi_e}^T &= [\varphi_e(1) \; \cdots \; \varphi_e(N)] \\ {\Phi_\zeta}^T &= [\zeta(1) \; \cdots \; \zeta(N)] \end{aligned}$$

the parameter estimate can be written as

$$\widehat{\theta}_N = (\frac{1}{N}\Phi_\zeta^T\Phi)^{-1}\frac{1}{N}\Phi_\zeta^T[\Phi\theta_0 + \Phi_e\theta_e + \mathsf{V}]$$

with $\mathsf{V} = [v(1) \; \cdots \; v(N)]^T$, or alternatively

$$\widehat{\theta}_N = \theta_0 + \Psi\Phi_e\theta_e + \Psi\mathsf{V} \tag{7.11}$$

where

$$\Psi = (\frac{1}{N}\Phi_\zeta^T\Phi)^{-1}\frac{1}{N}\Phi_\zeta^T.$$

Writing

$$G_0(q) = G(q,\theta_0) + G_a(q)$$

it follows from (7.11) that

$$\widehat{G}(e^{i\omega}) - G_0(e^{i\omega}) = G_a(e^{i\omega}) + \Gamma_n^T(e^{i\omega})[\Psi\Phi_e\theta_e + \Psi\mathsf{V}]$$

and with $G_a(e^{i\omega}) = \Gamma_{n+1,\infty}^T(e^{i\omega})\theta_e$:

$$\begin{aligned}\widehat{G}(e^{i\omega}) - G_0(e^{i\omega}) &= [\Gamma_{n+1,\infty}^T(e^{i\omega}) + \Gamma_n^T(e^{i\omega})\Psi\Phi_e]\theta_e + \Gamma_n^T(e^{i\omega})\Psi\mathsf{V} \\ &= B_n^T(e^{i\omega})\theta_e + \Gamma_n^T(e^{i\omega})\Psi\mathsf{V}.\end{aligned}$$

In this expression, the first term with $B_n^T(e^{i\omega}) = \Gamma_{n+1,\infty}^T(e^{i\omega}) + \Gamma_n^T(e^{i\omega})\Psi\Phi_e$ refers to two bias errors in the model: one error due to unmodelled dynamics, and one term due to a bias in the estimated parameters.
Separating the real and imaginary parts of the equation results in

$$\begin{bmatrix}\mathrm{Re}(\widehat{G}(e^{i\omega}) - G_0(e^{i\omega})) \\ \mathrm{Im}(\widehat{G}(e^{i\omega}) - G_0(e^{i\omega}))\end{bmatrix} - \begin{bmatrix}\mathrm{Re}(B_n^T(e^{i\omega})\theta_e) \\ \mathrm{Im}(B_n^T(e^{i\omega})\theta_e)\end{bmatrix} \in \mathcal{N}(0, P), \tag{7.12}$$

with

$$P = T^T(e^{i\omega})\Psi\Lambda_v\Psi^T T(e^{i\omega}),$$

with $T(e^{i\omega})$ as defined in Section 7.2 and where Λ_v is the covariance matrix of the disturbance v, *i.e.* $\Lambda_v = \mathsf{E}\{\mathsf{V}\mathsf{V}^T\}$.
Exploiting again Lemma 7.1, a statistical bound can be derived for the (two-dimensional) random variable

$$\begin{bmatrix}\mathrm{Re}(\widehat{G}(e^{i\omega}) - G_0(e^{i\omega})) \\ \mathrm{Im}(\widehat{G}(e^{i\omega}) - G_0(e^{i\omega}))\end{bmatrix} \tag{7.13}$$

while for constructing the bound, it is necessary to have an a overbound of the expression

$$\left\|\begin{matrix}\mathrm{Re}(B_n^T(e^{i\omega})\theta_e) \\ \mathrm{Im}(B_n^T(e^{i\omega})\theta_e)\end{matrix}\right\|_2.$$

Moreover, it is also possible to consider the random variable (7.13) in a sequence of frequencies ω_j simultaneously, thus increasing the row dimension of (7.13) and adjusting the appropriate dimensions in (7.12). Note that the dimension of P then also increases by adjusting the column dimension of matrix $T(e^{i\omega})$.
More details on this alternative formulation are provided in [81].

7.3.3 A Mixed Probabilistic Worst-case Approach with Periodic Inputs

In a second approach to model uncertainty bounding in the prediction error framework, similar starting points are taken as in the procedure of the previous section, *i.e.*

- disturbance signals in the data are assumed to be realizations of stationary stochastic processes, and
- unmodelled dynamics, due to a model structure that does not capture all process dynamics, is a deterministic phenomenon that can be bounded in a worst-case sense.

In order to separate the contributions of disturbance signals and unmodelled dynamics, use is made of the principle that when repeating experiments using the same input signal, the unmodelled dynamics terms in the estimated models should be the same, while the random noise contributions should average out. This property of experiment repetition, reflected by the choice of periodic input signals in a single experiment, is exploited to separate the model uncertainty terms originating from the two different sources, mentioned above. It is again assumed that $G_0(q)$ allows a series expansion

$$G_0(z) = \sum_{k=1}^{\infty} g_0(k) F_k(z)$$

for an orthonormal basis induced by $\{F_k(z)\}_{k=1,\cdots\infty}$, while there exists an *a priori* known bound, reflected by real-valued constants K, η with $\eta < 1$ such that $|g_0(k)| \leq K\eta^k$.
The basic approach, presented in [69, 70], contains the following principal steps:

1. Perform an identification experiment with a periodic input signal of length N, composed of r periods of length N_0, $r > 1$.
2. For each of the r periods, identify a non-parametric empirical transfer function estimate (ETFE), according to

$$\widehat{G}_j(e^{i\omega}) = \frac{Y_j(e^{i\omega})}{U_j(e^{i\omega})}$$

for all values of ω in the DFT-frequency grid

$$\Omega := \{\frac{2\pi k}{N_0}, k = 0, \cdots N_0 - 1 \text{ restricted to } | \ |U_j(e^{\frac{2\pi k}{N_0}})| \neq 0\},$$

where $U_j(e^{i\omega})$ is the discrete Fourier transform of the input signal over period number j,

$$U_j(e^{i\omega}) = \frac{1}{\sqrt{N_0}} \sum_{t=0}^{N_0-1} u_j(t) e^{-i\omega t}.$$

3. For each of the r frequency response estimates $\{\widehat{G}_j(e^{i\omega}), \omega \in \Omega\}$, construct a parametric model by fitting an orthonormal basis function model structure $\Gamma_n^T(e^{i\omega})\theta$ to the frequency response data, with $\Gamma_n(e^{i\omega})$ as defined in (7.5). The resulting parameter estimate for interval j is denoted by $\widehat{\theta}_j$.
4. For each of the r parameter estimates $\widehat{\theta}_j$ a frequency response $\widehat{G}_j^f(e^{i\omega})$ is calculated according to $\widehat{G}_j^f(e^{i\omega}) = \Gamma_n^T(e^{i\omega})\widehat{\theta}_j$.

5. Error bounds are constructed for the averaged frequency response

$$\widehat{G}^f(e^{i\omega}) = \frac{1}{r}\sum_{j=1}^{r} \widehat{G}^f_j(e^{i\omega})$$

partly on the basis of a sample covariance estimate of the variability in $\widehat{G}^f(e^{i\omega})$.
The constructed uncertainty bounds contain
   - A boxed area per frequency induced by the bias / unmodelled dynamics;
   - An ellipsoidal area per frequency delivering a probabilistic confidence interval induced by the variance component in the model.

In the parametric identification, use is made of the linear regression model structure

$$G(q,\theta) = \sum_{k=1}^{n} c_k F_k(q)$$

with $\theta = [c_1 \; c_2 \; \cdots \; c_n]^T$, and a parameter estimate $\widehat{\theta}_j$ is determined for every period of the input data

$$\widehat{\theta}_j = \arg\min_{\theta} \|W(\widehat{\mathsf{G}}_{\mathsf{j}} - \Gamma\theta)\|_2^2$$

with

$$\begin{aligned} \widehat{\mathsf{G}}_{\mathsf{j}} &= [\widehat{G}_j(e^{i\zeta_1}) \;\; \widehat{G}_j(e^{i\zeta_2}) \;\; \cdots \;\; \widehat{G}_j(e^{i\zeta_{N_p}})]^T \\ \Gamma &= [\Gamma_n(e^{i\zeta_1}) \;\; \Gamma_n(e^{i\zeta_2}) \;\; \cdots \;\; \Gamma_n(e^{i\zeta_{N_p}})]^T \\ \widehat{G}_j(e^{i\zeta_m}) &= \frac{Y_j(e^{i\zeta_m})}{U_j(e^{i\zeta_m})} \qquad m = 1, \cdots N_p, \end{aligned}$$

where the complex numbers $e^{i\zeta_j}$ occur in complex conjugate pairs and ζ_j are chosen from the frequency grid Ω.
The weighting matrix W can be used to minimize the variance of the estimated parameters or to affect the bias distribution over frequency. The separate parameter estimates can be written as

$$\widehat{\theta}_j = \Psi\widehat{\mathsf{G}}_{\mathsf{j}} \;\text{ with } \Psi = (\Gamma^* W^* W \Gamma)^{-1}\Gamma^* W^* W. \tag{7.14}$$

and with the final averaged parameter

$$\widehat{\theta} = \frac{1}{r}\sum_{j=1}^{r}\widehat{\theta}_j,$$

it follows that because of the linear structure of the parameter estimates:

$$\widehat{\theta} = \Psi\widehat{\mathsf{G}} \;\text{ with } \widehat{\mathsf{G}} = \frac{1}{r}\sum_{j=1}^{r}\widehat{\mathsf{G}}_{\mathsf{j}}.$$

Employing the system's equations, it follows that

$$\widehat{G}_j(e^{i\omega}) = G_0(e^{i\omega}) + S_j(e^{i\omega}) + \frac{V_j(e^{i\omega})}{U_j(e^{i\omega})} \tag{7.15}$$

where $S_j(e^{i\omega})$ is reflecting the effect of unknown past inputs, preceding the time interval that is considered. This term will be discarded, as it will be assumed that all transients have decayed to 0.
For the data generating system G_0, we additionally write

$$\begin{aligned}
\mathsf{G}_0 &= [G_0(e^{i\zeta_1}) \;\; G_0(e^{i\zeta_2}) \;\cdots\; G_0(e^{i\zeta_{N_p}})]^T \\
\Gamma_e &= [\Gamma_{n+1,\infty}(e^{i\zeta_1}) \;\; \Gamma_{n+1,\infty}(e^{i\zeta_2}) \;\cdots\; \Gamma_{n+1,\infty}(e^{i\zeta_{N_p}})]^T, \\
\mathsf{N} &= \left[\frac{\frac{1}{r}\sum_{j=1}^r V_j(e^{i\zeta_1})}{U_j(e^{i\zeta_1})} \;\cdots\; \frac{\frac{1}{r}\sum_{j=1}^r V_j(e^{i\zeta_{N_p}})}{U_j(e^{i\zeta_{N_p}})}\right]^T
\end{aligned}$$

so that

$$\widehat{\mathsf{G}} = \Gamma\theta_0 + \Gamma_e\theta_e + \mathsf{N}.$$

Accordingly

$$\widehat{\theta} = \Psi\widehat{\mathsf{G}} = \theta_0 + \Psi\Gamma_e\theta_e + \Psi\mathsf{N}.$$

This leads to a bias term in the estimated frequency response

$$Z^f(e^{i\omega}) = \Gamma_n^T(e^{i\omega})\Psi\Gamma_e\theta_e,$$

while the unmodelled dynamics term is given by

$$G_a(e^{i\omega}) = \sum_{k=n+1}^{\infty} g_0(k)F_k(e^{i\omega}) = \Gamma_{n+1,\infty}^T(e^{i\omega})\theta_e,$$

constituting:

$$\mathsf{E}\{\widehat{G}^f(e^{i\omega}) - G_0(e^{i\omega})\} = Z^f(e^{i\omega}) - G_a(e^{i\omega}).$$

Bounding Bias and Unmodelled Dynamics

In [69] the two bias terms are bounded simultaneously, using the *a priori* assumed exponential bound on the system's expansion coefficients.

$$|\mathrm{Re}\{G_a(e^{i\omega}) - Z^f(e^{i\omega})\}| \leq \bar{\beta}_R(\omega)$$

with

$$\begin{aligned}
\bar{\beta}_R(\omega) := K\sum_{k=n+1}^{n_h} \eta^{k-1}\left|\mathrm{Re}\{F_k(e^{i\omega}) - \sum_{\ell=1}^{N_p}\Upsilon(\ell)F_k(e^{i\zeta_\ell})\}\right| \\
+K\frac{\eta^{n_h}}{1-\eta}\left[|F_1(e^{i\omega})| + \sum_{\ell=1}^{N_p}|\Upsilon(\ell)|F_1(e^{i\zeta_\ell})|\right]
\end{aligned}$$

for any $n_h > n$, and $\Upsilon = \Gamma_n^T(e^{i\omega})\Psi$. A similar bound $\bar{\beta}_I(\omega)$ can be derived for the imaginary part.
Note that $G_a(e^{i\omega})$ and $Z^f(e^{i\omega})$ are contributions of, respectively, the under-modelling term and the bias term in the parameters. They are bounded in a way that is very similar to the bounds that were constructed for the previous method considered in Section 7.3.2.

Bounding the Model Uncertainty

We can now formulate the following result on the uncertainty bound of the frequency response, $\widehat{G}^f(e^{i\omega})$.

Proposition 7.3 ([69]). *Consider the experimental setup with a periodic input signal containing $r > 1$ periods of N_0 data points, and assume that all transients in the data have decayed to 0. For the estimated frequency response $\widehat{G}^f(e^{i\omega})$ as considered above, it holds that, asymptotically in N_0, for all $\omega \in [0, 2\pi)$,*

$$G_0(e^{i\omega}) - \widehat{G}^f(e^{i\omega}) \in \mathcal{A}(e^{i\omega}) \quad w.p. \geq F_\alpha(2, r-2) \tag{7.16}$$

where the set $\mathcal{A}(e^{i\omega})$ is given by all $\Delta \in \mathbb{C}$, $\Delta(e^{i\omega}) = \Delta_1(e^{i\omega}) + \Delta_2(e^{i\omega})$, $\Delta_1, \Delta_2 \in \mathbb{C}$, satisfying

$$\begin{bmatrix} \mathrm{Re}\{\Delta_1(e^{i\omega})\} \\ \mathrm{Im}\{\Delta_1(e^{i\omega})\} \end{bmatrix}^T \widehat{\Sigma}_r^{-1} \begin{bmatrix} \mathrm{Re}\{\Delta_1(e^{i\omega})\} \\ \mathrm{Im}\{\Delta_1(e^{i\omega})\} \end{bmatrix} \leq \frac{2(r-1)}{r-2}\alpha$$

$$|\mathrm{Re}\{\Delta_2(e^{i\omega})\}| \leq \bar{\beta}_R(\omega)$$

$$|\mathrm{Im}\{\Delta_2(e^{i\omega})\}| \leq \bar{\beta}_I(\omega)$$

where the estimated covariance matrix is given by

$$\widehat{\Sigma}_r := \frac{1}{r(r-1)} \sum_{i=1}^{r} \begin{bmatrix} \mathrm{Re}\{\widehat{G}_i^f(e^{i\omega}) - \widehat{G}^f(e^{i\omega})\} \\ \mathrm{Im}\{\widehat{G}_i^f(e^{i\omega}) - \widehat{G}^f(e^{i\omega})\} \end{bmatrix} \begin{bmatrix} \mathrm{Re}\{\widehat{G}_i^f(e^{i\omega}) - \widehat{G}^f(e^{i\omega})\} \\ \mathrm{Im}\{\widehat{G}_i^f(e^{i\omega}) - \widehat{G}^f(e^{i\omega})\} \end{bmatrix}^T.$$

The uncertainty region $\mathcal{A}(\omega)$ in the proposition is composed as a summation of a (soft-bounded) ellipsoidal area and a (hard-bounded) boxed area.
The particular F-distributions in this proposition result from Hotelling's T^2 statistic, see *e.g.* [69]. The F-distribution appears as the result of the fact that in the current situation, the variance of the frequency response model is estimated on the basis of the same data that is used for model identification. The particular advantage of this approach in relation to the method of Section 7.3.2 is that the periodic nature of the input signal allows the estimation of the necessary noise statistics, that are required for specification of the variance

part of the uncertainty bounds directly from the identification data. In the previous method, an independent data sequence is used for estimating the noise covariance matrix.
The model uncertainty bound for this method is specified slightly different from the one in Section 7.3.2. Note however that also in the result of Proposition 7.3, the structural property of Lemma 7.1 can be employed, so as to maintain an ellipsoidal uncertainty region in the frequency domain rather than a summation of an ellipsoid and a box.

In the methods of this section and the previous one, an important step is the specification of the *a priori* chosen exponential bound on the system's expansions coefficients. As explained in Section 7.3.2, the selection of this bound can be done iteratively in order to adjust the bound to the decay rate of the estimated expansion coefficient uncertainty bounds, so as to avoid unnecessarily conservative results.

The method considered in this section is further explored and analysed, incorporating non-zero transient conditions, in [69, 70].

7.3.4 Stochastic Embedding

In the previous two methods for uncertainty bounding, variance errors were considered as random variables, whereas bias and under-modelling errors were bounded deterministically in a worst-case setting. In the stochastic embedding method, as introduced in [105], the under-modelling and bias errors are also modelled as stochastic processes, and the parameters describing their properties are estimated from data. The result is an ellipsoidal uncertainty region for the estimated frequency response.
The method is very closely related to the computational approach of Hakvoort, as described in Section 7.3.2, and so we will stay very close to the exposition presented there.
Again, the system is assumed to be written as

$$y(t) = G_0(q)u(t) + v(t) \tag{7.17}$$

with $v(t)$ a Gaussian stationary stochastic process, while G_0 is assumed to allow a decomposition:

$$G_0(q) = \sum_{k=1}^{n} g_0(k)F_k(q) + G_\Delta(q)$$

with $G_\Delta(e^{i\omega})$ a zero-mean Gaussian stochastic process, written as

$$G_\Delta(q) = \sum_{k=1}^{L} \eta(k)F_k(q)$$

while the stochastic properties of G_Δ are characterized by

$$\mathsf{E}\{\eta(k)^2\} = \alpha\lambda^k \tag{7.18}$$

and the real-valued coefficients α and λ act as unknown parameters to be estimated from data. It is further assumed that the stochastic processes $v(t)$ and G_Δ are independent.
The stochastic embedding approach was originally derived for FIR representations, reflected by $F_k(z) = z^{-k}$, but actually work for any choice of fixed denominator or basis function model structure, as was already observed in [105].

Writing the output signal of the system as

$$y(t) = \varphi^T(t)\theta_0 + \varphi_e^T(t)\eta + v(t)$$

with similar notation as in Section 7.3.2, but now with

$$\varphi_e(t) = [F_1(q)u(t) \ \cdots \ F_n(q)u(t)]$$

and Φ and Φ_e as defined before, it follows that the least squares estimate $\widehat{\theta}_N$ is specified by

$$\begin{aligned}\widehat{\theta}_N &= (\frac{1}{N}\Phi\Phi^T)^{-1}\frac{1}{N}\Phi^T\mathsf{Y} \\ &= \theta_0 + \Psi\Phi_e\eta + \Psi\mathsf{V}\end{aligned} \tag{7.19}$$

with $\mathsf{V} = [v(1) \ \cdots v(N)]^T$, and

$$\Psi = (\frac{1}{N}\Phi\Phi^T)^{-1}\frac{1}{N}\Phi^T.$$

Because the difference between $G(\widehat{\theta}_N)$ and G_0 is constituted by the parameter vector

$$\begin{pmatrix}\widehat{\theta}_N - \theta_0 \\ -\eta\end{pmatrix}$$

we analyse this parameter vector to conclude from (7.19) that

$$\mathrm{Cov}\begin{pmatrix}\widehat{\theta}_N - \theta_0 \\ -\eta\end{pmatrix} = P_z,$$

with

$$P_z = \begin{bmatrix}\Psi[\Phi_e\Lambda_\eta{\Phi_e}^T + \Lambda_v]\Psi^T & -\Psi\Phi_e\Lambda_\eta \\ -\Lambda_\eta{\Phi_e}^T\Psi^T & \Lambda_\eta\end{bmatrix},$$

while $\Lambda_\eta := \mathsf{E}\{\eta\eta^T\}$ and $\Lambda_v := \mathsf{E}\{\mathsf{V}\mathsf{V}^T\}$.
Writing

$$\widehat{G}(e^{i\omega}) - G_0(e^{i\omega}) = [\Gamma_n^T(e^{i\omega}) \ \ \Gamma_e^T(e^{i\omega})]\begin{pmatrix}\widehat{\theta}_N - \theta_0 \\ -\eta\end{pmatrix}$$

with $\Gamma_e(e^{i\omega}) = [F_1(e^{i\omega}) \ \cdots \ F_L(e^{i\omega})]^T$, and denoting

$$\Pi(e^{i\omega}) := \begin{bmatrix} \text{Re}\{[\Gamma_n^T(e^{i\omega}) \;\; \Gamma_e^T(e^{i\omega})]\} \\ \text{Im}\{[\Gamma_n^T(e^{i\omega}) \;\; \Gamma_e^T(e^{i\omega})]\} \end{bmatrix}$$

it follows that

$$\text{Cov}\left\{\begin{bmatrix} \text{Re}\left(\widehat{G}(e^{i\omega}) - G_0(e^{i\omega})\right) \\ \text{Im}\left(\widehat{G}(e^{i\omega}) - G_0(e^{i\omega})\right) \end{bmatrix}\right\} = \Pi(e^{i\omega}) P_z \Pi^T(e^{i\omega}).$$

If the prior distributions of η and $v(t)$ are chosen to be Gaussian, then $\widehat{G}(e^{i\omega}) - G_0(e^{i\omega})$ is Gaussian distributed and hence quadratic functions have χ^2 distributions, so that the model uncertainty can be bounded accordingly by the following expression.

Proposition 7.4 ([105]). *Consider the assumptions on the data generating system, the under-modelling error and the noise as formulated in (7.17) – (7.18). Then*

$$\begin{bmatrix} \text{Re}\left(\widehat{G}(e^{i\omega}) - G_0(e^{i\omega})\right) \\ \text{Im}\left(\widehat{G}(e^{i\omega}) - G_0(e^{i\omega})\right) \end{bmatrix}^T P_\omega^{-1} \begin{bmatrix} \text{Re}\left(\widehat{G}(e^{i\omega}) - G_0(e^{i\omega})\right) \\ \text{Im}\left(\widehat{G}(e^{i\omega}) - G_0(e^{i\omega})\right) \end{bmatrix} \leq c_{\chi,\alpha}/N$$

with probability $\geq \alpha$, where $P_\omega = \Pi(e^{i\omega}) P_z \Pi^T(e^{i\omega})$ and $c_{\chi,\alpha}$ corresponds to a probability α in the χ^2-distribution with 2 degrees of freedom.

The only thing missing in the result of the proposition is the specification of the components Λ_η and Λ_v that occur in the covariance matrix P_z and thus in the covariance matrix P_ω. The estimation of these quantities is considered next.

Estimating the Parameters of Under-modelling and Noise

For estimating the parameters of the covariance matrices Λ_η and Λ_v, we consider the N-vector of residuals:

$$\varepsilon = \mathsf{Y} - \Phi\widehat{\theta}_N = [I - \Phi(\Phi^T\Phi)^{-1}\Phi^T]\mathsf{Y}.$$

Because the matrix between brackets on the right-hand side is singular, the vector ε will not contain N independent terms but will have a singular distribution. A vector of $N-n$ independent terms can be constructed by considering a $N \times (N-n)$ matrix R, such that $R^T\Phi = 0$, and thus

$$W := R^T\varepsilon = R^T\mathsf{Y} = R^T\Phi_e\eta + R^T\mathsf{V}.$$

Because R and Φ_e depend on input signals only, the probability density function of W can be calculated, conditioned on the parameters of Λ_η and Λ_v. For the situation that

$$\eta \sim \mathcal{N}(0, \Lambda_\eta), \qquad \Lambda_\eta = diag_{1 \le k \le L}\{\alpha\lambda^k\}$$
$$v(t) \sim \mathcal{N}(0, \sigma_v^2)$$

where v is assumed to be a white noise process, and all distributions are Gaussian, maximum likelihood estimates for the parameters $(\alpha, \lambda, \sigma_v^2)$ can be obtained by constructing the likelihood function of W:

$$L(\alpha, \lambda, \sigma_v^2) = -\frac{1}{2} \ln \det \Sigma - \frac{1}{2} W^T \Sigma^{-1} W + constant \tag{7.20}$$

where

$$\Sigma = R^T \Phi_e \Lambda_\eta(\alpha, \lambda) \Phi_e{}^T R + \sigma_v^2 R^T R$$
$$\Lambda_\eta(\alpha, \lambda) = diag(\alpha\lambda, \alpha\lambda^2, \cdots, \alpha\lambda^L).$$

Minimization of (7.20) with respect to the considered parameters leads to estimated values $\widehat{\alpha}, \widehat{\lambda}, \widehat{\sigma}_v^2$ and correspondingly estimated covariance matrices

$$\widehat{\Lambda}_\eta = diag(\widehat{\alpha}\widehat{\lambda}, \widehat{\alpha}\widehat{\lambda}^2, \cdots, \widehat{\alpha}\widehat{\lambda}^L)$$
$$\widehat{\Lambda}_v = \widehat{\sigma}_v^2 \cdot I_N$$

that are substituted in P_z and accordingly in P_ω to specify the uncertainty bounds as formulated in Proposition 7.4.
For non-white noise disturbances, the parameterization of Λ_v can become considerably more complex, and therefore white disturbance signals were considered in the original presentation of this method.
As the other methods discussed in this section, the stochastic embedding procedure leads to a model uncertainty region that, for a fixed value of ω, is specified by an ellipsoid in the Nyquist plane that contains the true system response at that frequency at a prespecified level of probability.

7.3.5 Model Error Modelling

There is one approach in model uncertainty bounding in which the undermodelling term is not bounded separately, but where the total model error is estimated by an auxiliary model. This auxiliary model should be of sufficiently high order to guarantee that unmodelled dynamics does not play a role when constructing the model uncertainty bounds. The so-called model error modelling approach is composed of the following steps.

- First, a parametric model $\widehat{G}(q)$ is estimated on the basis of measured data. The particular choice of model structure and identification method is not essential here.
- The residual simulation error is constructed according to

$$\varepsilon(t) = y(t) - \widehat{G}(q)u(t).$$

- Subsequently a model error model is identified by considering the transfer function between input u and 'output' ε, and by estimating this transfer through an FIR model of sufficiently high order. The estimated transfer function is denoted by $G_{mem}(q)$.
- The (frequency domain) covariance matrix of $G_{mem}(e^{i\omega})$, indicated by $P_{mem}(\omega)$, is constructed by applying the analysis of Section 7.2, assuming a full-order model structure.
- Bounds on the error $\widehat{G}(e^{i\omega}) - G_0(e^{i\omega})$ can now be constructed based on the bias term $G_{mem}(e^{i\omega})$ and the variance represented by $P_{mem}(\omega)$.

In light of the other methods discussed in this chapter, the model uncertainty set can be constructed on the basis of the property that

$$\begin{bmatrix} \text{Re}\left(\widehat{G}(e^{i\omega}) + G_{mem}(e^{i\omega}) - G_0(e^{i\omega})\right) \\ \text{Im}\left(\widehat{G}(e^{i\omega}) + G_{mem}(e^{i\omega}) - G_0(e^{i\omega})\right) \end{bmatrix}^T P_{mem}^{-1} \cdot \\ \cdot \begin{bmatrix} \text{Re}\left(\widehat{G}(e^{i\omega}) + G_{mem}(e^{i\omega}) - G_0(e^{i\omega})\right) \\ \text{Im}\left(\widehat{G}(e^{i\omega}) + G_{mem}(e^{i\omega}) - G_0(e^{i\omega})\right) \end{bmatrix} \leq c_{\chi,\alpha}/N$$

with probability $\geq \alpha$, and $c_{\chi,\alpha}$ corresponding to a probability α in the χ^2-distribution with 2 degrees of freedom.
Analysis of an uncertainty set based on $\widehat{G}(e^{i\omega}) - G_0(e^{i\omega})$ now follows along similar lines as for the methods discussed in Sections 7.3.2 and 7.3.3, where $G_{mem}(q)$ is now considered a fixed under-modelling term that by applying Lemma 7.1 can be incorporated when specifying ellipsoidal bounds on $\widehat{G}(e^{i\omega}) - G_0(e^{i\omega})$.

In the original model error modelling approach [163], an FIR model is used as a carrier for the estimated model uncertainty regions. Extension of this model structure to orthonormal basis function structures is straightforward and has the obvious advantage that the flexibility of the model structures can be exploited to limit the number of parameters that need to be estimated. Examples that illustrate the relations between several model uncertainty bounding procedures are provided in [255].

7.4 Bounded Error Identification

As an alternative to the classical stochastic prediction error approach, and in particular in order to replace the probabilistic model uncertainty bounds by deterministic hard-bounded model uncertainties, the so-called bounded error models have been extensively given attention, see *e.g.* [194]. In this approach, the framework is

$$y(t) = G_0(q)u(t) + v(t)$$

where v is any disturbance signal acting on the measurement data, reflecting all kinds of deviations between measurement data and real system. A model $\widehat{G}(q)$ is said to be unfalsified by the data if the residual signal

$$v'(t) = y(t) - \widehat{G}(q)u(t)$$

belongs to some hypothesized class $\mathcal{V}$ of disturbance signals v. The set of models that are unfalsified by the data is dependent on the choice of the class $\mathcal{V}$. Several alternatives for $\mathcal{V}$ have been considered to replace the standard assumption in prediction error identification of v being a realization of a stationary stochastic process with rational spectral density function. Alternatives include

- Bounded error disturbances $|v(t)| \leq c \in \mathbb{R}$ for all t;
- Bounded power disturbances $\frac{1}{N}\sum_{t=1}^{N} v^2(t) \leq c \in \mathbb{R}$, with N the length of the measured data sequence

while related approaches are also pursued in a frequency domain setting (for an overview, see *e.g.* [48, 212] and Chapter 9). The ***assumption*** on the noise disturbance class $\mathcal{V}$ will have a high influence on the set of models that is unfalsified by the data. In bounded error models, the chosen set $\mathcal{V}$ is hard-bounded, and the resulting set of unfalsified models will necessarily also be hard-bounded, leading to a deterministic specification of model uncertainty results; this identification approach is also referred to as set membership identification.

In practically all methods for bounded error modelling, the models are parameterized linearly in the unknown parameters to provide identification results that are computationally feasible. Resulting polytopic areas in the parameter space are obtained by solving linear programming (LP) problems.

The principal problem and challenge in the bounded error approach is the choice of the disturbance class $\mathcal{V}$. If this class is chosen too large in comparison with the actual disturbances on the data, the results will be unnecessarily conservative, *i.e.* the uncertainty sets will be hard-bounded but unnecessarily large. The information that is retrieved from the identified uncertainty set is very limited then. This happens in particular if the disturbance signals are allowed to be correlated with the input signal (which is *e.g.* the case when only considering amplitude bounds or power bounds on v). The incorporation of constraints on the correlation between u and v can then be added to limit this conservatism. In [118] it is shown that consistent identification results are obtained when incorporating a bound on the cross-covariance between noise and input:

$$c_l(p) \leq \frac{1}{\sqrt{N}} \sum_{t=1}^{N} v(t) r_p(t) \leq c_u(p), \quad p = 1, \cdots n_p$$

where $r_p(t)$ is a signal that is typically correlated with the input signal and uncorrelated with the noise. The most common choice of $r_p(t)$ is to be composed of delayed versions of the input u, *i.e.* $r_p(t) = u(t - p)$.
This approach is extended to a full model uncertainty bounding procedure in which several linear constraints are used to specify the disturbance class $\mathcal{V}$, while together with the linear model parameterizations in terms of orthonormal basis functions, the computation of uncertainty bounds in parameter space comes down to solving a set of linear programming problems. The resulting hard-bounded uncertainty sets actually come very close to the soft-bounded uncertainty sets discussed in the first part of this chapter. For a more extensive discussion, see [115, 116].

7.5 Summary

In this chapter, several methods have been discussed for model uncertainty quantification on the basis of measurement data, while including the possibility of biased (approximate) models. In the several techniques, linear model parameterizations play a crucial role in enabling analytical expressions as well as feasible computational tools. The prediction error methods considered in this chapter show some clear resemblances. They distinguish between the several components that compose the model uncertainty bounds: (a) bias effects due to unmodelled dynamics, (b) bias effects in parametric estimates, and (c) variance errors due to noise. The presented methods differ in the way the under-modelling terms are bounded: either deterministic (worst-case bounded) or probabilistic.
The linearly parameterized output error orthonormal basis functions have proven to be instrumental in both the analysis and implementation of the considered methods. Principle advantages of the basis function structures are, besides the linear structure, the fact that when chosen properly, the number of unknown parameters considerably decreases, allowing a reduced variance in model estimates and related attractive model uncertainty bounds.

Acknowledgement

The author acknowledges the contributions of and discussions with Sippe Douma that were helpful in developing the current presentation of the methods considered in this chapter.

8

Frequency-domain Identification in $\mathcal{H}_2$

József Bokor and Zoltan Szabó

Hungarian Academy of Sciences, Budapest, Hungary

8.1 Introduction

This chapter discusses results in the frequency domain related to approximate identification in $\mathcal{H}_2$ by using rational orthonormal parameterization of the transfer function. The criteria for modelling and identification are formulated in terms of L_2 norms.

The identification of models on the basis of frequency-domain data is a subject that attracts a growing number of researchers and engineers. Especially in application areas where experimental data of a process with (partly) unknown dynamics can be taken relatively cheaply, the excitation of the process with periodic signals, *e.g.* sinusoids, is an attractive way of extracting accurate information of the process dynamics from experiments.

Due to the commercial availability of frequency analysers that can handle huge amounts of data by special-purpose hardware, the experimental determination of frequency responses of dynamical systems has gained increasing interest in application areas as the modelling of *e.g.* mechanical servo systems and flexible space structures.

Identification on the basis of frequency-domain data can have a number of advantages when compared to the 'classical' time-domain approach, such as:

- Easy noise reduction. The non-excited (noisy) frequency lines are eliminated.
- Data reduction. A large number of time-domain samples are replaced by a small number of spectral lines, thus enabling us to handle huge amounts of time-domain data with a corresponding reduced variance of the model estimates.
- When using a discrete Fourier transform (DFT) to calculate the spectra, the frequency-domain noise is asymptotically (number of time-domain samples going to infinity) complex normal distributed.
- Model validation. Using periodic excitations, one has very good point estimates of the frequency response function. The formulation of an identi-

fication criterion in the frequency domain can be beneficial, especially in those situations where the application of the model dictates a performance evaluation in terms of frequency-domain properties. This last situation occurs often when the identified models are used as a basis for model-based control design.

- It is very easy to combine data from different experiments.

For a very nice overview of these arguments, the reader is referred to [253].
A model of a linear time invariant system can be written as

$$y(t) = \sum_{k=1}^{\infty} g_k u(t-k),$$

which, by using the delay operator $q^{-1}u(t) = u(t-1)$, can also be written as

$$y(t) = G(q)u(t),$$

with $G(q) = \sum_{k=1}^{\infty} g_k q^{-k}$. The complex valued function

$$G(e^{i\omega}) = \sum_{k=1}^{\infty} g_k e^{-ik\omega}, \quad -\pi \leq \omega \leq \pi$$

is the transfer function associated to the model.
The common way of formulating an identification problem in the frequency domain is by assuming the availability of the exact frequency response $G(e^{i\omega})$ of the (unknown) linear system in the points of a frequency grid, *i.e.* $\omega \in \mathcal{W}_N$,

$$\mathcal{W}_N := \{\omega_1, \cdots, \omega_N\},$$

disturbed by some additive (frequency-domain) noise with specific properties, *e.g.* independence among the several frequencies, *i.e.*

$$G(e^{i\omega}) = G_0(e^{i\omega}) + \eta(\omega), \quad \omega \in \mathcal{W}_N. \tag{8.1}$$

Data (8.1) can be obtained by direct measurements, *e.g.* by using frequency analysers or by an indirect method using some data handling/processing mechanism that starts off with time-domain data. Estimating a function is basically a nonparametric problem; however, one has to carry out the estimation via a finite-dimensional parameter vector. At this point, the parameters are only vehicles for arriving at a transfer function estimate.
The estimates are found by minimizing a quadratic-like cost function

$$\hat{\theta}_N = \arg\min_{\theta \in \Theta} \frac{1}{N} \sum_{z_k \in \mathcal{W}_N} ||G(z_k) - G(z_k, \theta)||^2.$$

The objective function in this general setting is a non-linear vector function of the measurements and the model parameters, which are usually constrained to be in a predefined set Θ. A large number of identification methods exist that are based on this least squares criteria [159,273,274,326]. These methods can be deterministic, when $\eta(\omega)$ are treated as bounded disturbances, or stochastic, when the sequence $\eta(\omega)$ is considered as a noise process with certain statistical properties. A nice overview of the different least squares approaches can be find in [252].

Maximum amplitude criteria are considered in [117,280], while special-purpose multi-variable algorithms are discussed, *e.g.* in [23, 160]. Recently, subspace algorithms have also been analysed for frequency-domain identification [184, 185]. Many more references and techniques can be found in [271].
A related approach to the problem, based on the discrete Fourier transforms of input and output data in [164], shows the close resemblance with results of the standard time-domain approach. As a framework for identification, it is assumed that a data sequence of input and output signals $\{u(t), y(t)\}_{t=1,\cdots N}$ is available, generated by a linear, time-invariant, discrete-time *data generating system*

$$y(t) = G_0(q)u(t) + v(t)$$

with $G_0 \in \mathcal{H}_\infty$, u a quasi-stationary signal and v a stationary stochastic process with rational spectral density, see [164].
These time-domain signals generate a (frequency-domain) empirical transfer function estimate (ETFE) which represents the data in a least squares identification criterion. Consider an equidistant frequency grid

$$\mathbb{U}_M := \{\frac{2\pi k}{M} \mid k = 0, 1, \cdots, M-1\} \tag{8.2}$$

and the discrete Fourier transform (DFT) of a signal $x(t)$ given by

$$X_M(e^{i\omega}) := \frac{1}{\sqrt{M}} \sum_{t=0}^{M-1} x(t)e^{-i\omega}, \quad \omega \in \mathbb{U}_M.$$

Then, the ETFE of the transfer function G_0 is given by

$$G(e^{i\omega}) := \frac{Y_M(e^{i\omega})}{U_M(e^{i\omega})}.$$

The main properties of this estimate can be summarized as follows:

$$G(e^{i\omega}) = G_0(e^{i\omega}) + \frac{R_M(e^{i\omega})}{U_M(e^{i\omega})} + \frac{V_M(e^{i\omega})}{U_M(e^{i\omega})},$$

where $|R_M(e^{i\omega})| \leq \frac{C_1}{\sqrt{M}}$, and, provided that $\mathsf{E}\{v\} = 0$, *i.e.* $\mathsf{E}\{V_M(e^{i\omega})\} = 0$,

$$\mathsf{E}\{G(e^{i\omega})\} = G_0(e^{i\omega}) + \frac{R_M(e^{i\omega})}{U_M(e^{i\omega})}. \tag{8.3}$$

For the covariance of the estimate one has the following result:

$$\mathsf{E}\{(G(e^{i\omega}) - G_0(e^{i\omega}))(G(e^{-i\xi}) - G_0(e^{-i\xi}))\} = \begin{cases} \frac{\Phi_v(\omega)+\rho(M)}{|U_M(e^{i\omega})|^2} & \omega = \xi, \\ \frac{\rho(M)}{U_M(e^{i\omega})U_M(e^{-i\xi})} & |\omega - \xi| = \frac{2k\pi}{M}, \end{cases}$$

where $\Phi_v(\omega)$ is the spectral function of the process v and where $|\rho(M)| \leq \frac{C_2}{M}$. For details see [164].

Results that concern frequency-domain identification based on an empirical transfer function estimate using generalized orthonormal basis functions can be found in [71].

The structure of this chapter will be the following: after introducing some basic notation, we recall a number of fundamental results about least squares identification using a parameterization with rational orthonormal functions. Section 8.3 gives the basic results concerning the approximation properties of rational orthonormal expansions. It is followed by a section about the discrete rational approximation, as an extension of the trigonometric interpolation results to the rational case.

The chapter is concluded by a section that shows the possible advantages of using data that are measured in dedicated frequency points, determined by the poles that generate the basis.

8.1.1 Basic Notation

In what follows, we will be mainly concerned with complex function theory on the unit disk. Therefore, to set the notation, let us denote by $\mathbb{R}$ the set of real numbers, by $\mathbb{C}$ the set of complex numbers and let $\mathbb{Z}$ be the set of integers. The open unit disc, its boundary, and its exterior will be denoted by

$$\mathbb{D} := \{z \in \mathbb{C} \,|\, |z| < 1\}, \quad \mathbb{T} := \{z \in \mathbb{C} \,|\, |z| = 1\}, \quad \text{and} \quad \mathbb{E} := \{z \in \mathbb{C} \,|\, |z| > 1\}.$$

We denote by $\mathfrak{I}$ the integral mean on $\mathbb{T}$, *i.e.*

$$\mathfrak{I}(f) := \frac{1}{2\pi}\int_{-\pi}^{\pi} f(e^{i\omega})d\omega. \tag{8.4}$$

By L_p, $1 \leq p \leq \infty$, we will denote the classical $L_p(\mathbb{T})$ Banach space endowed with the norm

$$||f||_p := (\frac{1}{2\pi}\int_{-\pi}^{\pi} |f(e^{i\omega})|^p d\omega)^{\frac{1}{p}} \quad f \in L_p, 1 \leq p < \infty$$

and

$$||f||_\infty := ess.\sup_{\omega\in\mathbb{T}} |f(e^{i\omega})| \quad \text{for} \quad f \in L_\infty, p = \infty.$$

The scalar product considered in L_2 is the usual one, *i.e.*

$$\langle f, g\rangle := \frac{1}{2\pi}\int_{-\pi}^{\pi} f(e^{i\omega})\overline{g}(e^{i\omega})d\omega = \mathfrak{I}(f\overline{g}) \quad f, g \in L_2.$$

$\mathcal{H}_2$ will be the Hardy space of square integrable functions on $\mathbb{T}$ with analytic continuation outside the unit disc, *i.e.* $\mathcal{H}_2$ can be identified with $\mathcal{H}_2(\mathbb{E})$. Its orthogonal complement in L_2 will be denoted by $\mathcal{H}_{2,\perp}$, *i.e.* $\mathcal{H}_{2,\perp}$ can be identified with $\mathcal{H}_2(\mathbb{D})$. Thus, the stable transfer functions are elements of $\mathcal{H}_2$. A classical introduction in $\mathcal{H}_p$ theory is given in *e.g.* [83, 135, 259]. A more advanced treatise of the topic can be found in [98].
As otherwise not stated, it will be supposed that $z \in \mathbb{T}$, *i.e.* $z := e^{i\omega}$, $\omega \in \mathbb{R}$.

8.2 Model Parameterization with Orthonormal Basis Functions

To formalize the identification problem in the frequency domain, assume that the 'measurements' are noise corrupted values of the transfer function G_0 at frequencies $\omega_k \in \mathcal{W}_N$:

$$G(e^{i\omega_k}) = G_0(e^{i\omega_k}) + \eta_k, \quad \omega_k \in \mathcal{W}_N, \tag{8.5}$$

where η_k denotes the noise. Assumptions about the noise $\eta_k \in \mathcal{N}$ will be made later on, *e.g.* a straightforward assumption being boundedness, *i.e.* $|\eta_k| \leq \epsilon$ where ϵ might be known or unknown in advance. Instead of the frequencies ω_k it will be considered points $z_k = e^{i\omega_k}$ on the unit circle, *i.e.* the starting point of the identification problem in the frequency domain will be the set of data given by $G(z_k)$, $z_k \in \mathcal{W}_N$. Here we consider a set of equidistant nodes, *i.e.* $\mathcal{W}_N = \mathbb{U}_N$.
As a parametric model structure, we will use a linear regression form, using a finite series expansion of the model transfer function in terms of very flexible orthonormal basis functions as introduced in [132] and the previous chapters of this book:

$$G(z) = \sum_{k=1}^{n} G_k V_k(z) \tag{8.6}$$

where the sequence of basis functions $\{V_k(z)\}_{k=1,\cdots n}$ can be constructed with one of the procedures presented in Chapter 2, specifying n_b pole locations $\alpha_{n_b} := (\alpha_1, \cdots, \alpha_{n_b})$.
The series expansion coefficients are collected in the parameter vector

$$\theta = [G_1 \; G_2 \; \cdots \; G_n]^T \in \mathbb{R}^{n_p},$$

with $n_p = n \cdot n_b$. Futhermore we denote

$$\varphi^T(z) := [V_1(z)^T \; \cdots \; V_n(z)^T].$$

Then the corresponding parameterized transfer function can be expressed as

$$G(z, \theta) := \varphi^T(z)\theta.$$

If $\mathcal{P}_k$ denotes the space of polynomials of degree at most k, and we denote by $\eta(z) := \prod_{i=1}^{n}(1-\overline{\alpha}_i z)$, and $\omega(z) := \prod_{i=1}^{n}(z-\alpha_i)$, then consider the sets

$$\mathcal{R}_n := \{\frac{p}{\omega} \,|\, p \in \mathcal{P}_{n-1}\}, \quad \mathcal{R}_{-n} := \{\frac{p}{\eta} \,|\, p \in \mathcal{P}_{n-1}\},$$

respectively. Accordingly, one can set

$$\mathcal{R}_{\pm n} := \{\frac{p}{\eta\omega} \,|\, p \in \mathcal{P}_{2n-1}\},$$

i.e. the orthogonal sum of $\mathcal{R}_{-n}$ and $\mathcal{R}_n$. If $\mathcal{R}_n$ includes the constant functions then it will be denoted by $\mathcal{R}_n^0$ and accordingly, $\mathcal{R}_{\pm n}^0 := \mathcal{R}_{-n} \oplus \mathcal{R}_n^0$. Observe that $\mathcal{R}_n$ and $\mathcal{R}_n^0$ are the sets of stable strictly proper transfer functions and proper transfer functions, respectively, that correspond to a fixed denominator structure, *i.e.* to a fixed set of poles.
The finite Blaschke product of order n with zeros that corresponds to $\boldsymbol{\alpha}_{n_b}$ will be denoted by G_b, *i.e.*

$$G_b(z) = \prod_{j=1}^{n_b} b_j(z) \quad \text{where} \quad b_j(z) = \frac{1-\overline{\alpha}_j z}{z-\alpha_j}, \quad |\alpha_j| < 1,$$

and the Blaschke product with poles that corresponds to $\boldsymbol{\alpha}_{n_b}$ will be denoted by G_b^*, *i.e.*

$$G_b^* = \prod_{j=1}^{n_b} \overline{b}_j \quad \text{where} \quad \overline{b}_j(z) = \frac{z-\alpha_j}{1-\overline{\alpha}_j z}, \quad |\alpha_j| < 1,$$

and it is clear that $G_b^*(e^{i\omega}) = \overline{G_b(e^{i\omega})}$, where the overbar denotes complex conjugation.
Throughout this chapter it is assumed that the Blaschke condition, *i.e.*

$$\sum_{i=1}^{\infty}(1-|\alpha_i|) = \infty,$$

is fulfilled.
If $\varphi_j(z) = \dfrac{d_j}{z-\alpha_j}$, where $d_j = \sqrt{1-|\alpha_j|^2}$, one has that the system

$$\Phi_n = \{\phi_j := \varphi_j B_{j-1},\, B_{j-1} := \prod_{k=1}^{j-1} b_j \,|\, j = 1, \cdots, n_b\}, \tag{8.7}$$

forms an orthonormal basis in $\mathcal{R}_{n_b}$. This is the so-called Takenaka-Malmquist system and it has the property, that $\phi_j \perp \mathcal{R}_{j-1}$, *i.e.* the elements of the basis can be obtained by a Gram-Schmidt orthogonalization process of the sequence of functions $\{\psi_j \,|\, j = 1, \cdots, n_b\}$, where $\psi_j(z) = \dfrac{d_j}{(z-\alpha_j)^k}$ if $\alpha_j \neq 0$

and $\psi_j(z) = z^{k-1}$ for $\alpha_j = 0$, where k is the multiplicity of α_j in the sequence $\alpha_1, \cdots, \alpha_j$. It has been seen that in the special case when all the poles are set in the origin, one has the classical Fourier series; when all the poles are equal and real, then one has the so-called Laguerre basis, and finally, when one complex pole pair is repeated periodically, one has the so-called Kautz system.
In some cases it is convenient to extend the system Φ_n for negative indices by setting

$$\varphi_{-k}(z) := \frac{d_k}{1 - \overline{\alpha}_j z} B_{-k+1}(z).$$

Then one has

$$\varphi_{-k}(z) = \overline{z\varphi_k(z)}. \tag{8.8}$$

With these notations, one can consider for example the parameterization with the system defined by

$$V_k(z)^T = [\ \varphi_1(z)G_b^{k-1}(z), \cdots, \varphi_{n_b}(z)G_b^{k-1}(z)\ ].$$

8.2.1 Least Squares Identification

Given data $\{\gamma_k = G(z_k)\}_{k=1,\cdots,N}$, we would like to obtain a parameter vector $\hat{\theta}_N$ related to the considered model structure that minimizes the error performance function

$$\hat{\theta}_N = \arg\min_{\theta \in \Theta} \frac{1}{N} \sum_{z_k \in W_N} \varepsilon^2(z_k, \theta), \tag{8.9}$$

with $\varepsilon(z_k, \theta) := |\gamma_k - G(z_k, \theta)|$.
Recall, that we have collected the expansion coefficients in the parameter vector

$$\theta = [G_1\ G_2\ \cdots\ G_n]^T \in \mathbb{R}^{n_p},$$

with $n_p = n \cdot n_b$, and the basis functions in the vector

$$\varphi^T(z) := [V_1(z)^T\ \cdots\ V_n(z)^T].$$

The parameterized model is then $G(z, \theta) = \varphi^T(z)\theta$.
Let us denote by Φ the matrix $\Phi = [\varphi^T(z_k)]_{z_k \in W_N}$ and by Λ the column vector $\Lambda = [\gamma_k]_{z_k \in W_N}$. Then, the least-squares parameter estimate is given by

$$\hat{\theta}_N = (\Phi^* \Phi)^{-1} \Phi^* \Lambda. \tag{8.10}$$

Let us suppose that $G_0(z) = \varphi^T(z)\theta_0 + T(z)$, then $\Lambda = \Phi\theta_0 + \tau + \eta$, with $\eta = [\eta_k]_{z_k \in W_N}$ and $\tau = [T(z_k)]_{z_k \in W_N}$.
It follows that one has the following expressions for the estimated parameter vector:

$$\hat{\theta}_N = \theta_0 + (\Phi^* \Phi)^{-1} \Phi^* \tau + (\Phi^* \Phi)^{-1} \Phi^* \eta, \tag{8.11}$$

and

$$G(z, \hat{\theta}_N) = \varphi^T(z)\theta_0 + \varphi^T(z)(\Phi^*\Phi)^{-1}\Phi^*\tau + \varphi^T(z)(\Phi^*\Phi)^{-1}\Phi^*\eta, \qquad (8.12)$$

respectively.

8.2.2 Asymptotic Bias Results

Using the fact that the system $\varphi(z)$ is an orthonormal system and recalling the definition of the definite integral, one has that:

Lemma 8.1. *For any measurable function $g(z)$ on the unit circle, one has*

$$\lim_{N\to\infty} \frac{1}{N}\Phi^*\gamma = \langle \varphi, g \rangle \,, \qquad (8.13)$$

hence,

$$\lim_{N\to\infty} \frac{1}{N}\Phi^*\Phi = \mathbb{I}, \qquad (8.14)$$

where $\gamma = [g(z_k)]_{z_k \in \mathcal{W}_N}$, provided that the norm of the partition on the unit circle defined by the points of $\mathcal{W}_N$ tends to 0 as $N \to \infty$.

Let us denote by $[\cdot\,,\cdot]_N$ the usual Euclidean scalar product in $\mathbb{C}^N$ and by $||\cdot||_{2,N}$ the induced norm. Then by the Cauchy-Schwartz inequality one has

$$|e_j^*\Phi^*\eta| = |\,[\Phi e_j, \eta]_N\,| \leq ||\Phi e_j||_{2,N}||\eta||_{2,N},$$

where e_j is the jth canonical unit vector in $\mathbb{C}^N$. It follows that

$$|\frac{1}{N}\Phi^*\eta| \leq \frac{||\eta||_{2,N}}{\sqrt{N}}.$$

Because $\langle \varphi, T(z) \rangle = 0$ one has that

Proposition 8.1. *If* $\lim_{N\to\infty} \frac{||\eta||_{2,N}}{\sqrt{N}} = 0$, *(or* $\lim_{N\to\infty} \frac{||\mathsf{E}\{\eta\}||_{2,N}}{\sqrt{N}} = 0$ *in the stochastic approach) then one has*

$$\lim_{N\to\infty} \hat{\theta}_N = \theta_0.$$

It follows that – under mild assumptions on the disturbances – the bias error

$$\mathcal{B}_N(z) := |G_0(z) - G(z, \mathsf{E}\{\hat{\theta}_N\})|$$

is asymptotically equal to $|G_0(z) - G(z, \theta_0)|$, *i.e.* the approximation error is determined by the choice of the poles.

8.2.3 Asymptotic Variance Results

The total estimation error $|G_0(z) - G(z, \hat{\theta}_N)|^2$ can be decomposed into a 'bias error' and a 'variance error' part:

$$|G_0(z) - G(z, \hat{\theta}_N)|^2 \leq |G_0(z) - G(z, \mathsf{E}\{\hat{\theta}_N\})|^2 + |G(z, \mathsf{E}\{\hat{\theta}_N\}) - G(z, \hat{\theta}_N)|^2.$$

The variance error term, defined as $\mathcal{V}(z) := |G(z, \mathsf{E}\{\hat{\theta}_N\}) - G(z, \hat{\theta}_N)|^2$ is given by the expression:

$$\mathcal{V}(z) = \varphi(z)^T \text{Cov}\{\hat{\theta}_N\}\overline{\varphi}(z), \tag{8.15}$$

where

$$\text{Cov}\{\hat{\theta}_N\} = (\Phi^*\Phi)^{-1}\Phi^*\text{Cov}\{\eta\}\Phi(\Phi^*\Phi)^{-*} \tag{8.16}$$

Under different assumptions on the disturbances, one can give certain asymptotic expressions for this latter covariance, see *e.g.* [332]. It is not our intention to present all these expressions here; however, to summarize all the results, we can state as a main result that asymptotically one has

$$\mathcal{V}(e^{i\omega}) \approx \frac{1}{N}\frac{\Phi_v(\omega)}{\Phi_u(\omega)}\sum_{k=1}^{n_p}|\varphi_k(e^{i\omega})|^2,$$

where $\Phi_v(\omega)$ and $\Phi_u(\omega)$ are the spectral functions of the noise $v(t)$ and input $u(t)$, respectively, see [71, 223, 224] and Chapter 5 of this book.

8.3 Approximation by Rational Orthonormal Functions on the Unit Circle

When the trigonometrical basis is used for model parameterization purposes, *i.e.* when ones build FIR models, and the measurement points are equidistant, *i.e.* $\mathbb{U}_M$, see (8.2), then the least squares problem leads to a classical trigonometric interpolation problem. In this case, there is an efficient method to compute the model parameters, namely, by using a fast Fourier transform (FFT).

The purpose of this section is to extend the classical results using approximation by trigonometrical polynomials on the unit circle to the case of the generalized rational case. The aim is to highlight the close relation of the rational case to the classical results and to show that for suitable interpolation nodes, *i.e.* frequency measurement points that are determined by the poles of the rational functions, one can construct an efficient algorithm that solves the interpolation problem. Moreover, asymptotic properties of this interpolation process are provided.

8.3.1 The β Function

In what follows, a central role will be played, in engineering terms, by the phase function of the Blaschke product, which will be called the 'β function' throughout this work. This function was introduced in [270] and was shown to be very useful in this context since then.
Let us denote the phase function of a single term by

$$\beta_j(\omega), \quad \text{i.e.,} \quad b_{\alpha_j}(e^{i\omega}) = e^{i\beta_j(\omega)}.$$

An explicit expression for this function can be given as follows: denote the poles (zeros) of the Blaschke product by $\alpha_j = \rho_j e^{i\theta_j}$, then

$$\beta_j(\omega) = \theta_j + \tau_{s(\rho_j)}(\omega - \theta),$$

where

$$\tau_s(\omega) = 2\arctan(s \tan \frac{\omega}{2}), \quad s(\rho_j) = \frac{1+\rho_j}{1-\rho_j}, \quad \omega \in [-\pi, \pi],$$

and it is extended periodically to $\mathbb{R}$ by

$$\tau_s(\omega + 2\pi) = \tau_s(\omega) + 2\pi,$$

see [267].
For the derivatives one has

$$\beta_j'(\omega) = \frac{1 - |\alpha_j|^2}{|1 - \overline{\alpha}_j e^{i\omega}|^2}. \tag{8.17}$$

Hence, it follows that $\beta_j : \mathbb{R} \to \mathbb{R}$ is a strictly increasing function. Let us mention here that $\beta_j'(\omega) = |\varphi_j(e^{it})|^2 = |\phi_j(e^{i\omega})|^2$.
As a result, for a finite Blaschke product B_n of order n, there exists a monotone increasing, invertible and differentiable function $\beta_{(n)}(\omega)$ mapping the space $\mathbb{R}$ onto itself, such that,

$$B_n(e^{i\omega}) = e^{in\beta_{(n)}(\omega)}, \tag{8.18}$$

where the function $\beta_{(n)}(\omega)$ can be expressed as

$$\beta_{(n)}(\omega) := \frac{1}{n} \sum_{k=1}^{n} \beta_k(\omega).$$

Because $1 - |\alpha_j| \leq |1 - \overline{\alpha}_j e^{i\omega}| \leq 1 + |\alpha_j|$, one can obtain the bounds

$$\frac{1 - |\alpha_j|}{1 + |\alpha_j|} \leq \beta_j'(\omega) \leq \frac{1 + |\alpha_j|}{1 - |\alpha_j|},$$

and hence

$$\frac{1}{n}\sum_{k=1}^{n}\frac{1-|\alpha_k|}{1+|\alpha_k|} \leq \beta'_{(n)}(\omega) \leq \frac{1}{n}\sum_{k=1}^{n}\frac{1+|\alpha_k|}{1-|\alpha_k|}.$$

It follows that the derivative of the inverse is bounded by

$$\frac{n}{\sum_{k=1}^{n}\frac{1+|\alpha_k|}{1-|\alpha_k|}} \leq \gamma'_{(n)}(\omega) \leq \frac{n}{\sum_{k=1}^{n}\frac{1-|\alpha_k|}{1+|\alpha_k|}},$$

where $\gamma_{(n)}(\omega) := \beta^{-1}_{(n)}(\omega)$.
If there is a constant $0 < c < 1$ such that $|\alpha_k| < c$, $k = 1, \cdots, n$ then one has the uniform bounds

$$\frac{1-c}{2} \leq \beta'_{(n)}(\omega) \leq \frac{2}{1-c}, \quad \text{and} \quad \frac{1-c}{2} \leq \gamma'_{(n)}(\omega) \leq \frac{2}{1-c}. \tag{8.19}$$

The shape of the argument-functions belonging to some values of parameter a (that correspond to different Laguerre system poles) is presented in Figure 8.1. The real parameter values $a = 0.5$, $a = 0.7$, $a = 0.9$, and $a = 0.99$ were considered.
The effect of the non-uniformly spaced frequency scale produced by the inverse argument-transform is presented in Figure 8.2. The Nyquist diagrams as well as Bode-like frequency diagrams containing the absolute value and phase belonging to the function $b_a(z) = \frac{1-\bar{a}z}{z-a}$ with a single real pole $a = 0.7$ are presented by using finite number of uniformly spaced samples on the left-side diagrams and those given by the argument-transform on the right-side ones. It can be observed that the Nyquist-plots belonging to the non-uniform scale interpolate the continuous function better, as the non-uniform scale applied on the domain results in uniform spacing of the image. On the frequency-diagrams the observation is different: the non-uniform scale results in denser

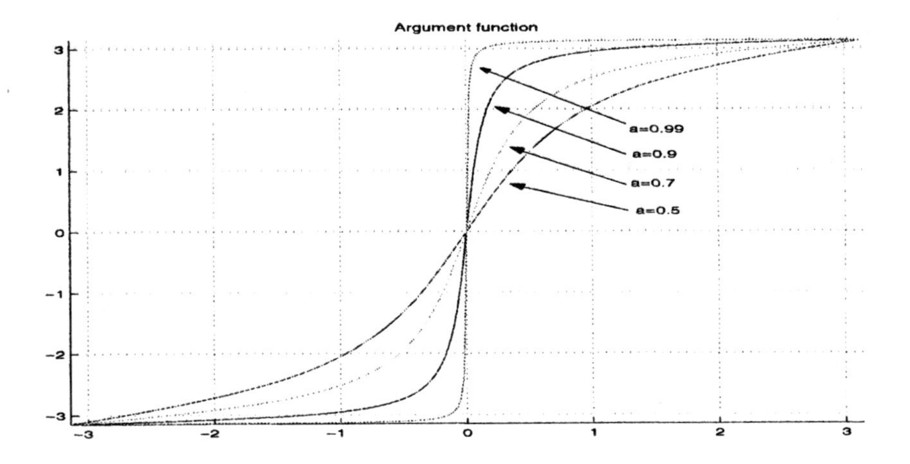

Fig. 8.1. Argument-functions belonging to real a values.

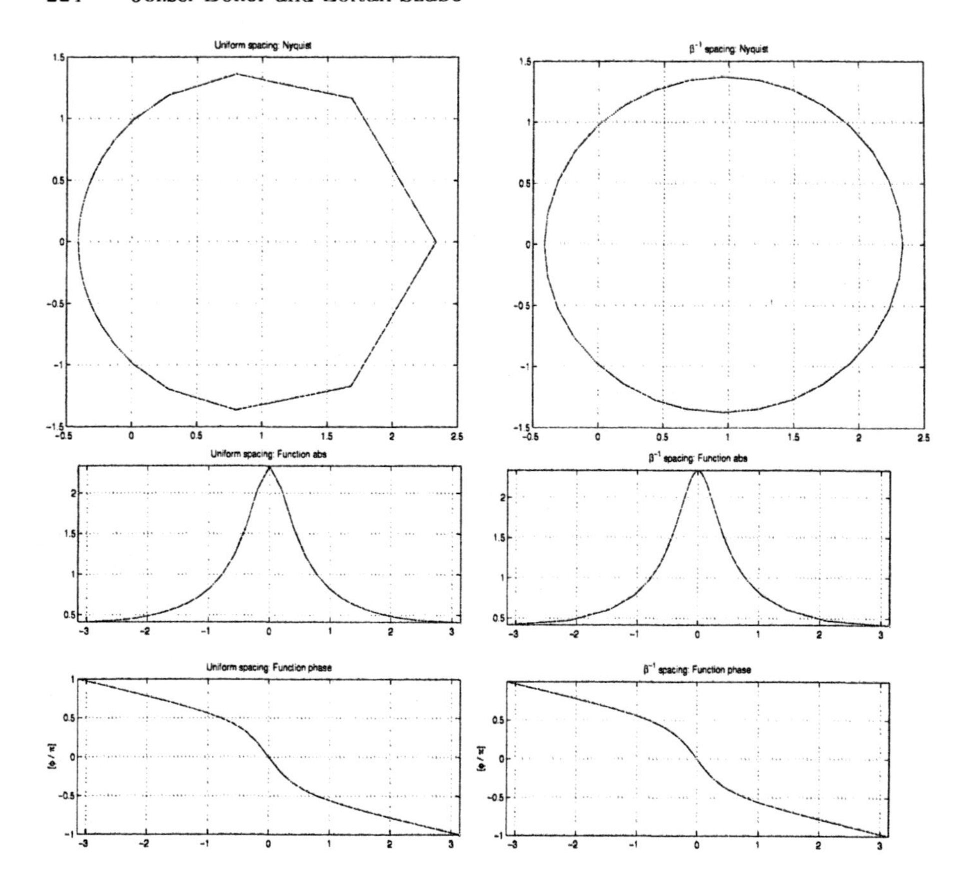

Fig. 8.2. Uniform and non-uniform spacing.

spacing on the sample points in the neighborhood of the peak and sparser on the places where "nothing can be observed".

The consequences of these observations and the role of the non-uniform spacing will be elaborated upon later on in this chapter in association with questions related to discrete interpolation of rational functions on the unit circle.

8.3.2 Reproducing Kernels

Fundamental concepts and basic tools in investigating approximation results concerning subspaces in Hilbert spaces are the concepts of reproducing property and reproducing kernels. Moreover, they play a central role in the construction of the interpolation operators that are very useful in building practical algorithms.

The reproducing kernel $K : \mathbb{T} \times \mathbb{T} \to \mathbb{C}$ of a subspace $\mathcal{V} \subset L_2$ is defined by its reproducing property, *i.e.*

$$\forall f \in \mathcal{V} \quad f(w) = \langle f, K(\cdot, w) \rangle, \qquad w \in \mathbb{T}.$$

If an orthonormal basis $\{\psi_j(z) \,|\, j = 1, \ldots, n\}$, for the $n < \infty$ dimensional subspace $\mathcal{V}$ is considered, then the reproducing kernel, or Dirichelet kernel, of the system is given by

$$K_n(z, w) = \sum_{k=1}^{n} \psi_k(z) \overline{\psi}_k(w), \qquad z, w \in \mathbb{T},$$

and it is independent of the choice of the orthonormal basis $\{\psi_j(z)\}$, see [19]. The orthogonal projection onto $\mathcal{V}$ is given by

$$\mathbf{P}_{\mathcal{V}} f(w) = \langle f, K_n(\cdot, w) \rangle, \quad f \in L_2.$$

For any set of distinct points $\mathbf{w}_n := \{w_1, w_2, \cdots, w_n\}$ on $\mathbb{T}$,

$$\langle K_n(\cdot, w_i), K_n(\cdot, w_j) \rangle = K_n(w_i, w_j)$$

and the matrix

$$\mathbf{K}_n(\mathbf{w}_n) := [K_n(w_i, w_j)]_{i,j=1}^{n}$$

is positive semidefinite.
Applying these properties to the subspace $\mathfrak{R}_n$, one can obtain the function $K_n(z, w) = \sum_{k=1}^{n} \phi_k(z) \overline{\phi}_k(w)$ as a reproducing kernel that can be expressed in a compact form by the the following Christoffel-Darboux formula, see [80,169]:

Lemma 8.2.

$$K_n(z, w) := \sum_{k=1}^{n} \phi_k(z) \overline{\phi}_k(w) = \frac{B_n(z) \overline{B}_n(w) - 1}{1 - z\overline{w}}, \quad z, \mu \in \mathbb{D} \cup \mathbb{T}, \tag{8.20}$$

where B_n is given in (8.18).

Using the expression for the derivative of the function $\beta_{(n)}$ derived from (8.7) and (8.17) one has

$$K_n(e^{i\omega}, e^{i\omega}) := \sum_{k=1}^{n} |\phi_k(e^{i\omega})|^2 = n\beta'_{(n)}(\omega). \tag{8.21}$$

Because the system Φ_n, see (8.7), forms a Haar system, see [62], *i.e.* the matrix

$$\mathbf{F}_n(\mathbf{w}_n) := \begin{bmatrix} \phi_1(w_1) & \cdots & \phi_1(w_n) \\ \vdots & & \vdots \\ \phi_n(w_1) & \cdots & \phi_n(w_n) \end{bmatrix}$$

is positive definite, one has the following lemma:

Lemma 8.3. *For any set of distinct points* $\mathbf{w}_n := \{w_1, w_2, \cdots, w_n\}$ *on* $\mathbb{T}$ *the system defined by* $\boldsymbol{\kappa}_n := \{\kappa_j(z) := K_n(z, w_j) \,|\, j = 1, \cdots, n$ *forms a basis for* $\mathcal{R}_n$.

As a consequence, the matrix $\mathbf{K}_n(\mathbf{w}_n)$ defined above is positive definite. The reproducing kernel of $\mathcal{R}_{\pm n}$ is

$$S_n(z, w) = (z\overline{w})^{-\frac{1}{2}} \frac{B_n(z)\overline{B}_n(w) - B_n(w)\overline{B}_n(z)}{(z\overline{w})^{\frac{1}{2}} - (w\overline{z})^{\frac{1}{2}}}.$$

Using the shorthand $S_n(\omega, \tau)$ for $S_n(e^{i\omega}, e^{i\tau})$, one has

$$S_n(\omega, \tau) = e^{-i\frac{\omega-\tau}{2}} \frac{\sin n(\beta_{(n)}(\omega) - \beta_{(n)}(\tau))}{\sin(\frac{\omega-\tau}{2})}. \tag{8.22}$$

Considering the subspace $\mathcal{R}^0_{\pm n}$, then its reproducing kernel is

$$S^0_n(\omega, \tau) = \frac{\sin\left(n(\beta_{(n)}(\omega) - \beta_{(n)}(\tau)) + \frac{1}{2}(\omega - \tau)\right)}{\sin(\frac{\omega-\tau}{2})}, \tag{8.23}$$

This can be seen as a generalization of the Dirichlet kernel for $\mathcal{P}_{\pm n}$,

$$D_n(\omega, \tau) = \frac{\sin\,(n + \frac{1}{2})(\omega - \tau)}{\sin\,(\frac{\omega-\tau}{2})}$$

that can be obtained when all of the poles are placed at the origin, *i.e.* $\beta_{(n)}(\omega) = \omega$.

8.3.3 L_p Norm Convergence of the Partial Sums

As was shown in the analysis of the asymptotical bias, a central question is the approximation property of the system that is used for parameterization. This section gives a generalization of the classical L_p-norm convergence theorem of partial sums of the Fourier series to partial sums in the rational orthonormal system generated by the sequence of poles $(\alpha_k \,|\, k \in \mathbb{Z})$.

Recall the completeness property of the Takenaka-Malmquist system in the $\mathcal{H}_p$ spaces, which follows from [135], pp. 64 and [98], pp. 53:

Theorem 8.1. *The system* $\Phi = \{\phi_k \,|\, k \in \mathbb{N}\}$ *is complete in* $\mathcal{H}_p$ *for* $0 < p < \infty$ *if and only if the Blaschke condition holds,* i.e.

$$\sum_{k=1}^{\infty}(1 - |\alpha_k|) = \infty.$$

Because the subspaces $\mathcal{R}_n$ are finite dimensional, there exist elements from these subspaces with the the best uniform approximation property. If $E_n(f) := \inf_{g\in\mathcal{R}_n} ||f-g||_\infty$ denotes the distance of $f \in \mathbb{A}(\mathbb{D})$ from the subspace $\mathcal{R}_n$, then by the preceding theorem it follows that $\lim_{n\to\infty} E_n(f) = 0$. A function f, continuous on $\mathbb{D}$, belongs to the disc algebra $\mathbb{A}(\mathbb{D})$ if f is analytic on $\mathbb{D}$. If one considers (8.8), it follows that the system $\Phi = \{\phi_k, \phi_{-k} \,|\, k \in \mathbb{N}\}$ is complete in L_p for $0 < p < \infty$ if and only if the Blaschke condition holds. Let us denote the partial sums of the expansion of a function f in the orthonormal system $\{\phi_k \,|\, k \in \mathbb{Z}\}$ by $\mathcal{S}_n f$ *i.e.* $\mathcal{S}_n f(w) = \langle f, S_n(\cdot, w)\rangle$.

Proposition 8.2. *For* $f \in L_p$, $\quad 1 < p < \infty$ *one has*

$$||\mathcal{S}_n f||_p \leq C_p ||f||_p \tag{8.24}$$

and if $\sum_{k=1}^\infty (1-|\alpha_k|) = \infty$ *then*

$$\lim_{n\to\infty} ||f - \mathcal{S}_n f||_p = 0. \tag{8.25}$$

8.4 A Discrete Rational Orthonormal System on the Unit Circle

In practice, one has to deal with a finite amount of data, and therefore it is necessary to construct methods that render sufficiently accurate approximations based only on the available finite information. Moreover, if one uses interpolation type methods, then it is possible to recover functions that are members of certain finite dimensional spaces by using only a finite amount of data. A well-known example for such methods is given by the operator that interpolates polynomials up to a certain degree, *e.g.* Lagrange-type interpolation operators. The properties of these algorithms depend heavily on the set of the points where interpolation is required. For example, polynomial-type interpolation defined on a uniform grid of the unit circle leads to FFT-type algorithms. In what follows, the theoretical background will be given for the construction of such discrete systems in the context of rational parameterizations.
Let us denote by $\mathbb{W}_n$ the set of images of the roots of unity through the $\gamma_{(n)} = \beta_{(n)}^{-1}$ function, *i.e.*

$$\mathbb{W}_n = \{\zeta_k = e^{i\gamma_k} \,|\, \gamma_k = \beta_{(n)}^{-1}(\eta_k),\ \eta_k \in \mathbb{U}_n\},$$

where $\mathbb{U}_n$ is given by (8.2).
Considering the set $\mathbb{W}_n$ as nodes, one can introduce the following rational interpolation operator:

$$(\mathcal{L}_n f)(z) := \sum_{\zeta\in\mathbb{W}_n} \frac{K_n(z,\zeta)}{K_n(\zeta,\zeta)} f(\zeta), \tag{8.26}$$

where f is a continuous function on $\mathbb{T}$ and $z \in \mathbb{T}$. Let us denote by

$$\mathbf{l}_{n,\zeta}(z) := \frac{K_n(z,\zeta)}{K_n(\zeta,\zeta)}, \quad \zeta \in \mathbb{W}_n.$$

From the definition of $\mathbb{W}_n$ and by (8.20), it follows that for $0 \le k, l < n,\ k \neq l$, one has:

$$\mathbf{l}_{n,\zeta_k}(\zeta_l) = \frac{1 - e^{in(\beta_n(\gamma_l) - \beta_n(\gamma_k))}}{K_n(\zeta_k,\zeta_k)(1-\zeta_l\overline{\zeta}_k)} = \frac{1 - e^{2\pi(l-k)}}{K_n(\zeta_k,\zeta_k)(1-\zeta_l\overline{\zeta}_k)} = 0.$$

Consequently, for $0 \le k, l < n$,

$$\mathbf{l}_{n,\zeta_k}(\zeta_l) = \delta_{k,l}, \tag{8.27}$$

i.e. $\mathbf{l}_{n,\zeta}$, $\zeta \in \mathbb{W}_n$, are the Lagrange functions corresponding to the system $\{\phi_i \mid i = 1, \cdots, n\}$.

This implies that $\mathfrak{L}_n f$ interpolates f at the points of $\mathbb{W}_n$, *i.e.* $\mathfrak{L}_n f(\zeta) = f(\zeta)$, $\zeta \in \mathbb{W}_n$. It is also clear that $\mathfrak{L}_n f = f$ for $f \in \mathfrak{R}_n$, and $\{\mathbf{l}_{n,\zeta} \mid \zeta \in \mathbb{W}_n\}$ is a basis in $\mathfrak{R}_n$.

Let us define the discrete scalar product

$$[f,g]_n := \sum_{\zeta \in \mathbb{W}_n} \frac{f(\zeta)\overline{g}(\zeta)}{K_n(\zeta,\zeta)} = \sum_{\zeta \in \mathbb{W}_n} \frac{f(\zeta)\overline{g}(\zeta)}{n\beta'_{(n)}(\gamma)}, \tag{8.28}$$

where $\zeta = e^{i\gamma}$.

For the classical case, *i.e.* when $\alpha_1 = \cdots = \alpha_n = 0$, the β function is the identity and this scalar product is exactly the discrete Fourier scalar product defined by the trigonometric interpolation, *i.e.*

$$[f,g]_n^f := \frac{1}{n} \sum_{\zeta \in \mathbb{U}_n} f(\zeta)\overline{g}(\zeta).$$

Using this discrete scalar product, the interpolation operator can be written as:

$$(\mathfrak{L}_n f)(z) = [f, K_n(\cdot, z)]_n,$$

for $f \in \mathbb{A}(\mathbb{D})$. Using this fact and by (8.27) it follows that for $\zeta, \xi \in \mathbb{W}_n$ one has

$$\mathfrak{I}(\mathbf{l}_{n,\zeta}\overline{\mathbf{l}_{n,\xi}}) = \delta_{\zeta,\xi}, \tag{8.29}$$

where the integral operator $\mathfrak{I}$ is given in (8.4).

Using the reproducing property of the kernel, it is easy to see that

$$\langle \mathfrak{L}_n f, \mathfrak{L}_n g \rangle = [f,g]_n,$$

and it follows that every orthonormal system $\{\psi_k \mid k = 1, \cdots, n\}$ on the subspace defined by the reproducing kernel is also discrete orthonormal, *i.e.*

$$[\psi_k, \psi_l]_n = \delta_{k,l} \quad \text{for} \quad 1 \le k, l \le n.$$

For details see [284–286].

8.4.1 A Quadrature Formula for Rational Functions on the Unit Circle

Using the interpolation operator $\mathfrak{L}_n$, see (8.26), one can introduce a quadrature formula as

$$\mathfrak{I}_n(f) := \sum_{\zeta \in \mathbb{W}_n} \rho_\zeta^{(n)} f(\zeta), \quad \text{where} \quad \rho_\zeta^{(n)} := \mathfrak{I}\left(\frac{K_n(\cdot,\zeta)}{K_n(\zeta,\zeta)}\right), \tag{8.30}$$

where $\mathfrak{I}$ denotes the integral mean operator on $\mathbb{T}$, see (8.4). Then it is clear that $\mathfrak{I}_n(f) = \mathfrak{I}(f)$ for all $f \in \mathfrak{R}_n$.
To get $\rho_\zeta^{(n)}$, we used the fact that for any $g \in \mathbb{A}(\mathbb{D})$ one has $\mathfrak{I}(g) = g(0)$. Thus by (8.20), for $\zeta \in \mathbb{W}_n$ one has

$$\mathfrak{I}(K_n(\cdot,\zeta)) = \sum_{k=1}^{n} \mathfrak{I}(\phi_k)\overline{\phi}_k(\zeta) = \sum_{k=1}^{n} \phi_k(0)\overline{\phi}_k(\zeta) = 1 - B_n(0)\overline{B}_n(\zeta) = 1 - B_n(0).$$

Consequently, by (8.30) the coefficients of the quadrature formula are of the form

$$\rho_\zeta^{(n)} := \frac{1 - B_n(0)}{K_n(\zeta,\zeta)} = \frac{1 - B_n(0)}{n\beta'_{(n)}(\gamma)}, \quad \text{where} \quad \zeta = e^{i\gamma}. \tag{8.31}$$

If one of the zeros of the Blaschke product is zero, say, $\alpha_n = 0$, hence $b_{\alpha_n}(z) = z$, one has $B_n(0) = 0$. It follows that in this case, the coefficients of the quadrature formula

$$\rho_\zeta^{(n)} = \frac{1}{n\beta'_{(n)}(\gamma)} > 0, \quad \text{where} \quad \zeta = e^{i\gamma},$$

are positive.
For every $g \in \mathfrak{R}_{-n}$ one has $\mathfrak{I}(g) = 0$, and $g = hB_{\overline{n}}$ for some $h \in \mathfrak{R}_n$. It follows, that $\mathfrak{I}_n(g) = \mathfrak{I}_n(h) = \mathfrak{I}(h) = h(0)$, *i.e.* in general one cannot expect $\mathfrak{I}(g) = \mathfrak{I}_n(g)$. But $\mathfrak{I}(g) = \mathfrak{I}_n(g)$ if $g \in \mathfrak{R}_{-n} \cap z\mathfrak{R}_{-n}$.
One can obtain a completely analogous result as for the polynomial case, if one chooses the quadrature formula induced by $\mathfrak{L}_n^0$, based on the interpolation nodes $\mathbb{W}_n^0$.

Proposition 8.3 ([285]). *Let us introduce the Gauss-type quadrature formula*

$$\mathfrak{I}_n^0(f) := \sum_{\zeta \in \mathbb{W}_n^0} \frac{f(\zeta)}{K_n^0(\zeta,\zeta)},$$

then $\mathfrak{I}_n^0(f) = \mathfrak{I}(f)$ *for all* $f \in \mathfrak{R}_{\pm n}^0$.

The asymptotic properties of the the quadrature formula induced by $\mathfrak{L}_n$ can be summarized as follows:

Proposition 8.4 ([285]). *For every $n \in \mathbb{N}$, $n \geq 2$,*

$$\sum_{\zeta \in \mathbf{W}_n} \frac{1}{K_n(\zeta, \zeta)} = \frac{1 - |B_n(0)|^2}{|1 - B_n(0)|^2} \tag{8.32}$$

and consequently for the norm of the functionals $\mathfrak{I}_n$ one has

$$||\mathfrak{I}_n|| = \sum_{\zeta \in \mathbf{W}_n} |\rho_\zeta^{(n)}| < \frac{2}{1 - |\alpha_1|}. \tag{8.33}$$

Moreover, if $\sum_{k=1}^{\infty}(1 - |\alpha_k|) = \infty$, then

$$\lim_{n \to \infty} \sum_{\zeta \in \mathbf{W}_n} \frac{1}{K_n(\zeta, \zeta)} = \lim_{n \to \infty} \sum_{\zeta \in \mathbf{W}_n} |\rho_\zeta^{(n)}| = 1. \tag{8.34}$$

As a corollary one has:

Proposition 8.5 ([286]). *If $\sum_{k=1}^{\infty}(1 - |\alpha_k|) = \infty$, then for every $f \in \mathbb{A}(\mathbb{D})$, one has*

$$\lim_{n \to \infty} \mathfrak{I}_n(f) = \mathfrak{I}(f).$$

Based on this result, one has the following generalization of the Erdős-Turán theorem for $\mathfrak{L}_n$ on $\mathbb{A}(\mathbb{D})$:

Proposition 8.6 ([286]). *Consider the interpolation operator*

$$(\mathfrak{L}_n f)(z) = \sum_{\zeta \in \mathbf{W}_n} \frac{K_n(z, \zeta)}{K_n(\zeta, \zeta)} f(\zeta).$$

If $\sum_{k=1}^{\infty}(1 - |\alpha_k|) = \infty$, then for every $f \in \mathbb{A}(\mathbb{D})$, one has

$$\lim_{n \to \infty} ||f - \mathfrak{L}_n f||_2 = 0.$$

8.4.2 L_p Norm Convergence of Certain Rational Interpolation Operators on the Unit Circle

In what follows, an extension of the Marcinkiewicz-Zygmund type inequalities will be given for the interpolation operator $\mathfrak{L}_n$ on $\mathbb{A}(\mathbb{D})$. Based on this result, the mean convergence of this interpolation operator will be proved.

Proposition 8.7 ([286]). *Let $f \in \mathfrak{R}_n$, and $1 - |\alpha_k| > \delta > 0$, $k \in \mathbb{N}$. Then there exist constants $C_1, C_2 > 0$ depending only on p, such that for $1 < p < \infty$ one has*

$$C_1 ||f||_p \leq [\mathfrak{I}_n(|f|^p)]^{\frac{1}{p}} \leq C_2 ||f||_p.$$

One can observe that for the case when $B(z) = z$, one can obtain the classical Marcinkiewicz theorems.
By using these results, one can prove the following mean convergence theorem:

Proposition 8.8 ([286]). *If* $1-|\alpha_k| > \delta > 0$, $k \in \mathbb{N}$, *then for every* $f \in \mathbb{A}(\mathbb{D})$ *and* $1 < p < \infty$ *one has*

$$||f - \mathfrak{L}_n f||_p \leq CE_n(f),$$

and consequently,

$$\lim_{n\to\infty} ||f - \mathfrak{L}_n f||_p = 0.$$

In a practical situation, one would like to recover all rational transfer functions from a given finite dimensional subspace and, at the same time, to have an approximation property for the entire $\mathcal{H}_2$ space in order to cope with the possible unmodelled dynamics. This property is not granted by all interpolatory type approximation methods. Proposition 8.8 shows that using the interpolation operator $\mathfrak{L}_n$ defined in (8.26), one has the property necessary in order for the required asymptotic bias results to hold.

8.5 Asymptotic Results for Adapted Nodes

Assume that the data of the identification problem is given in the nodes of $\mathcal{W}_n$. Denote by $\tilde{\Lambda}$ the column vector $\tilde{\Lambda} = [G(\xi)]_{\xi\in\mathcal{W}_n}$ and by $\tilde{\Phi}$ the matrix $\tilde{\Phi} = [\varphi^T(\xi)]_{\xi\in\mathcal{W}_n}$.
Recall that the least squares parameter estimation is given by

$$\hat{\theta}_n = (\tilde{\Phi}^*\tilde{\Phi})^{-1}\tilde{\Phi}^*\tilde{\Lambda}.$$

Because the system φ is orthonormal with respect to the discrete scalar product defined by (8.28), *i.e.*

$$[\varphi_i, \varphi_j]_n := \sum_{\xi\in\mathbf{W}_n} \frac{\varphi_i(\xi)\overline{\varphi}_i(\xi)}{K_n(\xi,\xi)} = \delta_{i,j}$$

one has that the system $\frac{1}{\sqrt{K_n(z,z)}}\varphi(z)$ is orthogonal for the product defined by

$$[f,g] := \sum_{\xi\in\mathbf{W}_n} f(\xi)\overline{g}(\xi).$$

Because

$$e_i\tilde{\Phi}^*\tilde{\Phi}e_j = \sum_{\xi\in\mathcal{W}_n} \varphi_j(\xi)\overline{\varphi}_i(\xi) = [\varphi_j, \varphi_i],$$

it follows that, by a slight abuse of notation,

$$(\tilde{\Phi}^*\tilde{\Phi})^{-1}\tilde{\Phi}^*\tilde{\Lambda} = diag(\frac{1}{K_n(\xi,\xi)})_{\xi\in\mathbf{W}_n}\tilde{\Phi}^*\tilde{\Lambda} = [G,\varphi]_n,$$

i.e. the parameter estimate θ_n can be computed as:

$$\hat{\theta}_n = [G, \varphi]_n. \tag{8.35}$$

The bias term is given by

$$G_0(z) - G(z, \mathsf{E}\{\hat{\theta}_n\}) = G_0(z) - \mathfrak{L}_n G_0(z) - \sum_{\zeta \in \mathbb{W}_n} \frac{K_n(z,\zeta)\mathsf{E}\{\eta_\zeta\}}{K_n(\zeta,\zeta)},$$

i.e. because $\langle \mathfrak{L}_n f, \mathfrak{L}_n g \rangle = [\mathfrak{L}_n f, \mathfrak{L}_n g]_n$,

$$||G_0(z) - G(z, \mathsf{E}\{\hat{\theta}_n\})||_2 \leq ||G_0 - \mathfrak{L}_n G_0||_2 + \sum_{\zeta \in \mathbb{W}_n} \frac{|\mathsf{E}\{\eta_\zeta\}|^2}{K_n(\zeta,\zeta)}.$$

If there is a $\delta > 0$ such that for the generating poles one has $|\alpha_k| > 1 - \delta$, then $C' n \leq K_n(\zeta, \zeta)$, *i.e.*

$$||G_0(z) - G(z, \mathsf{E}\{\hat{\theta}_n\})||_2 \leq ||G_0 - \mathfrak{L}_n G_0||_2 + C\frac{||\mathsf{E}\{\eta\}||_{2,n}^2}{n}, \tag{8.36}$$

where C', C are constants depending on δ.
As in the case with uniformly distributed nodes, the bias error asymptotically tends to the approximation error $||G_0 - \mathfrak{L}_n G_0||_2$, but in contrast with the equidistant case, this error vanishes if $G_0 \in \mathfrak{R}_m$ for an $m << n$.
For the variance one has:

$$||G(z, \hat{\theta}_n) - G(z, \mathsf{E}\{\hat{\theta}_n\})||_2^2 = || \sum_{\zeta \in \mathbb{W}_n} \frac{K_n(z,\zeta)(\eta_\zeta - \mathsf{E}\{\eta_\zeta\})}{K_n(\zeta,\zeta)} ||_2^2,$$

i.e.

$$||G(z, \hat{\theta}_n) - G(z, \mathsf{E}\{\hat{\theta}_n\})||_2^2 \leq C\frac{\mathrm{Tr}(\mathrm{Cov}(\eta))}{n}. \tag{8.37}$$

8.6 Computational Algorithm

To conclude this chapter, we summarize the case when the data points are supposed to be known at frequencies adapted to the poles that generate the rational orthonormal basis:

- Select a set of poles $\{\alpha_j \mid j = 1, \cdots n\}$ to define the Blaschke product $G_b(z)$ and the generalized rational orthonormal basis, say $\{\varphi_j(z) \mid j = 1, \cdots n\}$.
- Compute the function $\beta(t)$ to obtain the sets of points defined by

$$\mathbb{W}_n = \{\zeta_k = e^{i\gamma_k} \mid \gamma_k = \beta^{-1}(\frac{2\pi k}{n}),\ k = 0, \ldots, n-1\},$$

  and compute $\mu_\xi := \beta'(\frac{2\pi k}{n})$.

- Obtain the frequency-response data sets $G(\xi_k)$, $\xi_k \in \mathbb{W}_N$ and compute

$$E_{G,i}^{\langle\alpha\rangle}(\xi) := \frac{G(\xi)\overline{\varphi}_i(\xi)}{\mu_\xi}.$$

- The least squares parameter estimate can be computed as given (8.35), *i.e.*

$$\hat{\theta}_{n,i} = [G, \varphi_i]_n = \sum_{\xi \in \mathbb{W}_n} \frac{E_{G,i}^{\langle\alpha\rangle}(\xi)}{n}.$$

As the final result, the estimate will be given by:

$$G(z, \hat{\theta}_{n,}) = \sum_{i=1}^{n} \varphi_i(z)\hat{\theta}_{n,i}.$$

8.7 Concluding Remarks

In this chapter we have discussed results in the frequency domain related to approximate identification in $\mathcal{H}_2$ by using a rational orthonormal parameterization of the transfer function. The criteria for modelling and identification were formulated in terms of L_2 norms. The presented method can be seen as a generalization of the results of the FIR modelling to the case of a generalized orthonormal rational basis.
It was shown that the approximation power of these rational basis functions concerning the entire $\mathcal{H}_2$ space is the same as for the FIR case, *i.e.* one has the asymptotical convergence of the partial sums formed using the expansion defined by a given set of generalized orthonormal rational basis functions.
From a practical point of view, it is important to prove that the discrete approximation operators, *i.e.* the methods that use only a finite amount of data, can also provide the required convergence properties necessary to establish the asymptotic bias results.
An interpolation operator was introduced, which is defined on a non-uniform grid, given by the image of the uniform grid through the inverse of a function determined only by the prescribed poles.
The properties of this interpolation method were investigated and it was shown that by using the operator, one can obtain a method that is convergent on $\mathcal{H}_2$ and at the same time interpolates the rational transfer functions of a subspace determined by the given poles. Thus it was proved that the knowledge about the poles of the transfer function can be efficiently exploited in practice by using an identification method that is based on rational orthonormal expansions rather than a conventional FIR modelling.

9

Frequency-domain Identification in $\mathcal{H}_\infty$

József Bokor and Zoltan Szabó

Hungarian Academy of Sciences, Budapest, Hungary

9.1 Introduction

In this chapter we discuss results related to approximate identification in $\mathcal{H}_\infty$. The criteria for modelling and identification are formulated in terms of L_∞ or $\mathcal{H}_\infty$ norms. Emphasis is made on the construction of a model set by specifying bases in function spaces L_2, $\mathcal{H}_2$ or in the disc algebra $\mathbb{A}(\mathbb{D})$. These bases include – besides the most widely used trigonometric basis – the recently introduced rational orthogonal bases and some wavelet bases. The construction of identification algorithms considered as bounded operators mapping measured noisy frequency response data to an element in the model space is discussed. Bounds on the operator norm effecting the approximation and the noise errors are given, too.

It has been known that control design for dynamic systems usually requires the knowledge of an appropriate model of the system. These models can, in many cases, be derived from first principles but in more realistic situations from data that are the measured input/output signals of a system. Model construction from measured data is usually called system identification.

The goal of system identification is to construct models from noise corrupted measured data such that the model and the system generating the data should be small under suitably chosen criteria. The choice of identification criteria and the parameterization of models should reflect the ultimate goal the model is intended to be used for and depends also on the available information about the experimental conditions, noise assumptions (stochastic, deterministic norm bounded), *etc.* The identification criteria can be formulated either in the time domain, for instance the prediction error criterion as in Chapter 4, or in the frequency domain with $\mathcal{H}_2$ and $\mathcal{H}_\infty$ criteria.

Traditionally, the main approach to system identification has been based on stochastic assumptions explaining the errors between the actual system and its models. Properly parameterizing the models, the elaborated results provide a point estimate for the parameters of a nominal model and additional statistical properties characterizing the estimated parameters and the goodness of fit. The control design was based on the nominal model (applying the certainty equivalence principle) disregarding most of the statistical information provided by the identification.

The appearance of the robust control paradigm in the past decade accompanied by the formal $\mathcal{H}_\infty$ analysis and design theory incorporated the modelling uncertainties into control design. This started as a completely deterministic approach with the design based on a family of models given by a nominal model and an uncertainty model, describing *e.g.* the modelling error (or the bound on the magnitude of the error) in the frequency domain of interest. The design has been usually formulated as an $\mathcal{H}_\infty$ optimization (*e.g.* minimization of certain operator norms) over a set of stabilization controllers.

It was soon realized that existing methods of system identification are not able to provide initial data for robust control, and this inspired intensive research on both fields. The first concepts for a solution were published in the early 1990s by [123, 124]. This non-stochastic approach, usually referred to as worst-case identification for robust control, proposed to identify a nominal LTI model from frequency response data.

The problems related to this research can be characterized as follows.

The first subject considered is the modelling of uncertain systems and approximation of uncertain systems by a low complexity nominal model. The nominal model set is usually chosen as a finite dimensional subspace spanned by specific basis of one of the spaces $\mathcal{H}_2$, L_2 or the disc algebra $\mathbb{A}(\mathbb{D})$; see their definitions below.

The second problem is the identification of the uncertain system by determining a nominal model and bounds on the error that characterizes the uncertainty.

The discrepancies between the system to be modelled and the nominal model is explained usually by two error sources. One of them is the approximation error and the second is the noise error. The approximation error is generally defined by the choice of approximation operator, whereas the noise error term is influenced both by the operator (more precisely the operator norm) and the norm of the noise that corrupts the measurements.

A third group of problems, called validation/invalidation of models, comes from the question: If one specifies a nominal model and its uncertainty model, how it can be decided if these are consistent with measured information on the system?

There are various approaches to the above problems both in the time and frequency domains. Recent overviews using information based complexity and the set-membership approach to modelling and identification can be found in [195–197] and [311].

Concerning the choice of identification criteria, for worst-case identification in l_1 see *e.g.* [50, 101, 114, 303]. Worst-case identification under $\mathcal{H}_\infty$ criterion appeared in a large number of papers including [123,124], [109,110], [246,247], [177], [179]. Time-domain approaches to this problem appeared in [49] and [343].

Closed-loop issues of identification for robust control were initiated by [100] using an LQG/LTR approach and were further discussed by [20, 305]. In [145, 146] a generic scheme was introduced for the joint identification/control design by showing that the identification and control errors are identical in this scheme. This allows to elaborate very powerful iterative tools to obtain high closed-loop performance.
Model invalidation was discussed *e.g.* by [276, 277].
The subject of this chapter is the most closely related to worst-case identification under the $\mathcal{H}_\infty$ criterion by showing some new concepts and results that appeared recently in approximate modelling and identification. Our approach to build up the necessary tools on this area is based on the following concepts. The system to be modelled or identified from the data is supposed to be stable. The models should approximate this system uniformly under the $\mathcal{H}_\infty$ or L_∞ criterion if bounded noise assumption is applied. The problem is to find operators mapping the data set to a nominal LTI model such that the operator norm should be bounded to ensure convergence of the $\mathcal{H}_\infty$ or L_∞ norm of the approximation error to zero by increasing model order. The noise error is required to be bounded, too. We put attention of choosing the model parameterization, *i.e.* by specifying the subspaces spanned by various bases that can be used to ensure the above requirements.
The structure of the chapter is the following. The next section discusses the problem of approximate worst-case modelling and identification under the $\mathcal{H}_\infty$ criterion and provides the basic definitions.
It is followed by a discussion about approximate modelling and identification using the most widely used trigonometric basis in $C_{2\pi}$, the space of 2π periodic functions continuous on the unit circle. This leads to model sets represented by weighted Fourier partial sum operators. These algorithms are called φ-summations of trigonometric Fourier series. Specific choices of the window function φ lead to well-known examples like the Fejér or de la Vallée-Poussin summations. The model parameters are computed from frequency response data by using FFT. If the data are corrupted by L_∞-norm bounded noise, the model set will be in a subspace of L_∞. If a stable rational model with transfer function from the space $\mathbb{A}(\mathbb{D})$ is needed, one has to solve a Nehari-approximation problem. This typically two-step approach was discussed *e.g.* by [109, 110] and [112], using Fejér-summation in the first step. For a thorough treatment of this classical approach, one can consult [48]. The properties and bounds of the operators generated by the general φ-summation are discussed in [266].
Section 9.3 discusses the use of some recently introduced rational orthogonal bases in l_2, $\mathcal{H}_2$ and L_2. These bases were proposed by the authors of [306], [213, 214], and by [270]. Special cases like the Laguerre and Kautz bases, discussed in [314, 315] and [111] will be also considered The new results are related to the discrete versions of the frequency-domain identification of these models. It will be shown that by introducing a special argument transform in the inner functions representing the generalization of the shift operator,

these basis constructions can be relatively simply related to the trigonometric bases. This was shown in [269]. The advantage coming from this property is that one can use the well-known FFT or DFT to compute the coefficients of the models.
This enables the extension of the results obtained for the trigonometric bases to obtain approximate models and approximate identification algorithms under the $\mathcal{H}_\infty$ criterion, too.
The next section discusses basis construction in the disc algebra (it is known that there is no basis in $\mathcal{H}_\infty$). The basis is derived from the Faber-Schauder or from the Franklin systems resulting in wavelet-like basis functions. This basis allows to use simple bounded linear operators to obtain approximate model sets. The parameters can be computed by biorthogonal functionals specified by the basis resulting in a very simple identification algorithm, see [269]. Wavelet bases derived from frames and special rational bases proposed by [325] are considered briefly as well.
This chapter is concluded by an application example of the frequency-domain identification of a MTI interferometer testbed.

9.1.1 Basic Notation and Definitions

In this chapter, we consider functions that belong to the disc algebra $\mathbb{A}(\mathbb{D})$. Recall that a function $f : \overline{\mathbb{D}} \to \mathbb{C}$ belongs to $\mathbb{A}(\mathbb{D})$ if f is analytic on $\mathbb{D}$ and continuous on $\overline{\mathbb{D}}$, where $\overline{\mathbb{D}} := \mathbb{D} \cup \mathbb{T}$ stands for the closure of $\mathbb{D}$. The set of the boundary functions $f(z) \quad (z \in \mathbb{T})$ can be considered as a subspace of the set of continuous functions $C(\mathbb{T})$. This set will be identified with the set of 2π-periodic functions $C_{2\pi}$.
Moreover by the maximum modulus principle the $\mathcal{H}_\infty$-norm of f

$$\|f\|_{\mathcal{H}_\infty} := \max\{|f(z)| : z \in \overline{\mathbb{D}}\}$$

and the max-norm of the boundary function

$$\|f\|_\infty := \max\{|f(z)| : z \in \mathbb{T}\},$$

coincide. The problem of approximate $\mathcal{H}_\infty$ identification can be described as follows. Assume that the frequency-response measurements generated by a stable 'system' with transfer function $f \in C(\mathbb{T})$ are given, where f may be infinite dimensional. The goal is to produce an approximate model that should usually be a finite, possible low dimensional stable rational model, such that a prescribed small $\mathcal{H}_\infty$ norm of the error between the system and the model can be achieved if the noise level is sufficiently small and the number of measurements is sufficiently large.
To formalize the above problem, assume that the measurements are corrupted by some bounded noise $|\eta_k| \leq \epsilon \, (k \in \mathbb{N})$, where ϵ is unknown.
This concept of the noise can describe *e.g.* unmodelled dynamics, *i.e.* unstructured stable perturbations when using a nominal stable rational model, see [178].

Denote the data set by

$$E_N^{f,\eta}(z_l) := f(z_l) + \eta(z_l) \quad (z_l \in T_N).$$

where the set $T_N \subset \mathbb{T}$ with N discrete points.
Our intention is to find an algorithm

$$A_N : \mathbb{C}^N \to \mathcal{H}_\infty$$

that maps the data to a model:

$$A_N : \left\{E_N^{f,\eta}(z_l) \quad : z_l \in T_N\right\} \to f_N^{id} \in \mathcal{H}_\infty$$

such that the identification error

$$e_N^{id} := \|f_N^{id} - f\|_\infty$$

satisfies:

$$\lim_{N\to\infty,\epsilon\to 0} \sup_{||\eta||<\epsilon} \sup_{f\in\mathcal{H}^\infty} ||e_N^{id}||_\infty \to 0.$$

The above algorithm A_N is said to be convergent.
In some papers, the subspace $\mathcal{H}_\infty(D_\rho, M) \subset \mathcal{H}_\infty$ is used, *i.e.* the set of functions analytic on the disc with radius $\rho > 1$ and with $\sup\{|f(z)|\,|\,|z| < \rho\} \leq M$ (a set of exponentially stable systems). Moreover, an algorithm is said to be *robustly convergent* when the above is true for all $\rho > 1$ and $M > 0$, and *untuned* if it does not depend on prior information about ρ, M and ϵ, while the others are *tuned.*
It was proved that there is no linear algorithm that satisfies the above requirement if $f_N^{id} \in \mathcal{H}_\infty$, see [247]. This leads to the elaboration of a family of two-step methods. Under the L_∞ norm bound assumption on the noise, the first step typically includes an L_∞ approximation and this is followed by a Nehari-approximation.
It can be shown that for a sequence of bounded linear operators, strongly convergent to the identity – *i.e.* an interpolation operator on $\mathcal{L}^\infty$ – the conditions for robust convergence are fulfilled. For this choice, one can drop the *worst-case* requirement $\sup_{||\eta||<\epsilon} \sup_{f\in L_\infty}$, as it has been already included in the concept of strong operator convergence – see one of the equivalent formulations of the Banach-Steinhaus uniform boundedness principle in [153, pp. 267]. Denote by $V_N : \mathbb{C}^N \to L_\infty$ the sequence of uniformly bounded linear operators, then the robust convergence in L_∞ means

$$\lim_{N\to\infty,\epsilon\to 0} ||V_N(f+\eta) - f||_\infty \to 0.$$

By linearity one has

$$\begin{aligned}
&\lim_{N\to\infty,\epsilon\to 0} ||V_N(f+\eta) - f||_\infty \leq \\
&\leq \lim_{N\to\infty,\epsilon\to 0}\{||V_N(f) - f||_\infty + ||V_N(\eta)||_\infty\} \leq \\
&\leq \lim_{N\to\infty,\epsilon\to 0}\{||V_N(f) - f||_\infty + ||V_N||\epsilon\}.
\end{aligned}$$

The first term tends to zero due to the convergence property of the approximation algorithm, regardless of the value of ϵ, the second term tends to zero if $\epsilon \to 0$.
The above discussions show that if one can find a sequence of uniformly bounded linear operators in the disc algebra $\mathbb{A}(\mathbb{D})$, such that

$$\lim_{N\to\infty} \{||V_N(f) - f||_\infty \to 0,$$

then a linear approximation algorithm, fulfilling the robust convergence property, has been found.
The existence of such operators when using trigonometric polynomials is a classical result. The Fejér-mean and the de la Vallée-Poussin operators were discussed *e.g.* in [178] for stable perturbations *i.e.* when $\eta \in \mathbb{A}(\mathbb{D})$. Generalization of this class of operators that are generated by the φ-summations are studied in [266] and will be summarized in the next paragraph.

9.2 Approximate Linear Models Generated by Discrete φ-Summation

Let us consider the continuous Fourier coefficients of a function $f \in C_{2\pi}$:

$$c_k(f) = \frac{1}{2\pi} \int_0^{2\pi} f(e^{i\omega}) e^{-ik\omega} d\omega,$$

and define the discrete Fourier transform of f as:

$$c_N(k) := \frac{1}{N} \sum_{\ell=0}^{N-1} f(\frac{2\pi\ell}{N}) e^{-2\pi i k\ell/N} \quad (k \in \mathbb{Z}).$$

The sequence $(c_N(k), k \in \mathbb{Z})$ is periodic with period N.
The partial sums of the Fourier expansion will have the form:

$$(S_n f)(\omega) := \sum_{k=-n}^{n} c_k(f) e^{ik\omega} \quad (n \in \mathbb{N}).$$

and the discrete partial sums will have the form:

$$(S_{n,N} f)(\omega) := \sum_{k=-n}^{n} c_N(k) e^{ik\omega} \quad (n \in \mathbb{N}),$$

where $n < N$ denotes the order of the truncation that is called model order, too.
The Fourier coefficients are obtained from the data as follows:

$$\hat{c}_N(k) := \frac{1}{N} \sum_{\ell=0}^{N-1} E_N^{f,\eta}(z_\ell) e^{2\pi i k\ell/N} = c_N(k) + \eta_N(k),$$

where

$$\eta_N(k) := \frac{1}{N}\sum_{\ell=0}^{N-1} \eta_\ell e^{2\pi i k\ell/N}.$$

A model could be obtained from the Fourier series as a partial sum

$$f_{n,N}(\omega) := \sum_{k=-n}^{n} \hat{c}_N(k) e^{ik\omega},$$

i.e.

$$f_{n,N}(\omega) = \big(S_{n,N} f\big)(\omega) + \big(S_{n,N}\eta\big)(\omega). \tag{9.1}$$

It is known, however, that even in the noiseless case the partial sum operators diverge at a point $\omega \in [0, 2\pi]$, as the operators in question are not uniformly bounded. This problem is usually avoided when replacing $S_{n,N}$ by a bounded linear operator obtained by applying summation over the 'windowed' Fourier series (see *e.g.* Fejér-summation), resulting in a model $f^w_{n,N} \in L_\infty$:

$$f^w_{n,N} := \sum_{k=-n}^{n} w_k \hat{c}_N(k) e^{ik\omega}.$$

In case of the Fejér-summation $w_k = 1 - \frac{2|k|}{n}$. We investigate a generalization of the 'windowing' idea that will be called φ-summation, *i.e.* a summation generated by a function φ as defined below. The continuous φ-summation was investigated in the book of [42] and [203]. The results for the discrete case are from [266]. For a thorough treatment see [295]
Let φ be a continuous, even, compactly supported function satisfying $\varphi(0) = 1$. Assume that the Fourier transform of φ is absolute integrable, *i.e.*

$$\hat{\varphi}(\omega) := \frac{1}{2\pi}\int_{-\infty}^{\infty} \varphi(t) e^{-i\omega t}\, dt \quad (\omega \in \mathbb{R})$$

and $\hat{\varphi} \in L^1(\mathbb{R})$.
Recall that the infinite series $\sum_{-\infty}^{\infty} g_k$ is called φ-summable if the sequence

$$U^\varphi_n g := \sum_{k=-\infty}^{\infty} \varphi\big(\frac{k}{n}\big) g_k \quad (n \in \mathbb{N}^*)$$

converges as $n \to \infty$. The limit is called the φ-sum of $\sum_{-\infty}^{\infty} g_k$.
Let us define the continuous φ-summation operator as:

$$\big(U^\varphi_n f\big)(e^{i\omega}) := \sum_{k=-\infty}^{\infty} \varphi\big(\frac{k}{n}\big) c_k(f) e^{ik\omega}. \tag{9.2}$$

The continuous φ-summation theorem can be stated as follows:

Proposition 9.1 ([266]). *Suppose that $\varphi : \mathbb{R} \to \mathbb{R}$ is a continuous even function supported in $[-1, 1]$, with $\varphi(0) = 1$, and such that its Fourier transform is Lebesgue integrable on $\mathbb{R}$. Then*

$$\lim_{n\to\infty} \|f - S_n^{\varphi} f\|_{\infty} = 0$$

for every continuous function on $\mathbb{T}$.

The model set used in the subsequent linear L_{∞} identification will be constructed from the φ-sum of the discrete Fourier series as

$$\mathcal{M}: \quad f_{n,N}^{\varphi} = \sum_{k=-\infty}^{\infty} \varphi(\frac{k}{n}) \hat{c}_N(k) e^{ik\omega},$$

i.e.

$$f_{n,N}^{\varphi} = U_{n,N}^{\varphi} f + U_{n,N}^{\varphi} \eta, \tag{9.3}$$

where

$$(U_{n,N}^{\varphi} f)(\omega) := \sum_{j=-\infty}^{\infty} \varphi(\frac{j}{N}) c_N(j) e^{ij\omega}$$

is the φ-sum of the noiseless discrete Fourier coefficients.

The following theorem shows that the sequence of operators $U_{n,N}^{\varphi}$ is uniformly bounded and that the algorithm which results in the models above is robustly convergent. Let us denote

$$\frac{2\pi}{N}\mathbb{Z} := \{\frac{2\pi\ell}{N} : \ell \in \mathbb{Z}\}.$$

Proposition 9.2 ([266]). *Suppose that φ is a continuous real function supported in the interval $[-s, s]$ for which $\varphi(0) = 1$ and $\hat{\varphi} \in L^*(\mathbb{R})$. Then for any discrete 2π-periodic function $f : \frac{2\pi}{N}\mathbb{Z} \to \mathbb{C}$ and any $x \in \mathbb{R}$*

$$(U_{n,N}^{\varphi} f)(x) = \frac{2\pi n}{N} \sum_{r \in \frac{2\pi}{N}\mathbb{Z}} f(r) \hat{\varphi}(n(r - x)). \tag{9.4}$$

Moreover

$$\|U_{n,N}^{\varphi}\| \leq \frac{n}{N} \|\varphi\|_1 + \|\hat{\varphi}\|_* \quad (n, N \in \mathbb{N}^*) \tag{9.5}$$

and for any sequence N_n $(n \in \mathbb{N}^*)$ with

$$\lim_{n\to\infty} N_n/n \geq s$$

we have

$$\lim_{n\to\infty} \|U_{n,N_n}^{\varphi} f - f\|_{\infty} = 0 \quad (f \in C_{2\pi}).$$

Remark 9.1.

1. In (9.5) the bound on the operator depends on the $\mathcal{L}^1 1$ norm of φ and on a norm of $\hat{\varphi}$ that is the Fourier-transform of φ. This latter norm is defined as follows. Assume that for the even function $\hat{\varphi} \in \mathcal{L}^1(\mathbb{R})$ there exists an even function $\gamma \in \mathcal{L}^1(\mathbb{R})$ such that

$$\text{i) } \gamma \quad \text{is decreasing in } [0, +\infty), \text{ and} \tag{9.6}$$

$$\text{ii) } |\hat{\varphi}(\omega)| \leq \gamma(\omega) \quad (\omega \in \mathbb{R}). \tag{9.7}$$

Denote by $\mathcal{L}^*(\mathbb{R})$ the set of even functions $\hat{\varphi} \in \mathcal{L}^1(\mathbb{R})$ for which there exists an even function $\gamma \in \mathcal{L}^1(\mathbb{R})$ satisfying (9.6). Obviously $\mathcal{L}^*(\mathbb{R}) \subset \mathcal{L}^1(\mathbb{R})$ and

$$\|\hat{\varphi}\|_* := \inf_\gamma \|\gamma\|_1$$

is a norm in $\mathcal{L}^*(\mathbb{R})$. Here the infimum is taken over all γ satisfying (9.6) and $\|\cdot\|_p$ denotes the $\mathcal{L}^p(\mathbb{R})$-norm $(1 \leq p \leq \infty)$.
Various choices of φ lead to specific algorithms, of which many have already been considered in the literature.

2. In the continuous case the operator norm in (9.5) is reduced simply to

$$\|U_n^\varphi\| \leq \|\hat{\varphi}\|_1.$$

Example 9.1.

1. In the special case when φ is supported in $[-1, 1]$ and linear on $[0, 1]$, *i.e.* if

$$\varphi_1(x) := 1 - x \quad (0 \leq x \leq 1),$$

then U_{n+1}^φ is the nth discrete Fejér operator, see *e.g.* [110]:

$$U_{n+1}^\varphi f = \sigma_n f = \frac{1}{n+1} \sum_{k=-n}^{n} S_{k,N} f \quad (n \in \mathbb{N}).$$

2. Let $\varphi_2(x) := 1$ on $[0, 1/2]$, linear on $[1/2, 1]$, even and supported in $[-1, 1]$. In this case we get the de la Vallée-Poussin means, discussed *e.g.* in [178], [110],

$$U_{2n+2}^\varphi f = \frac{S_{n+1,N} f + \cdots + S_{2n+1,N} f}{2(n+1)} = $$
$$= 2\sigma_{2n+1} f - \sigma_n f.$$

3. When $\varphi_3(x) := \cos \frac{\pi x}{2} \quad (-1 \leq x \leq 1)$, the $U_{n,N}^\varphi$'s are the Rogosinski operators,

$$\hat{\varphi}_3(x) = \frac{\sin(x - \pi/2)}{2(x^2 - (\pi/2)^2)} \quad (x \in \mathbb{R}).$$

4. If $\varphi_4(x) := 1 - x^s \quad (0 \leq x \leq 1)$ for some $s > 1$ then we get the Marcel Riesz means.

5. An important class of φ-functions can be obtained using cardinal B-splines. Let
$$M_1(x) := \chi_{[0,1)}(x) \quad (x \in \mathbb{R})$$
be the characteristic function of $[0,1)$ and define M_m recursively by
$$\begin{aligned} M_m(x) &:= \int_{-\infty}^{\infty} M_1(t) M_{m-1}(x-t)dt \\ &= \int_0^1 M_{m-1}(x-t)\,dt \quad (x \in \mathbb{R}, m \geq 2). \end{aligned}$$
The cardinal B-spline M_m is symmetric with respect to the centre of its support, consequently the continuous function
$$\psi_m(x) := M_m(\frac{m}{2} + \frac{mx}{2})/M_m(\frac{m}{2})$$
$(x \in \mathbb{R}, m \geq 1)$ is supported in $[-1,1]$ and satisfies $\psi_m(0) = 1$.
For example, if $m = 4$ then the Fourier transform of ψ_4 is the Jackson kernel.

In the previous section it was mentioned that there is no linear algorithm that satisfies the robust convergence requirement. This makes it necessary to elaborate on an efficient non-linear method *even if the noise is known to be continuous.* Based on the celebrated Adamjan-Arov-Krein (AAK) theorem in a Hankel operator setting, this problem has a solution. The algorithm has typically two parts, see *e.g.* [109].

By using the discrete Fourier transform coefficients and the φ-summation method, one can obtain the desired linear algorithm $V_N : \mathbf{C}^N \rightarrow C(\mathrm{T})$ for the first step.
In the second, non-linear, step one can get for the continuous function f its best $\mathcal{H}_\infty$ approximation, *i.e.* one has to find $A : L_\infty \rightarrow \mathcal{H}_\infty$ such that:
$$||f - A(f)||_\infty = \inf\{||f - h||_\infty \,|\, h \in \mathcal{H}_\infty\}.$$
This problem is usually called a Nehari-approximation problem and its solution is given as a special case of the AAK theorem, as follows:
Consider a function $f = \sum_{|k| \leq n} c_k z^k$ and the Hankel matrix
$$\Gamma_f = \begin{pmatrix} c_{-1} & c_{-2} & \dots & c_{-n} \\ c_{-2} & c_{-3} & \dots & 0 \\ \dots & & & \\ c_{-n} & 0 & \dots & 0 \end{pmatrix}.$$

For the vectors v, w from $\mathbf{C}^m$ that $||w|| = ||\Gamma_f||||v||$ and $w = \Gamma_f v$ one has the rational function

$$h = \sum_{|k|\leq n} c_k z^k - \frac{\sum_{j=1}^n w_j z^{-j}}{\sum_{j=1}^n v_j z^{j-1}},$$

which solves the optimization problem see [246, 247]. For the error of the method one can get the bound [246,247]:

$$\begin{aligned} &||f - AV_N(f,\eta)||_\infty \leq \\ &2(||f - V_N(f,0)||_\infty + ||V_N(0,\eta)||_\infty). \end{aligned}$$

Both of the errors on the left-hand side, *i.e.* the L_∞ approximation error for the plant nominal model and the amplification factor of the noise by the uniformly bounded factor $||V_N||$ depend on the actual form of the approximation process V_N, *i.e.* depend on the summation technique.

9.3 Approximate Modelling Using Rational Orthogonal Bases

In the previous chapters, the reader has seen some examples for the parameterization of the transfer functions using rational orthonormal bases. In what follows we restrict this modelling to the situation where the set of poles that generate the orthonormal system is formed by a periodic repetition of the same finite sequence $\alpha_d = (\alpha_k \,|\, k = 1, \ldots, d)$.
As we have already seen, see *e.g.* Proposition 8.2, if one considers the finite Blaschke product $G_b(z) = \prod_{k=1}^d \frac{1-\overline{\alpha}_k z}{z-\alpha_k}$, and an orthonormal basis $\{\varphi_l \,|\, l = 1, \ldots, d\}$ in the subspace $\mathfrak{R}_d$, then the system $\phi_{l+kd} = \varphi_l G_b^k, \quad k \in \mathbb{Z}$ forms an orthonormal basis of L_2. For a proof based on the properties of the shift operator induced by the multiplication by the inner function G_b on $\mathcal{H}_2$ see [270, 290].

9.3.1 Summation Theorems for Rational Expansions

It is known that for classical Fourier series, Proposition 8.2 fails to be true for $f \in \mathbb{A}(\mathbb{D})$ and the uniform norm, as the operator norm of the Dirichlet kernels are unbounded. An analogous result holds for the rational case, too, *i.e.*

Proposition 9.3 ([285]). *If* $1 - |\alpha_k| > \delta > 0$, $k \in \mathbb{N}$, *then*

$$C_1 \log n \leq \sup_{\tau \in [-\pi,\pi]} \mathfrak{I}(|S_n^0(\omega,\tau))|) \leq C_2 \log n, \quad C_1, C_2 > 0 \tag{9.8}$$

Let $S_n(\omega,\tau)$ be generated by the sequence of poles $\alpha_1, \alpha_2, \cdots$ and $S_n^0(\omega,\tau)$ by the sequence $0, \alpha_1, \alpha_2, \cdots$ then, using (8.7) and the definition of the kernels one has $S_n^0(\omega,\tau) = 1 + e^{i(\omega-\tau)} S_n(\omega,\tau)$. It follows that

$$|S_n(\omega,\tau)| = |S_n^0(\omega,\tau)| + O(1),$$

i.e. Proposition 9.3 is valid for S_n, too.

Let us consider the 'block' analogy of the Fejér summation, *i.e.* $\mathcal{F}_n = \frac{1}{n}\sum_{k=1}^{n} \mathcal{S}_{kd}$, where $\mathcal{S}_n f$ denotes the partial sums of the expansion of a function f in the orthonormal system $\{\phi_k \mid k \in \mathbb{Z}\}$.

Proposition 9.4 ([285]). *For $f \in L_\infty$ it holds that*

$$||\mathcal{F}_n f||_\infty \leq C||f||_\infty \tag{9.9}$$

and for all continuous f one has

$$\lim_{n\to\infty} ||f - \mathcal{F}_n f||_\infty = 0. \tag{9.10}$$

Using this theorem, one can prove that the classical, *i.e.* not 'block', summation for the periodic case is also convergent, *i.e.*

Proposition 9.5 ([285]). *For $f \in L_\infty$ and $\mathcal{F}_N^c := \frac{1}{N}\sum_{k=1}^{N} \mathcal{S}_k$ then one has*

$$||\mathcal{F}_N^c f||_\infty \leq C||f||_\infty$$

and for all of the continuous f one has

$$\lim_{N\to\infty} ||f - \mathcal{F}_N^c f||_\infty = 0.$$

One can introduce the generalization of the de la Vallée-Poussin operators as $\mathcal{V}_n = \frac{1}{2n}\sum_{k=n+1}^{2n} \mathcal{S}_k = 2\mathcal{F}_{2n} - \mathcal{F}_n$, that has the same convergence properties. Let us mention here that getting the square, [222], or the fourth power, [169], of the absolute value of S_{α_n} as kernels leads also to uniformly bounded operators, but these operators cannot be associated with simple summation processes. For the sake of completeness and for further reference let us recall here the result from [222]:

Theorem 9.1. *Let us introduce the kernel*

$$T_n(t,\tau) = \frac{|K_n(e^{i\omega}, e^{i\tau})|^2}{K_n(e^{i\omega}, e^{i\omega})}, \tag{9.11}$$

and the operator

$$\mathcal{T}_n f(\omega) = \mathcal{I}(f(\tau)T_n(\omega,\tau)). \tag{9.12}$$

If $\sum_{k=1}^{\infty}(1 - |\alpha_k|) = \infty$ then for every f continuous on $\mathbb{T}$ one has

$$\lim_{n\to\infty} ||f - \mathcal{T}_n f||_\infty = 0. \tag{9.13}$$

The φ-summation theorem was phrased in Section 9.2 in the context of the classical Fourier series. In what follows, a proof will be given for the extension of this results for rational expansions, too. The presented ideas leads to a possible numerical algorithm to compute the expansion coefficients and to the discrete version of the theorem.

Let us consider the following operator:

$$Pf := \sum_{l \in \mathbb{Z}} \langle f, G_b^l \rangle G_b^l$$

and the operators

$$P_r f := P(f \overline{\varphi}_r).$$

One has

$$P_r f = \sum_{l \in \mathbb{Z}} \langle f \overline{\varphi}_r, G_b^l \rangle G_b^l = \sum_{l \in \mathbb{Z}} \langle f, \phi_{r+dl} \rangle G_b^l,$$

and is clear that $f = \sum_{r=1}^{d} \varphi_r P_r f$.
Because $G_b(e^{i\gamma(s)}) = e^{ids}$, it follows that $Pf(e^{i\gamma(s)}) = \sum_{l \in \mathbb{Z}} \langle f, G_b^l \rangle \epsilon^{ld}$, *i.e.*

$$Pf(e^{i\gamma(s)}) = \sum_{l \in \mathbb{Z}} \langle \tilde{f}, \epsilon^{dl} \rangle \epsilon^{ld},$$

where $\epsilon(\omega) = e^{i\omega}$ and $\tilde{f}(e^{is}) = f(e^{i\gamma(s)})\gamma'(s)$, and where γ denotes the function β^{-1}.

9.3.2 The Conditional Expectations E_d

Let us consider the following operator:

$$E_d f(\omega) = \frac{1}{d} \sum_{r=0}^{d-1} f(\omega + \frac{2\pi}{d} r).$$

The trigonometric system $\epsilon_l(\omega) = \epsilon^l(\omega)$ is E_d orthogonal, as:

$$E_d(\epsilon_k \overline{\epsilon}_l) = \delta_{kl},$$

in particular $E_d(\epsilon_r) = 0$, $r = 1, \dots d-1$. For a function $g = \sum_{k \in \mathbb{Z}} g_k \epsilon^{dk}$, $g_k \in \mathbb{C}$ one has $E_d g = g$, so, one has:

$$f = \sum_{r=0}^{d-1} E_d(f \overline{\epsilon}_r) \epsilon_r,$$

and the operators $E_d(f \overline{\epsilon}_r)\epsilon_r$ are pairwise orthogonal projections.

Lemma 9.1. *For all $f \in L_2$ one has*

$$E_d f = \sum_{k\in\mathbf{Z}} \langle f, \epsilon_{dk}\rangle \epsilon_{dk}.$$

Proof. It is clear that $f = \sum_{k\in\mathbf{Z}} f_k \epsilon_{dk}$, where $f_k \in \mathfrak{P}_d$. Now,

$$E_d f = \sum_{k\in\mathbf{Z}} E_d(f_k)\epsilon_{dk} = \sum_{k\in\mathbf{Z}} \langle f, \epsilon_{dk}\rangle \epsilon_{dk},$$

as $E_d(f_k) = \langle f, \epsilon_{dk}\rangle$.

■

Now, because $Pf \circ \gamma = \sum_{l\in\mathbf{Z}} \langle \tilde{f}, \epsilon_{dl}\rangle \epsilon_{dl}$, it follows that:

Lemma 9.2.

$$Pf \circ \gamma = (E_d \tilde{f})(e^{i\omega}).$$

Because $Pf = E_d \tilde{f} \circ \beta$, it follows that $||Pf||_\infty \leq ||E_d\tilde{f}||_\infty \leq ||\tilde{f}||_\infty \leq ||f||_\infty ||\gamma'||_\infty$. Then

Lemma 9.3.

$$||Pf||_\infty \leq ||\gamma'||_\infty ||f||_\infty.$$

It follows, that the projection operator P_r is bounded, *i.e.* $||P_r f||_\infty = ||P(f\overline{\varphi}_r)||_\infty \leq C||f||_\infty$.
Let us recall that the expansion coefficients of f, *i.e.* $c_{kd+l}(f)$, can be computed as

$$c_{kd+l}(f) = \langle f, \phi_{r+dl}\rangle = \langle E_d(f\tilde{\overline{\varphi}}_r), \epsilon_{dl}\rangle.$$

Let us define the generalized continuous φ-summation operator as:

$$(R_n^\varphi f)(z) := \sum_{k=-\infty}^{\infty} \sum_{l=1}^{d} \varphi\left(\frac{k}{n}\right) c_{kd+l}(f)\varphi_l(z) G_b^k(z). \tag{9.14}$$

Because $P_l f \circ \gamma = \sum_{k\in\mathbf{Z}} \langle f\tilde{\overline{\varphi}}_l, \epsilon_{dk}\rangle \epsilon_{dk} = \sum_{k\in\mathbf{Z}} c_{kd+l}(f)\epsilon_{dk}$, it follows that

$$S_{nd}^\varphi(P_l f \circ \gamma) = \sum_{k=-\infty}^{\infty} \varphi\left(\frac{k}{n}\right) c_{kd+l}(f)\epsilon_{dk},$$

i.e.

$$(R_n^\varphi f)(e^{i\omega}) = \sum_{l=1}^{d} \varphi_l(e^{i\omega}) S_{nd}^\varphi(P_l f \circ \gamma) \circ \beta(\omega)$$

Because the projection operator P_l is bounded, the classical φ-summation result stated in Proposition 9.1 is applicable, *i.e.* one has that

Proposition 9.6 ([285]). *Suppose that $\varphi : \mathbb{R} \to \mathbb{R}$ is a continuous even function supported in $[-1,1]$, with $\varphi(0) = 1$, and such that its Fourier transform is Lebesgue integrable on $\mathbb{R}$. Then*

$$\lim_{n\to\infty} \|f - R_n^\varphi f\|_\infty = 0$$

for every continuous function on $\mathbb{T}$.

9.3.3 Discrete φ-Summation for Rational Expansions

In this section, we would like to give the discrete version of the summation results, already presented in the classical Fourier context in Proposition 9.2, for rational orthonormal expansions.
Let us recall the discrete rational orthonormal system, introduced in Section 8.4, defined using the nodes

$$\mathbb{W}_n = \{\zeta_k = e^{i\gamma_k} \mid \gamma_k = \beta_{(n)}^{-1}(\nu_k),\ k = 0, \ldots, n-1\},$$

i.e. the images of the roots of unity through the $\gamma_{(n)} = \beta_{(n)}^{-1}$ function. Considering as nodes the set $\mathbb{W}_n$, it was introduced the rational interpolation operator:

$$(\mathfrak{L}_n f)(z) := \sum_{\zeta\in\mathbb{W}_n} \frac{K_n(z,\zeta)}{K_n(\zeta,\zeta)} f(\zeta),$$

where f is a continuous function on $\mathbb{T}$ and $z \in \mathbb{T}$, and defined the discrete scalar product

$$[f,g]_n := \sum_{\zeta\in\mathbb{W}_n} \frac{f(\zeta)\overline{g}(\zeta)}{K_n(\zeta,\zeta)} = \sum_{\zeta\in\mathbb{W}_n} \frac{f(\zeta)\overline{g}(\zeta)}{n\beta_{(n)}'(\gamma)},$$

where $\zeta = e^{i\gamma}$, with the property

$$\langle \mathfrak{L}_n f, \mathfrak{L}_n g\rangle = [f,g]_n.$$

Let us define the discrete rational transform of f as:

$$c_N(kd + l) := [f, \varphi_l G_b^k]_N, \tag{9.15}$$

and let us observe that if the poles a placed in the origin, then these coefficients are exactly those given by the discrete Fourier transform. Moreover, let us observe, that

$$c_N(kd + l) = [f\tilde{\overline{\varphi}}_l, e_{dk}]_N^f,$$

i.e. $c_N(kd+l)$ is the sampled discrete Fourier coefficient of $f\tilde{\overline{\varphi}}_l$.
Using these coefficients, let us introduce the discrete partial sums as:

$$(R_{n,N}f)(z) := \sum_{k=-n}^{n} \sum_{l=1}^{d} c_N(kd+l)\varphi_l(z)G_b^k(z) \quad (n \in \mathbb{N}),$$

and the φ-summation of the discrete rational expansion as:

$$(W_{n,N}^{\varphi}f)(z) := \sum_{j=-\infty}^{\infty} \sum_{l=1}^{d} \varphi(\frac{j}{n})c_N(jd+l)\varphi_l(z)G_b^j(z).$$

Proposition 9.7 ([266]). *Suppose that φ is a continuous real function supported in the interval $[-1,1]$ for which $\varphi(0) = 1$ and $\hat{\varphi} \in L^*(\mathbb{R})$. Then for any discrete continuous function f on $\mathbb{T}$ one has*

$$\lim_{n\to\infty} \|W_{n,N_n}^{\varphi}f - f\|_\infty = 0$$

for any sequence N_n $(n \in \mathbb{N}^*)$ with

$$\lim_{n\to\infty} N_n/n \geq 1.$$

9.3.4 Optimal Non-linear Rational Approximation

To obtain the best approximate model from $\mathcal{H}_\infty$, one has to solve a Nehari-problem. The generalization of the results given for the trigonometric system to the generalized orthonormal expansion can be found in [289].
In Section 2, it was mentioned that there is no linear algorithm that satisfies the robust convergence requirement. This makes necessary to elaborate an efficient non-linear method *even if the noise is known to be continuous.* Based on the celebrated Adamjan-Arov-Krein (AAK) theorem in a Hankel operator setting, this problem has a solution. The algorithm has typically two parts, see *e.g.* [109].
By using the the discrete Fourier transform coefficients and the φ-summation method, one can obtain the desired linear algorithm $V_N : \mathbf{C}^N \to C(\mathrm{T})$ for the first step.
In the second, non-linear step one can get for the continuous function f its best $\mathcal{H}_\infty$ approximation, *i.e.* one has to find $A : L_\infty \to \mathcal{H}_\infty$ such that:

$$||f - A(f)||_\infty = \inf\{||f - h||_\infty \mid h \in \mathcal{H}_\infty\}.$$

This problem is usually called a Nehari-approximation problem and its solution is given as a special case of the AAK theorem, as follows:
Consider a function $f = \sum_{|k|\leq n} c_k z^k$ and the Hankel matrix

$$\Gamma_f = \begin{pmatrix} c_{-1} & c_{-2} & \dots & c_{-n} \\ c_{-2} & c_{-3} & \dots & 0 \\ \dots & & & \\ c_{-n} & 0 & \dots & 0 \end{pmatrix}.$$

For the vectors v, w from $\mathbf{C}^m$ that $||w|| = ||\Gamma_f|| ||v||$ and $w = \Gamma_f v$ one has the rational function

$$h = \sum_{|k|\leq n} c_k z^k - \frac{\sum_{j=1}^n w_j z^{-j}}{\sum_{j=1}^n v_j z^{j-1}},$$

which solves the optimization problem see [246, 247]. For the error of the method one can get the bound [246, 247]:

$$||f - AV_N(f,\eta)||_\infty \leq 2(||f - V_N(f,0)||_\infty + ||V_N(0,\eta)||_\infty).$$

Both of the errors on the left-hand side, *i.e.* the L_∞ approximation error for the plant nominal model and the amplification factor of the noise by the uniformly bounded factor $||V_N||$ depend on the actual form of the approximation process V_N, *i.e.* depend on the summation technique.

The second, non-linear step of the approximation method presented above can be extended to generalized orthonormal representations, too. To do this, the Nehari-problem is reduced to an $\mathcal{H}_\infty$ optimization one, as follows: let us denote by $\mathbf{H}(G_b) = \mathcal{H}_2 \ominus G_b\mathcal{H}_2$. and let us consider a function $f \in \mathbf{H}(G_b)$, then

$$||f\overline{G}_b - f_*||_\infty = \inf_{g\in\mathcal{H}_\infty} ||f\overline{G}_b - g||_\infty = \inf_{g\in\mathcal{H}_\infty} ||f - G_b g||_\infty,$$

using the fact that G_b is inner, *i.e.* $|G_b| = 1$ on the unit circle.

Let us consider the so called Sarason operators [262], *i.e.*:

$$\mathtt{A}_{f,G_b} : \mathbf{H}(G_b) \to \mathbf{H}(G_b), \quad \mathtt{A}_{f,G_b} = \mathtt{P}_{\mathbf{H}(G_b)}\mathtt{M}_f,$$

where $\mathtt{M}_f$ is the multiplication operator by f. For a given inner function G_b we denote $\mathtt{A}_{f,G_b} = \mathtt{A}_f$. One has the following proposition due to [262]:

Proposition 9.8. *One can always choose an optimal $f_* \in \mathcal{H}_\infty$ such that $\mathtt{A} = \mathtt{A}_{f_*}$, $||\mathtt{A}|| = ||f_*||_\infty$. For any function $f \in \mathcal{H}_\infty$ there exists an optimal $\psi_* \in \mathcal{H}_\infty$ solving the following optimization problem:*

$$||f - G_b\psi_*|| = \inf_{\psi\in\mathcal{H}_\infty} ||f - G_b\psi|| = ||\mathtt{A}_f||.$$

Assume that x is a non-zero vector in $\mathbf{H}(G_b)$ satisfying $||\mathtt{A}x|| = ||\mathtt{A}|| ||x||$ and set $y = \mathtt{A}x$. Then $f_* = y/x$ is an inner function and this is an optimal one, moreover $\psi_* = G_b^*(f - y/x)$.

The algorithm that solves the problem

$$||fG_b^* - f_*||_\infty = \inf_{g\in\mathcal{H}_\infty} ||f - G_b g||_\infty,$$

can be summarized as follows: one has to construct the matrix $\mathtt{C}_\phi$ determined by G_b and f that represents the operator $\mathtt{A}_f$ in the orthonormal basis $\phi =$

$\{\phi_j, j=\overline{1,d}\}$ of $\mathbf{H}(G_b)$, then one has to compute some maximal vectors *i.e.* any non-zero vector x satisfying $||\mathtt{C}x|| = ||\mathtt{C}||||x||$, and $y = \mathtt{C}x$. The solution of the problem is given by $f_* = G_b^*(f - \frac{\mathtt{W}_\phi y}{\mathtt{W}_\phi x})$, where $\mathbf{W}_\phi$ denotes the unitary operator $\mathbf{C}^d \to \mathbf{H}(G_b)$ fixed by the (orthonormal)basis ϕ, *i.e.* the coordinate map.
The entries of $\mathtt{C}$ are given by

$$\mathtt{C}_{i,j} =< \mathtt{A}(f)\phi_j, \phi_i >=< \mathtt{P}_{\mathbf{H}(G_b)}(f\phi_j), \phi_i >= \sum_{k=1}^{d} f_k < \phi_k\phi_j, \phi_i > .$$

It can be seen that $\mathtt{C}_{i,j} = 0, \quad j > i$.
If one considers $f = \sum_{k=0}^{N-1} f_k G_{b,0}^k$, where $f_k \in \mathbf{H}(G_{b,0})$ and $G_{b,0}$ is a finite Blaschke product of order d with poles having multiplicity one, the matrix C has a block Toeplitz structure where the d dimensional matrices C_j have as rows the coefficients of the projection of $f_j\phi_i$ on $\mathbf{H}(G_{b,0})$ where $\{\phi_i\}$ is an orthonormal basis in $\mathbf{H}(G_{b,0})$. A possible way to compute these coefficients for a function h is to use the recursion:

$$h_1 = h, \quad h_{k+1}(z) = h_k(z)(1 - \frac{\overline{\alpha}_k z}{z - \alpha_k}) - h_k(\alpha_k)\frac{d_k^2}{z - \alpha_k},$$

where the coefficients are given by $h_k(\alpha_k)d_k$.
Putting these facts together, the solution of the problem

$$||fG_b^* - f_*||_\infty = \inf_{g \in \mathcal{H}_\infty} ||f - G_b g||_\infty,$$

is given by

$$f_* = fG_b^* - \frac{G_b^*\mathtt{W}_\phi y}{\mathtt{W}_\phi x}.$$

The generalization of the results given for the trigonometric system to the generalized orthonormal expansion can be found in [289].

9.3.5 Computational Algorithm

The computational algorithm for identifying a system in the generalized rational orthonormal basis can be described as follows:

- Select a set of poles $\{\alpha_j \mid j = 1, \cdots n_b\}$ to define the Blaschke product $G_b(z)$ and the generalized rational orthonormal basis, say $\{\varphi_j(z)G_b^k(z) \mid j = 1, \cdots n_b, \, k \in \mathbb{Z}\}$.
- Compute the function $\beta(t)$ to obtain the sets of points defined by

$$\mathbb{W}_N = \{\zeta_k = e^{i\gamma_k} \mid \gamma_k = \beta^{-1}(\frac{2\pi k}{N}), \, k = 0, \ldots, N-1\},$$

  and compute $\mu_k := \beta'(\frac{2\pi k}{N})$.

- Obtain the frequency response data sets $E_N^{F,\eta}(\xi_k)$, $\xi_k \in \mathbb{W}_N$ and compute $E_{F,j}^{\langle\alpha\rangle}(k) = \frac{G_b^n(\xi_k)E_N^{F,\eta}(\xi_k)\overline{\varphi}_j(\xi_k)}{\mu_k}$.
- Using FFT compute the trigonometric Fourier coefficients, *i.e.*

$$\tilde{c}_{k,j} = \frac{1}{N}\sum_{l=0}^{N-1} E_{F,j}^{\langle\alpha\rangle}(l)e^{i2\pi kl/N}$$

  and keep every dth coefficient, *i.e.* $c_{k-n,j} = \tilde{c}_{kd,j}$, $k = 0, \cdots, 2n$.
- Choose the window function φ and set the identified model as

$$F_{n,N}^{lin}(z) = \left(\mathcal{W}_{n,N}^{\varphi}E_N^{F,\eta}\right)(z) = \sum_{k=-n}^{n}\sum_{j=1}^{n_b}\varphi(\frac{k}{n})c_{k,j}\varphi_j(z)G_b^k(z)$$

- Apply the non-linear method described in Section 9.3.4. to obtain the model $F_{n,N}^{nlin}(z)$.

9.3.6 Robust Approximation Using Disc Algebra Basis and Wavelets

Recently, there has been a great interest in rational approximation of the transfer functions using basis or basis-like decompositions for purposes of system identification and model reduction.
The basic approach followed in these methods is to represent a transfer function f by a series of the form

$$f(z) = \sum_{k\in\mathbb{N}} c_k(f)\phi_k(z)$$

where $\{\phi_k(z)\}$ is a dictionary of fixed 'basis' functions – usually rational functions – and then construct rational approximations to f by taking finite truncations of the series.
A class of algorithms is based on the fact that $\mathbb{A}(\mathbb{D})$ admits a basis. As it is known, a Schauder basis in a Banach space $(X, \|\ \|)$ is a sequence $\{\phi_k \in X\}$ such that for all $f \in X$ there is a unique expansion $f = \sum_{k\in\mathbb{N}} g_k\phi_k$, *i.e.* the sequence of partial sums $S_N^f f = \sum_{k=0}^{N} f_k\phi_k$ converges to f in the norm of the space. The coordinates g_k are given by coordinate functionals, *i.e.* bounded linear functionals $\phi_k^* \in X^*$ such that $\phi_k^*(\phi_l) = \delta_{kl}$. Such a sequence is called a biorthogonal sequence and is uniquely determined by the basis. Usually these functionals are integrals by properly chosen elements.
Because Hilbert spaces are reflexive, *i.e.* $\mathcal{H}$ is isomorph with $\mathcal{H}^*$, the functionals correspond by the canonical map to a sequence of elements $\phi_k^* \in \mathcal{H}$. When $\phi_k = \phi_k^*$ one has an orthonormal basis.
Because the disc algebra can be described as

$$\mathbb{A}(\mathbb{D}) = \{f + i\tilde{f} | f, \tilde{f} \in C_{2\pi}\},$$

if one has a basis-like expansion in $C_{2\pi}$ and the trigonometric conjugate of the basis functions are also continuous, that algorithm will be suitable for approximation in $\mathbb{A}(\mathbb{D})$ also. Following this idea based on some results of Bochkarev, a basis and an algorithm based on the partial sums were given in [269]. Starting from the Faber-Schauder system $\{\xi_k\}$ on the $[0,\pi]$, *i.e.* a wavelet-like system generated by the dyadic dilations and translations of the hat function:

$$\begin{aligned}&\xi_0 = 1 \quad \xi_1(x) = 1 - \tfrac{x}{\pi}\\ &\xi_{2^m+l}(x) = \psi(2^m x - (l-1)\pi),\\ &m = 0, 1, \ldots; l = 1, \ldots, 2^m,\end{aligned}$$

where

$$\psi(x) = \begin{cases} \frac{2x}{\pi} & x \in [0, \frac{\pi}{2}] \\ 2 - \frac{2x}{\pi} & x \in [\frac{\pi}{2}, \pi] \\ 0 & x \in \mathbb{R} \setminus [0, \pi] \end{cases},$$

one can obtain an orthonormal system by a Gram-Schmidt procedure, called Franklin system

$$\varphi_k(x) = \sum_{i=0}^{k} \alpha_{ki}\xi_i(x).$$

Let $f \in C[0,\pi]$, then the partial sum of the biorthogonal expansion with respect to the Faber-Schauder system is given by:

$$S^{\xi}_{2^{m+1}} f = \sum_{k=0}^{2^{m+1}} c_k(f)\xi_k.$$

The coefficients of a Faber-Schauder expansion can be obtained as

$$\begin{aligned}&c_0(f) = f(0), \quad c_1(f) = f(\pi) - f(0),\\ &c_{2^m+l}(f) = f(\pi \tfrac{2l+1}{2^{m+1}}) - \tfrac{f(\pi \frac{l}{2^m}) + f(\pi \frac{l+1}{2^m})}{2}.\end{aligned}$$

The Franklin system has the property that its trigonometric conjugate is also continuous, *i.e.* $\varphi_k + i\tilde{\varphi}_k \in \mathbb{A}(\mathbb{D})$, moreover they form a basis in the disc algebra. Extend the Franklin and conjugate Franklin system to $[-\pi, 0)$ as follows,

$$\varphi_k(x) = \varphi_k(-x), \quad \tilde{\varphi}_k(x) = -\tilde{\varphi}_k(-x).$$

It can be proved that the system of functions $\{\Phi_k(z), k \in \mathbb{N}\}$ defined as

$$\begin{aligned}&\Phi_k(re^{i\vartheta}) = [\varphi_k(\vartheta) + i\tilde{\varphi}_k(\vartheta)] * P(r, \vartheta),\\ &\vartheta \in [-\pi, \pi), \quad r \in [0, 1],\end{aligned}$$

where

$$P(r, \vartheta) = \frac{1 - r^2}{1 + r^2 - 2r\cos\vartheta}$$

is the Poisson kernel, form a basis in $\mathbb{A}(\mathbb{D})$, see [269]. The biorthogonal expansion of $f \in \mathbb{A}(\mathbb{D})$ can be expressed as

$$f(z) = f(0) + \sum_{k=1}^{\infty} c_k \Phi_k(z),$$

where

$$c_k = \int_{-\pi}^{\pi} Re\{f(e^{i\vartheta})\}\varphi_k(\vartheta)d\vartheta.$$

The details of the algorithm can be found in [269].
The functions in the previous construction or the Faber-Schauder functions are closely related to the wavelets. There exists an analogous construction for basis in the disc algebra starting from certain wavelet systems – the so called Littlewood-Paley wavelets – due to Bochkarev. Details can be found in [193, pp. 198–204].
The difficulty in these constructions is that one can hardly obtain an explicit formula for the conjugate functions, *i.e.* for practical computations one has to use their truncated Fourier series expansion only.
A possible approach to wavelet theory is an introduction through frame operators [248]. The concept of the frame is a generalization of the idea of the basis, *i.e.* a frame is a possible redundant representation of the elements using a fixed set of functions. In other words, the coordinates are no more given as values of unique biorthogonal functionals.
Let us consider a system of bounded linear functionals on a Banach space $\phi = \{\phi_k \in X^*\}$ and a system of elements $\mathcal{F} = \{f_k \in X\}$. The pair $\{\phi, \mathcal{F}\}$ is called a frame, if the map – frame operator – $F(x) = \sum_{k\in\mathbb{N}} \phi_k(x) f_k$ is an one to one map on X. For a frame one has

$$m||x|| \leq \sup_n || \sum_{|k|<n} \phi_k(x) f_k|| \leq M||x||,$$

for $0 < m \leq M < \infty$ and for all $x \in X$. For Hilbert spaces $\mathcal{H}$ the converse is also true, *i.e.* there exist $0 < A \leq B < \infty$, such that

$$A||x||^2 \leq \sum_{k\in\mathbb{N}} |\phi_k(x)|^2 \leq B||x||^2,$$

for all $x \in \mathcal{H}$.
One can define the inverse of the frame operator and the inverse frame, obtaining an expansion analogous to the biorthogonal expansions:

$$x = \sum_{k\in\mathbb{N}} \phi_k(x) F^{-1}(f_k) = \sum_{k\in\mathbb{N}} (F^{-1})^* \phi_k(x) f_k.$$

The inverse frame operator can be obtained by a fix point iteration for example.
An approach using frame operators on $\mathcal{H}_2$ is based on affine frames generated by dilates and translates of a single function ψ – the analyzing wavelet or mother wavelet. The frames generated by $\{\psi_{m,n}(x) = a_0^{\frac{m}{2}} \psi(a_0^m x - nb_0)\}$,

where a_0, b_0 are fixed constants, and the rational function ψ that fulfills the so-called admissibility condition $\int_{\mathbb{R}} \frac{|\hat{\psi}_x(t)|^2}{|t|} dt < \infty$, $x > 0$, where ψ_x denotes the restriction of ψ to the vertical line $Re(s) = x$. Any $f \in \mathcal{H}_2$ can be represented in the form $f = \sum_{m,n \in \mathbf{N}} f_{m,n}$, where

$$f_{m,0} = < f, F^{-1}\psi_{m,0} > \psi_{m,0},$$
$$f_{m,n} = < f, F^{-1}\psi_{m,n} > \psi_{m,n} + \overline{< f, F^{-1}\psi_{m,n} >}\psi_{m,-n},$$

and F is the frame operator associated with $\{\psi_{m,n}\}$, [248].
For the approximation of functions in the disc algebra, there exists another approach that exploits embedding theorems of Hardy-Sobolev classes while maintaining the advantages of the properties of the Hilbert spaces. The Hardy-Sobolev class $\mathbb{H}^{2,1}$ is defined as

$$\mathbb{H}^{2,1} = \{f \in \mathcal{H}_2,\ f' \in \mathcal{H}_2\}.$$

The disc algebra is a dense set in $\mathbb{H}^{2,1}$ and one has the embedding theorem

$$||f||_\infty \leq C||f||_{\mathbb{H}^{2,1}}.$$

For details see [324].

The two main advantages of all these methods are that the model is linear in the coefficient parameters $c_k(f)$, and it is often possible to incorporate various forms of *a priori* knowledge into the approximation problem through a suitable choice of the functions ϕ_k.
[325] proposed an approach to robust identification that is based on rational wavelets where the coefficients are no more computed by linear functionals. Starting from the lattice

$$\mathcal{L} = \{\alpha_{n,k} = (1 - \tfrac{1}{2^n})e^{\frac{2k\pi i}{2^n}}, \quad k = 0, ..2^n - 1,\ n \in \mathbb{N}^*\}$$

it was shown that the class of Cauchy kernels

$$\{C_w(z) = \frac{1}{1 - \overline{w}z} | w \in \mathcal{L}\}$$

is a fundamental set for $\mathbb{A}(\mathbb{D})$, *i.e.* for any $f \in \mathbb{A}(\mathbb{D})$ and $\epsilon > 0$ there exist coefficients and lattice elements such that

$$||f(z) - \sum_{k=1}^{N} \lambda_k C_{w_k}(z)||_\infty < \epsilon.$$

It can be also shown that for the finite dimensional subspaces

$$X_p = \mathrm{Sp}\{C_{\alpha_{m,k}}, \quad m = 0, .., p,\ k = 0, .., 2^m - 1\},$$

and a set $\{z_k | k = 1, \ldots, n\}$ equally distributed on the unit circle with $n \geq 2\pi 2^{2(p+1)}$, the following algorithm is robustly convergent: for every X_p let f_p denote the solution of the minimax problem

$$f_p = \min_{g \in X_p} \max_{1 \leq k \leq n} |g(z_k) - a_k|,$$

where $a_k = f(z_k) + \eta_k$ and $|\eta_k| \leq \epsilon$.
The result in [325] can also be applied to give conditions on the choice of the sequence of poles in the basis proposed by [214]. Denote

$$X_p = \mathrm{Sp}\{B_n(z), n = 0, .., p\}$$

and let $r_p = \max_{0 \leq k \leq p} |\alpha_k|$. Suppose that the set of points $\{z_k\}$ is dense on T. Then, with this choice of basis in X_p, the algorithm above is robustly convergent provided that for Δ_n, the maximum angular gap between the first n points, the condition $\Delta_N \leq \frac{1}{p+1} \frac{1-r_p}{1+r_p}$ is satisfied.

To conclude this section, another wavelet-like construction will be presented that uses product systems. The multiplication by $e_1(z) = z$ can be identified as the shift on $\mathcal{H}_2$. The power functions $(z) = z^m$ and the powers of the shift $\mathcal{S}^m$ can be generated by the functions $\phi_n := e_{2^n}$ $(n \in \mathbb{N})$ and by the corresponding subsystem $\mathfrak{T}_n := \mathfrak{S}^{2^n}$ $(n \in \mathbb{N})$. If the numbers $m \in \mathbb{N}$ are represented in binary form

$$m = \sum_{n=0}^{\infty} m_n 2^n \quad (m_n \in \{0, 1\}), \tag{9.16}$$

then $\mathcal{S}^m$ can be written as

$$e_m = \prod_{n=0}^{\infty} \phi_n^{m_n}, \quad \mathfrak{S}^m = \prod_{n=0}^{\infty} \mathfrak{T}_n^{m_n}. \tag{9.17}$$

The system of functions $(e_m, m \in \mathbb{N})$ is usually called the product system of $(\phi_n, n \in \mathbb{N})$. There is another important property - that can be used in an FFT-type algorithm - of the system $(e_m, m \in \mathbb{N})$; namely, that the generating $(\phi_n, n \in \mathbb{N})$ functions can be obtained by function compositions starting from the function $A(z) := \phi_1(z) = z^2$ $(z \in \mathbb{C})$ as

$$\phi_0(z) = z, \quad \phi_n = \phi_{n-1} \circ A = A \circ A \circ \cdots \circ A \tag{9.18}$$

where $n \in \mathbb{N}^* := \{1, 2, \cdots\}$.

We proceed with a general procedure for constructing discrete orthogonal product systems based on a series of two-folded mappings. Let X be a non-empty set and $A : X \to X$ a mapping on X. The mapping A is *two-folded* if $\forall x \in X$ the primordial element with respect to A is a two-element set: $A^{-1}(x) = \{x_0, x_1\}$. In other words, $\forall x \in X$ there are exactly to elements

$x_0, x_1 \in X$, for which $A(x_0) = A(x_1) = x$ hold. Mapping $\varphi : X \to \mathbb{C}$ is said to be *odd with respect to mapping* A, if $\forall x \in X$: $\varphi(x_0) = -\varphi(x_1)$, where x_0, x_1 are the primordial elements created by A.
Let us consider the Blaschke-term

$$B_b(z) := \epsilon(b)\frac{z-b}{1-\bar{b}z} \quad \left(\epsilon(b) := \frac{1-\bar{b}}{1-b},\ z, b \in \mathbb{C}\right), \tag{9.19}$$

where the scaling factor $\epsilon(b)$ results in the normalizing condition $B_b(1) = 1$. It can be easily verified that the two-factored Blaschke products, $A := B_{b_1}B_{b_2}$ $(b_1, b_2 \in \mathbb{D})$, constitute two-folded mappings from $\mathbb{T}$ to oneself and that the identity mapping $\phi(z) = z$ $(z \in \mathbb{T})$ is an *odd function* with respect to T, if $b_1 = -b_2 = b$. In this case, $A(z) = B_b(z)B_{-b}(z) = B_{b^2}(z^2)$ $(z \in \mathbb{C})$ and therefore the identity mapping is really odd with respect to A.
In general, it can be shown that the two-factored Blaschke products $B_{b_1}B_{b_2}$ $(b_1, b_2 \in \mathbb{D})$ can be generated in the form of

$$A := B_{b_1}B_{b_2} = B_a \circ B_0^2 \circ B_c,$$

where $a, c \in \mathbb{D}$. Moreover, the function $\varphi = B_c$ is odd with respect to A; namely, primordial elements of $x \in \mathbb{D}$, produced by A, can be written as

$$x_1 = B_c^{-1}(\sqrt{B_a^{-1}(x)}), \quad x_2 = B_c^{-1}(-\sqrt{B_a^{-1}(x)}),$$

and therefore $B_c(x_1) = -B_c(x_2)$ holds.

To define a point set for discretization, let us fix a series $A_n : X \to X$ $(n \in \mathbb{N}^*)$ of two-folded mappings on set X, and let $A_0 : X \to X$ be a bijection. Using function composition, let us introduce the following mappings of set X

$$\begin{aligned} T_0 &:= A_0, \\ &\vdots \\ T_n &:= A_n \circ A_{n-1} \circ \cdots \circ A_1 \circ A_0 = A_n \circ T_{n-1} \quad (n \in \mathbb{N}^*). \end{aligned} \tag{9.20}$$

Further, fix a point $x_0 \in X$ and take the primordial elements of x_0 created by T_n. Introduce the following sets

$$X_n := T_n^{-1}(x_0) = T_{n-1}^{-1}(A_n^{-1}(x_0)) \quad (n \in \mathbb{N}^*). \tag{9.21}$$

Now the number of elements in X_n is 2^n. If the point x_0 is a fix point of all mappings A_n, *i.e.* $A_n(x_0) = x_0$, then, using (9.21), we arrive at

$$X_n = T_{n-1}^{-1}(x_0) \cup T_{n-1}^{-1}(x_n) = X_{n-1} \cup X'_{n-1} \quad (A_n(x_0) = x_0)$$

In this case, the set X_n can be derived from X_{n-1} by adding a 2^{n-1}-element point set.

In order to interpret the discrete orthogonal system, let us start from functions $\varphi_n : X \to \mathbb{T}$ $(n \in \mathbb{N})$ that are odd with respect to mapping A_{n+1} and formulate the function system

$$\phi_n := \varphi_n \circ T_n \quad (n \in \mathbb{N}^*). \tag{9.22}$$

Denote the product system by

$$\psi_m := \prod_{j=0}^{\infty} \phi_j^{m_j} \quad \Big(m = \sum_{j=0}^{\infty} m_j 2^j \in \mathbb{N}\Big). \tag{9.23}$$

The product system ψ_m $(0 \leqq m < 2^N)$ is orthonormal with respect to scalar products

$$[f, g]_N := \int_{X_N} f\overline{g}\, d\mu_N = 2^{-N} \sum_{x \in X_N} f(x)\overline{g(x)}, \tag{9.24}$$

where μ_N is a discrete measure on the set X_N and defined as $\mu_N(\{x\}) := 2^{-N}$. Calculation of Fourier coefficients and reconstruction of discrete functions defined on X_N (*i.e.* the Fourier analysis and synthesis based on the product system) can be carried out using a fast algorithm similar to FFT.

Several rational systems applied in control theory can be derived on the basis of the above procedure. The so-called *trigonometric system* can be formulated with choices $A_0(z) = z$, $A_n(z) := z^2$, $(z \in \mathbb{T}, n \in \mathbb{N}^*)$ and $\phi_n(z) = z$ $(z \in \mathbb{T}, n \in \mathbb{N})$

$$\psi_m(z) = z^m \quad (m \in \mathbb{N})$$

The *discrete Laguerre system*, corresponding to parameter $a \in \mathbb{D}$, - not considering the factor $\sqrt{1 - |a|^2}/(1 - \bar{a}z)$ - is similar to the product system with choices $A_0 := B_a$, $A_n(z) := z^2$ $(z \in \mathbb{T}, n \in \mathbb{N}^*)$

$$\psi_m(z) = (z - a)^m/(1 - \bar{a}z)^m \quad (m \in \mathbb{N}, z \in \mathbb{T}).$$

In the general case, let us start from series a_n, $c_n \in \mathbb{D}$ $(n \in \mathbb{N})$ and construct functions

$$A_0 := B_{a_0}, \quad A_n := B_{a_n} \circ B_0^2 \circ B_{c_n} \quad (n \in \mathbb{N}^*).$$

Now

$$\varphi_n := B_{c_{n+1}} \quad (n \in \mathbb{N})$$

is odd with respect to A_{n+1}.

Similarly, the *Malmquist-Takenaka system* generated by the periodic zeros $a_n \in \mathbb{D}$, $a_{n+N} = a_n$ $(n \in \mathbb{N})$ can be decomposed into N product systems that can be produced separately using the above procedure. Specifically, this also holds for *Kautz systems*.

The following theorem summarizes the basic properties of product systems.

Proposition 9.9. *The reproduction*

$$D_{2^N}(x,t) = \prod_{j=0}^{N-1} (1 + \phi_j(x)\overline{\phi_j(t)})$$

holds for the kernel functions of index 2^N of any product system. If $\forall n \in \mathbb{N}$: the function φ_n is odd with respect to A_{n+1}, then the product system (9.23) is orthogonal to the scalar product (9.24). Further, the kernel functions can be written as

$$D_{2^N}(x,t) = \begin{cases} 2^N & (x = t, x, t \in X_N), \\ 0 & (x \neq t, x, t \in X_N). \end{cases} \tag{9.25}$$

Partial sums, corresponding to indices 2^N in the series expansion (using a discrete product system) of an arbitrary function $f : X \to X$, generate the function f in points of X_N

$$(S_{2^N} f)(x) = f(x) \quad (x \in X_N). \tag{9.26}$$

Equation (9.26) expresses the fact that in the points of X_N the partial sums $S_{2^N} f$ are interpolating the function f.

For computing Fourier coefficients

$$c_n := \left[f, \psi_n\right]_N = 2^{-N} \sum_{x \in X_N} f(x)\overline{\psi_n(x)}$$
$$(n = 0, 1, \cdots, 2^N - 1)$$

one needs to perform $2^N 2^N = 2^{2N}$ multiplications and $2^N(2^N - 1)$ additions (neglecting the calculation of function values and the normalization by 2^N). The function f can be reconstructed in the points of X_N based on Fourier coefficients and using the following formula that is equivalent to (9.26)

$$f(x) = \sum_{n=0}^{2^N - 1} c_n \psi_n(x) \quad (x \in X_N).$$

The number of necessary calculations is just the same as in the preceding case. One can construct an algorithm for computing the Fourier coefficients and reconstructing the function f in $O(N2^N)$ steps:

Proposition 9.10. *Starting from arbitrary functions $\varphi_n : X \to \mathbb{C}$ ($n = 1, 2, \cdots, N$) and two-folded mappings $A_n : X \to X$ ($n = 1, 2, \cdots, N$) let us introduce the system ϕ_n using Equations (9.22) and (9.23). For parameter $n = 0$, the initial value is*

$$Y_{k,0}^0 := f(x_k^0) \quad (0 \leqq k < 2^N). \tag{9.27}$$

Define a recursive (in n) series by

$$Y^n_{k,j2^{n-1}+\ell} = \frac{1}{2}\Big(\varphi_{n-1}(x^{n-1}_{2k})^j Y^{n-1}_{2k,\ell} + \varphi_{n-1}(x^{n-1}_{2k+1})^j Y^{n-1}_{2k+1,\ell}\Big), \tag{9.28}$$

where

$$0 \leqq k < 2^{N-n}, 0 \leqq \ell < 2^{n-1}, j = 0,1;\ n = 1,2,\ldots,N.$$

In the Nth step we arrive at the Fourier coefficients

$$Y^N_{0,m} = \Big[f, \psi_m\Big]_N \quad (m = 0,1,\cdots,2^N - 1). \tag{9.29}$$

Using two-factored Blaschke products, one can obtain the discrete orthogonal systems composed from true rational functions.

Proposition 9.11. *Let a series $b_n \in \mathbb{D}$ $(n \in \mathbb{N})$ be given and define the functions $A_0 := B_{b_0}$, $A_n := B_{b_n} B_{-b_n}$ $(n \in \mathbb{N}^*)$ that map the torus $\mathbb{T}$ onto itself. The identity mapping $\varphi_n(z) := z$ $(z \in \mathbb{T})$ is odd with respect to the functions A_{n+1} $(n \in \mathbb{N}^*)$. The product system of the system $\{\phi_n\}$, given by $\phi_n := A_n \circ \cdots \circ A_0$ $(0 \leqq n < N)$, is orthogonal with respect to the scalar product (9.24) defined on the base set $X_N := \phi_N^{-1}(1) \subset \mathbb{T}$. For an arbitrary function $f : \mathbb{T} \to \mathbb{T}$ the partial sum $S_{2^N} f$ of the Fourier expansion corresponding to the product system is generating the function f in the points of set X_N. If the function $f : \mathbb{C} \to \mathbb{C}$ is a true rational function with all of its poles contained in the poles of ϕ_k $(k = 0,1,\ldots,N-1)$, then $S_{2^N} f = f$ on the whole complex plane. The discrete Fourier coefficients of f can be computed by the fast algorithm given in (9.27)-(9.29).*

Details of this approach can be found in [268].

9.3.7 An Application Example – The MIT Interferometer Testbed

In the case of control for large flexible structures, system identification takes on a significant role due to the potential high complexity of finite element models for these structures. The approach that uses rational orthonormal basis functions, *i.e.* the approach that uses *a priori* knowledge about the system, like location of poles, by incorporating this information into the choice of the basis and also by specifying the distribution of measurement points (frequencies), can be especially useful when identifying lightly damped structures.
Such a specific structure is an interferometer. The MIT Space Engineering Research Center (SERC) has developed a controlled structures technology (CST) testbed based on one design for a space-based optical interferometer. The role of the testbed was to provide a versatile platform for experimental investigation and discovery of CST approaches. In particular, it served as the focus for experimental verification of CSI methodologies and control strategies

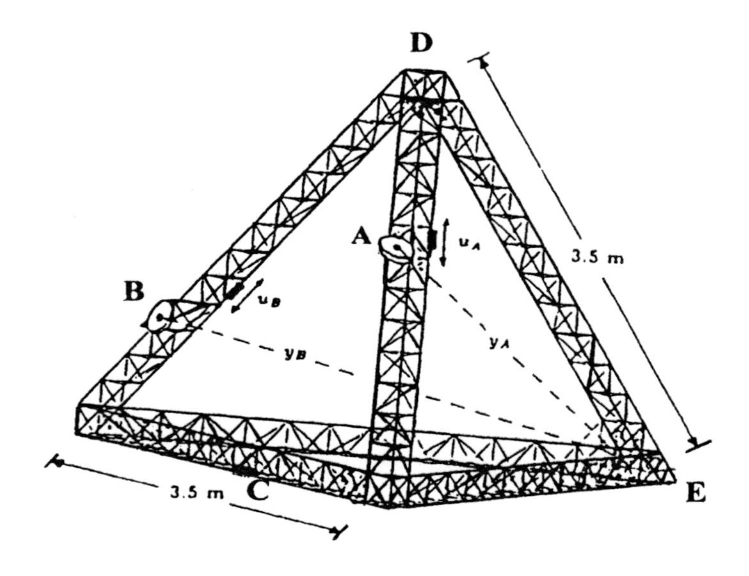

Fig. 9.1. MIT interferometer testbed.

at SERC. The testbed program had an emphasis on experimental CST, incorporating a broad suite of actuators and sensors, active struts, system identification, passive damping, active mirror mounts, and precision-component characterization.

The SERC testbed represents a one-tenth scaled version of an optical interferometer concept based on an inherently rigid tetrahedral configuration with collecting apertures on one face. The testbed consists of six 3.5-meter-long truss legs joined at four vertices and is suspended with attachment points at three vertices. Each aluminum leg has a 0.2 m by 0.2 m by 0.25 m triangular cross section. The structure has a first flexible mode at 31 Hz and has more than 50 global modes below 200 Hz. The stiff tetrahedral design differs from similar testbeds (such as the JPL Phase B) in that the structural topology is closed. The testbed has three mock siderostats located at points A, B, C. Located at point D is a disturbance source that simulates disturbances that would be encountered by a full scale space structure, while a laser assembly is mounted at a vertex shown as point E. At the mock siderostat locations are three-axis active mirror mounts for the use with the multi-axis laser metrology system. The tetrahedral design minimizes structural deflections at the vertices (site of optical components for maximum baseline) resulting in reduced stroke requirements for isolation and pointing of optics. Typical total light path length stability goals are of the order $\frac{\lambda}{20}$, with a wavelength of light, lambda, of roughly 500 nanometers.

The performance objective for the testbed was to maintain the internal path length errors between multiple points on the tetrahedron to within stringent

tolerances – to less than 50 nanometers from the nominal – in the presence of a disturbance source that cause the system to vibrate.
For more details about the interferometer hardware and performance objectives see [24]. A multi-variable identification of the SERC interferometer testbed is presented in [28, 102]. An $\mathcal{H}_2$ approach to the multivariable control design problem can be found in [170].

The control task for the system is to enhance its structurally designed disturbance rejection property, *i.e.* to minimize the differential path length when worst-case disturbance acts on the system. This control task leads to an $\mathcal{H}_\infty$ design method where due to modelling errors robustness issues have to be taken into account as well [171].
The identification experiments were performed by applying predesigned excitation signals like sine sweeps designed to cover the frequency range of interest. Both single-input–single-output (SISO) and multi-input–multi-output (MIMO) measurements were obtained in the frequency range 10–100 Hz. In the experiments two inputs and two outputs – the differential path length –

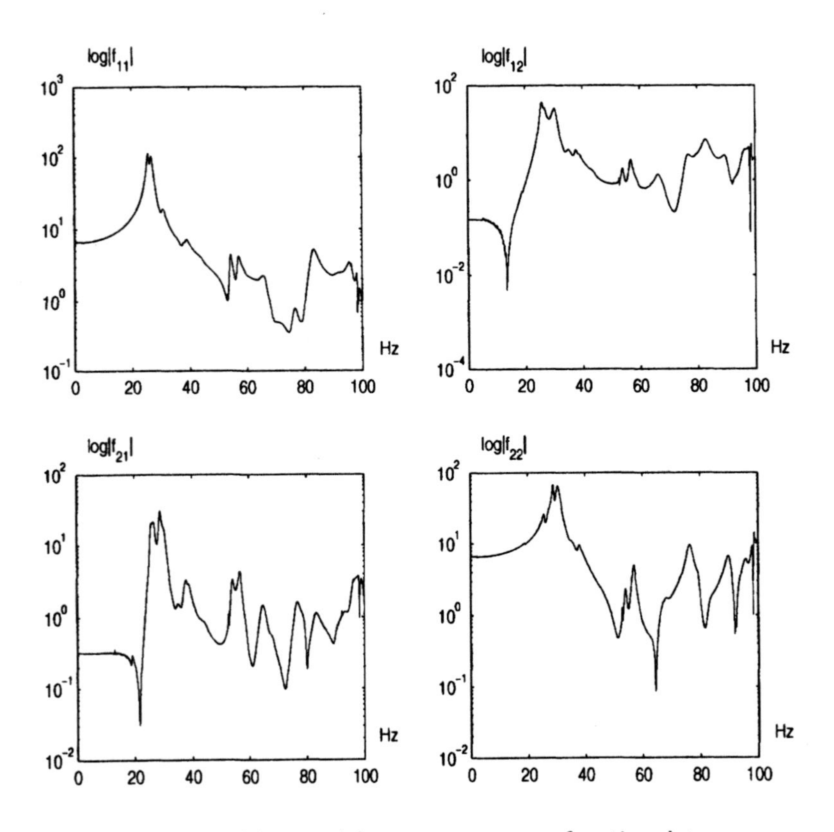

Fig. 9.2. Measured frequency-response function data.

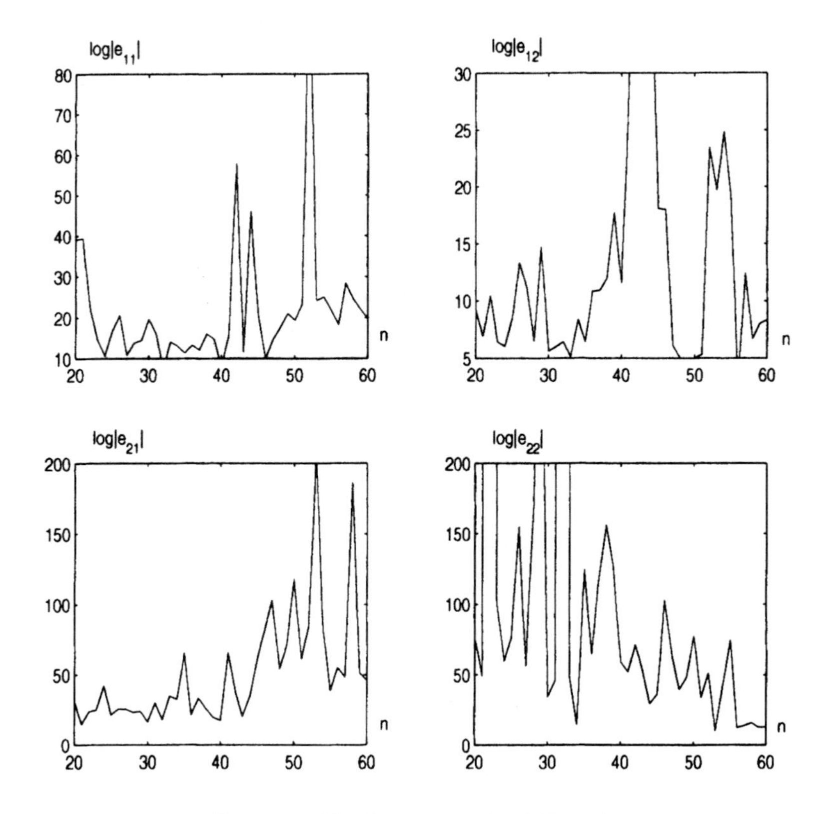

Fig. 9.3. Absolute errors for LS models

were used and a 2×2 transfer function matrix was to be identified. Figure 9.2 shows the four frequency-response functions ($f_{i,j}$, $i, j = 1, 2$) resulting from single-input excitations while measuring both outputs.

The approach of this chapter was applied to the identification of a model in an orthogonal basis generated by using the approximate knowledge about of the system poles. A frequency-domain least-squares method was also applied to SISO data and it was found that very high order models (in the SISO case with degree larger than 40) were needed. In Figure 9.3, the absolute errors of the least-squares models of orders $n = 20$ to 60 per channel, *i.e.* $e_{i,j} = ||f_{i,j} - f_{i,j}^{LS}||_2$, are shown. One can observe that there is no monotonicity in the absolute errors.

The first step in applying the system-based orthonormal functions is to obtain the approximate location of the poles of the system that will be identified. A linear least-squares method was used to obtain the *a priori* knowledge of the poles.

Models were fitted to the individual experimental transfer functions. The 'floating' poles were discarded and the fixed ones were kept as representatives of the poles of the real system. Based on this information, twelve pole pairs were chosen, as follows:

$$\begin{bmatrix} 0.6830 \pm 0.7133i & 0.6555 \pm 0.7328i & 0.6094 \pm 0.7789i \\ 0.5616 \pm 0.7955i & 0.3652 \pm 0.9184i & -0.1253 \pm 0.9788i \\ -0.2086 \pm 0.9607i & -0.4275 \pm 0.8637i & -0.7146 \pm 0.6504i \\ -0.7805 \pm 0.5884i & -0.8233 \pm 0.4914i & -0.9913 \pm 0.0507i \end{bmatrix}$$

The model set considered in the identification process was:

$$F_{N,d}^{GOBF}(z) = \sum_{k=0}^{N_{GOBF}} \sum_{l=1}^{d} C_{kd+l}\phi_l(z)m^k(z),$$

where $C_{kd+l} \in \mathbb{R}^{2\times 2}$. In our identification experiment, $N_{GOBF} = 5$ and $d = 24$. For comparison with FIR models,

$$F_N^{FIR}(z) = \sum_{k=0}^{N_{FIR}} C_k z^k,$$

having the same number of parameters, its absolute errors are also included.

Tables 9.1, 9.2, and 9.3 show the absolute errors, *i.e.* $e = \|F - F_{N,d}^{GOBF}\|_\infty$ between the measured data and the identified model functions using different type of summations. Because the effect of the noise was minor in this experiment, the effect of the non-analytic terms was not significant.

Table 9.1. Absolute error, no summation

N_{FIR}	$\|f_{11} - f_{11}^N\|$	$\|f_{12} - f_{12}^N\|$	$\|f_{21} - f_{21}^N\|$	$\|f_{22} - f_{22}^N\|$
143	18.1110	6.9959	7.5016	9.4925
119	21.3622	8.1301	10.6749	13.9060
95	22.5454	11.0616	12.1718	17.8260
71	23.7997	14.5813	15.1471	19.8204
47	36.3400	22.0595	26.5094	18.8657
23	65.1312	31.8669	31.7667	29.0462
N_{GOBF}	$\|f_{11} - f_{11}^N\|$	$\|f_{12} - f_{12}^N\|$	$\|f_{21} - f_{21}^N\|$	$\|f_{22} - f_{22}^N\|$
5	1.9875	2.7330	2.2730	2.9901
4	2.0025	3.1628	2.1566	3.1751
3	2.1066	3.5438	1.9912	3.2558
2	2.2635	3.8385	2.2085	3.3497
1	2.6253	4.0297	2.9097	7.1725
0	3.9750	4.4916	3.3972	10.7562

Table 9.2. Absolute error, Fejér summation

N_{FIR}	$\lvert f_{11} - f_{11}^N \rvert$	$\lvert f_{12} - f_{12}^N \rvert$	$\lvert f_{21} - f_{21}^N \rvert$	$\lvert f_{22} - f_{22}^N \rvert$
143	37.4324	18.7078	19.1123	21.1659
119	41.2310	20.9735	21.1323	23.0893
95	46.6497	23.8764	23.5248	24.8494
71	55.1782	27.8036	27.1100	27.1624
47	68.3843	32.5009	30.5609	32.1820
23	86.3988	39.7451	31.2778	43.2593
N_{GOBF}	$\lvert f_{11} - f_{11}^N \rvert$	$\lvert f_{12} - f_{12}^N \rvert$	$\lvert f_{21} - f_{21}^N \rvert$	$\lvert f_{22} - f_{22}^N \rvert$
5	1.9948	3.5262	2.0403	3.6347
4	2.0491	3.7053	2.1357	3.8992
3	2.1636	3.8562	2.2496	4.6656
2	2.3676	3.9667	2.5658	6.0013
1	2.8076	4.0309	3.0295	8.2421
0	3.9750	4.4916	3.3972	10.7562

Consulting these tables, it seems that the de la Vallée-Poussin summation performs better than the summation with the Fejér kernel. To obtain a state space model from the identified coefficients, a generalized Ho-Kalman algorithm was applied. To remain between bounds in processing time, we considered $N_{GOBF} = 1$ and $d = 24$, *i.e.* at most two 'coefficients'.

Figure 9.4 shows the error between the maximal (minimal) singular value for the model with state-space dimension 48 obtained for $N_{GOBF} = 0$, *i.e.* for the first coefficient only, and Figure 9.5 shows the error that corresponds to the reduced model obtained using $N_{GOBF} = 1$ and $d = 24$.

Table 9.3. Absolute error, de la Vallée-Poussin summation

N_{FIR}	$\lvert f_{11} - f_{11}^N \rvert$	$\lvert f_{12} - f_{12}^N \rvert$	$\lvert f_{21} - f_{21}^N \rvert$	$\lvert f_{22} - f_{22}^N \rvert$
143	21.3896	9.7197	11.1185	15.3736
119	22.5940	12.0058	13.2027	17.8108
95	24.8002	15.3877	16.4646	19.1765
71	33.1570	20.0021	22.9477	19.0988
47	50.1689	25.3497	29.8744	20.9482
23	75.0029	36.7596	31.4495	34.7482
N_{GOBF}	$\lvert f_{11} - f_{11}^N \rvert$	$\lvert f_{12} - f_{12}^N \rvert$	$\lvert f_{21} - f_{21}^N \rvert$	$\lvert f_{22} - f_{22}^N \rvert$
5	1.9880	3.1316	2.1384	3.1271
4	1.9749	3.4364	1.9978	3.2192
3	2.0913	3.6855	2.0196	3.2691
2	2.3281	3.9014	2.3314	4.1486
1	2.6253	4.0297	2.9097	7.1725
0	3.9750	4.4916	3.3972	10.7562

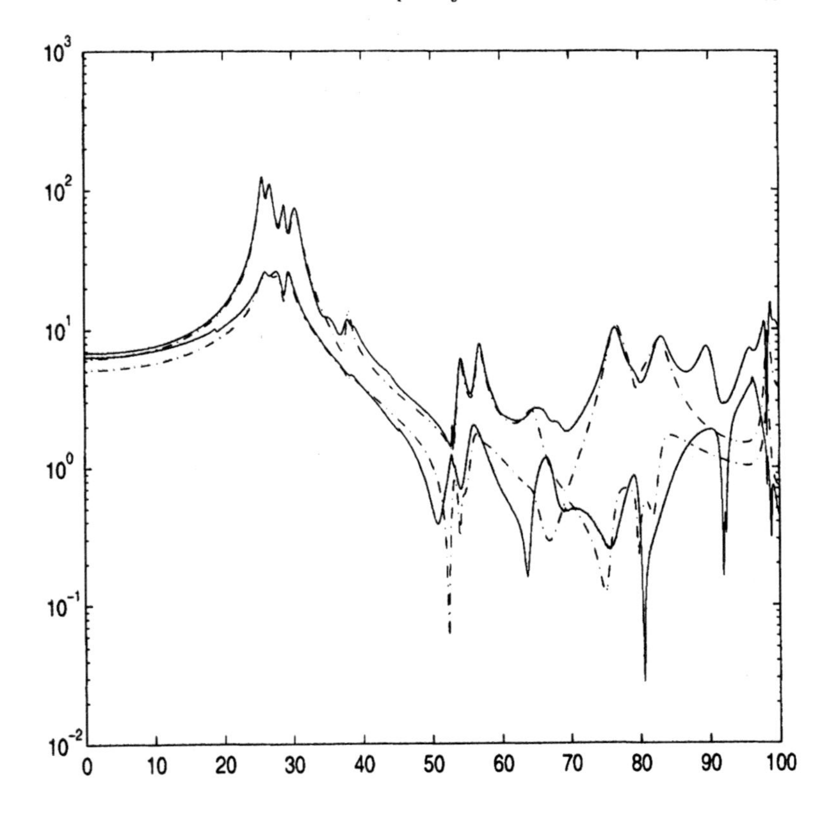

Fig. 9.4. The maximal and minimal singular values for the model obtained for $N_{GOBF} = 0$, $d = 24$.

9.4 Conclusions

In this chapter, we made an attempt to characterize some approaches and results that appeared very recently in the area of approximate modelling and identification under the $\mathcal{H}_\infty$ criterion. The use of this criterion has been motivated by demands of robust control design formulated as an $\mathcal{H}_\infty$ optimization.

The approach followed here is a frequency-domain one. Emphasis was made on the construction of model sets and the related identification algorithms, considered as bounded functionals, that map the noisy data to an approximate model. These operators are related to the choice of the basis in the spaces l_2, $\mathcal{H}_2$, L_2 or in the disc algebra $\mathbb{A}(\mathbb{D})$. The possible choices of bases, like trigonometric, rational orthogonal, disc algebra bases, and wavelets were studied. The norm of operators related to a specific basis influences the bound on the approximation and on the noise error. The estimates of the norm bounds of the operators related to a choice of a basis were also discussed.

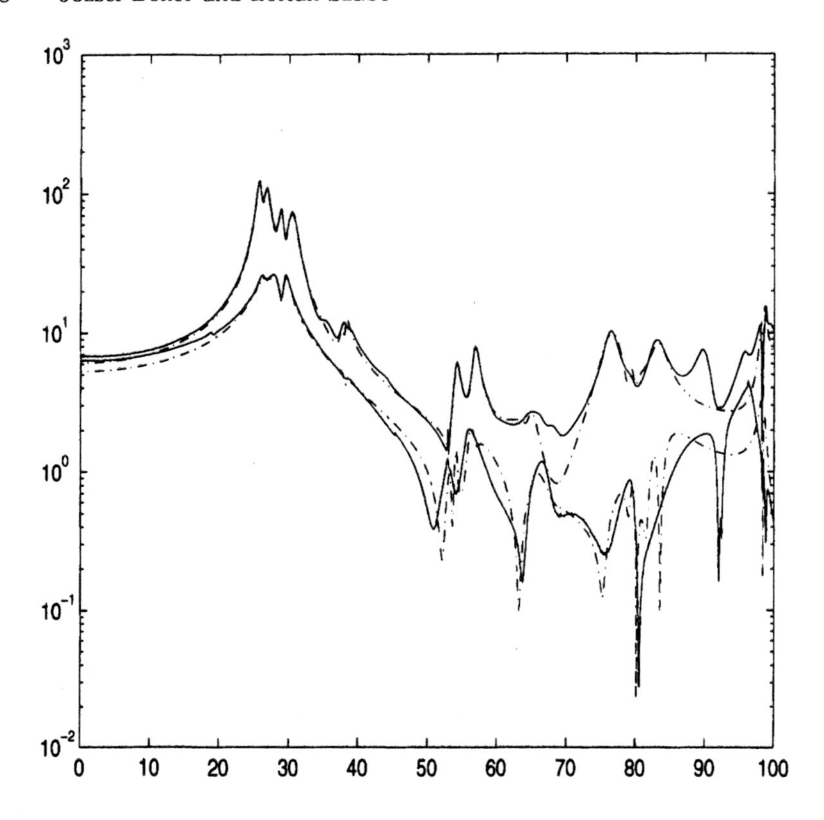

Fig. 9.5. The maximal and minimal singular values for the model obtained for $N_{GOBF} = 1$, $d = 24$.

It is expected that this summary can serve to get a coherent picture on these new approaches by calling attention to some basic aspects important for further investigation of the theory and for possible application of the results.

10

Design Issues

Peter Heuberger

Delft University of Technology, Delft, The Netherlands

10.1 Introduction

The previous chapters have concentrated on theoretical principles and results that underpin the importance of orthonormal bases in system identification as well as broader system theoretic settings. By way of contrast, this chapter devotes itself to practical and implementation issues that specifically pertain to the use of an orthonormal parameterization of system models.

There are many issues that can be addressed under this umbrella. This chapter in fact focuses on what, via experience, have emerged to be of greatest relevance – finite data/unknown initial conditions, state-space formulations, appropriate pole locations and multi-variable extensions.

In this chapter, we will consistently use the notation G_b for the basis generating inner function with McMillan degree n_b and minimal balanced state-space realization (A_b, B_b, C_b, D_b). As in the previous chapters, we will use the symbol $V_k(z)$ for the vectorized Hambo functions in the frequency domain,

$$V_k(z) = [zI - A]^{-1} BG_b^{k-1}(z)$$

as defined in Section 2.5, and $v_k(t)$ for their time-domain equivalent.

10.2 Choice of Basis Functions

The preceding chapters introduced both the Takenaka-Malmquist functions and the Hambo functions. The latter are a special case of the former, using a repetitive basis pole sequence. For practical application, the question arises which type of functions should be used. This issue is closely related to the prior information available, specifically in terms of uncertainty regions for the location of the 'true poles'. A detailed account of the choice of basis polesis given in Chapter 11. See also the bias discussion in Chapter 4.

Assume that the 'true poles' of the system at hand are contained in a region Ω, which can be represented by

$$\Omega = \{z \in \mathbb{C} \mid |G_b(z^{-1})| \leq \rho\},$$

where G_b is an inner function with poles $\{\xi_1, \cdots, \xi_{n_b}\}$. It is shown in Chapter 11 that in this case the Hambo functions with these poles are optimal in a Kolmogorov N-width sense (with $N = n \times n_b$ and n the number of pole repetitions), which simply stated means that the functions $\{V_1(z), \cdots, V_n(z)\}$ render the best worst-case approximation error[1] over all possible sets of $n \times n_b$ functions. The value of ρ (decay rate) is directly related to this error, in the sense that smaller values render smaller errors. In practice, it will often be cumbersome to represent the uncertainty about the pole locations in this manner, and one will have to approximate the uncertainty region with ρ as small as possible (see Chapter 11 for some examples of these regions). An illustrative example is the case where the uncertainty information is given by the knowledge that the true poles are real valued and in a specific interval, say $\Omega = \{z \in \mathbb{R} \mid 0.3 \leq z \leq 0.7\}$. We try to approximate this region for different values of n_b. So we search values ξ_i, ρ, with ρ as small as possible such that

$$\Omega \subset \{z \in \mathbb{C} \mid \prod_{i=1}^{n_b} |\frac{z - \xi_i}{1 - \xi_i z}| \leq \rho\}.$$

Table 10.1 gives an overview of the optimal basis pole locations and the corresponding minimal value of ρ, when we approximate Ω with $1, 2 \cdots, 5$ poles. The optimal uncertainty region $\{z \in \mathbb{C} | \; |G_b(z^{-1})| \leq \rho\}$ is depicted in Figure 10.1 for the cases $n_b = 1, 2, 3, 5$. It is clear that the optimal poles are distinct and distributed over the interval [0.3 0.7]. In this case, it will only be possible to exactly describe Ω with an infinite number of poles. Note that this discussion is based on minimization of the worst-case approximation error. If one would aim for instance at minimization of the average approximation error, the results will be quite different.

Table 10.1. Resulting radii and basis poles for the approximation of the uncertainty region $\Omega = \{z \in \mathbb{R} \mid 0.3 \leq z \leq 0.7\}$ with $1, 2, \cdots, 5$ basis poles

n_b	ρ	ξ_i
1	0.271879	0.529
2	0.037009	0.374, 0.655
3	0.005067	0.335, 0.529, 0.679
4	0.000690	0.320, 0.450, 0.599, 0.688
5	0.000102	0.313, 0.405, 0.532, 0.640, 0.698

[1] Best here means that the approximation will have the smallest possible squared error for the worst possible 'true' model with a given unit pulse response energy.

An important aspect in this regard is that the determination of the optimal basis pole locations involves a tedious optimization problem, which becomes increasingly more difficult to solve with growing n_b (see Chapter 11 for a detailed discussion). In practice, it is therefore advisable to keep the number of basis poles relatively low. In the example above, the optimal value of ρ for $n_b = 4$ is equal to $\rho = 0.00069$. Now suppose that we use the resulting poles for $n_b = 2$ and repeat them once. The resulting non-optimal function G_b with 4 poles then has a decay rate $\rho = 0.037009^2 = 0.001369$, about twice the value of the optimal ρ. However, if the optimization routine for $n_b = 4$ would get stuck in a local optimum, this might lead to a much worse 'optimum'.

A second argument for using the Hambo function construction lies in the fact that the corresponding expansion coefficient vectors can be interpreted as Markov parameters and can be used in a realization/interpolation context. This is explained in detail in Chapter 13 and is a generalization of the classical realization theory where the Ho-Kalman algorithm is used to infer a minimal state-space description of the underlying system from the expansion coefficients of an (over-parameterized) finite impulse response (FIR) model. Later in this chapter, we will discuss a procedure to update the basis poles in an iterative scheme.

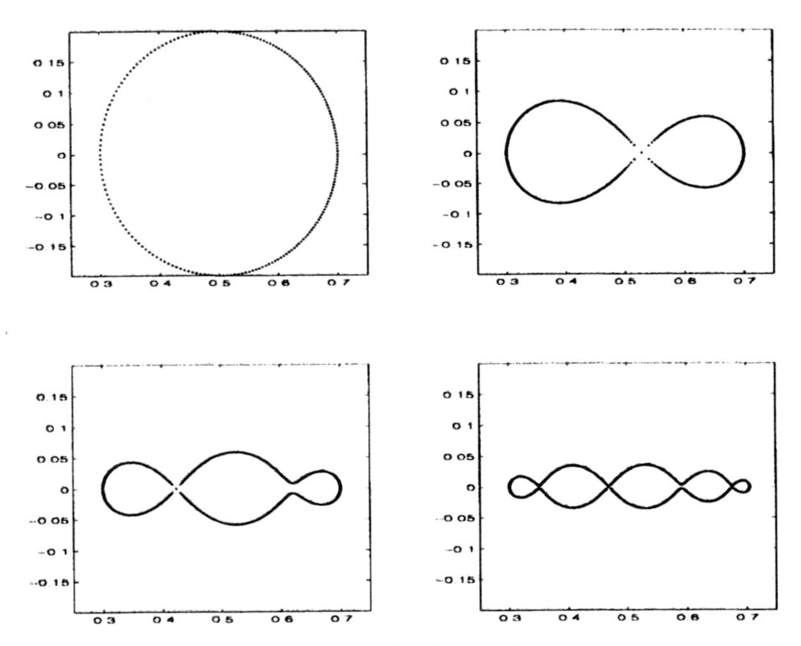

Fig. 10.1. Uncertainty regions for the data in Table 10.1 for $n_b = 1, 2, 3, 5$.

10.3 Finite Data Considerations

In a practical identification setting, the number of data points will always be finite, and this may directly impose limits on the choice of basis functions to be taken into account. There are two major features that should be addressed in this respect. First of all the length of the data sequence limits the lowest possible frequency present in the data and henceforth also limits the time-constants of the basis functions used for identification. Second, the influence of initial conditions cannot always be neglected. In this section, some guidelines are given for the design issue of choosing poles and the number of basis functions (*i.e.* parameters to be estimated). Furthermore, it is shown how the effect of initial conditions can be included in the estimation problem without losing the linear-regression structure. An equivalent feature can be applied for the inclusion of static-gain information in the problem.

10.3.1 Pole Locations and Number of Functions

The most important design issue when dealing with generalized orthonormal basis functions (GOBFs) is obviously the choice of the basis functions, which in the scalar case boils down to a choice of pole locations, see also Section 10.2 and Chapter 11. This choice can for instance be made on the basis of prior knowledge, as settling times, spectral estimates *etc.*, including uncertainty information. In this section, attention will be restricted to the situation where only short data sets are available, as is for instance often the case in process industry or environmental applications, where experiments and/or measurements are expensive or difficult to obtain.

Because the basis functions $\{f_k(t)\}_{k=1}^{\infty}$ constitute a basis for ℓ_2, it is immediate that with growing k, the energy distribution of $f_k(t)$ is shifted towards $+\infty$. An example of this feature is depicted in Figure 10.2, where the Laguerre functions $\{f_k, k = 5, 20, 60\}$ with a pole in 0.92 are plotted, together with the cumulative energy of these functions, *i.e.* the functions $C_k(t) = \sum_{s=1}^{t} f_k^2(s)$. Orthonormality of the basis functions implies that $C_k(t) \overset{t\to\infty}{\longrightarrow} 1$.

In this respect, it is important that the frequency response of the Laguerre functions $F_k(z)$ for low frequencies ω can be approximated as

$$F_k(e^{i\omega}) = \frac{\sqrt{1-a^2}}{e^{i\omega}-a}\left[\frac{1-ae^{i\omega}}{e^{i\omega}-a}\right]^{k-1} \approx \frac{\sqrt{1-a^2}}{e^{i\omega}-a}\, e^{-T_a i\omega(k-1)}, \quad T_a = \frac{1+a}{1-a}.$$

This approximation follows from the fact that the first derivative (the slope) of $(1-ae^{iw})/(e^{iw}-a)$ for $w \to 0$ is $-i(1+a)/(1-a)$, which gives

$$\frac{1-ae^{iw}}{e^{iw}-a} \approx e^{-iw(1+a)/(1-a)}, \quad \text{for small } \omega.$$

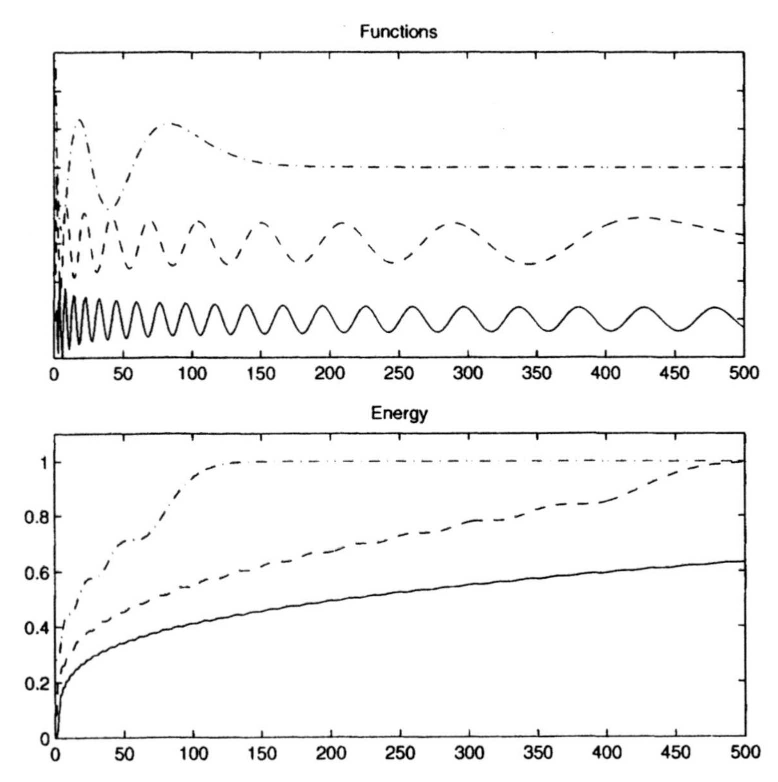

Fig. 10.2. Laguerre functions with a pole in $a = 0.92$. Upper: functions no. 5 (dash-dot), 20 (dash), 60 (solid). The functions are vertically shifted for clarity. Lower: the cumulative energy of the same functions.

Hence for low frequencies, the Laguerre filters act as time-delayed first-order systems, where the time delay approximately equals $(k-1)(1+a)/(1-a)$. This time-delay is increasing by larger a, which means that the Laguerre shift gives more phase-shift for lower frequencies that the ordinary delay shift.
For the example described above, where $a = 0.92$, it follows that $T_a = 1.92/0.08 = 24$. Hence for a time axis $[0 \quad 500]$, one can go up to around $(k-1) = 500/24$, or $k \approx 22$, which is in agreement with Figure 10.2.

In general, one can say that for a basis function

$$F_k(z) = \frac{\sqrt{1-|\xi_k|^2}}{z-\xi_k} \prod_{j=1}^{k-1} \left(\frac{1-\xi_j^* z}{z-\xi_j} \right) \tag{10.1}$$

the number

$$T = \sum_{j=1}^{k} \frac{1+\xi_j}{1-\xi_j} \tag{10.2}$$

gives a good rule of thumb of the settling time.

These arguments illustrate that the orthonormality of the functions will be violated if they are evaluated on a finite time axis. This feature may be of importance in an identification setup, where a system is modelled by

$$y(t) = \sum_{k=1}^{n} c_k F_k(q) u(t) + v(t).$$

If the corresponding time-domain functions f_k are not sufficiently damped out on the time-axis, imposed by the available measurements, the resulting regression equation can be bad conditioned. Consider for example the Laguerre functions discussed above. The upper part of Figure 10.3 depicts the singular values of the matrix $\mathbf{L}$ with the first 100 Laguerre functions on the time interval $[1 \quad 500]$, *i.e.*

$$\mathbf{L}_{ij} = f_i(j) \quad i = 1, \cdots, 100 \quad j = 1, \cdots, 500.$$

In the lower part, the singular values are shown of the matrix $\mathbf{X}$ created by using these 100 functions to filter a pseudo-random binary signal (PRBS) $u(t)$, *i.e.* the matrix with state signals $x_i(t) = F_i(q)u(t)$.

$$\mathbf{X}_{ij} = x_i(j), \quad i = 1, \cdots, 100 \quad j = 1, \cdots, 500$$

Small singular values indicate bad conditioning, hence these figures show that around 60 Laguerre functions is the maximum that can be used for this number of data points and the given pole location. This number could be smaller for other input signals.
The basic idea here is that the number of functions that can be used is limited and the actual number of functions that can be used depends strongly on the 'slowest' pole of $\{F_k(z)\}$. Obviously, the quality of the input signal plays a major role in this phenomenon as well.
See also [41, 307] for a discussion and possible solutions for these numerical problems, for instance by using alternative measures. See furthermore the discussion on numerical conditioning in Chapter 6.

Bias

Although one of the guidelines in using GOBFs is to choose the basis poles close to the actual poles in order to reduce the bias (see Chapter 4), this may become troublesome if the system at hand has slow modes and the data set is short. Especially when using the Hambo functions $V_k(z) = V_1(z)G_b^{k-1}(z)$, the number of functions to be used is limited due to the repetition of all poles,

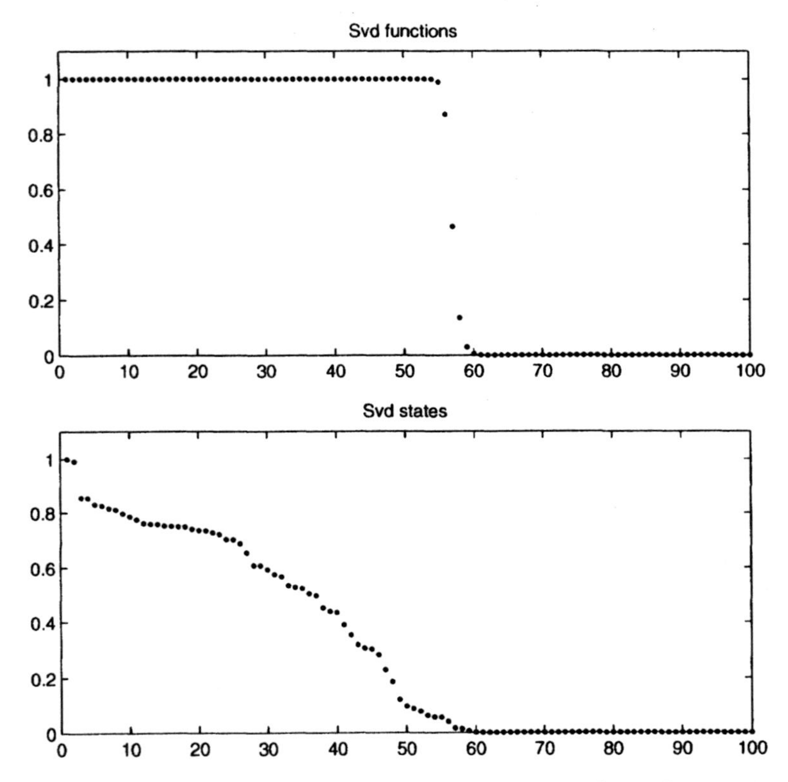

Fig. 10.3. Upper: scaled singular values of the matrix **L** with the first 100 Laguerre functions ($a = 0.92$) on the time interval [1 500]. Lower: scaled singular values of the matrix **X** with the states resulting from filtering a PRBS signal with these 100 functions.

which is the effect of multiplication with $G_b(z)$. An obvious solution to this problem is to use the general Takenaka-Malmquist functions $F_k(z)$, avoiding the repetition of slow poles. Another alternative is the use of 'faster' poles to be able to estimate a larger number of coefficients or to improve the conditioning of the regression problem.

Hence, the aim to reduce the bias by minimization of the neglected tail as in Chapters 4 and 11, which leads to the min-max problem

$$\min_{\xi_1,\cdots,\xi_n} \quad \max_{p\in\Omega} \quad \prod_{j=1}^{n} \left| \frac{p-\xi_j}{1-\xi_j p} \right| \tag{10.3}$$

where Ω is the region that contains the true poles and where $\{\xi_i\}$ are the basis poles, is more or less weighted against the aim to reduce the norm of the neglected coefficients by increasing the number of estimated coefficients. Note however that this latter approach may increase the variance of the estimated parameters.

Variance

A second feature of the pole locations is that they have a strong influence on the asymptotic variance through Expression (4.54), which stated that for $n, N \to \infty,\ n << N$

$$\mathrm{Var}\{G(e^{i\omega}, \widehat{\theta}_N\} \to \frac{1}{N}\frac{\Phi_v(e^{i\omega})}{\Phi_u(e^{i\omega})} \cdot \sum_{k=1}^{n} |F_k(e^{i\omega})|^2,$$

where n, N are respectively the order of the model (*i.e.* the number of basis functions) and the number of data points.
This expression can be seen as a grip to control the variance through the choice of the basis poles $\{\xi_1, \cdots, \xi_n\}$. If we denote by G_b the all-pass function

$$G_b(z) = \prod_{i=1}^{n} \frac{1 - \xi_i^* z}{z - \xi_i}$$

then it can be shown that for $z = e^{i\omega}$, *i.e.* $|z| = 1$,

$$\sum_{k=1}^{n} |F_k(z)|^2) = z\frac{d}{dz}\left(G_b^T(z^{-1})\right) G_b(z) = -zG_b^T(z^{-1})\frac{d}{dz}G_b(z)$$

which comes down to (see also Chapter 2):

$$\sum_{k=1}^{n} |F_k(z)|^2 = z \cdot \sum_{i=1}^{n} \frac{1 - |\xi_i|^2}{(1 - \xi_i^* z)(z - \xi_i)} = \sum_{i=1}^{n} \left(\frac{1}{1 - \xi_i^* z} + \frac{\xi_i}{z - \xi_i} \right), \quad (10.4)$$

giving a means to *shape* the weighting of the variance.

In this regard, it should be noted that this variance weighting has a so-called water-bed effect, as it redistributes the 'standard' weighting over the frequency axis. This follows from the fact that for the FIR case the asymptotic variance is given by:

$$\mathrm{Var}\{G(e^{i\omega}, \widehat{\theta}_N)\} \to \frac{n}{N}\frac{\Phi_v(\omega)}{\Phi_u(\omega)} \quad \text{as} \quad n, N \to \infty, n/N \to 0.$$

Now it is easy to derive that integrating (10.4) over the unit circle amounts to the model order n:

$$\frac{1}{2\pi}\int_{-\pi}^{\pi}\sum_{k=1}^{n}|F_k(e^{i\omega})|^2 d\omega = n$$

which shows that the integrals of the weightings match.

In Figure 10.4, some examples are given of the frequency weighting imposed by different choices of basis poles. See Chapter 5 for a detailed analysis on the role of orthonormal bases in relation to variance.

The expressions (10.3) and (10.4) also explicitly show the bias-variance trade-off, as choosing a basis pole close to a 'true' pole p will decrease the bias but will cause a peak in the variance near Im(p). Note that (10.3) and (10.4) are *asymptotic* expressions and should therefore in the case of short data sets only been seen as a rough directive.

10.3.2 Initial Conditions

Another important feature when dealing with relatively short data sequences is the influence of 'past' – unknown – input signals on the measured output signals, often referred to as the effect of *initial conditions* on the output signals. From an input/output perspective, representing the relation between signals u and y by

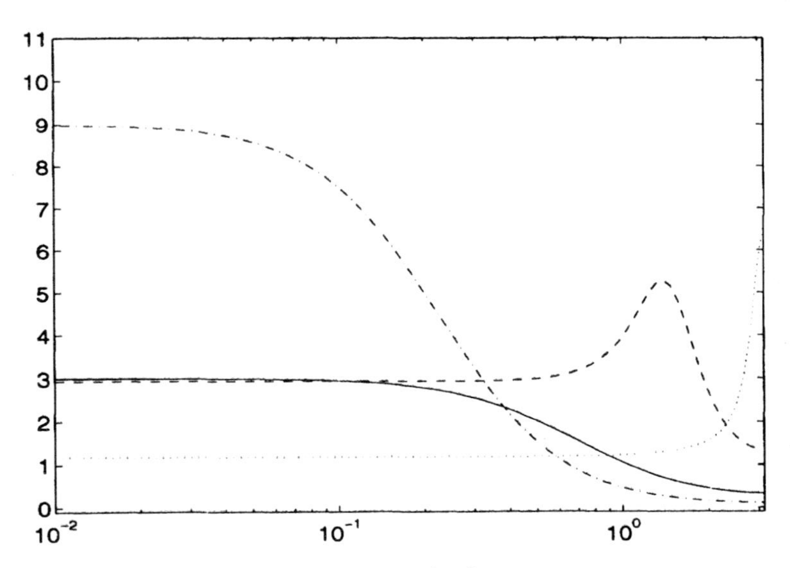

Fig. 10.4. Frequency weighting $\sum_{k=1}^{n}|F_k(e^{i\omega})|^2$ on the asymptotic variance with basis functions generated by poles in $\xi = 0.5$ (solid), $\xi = 0.8$ (dash-dot), $\xi_{1,2} = \{-.7, 0\}$ (dotted) and $\xi_{1,2,3} = \{.3, .1 \pm .6i\}$ (dash-dash).

$$y(t) = G(q)u(t) + e(t),$$

where e is an error signal, this influence can be formulated as follows: Define the 'past' and 'future' input signals u_1, u_2 as

$$u_1(s) = \begin{cases} u(s) \; ; s < t_0 \\ 0 \quad ; s \geq t_0 \end{cases} \qquad u_2(s) = \begin{cases} 0 \quad ; s < t_0 \\ u(s) \; ; s \geq t_0 \end{cases} \tag{10.5}$$

such that $u(t) = u_1(t) + u_2(t)$. Then we can decompose the output signal y as

$$y(t) = y_1(t) + y_2(t) + e(t),$$

where $y_i(t) = G(q)u_i(t), \quad i = 1, 2.$

If a data set $\{u(s), y(s), s = t_0, \cdots, t_N\}$ is to be used in an identification context to infer knowledge about the system, one should therefore be aware of the autonomous part y_1 in the output, which is completely independent of the measured input signal.

If the system G is represented in a state-space realization (assuming for simplicity that G has no feedthrough term):

$$\begin{aligned} x(t+1) &= Ax(t) + Bu(t) \\ y(t) &= Cx(t), \end{aligned}$$

then the effect of the past input signals can be reduced to an *initial condition* vector x_0 as:

$$\begin{aligned} x(t_0) &= x_0 \\ x_0 &= Bu(t_0 - 1) + ABu(t_0 - 2) + \cdots \\ &= \sum_{k=1}^{\infty} A^{k-1} Bu(t_0 - k). \end{aligned}$$

Under the assumption that $u \in \ell_2(\mathbb{Z})$ and $G \in \mathcal{H}_2(\mathbb{E})$, the existence of this vector x_0 is guaranteed. It follows that the effect on the output signal can be written as an impulse response of the system $H(z) = C[zI - A]^{-1}x_0$, that results from a pulse at time $t = t_0 - 1$:

$$y_2(t) = CA^{t-t_0}x_0 \qquad t \geq t_0.$$

The latter equation shows clearly that if the system has dominant slow poles (*i.e.* the matrix A has eigenvalues near the unit circle), the effect of the initial conditions may be substantial.

In the classical identification approach, when estimating FIR models, this feature is generally dealt with by omission of the first n output samples in the regression equation, where n is the order of the FIR model [164].

$$y(t) = \sum_{k=1}^{n} g_k u(t-k) + e(t)$$

If measurements are available for $t = t_0, \ldots, t_N$, then it readily follows that for $t = t_0, \cdots, t_n - 1$ the model output $y(t)$ depends on unknown 'past' values of $u(t)$. For $t \geq t_n$ all involved data are available.

When applying GOBFs, this is no longer valid, as GOBFs have an infinite energy horizon, *i.e.* for every time instant t the value $y(t)$ depends on $u(s)$ for $s = -\infty, \cdots, t$. If the contribution of the unknown inputs $u_1(s)$ is substantial, it may therefore be necessary to include initial conditions in the identification setup. This is especially the case when the model contains 'slow' modes.

Estimation Through State-Space Models

Estimating initial conditions can be included in GOBF modeling without violating the property that the parameters appear linearly in the model. For this purpose the model

$$y(t) = G(q)u(t) + e(t) = \sum_{k=1}^{n} c_k F_k(q) u(t) + e(t) \tag{10.6}$$

is represented for $t \geq t_0$ in state-space form,

$$x(t+1) = Ax(t) + Bu(t) \tag{10.7}$$

$$y(t) = Cx(t) + e(t) \tag{10.8}$$

$$x(t_0) = x_0, \tag{10.9}$$

where $C = [c_1 \cdots c_n]$ and x_0 can be interpreted as $x_0 = \sum_{k=1}^{\infty} A^{k-1} B u(t_0 - k)$, clearly representing the influence of the unknown 'past' inputs.

The conversion from (10.6) to (10.7–10.9) can easily be accomplished by using the state-space realizations for first and second order inner functions, given in Chapter 2, and the property that – given two all-pass functions $G_1(z)$ and $G_2(z)$ with minimal balanced realizations (A_i, B_i, C_i, D_i) $i = 1, 2$ – the product $G_2(z)G_1(z)$ is again an all-pass function with minimal balanced realization

$$\left(\begin{pmatrix} A_1 & 0 \\ B_2 C_1 \end{pmatrix}, \begin{pmatrix} B_1 \\ B_2 D_1 \end{pmatrix}, \begin{pmatrix} D_2 C_1 & C_2 \end{pmatrix}, D_2 D_1 \right).$$

Because any inner function can be written as the product of first and second order systems, this procedure can be used to create a state-space realization (A_b, B_b, C_b, D_b) for the inner function G_b related to the functions $\{F_k\}$. From this realization we can use the matrices $A = A_b$ and $B = B_b$ in Equation

(10.7), while the unknown parameters are contained in the matrix C and the vector x_0 in Equations (10.8, 10.9).

To simplify the discussion, we assume that the system at hand is a scalar system (*i.e.* $G = G^T$), showing that we can also assume that the fixed parameters are contained in the A and C matrices, while the matrix B and the vector x_0 contain the unknown parameters.
Using this setup we can write the following (regression) equation:

$$\begin{bmatrix} y(t_0) \\ y(t_1) \\ y(t_2) \\ \vdots \\ y(t_N) \end{bmatrix} = \begin{bmatrix} C \\ CA \\ CA^2 \\ \vdots \\ CA^{N-1} \end{bmatrix} x_0 + \begin{bmatrix} 0 \\ u(t_0)C \\ u(t_0)CA + u(t_1)C \\ \vdots \\ u(t_0)CA^{N-2} + \cdots + u(t_{N-1})C \end{bmatrix} B + \begin{bmatrix} e(t_0) \\ e(t_1) \\ e(t_2) \\ \vdots \\ e(t_N) \end{bmatrix}$$

where only the quantities x_0 and B are unknown. Rewriting this equation as:

$$\mathcal{Y} = \Phi \begin{bmatrix} x_0 \\ B \end{bmatrix} + \mathcal{E},$$

it follows immediately that $\{x_0, B\}$ can be determined analytically when the latter equation is solved using a least squares criterion

$$\begin{bmatrix} \hat{x}_0 \\ \hat{B} \end{bmatrix} = \left(\Phi^T \Phi\right)^{-1} \Phi^T \mathcal{Y}.$$

The key trick here is of course that the output is linear in the parameters B and x_0.

Variance and Identifiability Considerations

Note that inclusion of the initial conditions x_0 adds n parameters to the estimation problem, *i.e.* doubles the number of parameters to be estimated. This may seriously increase the variance of the 'dynamic' parameters in B. This can partly be dealt with by omitting irrelevant initial conditions. Thereto the regression matrix Φ can be analyzed on the contribution of the initial conditions. Write:

$$\Phi = [\Phi_x \Phi_u]$$

If columns of Φ_x are (nearly) linear independent of the columns of Φ_u, then the corresponding initial conditions will (hardly) influence the estimation of the dynamic parameters. On the other hand, dependence of a Φ_x column on Φ_u leads to rank deficiency of the regression matrix and reveals an identifiability problem, because in this case it is impossible to distinguish the effect of the initial condition from the dynamic effect. This might be solved by omission of the corresponding initial value, but this is a deceitful operation, since there is no criterion to discriminate between the initial or dynamic effect. Instead one should use a different model structure, with a different set of orthonormal functions, or a different data set.

As an example consider the system

$$G(z) = \frac{z^2 - 1.9z + 0.9104}{z^3 - 2.75z^2 + 2.56z - 0.8075} \tag{10.10}$$

with poles in $\{0.95, 0.9 \pm 0.2i\}$, which was excited with white noise and an initial condition, resulting in the system response depicted in Figure 10.5. A sixth-order GOBF model with poles in $\{0.7 \pm 0.2i, 0.9\}$, each with multiplicity 2 (*i.e.* a Hambo construction with $n_b = 3$ and 1 repetition) was estimated using $N = 100$ data points. Figure 10.6 shows the frequency responses of the resulting estimate, with and without initial state estimation. It is evident that the estimation of the initial conditions improves the overall estimate. Both models are not able to capture the static gain, which is caused by the very small number of data points. Thus the input signal does not contain enough low-frequency information.

Remark 10.1. From the discussion above it follows that – for the scalar case – the effect of past input signals can also be formulated in an input/output (transfer function) setting with a finite number of 'initial conditions', by writing

$$y(t) = \sum_{k=1}^{n} c_k F_k(q) u_2(t) + \sum_{k=1}^{n} d_k F_k(q) \delta(t - t_0 + 1) + e(t) \quad t \geq t_0,$$

where u_2 is defined as in Equation (10.5).

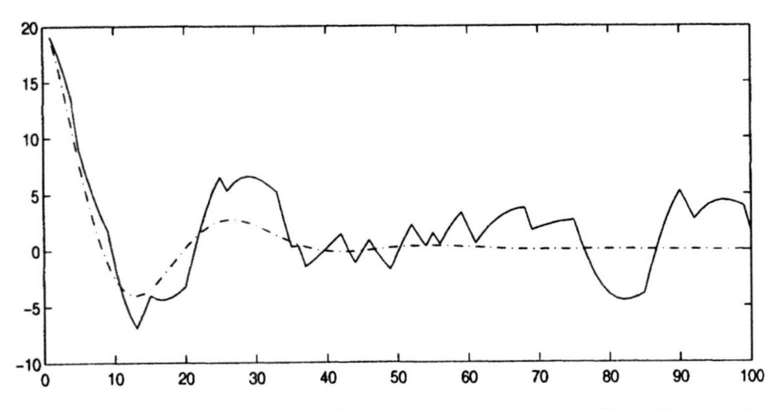

Fig. 10.5. Example output (solid) and initial condition effect (dash dotted).

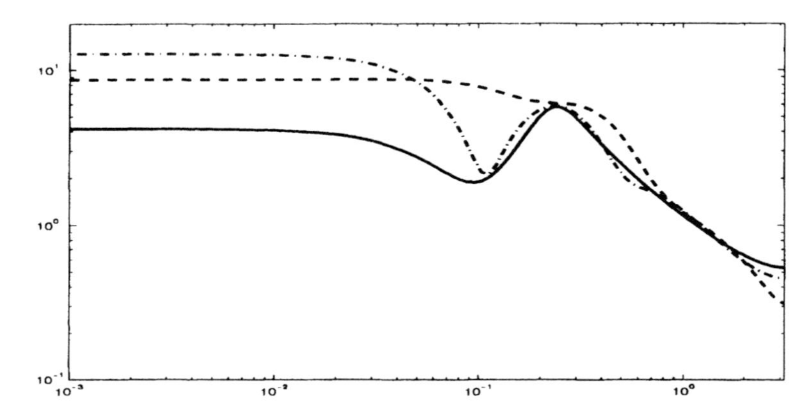

Fig. 10.6. Frequency response of example system (solid) and estimates with (dash-dotted) and without (dashed) initial state estimation.

10.3.3 Static-gain Enforcement

In the previous paragraph, it was shown how the estimation of initial conditions can be included in the GOBF formulation without violating the linear-in-the-parameters property. A similar phenomenon occurs if one wants to include static-gain information in the estimation problem.

Assuming the static-gain value to be equal to $\mathbf{S}$, *i.e.* $\widehat{G}(1) = \mathbf{S}$, then this adds a constraint to the estimation problem

$$y(t) = \sum_{k=1}^{n} c_k F_k(q) u(t) + e(t), \tag{10.11}$$

in the form of

$$\sum_{k=1}^{n} c_k F_k(1) = \mathbf{S} \tag{10.12}$$

or in state-space form

$$C\left[I - A\right]^{-1} B = \mathbf{S}$$

There are two ways to consider this constraint, treating it as a *hard* or a *soft* constraint, where a hard constraint should always be fulfilled, *i.e.* the resulting model will obey Equation (10.12), whereas a soft constraint is merely treated as an extra constraint in the regression equation. This is formalized as follows:

Let the model be given by

$$y(t) = \sum_{k=1}^{n} c_k F_k(q) u(t) + e(t) = \sum_{k=1}^{n} c_k x(t) + e(t),$$

where the states $\{x_k\}$ are defined by

$$x_k(t) = F_k(q)u(t).$$

Then the resulting regression equation is given by

$$\begin{bmatrix} y(t_0) \\ y(t_1) \\ \vdots \\ y(t_N) \end{bmatrix} = \begin{bmatrix} x_1(t_0) & \cdots & x_n(t_0) \\ x_1(t_1) & \cdots & x_n(t_1) \\ \vdots & \vdots & \vdots \\ x_1(t_N) & \cdots & x_n(t_N) \end{bmatrix} \begin{bmatrix} c_1 \\ c_2 \\ \vdots \\ c_n \end{bmatrix} + \begin{bmatrix} e(t_0) \\ e(t_1) \\ \vdots \\ e(t_N) \end{bmatrix}$$

or in compact notation

$$\mathcal{Y} = \mathcal{X} \cdot \mathcal{C} + \mathcal{E}. \tag{10.13}$$

When we use a least-squares criterion

$$V(\theta) = ||\mathcal{E}||_2,$$

this leads to the *unconstrained* parameter estimate

$$\widehat{\mathcal{C}} = \left(\mathcal{X}^T\mathcal{X}\right)^{-1}\mathcal{X}^T\mathcal{Y}. \tag{10.14}$$

Soft Constraints

A soft constraint is entered in this formulation by extending Equation (10.13) with a static-gain equation as given by Equation (10.12).
Defining

$$\mathcal{F} := \left[F_1(1) \cdots F_n(1)\right],$$

we can write

$$\begin{bmatrix} \mathcal{Y} \\ \mathbf{S} \end{bmatrix} = \begin{bmatrix} \mathcal{X} \\ \mathcal{F} \end{bmatrix} \mathcal{C} + \begin{bmatrix} \mathcal{E} \\ e_s \end{bmatrix}. \tag{10.15}$$

Now we apply a weight w_s in the criterion function

$$V(\theta) = ||\mathcal{E}||_2 + w_s||e_s||_2,$$

where the weight $w_s \in [0\ \infty)$ is a design variable, reflecting the confidence in the static-gain information. This extra constraint results in a parameter estimate

$$\widehat{\mathcal{C}} = \left(\mathcal{X}^T\mathcal{X} + w_s\mathcal{F}^T\mathcal{F}\right)^{-1}\left(\mathcal{X}^T\mathcal{Y} + w_s\mathcal{F}^T\mathbf{S}\right). \tag{10.16}$$

Comparing this estimate with the unconstrained estimate given by Equation (10.14) reveals the role of the weight w_s.

Hard Constraints

A hard constraint is substituted in the so-called normal equation instead of the regression equation. The normal equation resulting from (10.13) is

$$\mathcal{X}^T \mathcal{Y} = \mathcal{X}^T \mathcal{X} \mathcal{C} + \mathcal{X}^T \mathcal{E}.$$

This equation is now extended to

$$\begin{pmatrix} \mathcal{X}^T \mathcal{Y} \\ \mathbf{S} \end{pmatrix} = \begin{pmatrix} \mathcal{X}^T \mathcal{X} & \mathcal{F}^T \\ \mathcal{F} & 0 \end{pmatrix} \begin{pmatrix} \mathcal{C} \\ \Lambda \end{pmatrix} + \begin{pmatrix} \mathcal{X}^T \mathcal{E} \\ 0 \end{pmatrix}. \tag{10.17}$$

where Λ is a vector of Lagrange multipliers. It is evident from this equation that the resulting model will exactly obey the static-gain constraint. As a result any misrepresentation of this information may seriously affect the quality of the resulting model.

A Unified Formulation

Both these constraint equations can be captured in one general formulation, by introducing a variable R and the following adapted normal equation:

$$\boxed{\begin{pmatrix} \mathcal{X}^T \mathcal{Y} \\ \mathbf{S} \end{pmatrix} = \begin{pmatrix} \mathcal{X}^T \mathcal{X} & \mathcal{F}^T \\ \mathcal{F} & R \end{pmatrix} \begin{pmatrix} \mathcal{C} \\ \Lambda \end{pmatrix} + \begin{pmatrix} \mathcal{X}^T \mathcal{E} \\ \mathcal{E}_r \end{pmatrix}}$$

It is easily verified that with $R = \frac{1}{w_s}$, $w_s > 0$, this leads to the soft-constrained estimate (10.16), whereas setting $R = 0$ reproduces the hard constraint of Equation (10.17). This reveals in a clear way the transition between soft and hard constraints and shows – as expected – that when $w_s \to \infty$ and thus $R \to 0$ the soft constraint becomes indeed a hard constraint. When $w_s \to 0$, this results in $\Lambda \to 0$ and we are back in the unconstrained situation.

To illustrate the effect of including a hard static-gain constraint, the example with system (10.10), as used in Figures 10.5 and 10.6 is extended with a correct hard constraint. The frequency response of the resulting estimates are shown in Figure 10.7, distinguishing again between estimation with and without initial conditions.

10.3.4 Time Delays

Another important aspect of the modelling procedure is the incorporation of time delays. If it is certain that the system at hand has a time delay of value k, then the system can be modelled as

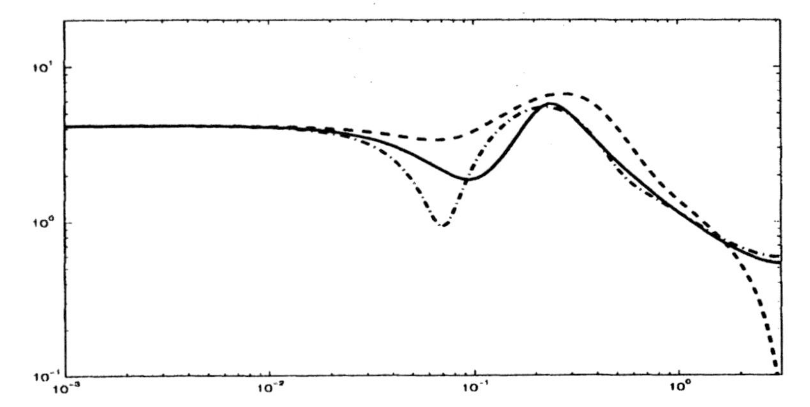

Fig. 10.7. Frequency response of example system (solid) and estimates with (dash-dotted) and without (dashed) initial state estimation, where a correct hard static gain is imposed.

$$G(z) = z^{-k}G_1(z),$$

where the system G_1 is modelled using a basis function description

$$G_1(z) = \sum_{i=1}^{\infty} c_i F_i(z).$$

In a practical setting this comes down to a structure

$$y(t) = \sum_{i=1}^{N} c_i F_i(q) u(t-k) + e(t).$$

See for instance [91].
If one assumes that the system *may* have a time delay, but the length is uncertain, then a reasonable option would be to start the basis generating process with k poles at $z = 0$, *i.e.*

$$F_i(z) = z^{-i}, \quad i = 1, \cdots, k,$$

thus enabling the estimation procedure to set

$$c_i = 0, \quad i = 1, \cdots, k_1 \leq k. \tag{10.18}$$

If in this last situation, there is (partial) confidence/knowledge about the length of the delay, then this information can be included in the estimation procedure, analogously to the static-gain situation in the previous section. This follows from the fact that the time-delay constraints (10.18) can again be incorporated in the linear-in-the-parameters framework. Hence one can use an equivalent procedure as in the static-gain case, expressing confidence in the value of the delay in terms of weighting functions.

10.4 Iterative Scheme

One of the major benefits of the GOBF approach is the possibility to make explicit use of prior knowledge in problems of estimation and approximation. If such knowledge is unavailable, however, the best choice of basis is the standard pulse basis $\{z^{-k}\}$ as will be shown in Chapter 11. It is of paramount importance that – once a model has been estimated or built – the amount of confidence in this model can be considered as prior information about the system at hand. Hence in a consecutive step a GOBF model can be estimated that uses this knowledge.
Of course there is no reason to stop after this step, so the confidence in the created GOBF model can again serve as prior knowledge. This line of reasoning motivates the use of a so-called *iterative approach*, where in each iteration step the amount of prior knowledge is updated. The approach can – to some extend – be compared to what is known as the Steiglitz-McBride iteration [282].

The basic idea is that by extracting this knowledge from the data, the complexity of the estimation problem will decrease, as is illustrated in Figure 10.8. This concept is motivated by the notion that in the ideal case, where the data are generated by a finite dimensional time invariant linear system G, the optimal choice of basis poles will obviously be the set of poles of G, as then the number of non-zero parameters would be reduced to the McMillan degree of G.
Hence the aim of the iterative procedure is the reduction of the number of parameters and henceforth a decrease of the variance of the estimated models. The flow diagram of Figure 10.9 illustrates the iterative approach

This scheme immediately leads to the question of how to construct the update mechanism. The answer to this question obviously depends on the purpose of the mechanism. Often the goal will be to decrease the value of the criterion function for a fixed number of basis functions, in which case a procedure is needed that, given data $\{y(t), u(t), t = 1, \cdots, N\}$ and an expansion

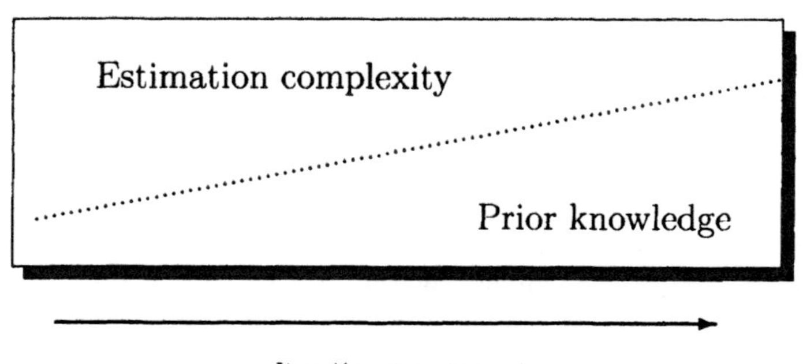

Fig. 10.8. Complexity versus prior knowledge in iterative estimation scheme.

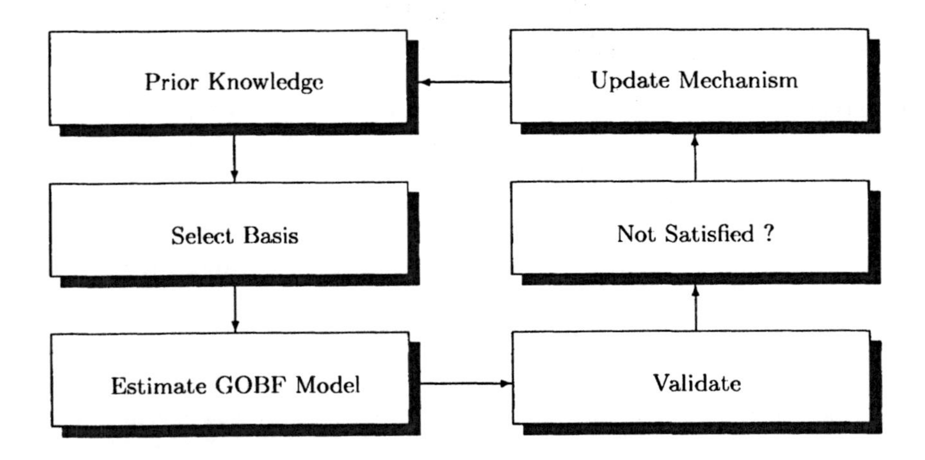

Fig. 10.9. Iterative scheme for GOBF modelling.

$$\widehat{G}(z) = \sum_{k=1}^{n} c_k F_k(z),$$

creates a new set of basis functions such that an expansion in this new basis

$$\widehat{G}'(z) = \sum_{k=1}^{n} c'_k F'_k(z),$$

obeys

$$V_N(G', y, u) \leq V_N(G, y, u),$$

where the criterion function V_N is for instance defined by

$$V_N(G, y, u) = \sum_{t=1}^{N} \varepsilon^2(t), \qquad \varepsilon(t) = y(t) - G(q)u(t).$$

Other aims of the procedure may be the construction of low-order models, reduction of uncertainty *etc.* In fact the question of the purpose of the mechanism is directly related to the intended purpose of the model to be identified. One approach that has been used in the past ([126,308,309]) is the application of model reduction techniques. This can be motivated by the observation that after estimating the coefficients in a GOBF model, the resulting model is generally of a high McMillan degree and over-parameterized. Applying model reduction will extract the dominant characteristics of the model. By adding new poles, for instance using the Hambo function approach, a new set of basis poles is generated.

Example 10.1. To illustrate this phenomenon, consider the first-order system $G(z) = \frac{z+0.4}{z-.85}$. A model of this system was calculated using Laguerre models, where in each iteration step 2 coefficients where calculated. Note that the resulting model will be an approximation, unless the Laguerre pole was chosen to be equal to 0.85. The resulting second-order approximative models were reduced to first-order models using balanced truncation [198]. The pole of the first-order model was used to generate Laguerre functions in the next iteration step. The initial basis pole was 0. The resulting poles are displayed in Table 10.2, which shows that this procedure converges indeed to the correct pole value.

The crucial issue here is of course the convergence of the iteration scheme. So far no theoretical results on this subject are known. Practical results using model reduction techniques based on balancing point at convergence in most of the cases, but also divergence has been observed. In [308] *weighted* balanced reduction was used, where the weighting was based on the inverse of the estimated variance, motivated by the notion that a relatively high variance reflects low confidence. This procedure shows a major improvement over the unweighted methods. However, there is a possible pitfall in this motivation, since high variance might be solely due to the noise spectrum and not related to the basis functions used. Thereby one might discard important poles if only variance information is used.
In Chapter 13 an alternative reduction method, generalizing the classical realization methods for the Hambo-GOBF case, is presented. See also [302] for a discussion on model reduction and identification.

Of course it should be noted that this iterative approach more or less gives up on the original motivation to circumvent non-linear optimization procedures, since the model reduction step is a non-linear addition. However, because both linear regression and model reduction are reasonably well defined and understood, one can still view the approach as a way to break down non-linear optimization problems in manageable steps.

Table 10.2. Example of basis convergence for the iteration scheme

Iteration number	Laguerre pole value
0	0
1	0.4310
2	0.7230
3	0.8351
4	0.8499
5	0.8500

10.5 MIMO Models

Most of the material presented so far has been for the scalar case, *i.e.* for single-input/single-output (SISO) systems. This section discusses the application of GOBF theory for multi-variable or multi-input/multi-output (MIMO) systems. It will be shown that there are two approaches to be considered. The first approach uses the theory on scalar basis functions as defined in Chapter 2, whereas the second approach is based on the construction of multi-variable basis functions.

10.5.1 Scalar Basis Functions

This approach deals with the MIMO aspect in accordance with the classical FIR approach, describing a $p \times m$ system G by

$$G(z) = \sum_{i=1}^{\infty} M_i F_i(z) \tag{10.19}$$

with $\{F_i(z)\}$ scalar basis functions and $M_i \in \mathbb{R}^{p\times m}$. For the FIR case $(F_i(z) = z^{-i})$, the matrices $\{M_i\}$ are the Markov parameters of the system. For identification or approximative modelling, the system (10.19) is approximated by a finite series expansion

$$\widehat{G}(z) = \sum_{i=1}^{n} \widehat{M_i} F_i(z) \tag{10.20}$$

where the parameters $\{M_i\}$ can again be estimated analytically, when using a least squares criterion. Details of this procedure can be found in [209,211]. It turns out that the bias and variance results, as discussed in Chapter 4 for scalar systems, can be extended rigorously to the multi-variable case [209].

In particular, (see [209] for more details), with $[\cdot]_{m,n}$ denoting the (m,n)th element of a matrix, and remembering that in the multi-variable case the input and noise spectral densities are possibly matrix valued quantities, then the variance error of using a multi-variable fixed-denominator model structure (whether or not it is orthonormal), may be quantified as:

$$\mathsf{E}\{|G_{m,r}(e^{j\omega}) - G_{m,r}(e^{j\omega},\theta_0)|^2\} \approx \frac{1}{N}\left[\Phi_u^{-1}(\omega)\right]_{r,r}[\Phi_\nu(\omega)]_{m,m} \times \sum_{k=1}^{n}|F_k(e^{j\omega})|^2$$

and the asymptotic bias error, for white input, can be quantified as

$$|G_{m,r}(e^{j\omega}) - \widehat{G}_{m,r}(e^{j\omega})| < \sum_{\ell=1}^{p}\left|\frac{\alpha_\ell^{m,r}}{e^{j\omega}-\gamma_\ell^{m,r}}\right|\prod_{k=1}^{n}\left|\frac{\gamma_\ell^{m,r}-\xi_k}{1-\overline{\xi_k}\gamma_\ell^{m,r}}\right|,$$

where the $\{\gamma_\ell^{m,r}\}$, $\ell = 1, 2, \cdots, p$ are the p true poles of the (m, r)th entry in $G(e^{j\omega})$. See Chapters 5 and 6 for a detailed discussion on fixed-denominator model structures.
Next to its intrinsic simplicity, a major advantage of using scalar basis functions to forming a model structure as in Equation (10.20), is that such a structure is very flexible in its possibilities to enforce specific (desired) structural knowledge into the model. Without problems, one can set specific elements of the M_i matrices to zero or pose other constraints. One can even use a different set of basis functions for each entry of $G(z)$.

Note that (without constraints) the number of parameters to be estimated in the model (10.20) is equal to $n \times p \times m$. This number can become quite large if the overall dynamic range is wide, since in that case a large number of basis functions is needed.
In [308] this approach is explored and analyzed for the situations with different dynamics in each row and/or column of the transfer function matrix.

10.5.2 Multi-variable Basis Functions

The use of MIMO orthonormal basis functions was first addressed in [126], where the Hambo functions were introduced for the MIMO case. The basic ingredient is the use of multi-variable, specifically square, all-pass functions. The derivation is the same as the state-space approach presented in Chapter 2, *i.e.* given an $m \times m$ stable all-pass function G_b with McMillan degree $n_b > 0$ and minimal balanced realization (A_b, B_b, C_b, D_b) with $||D_b||_2 < 1$, then the set of vector functions $\{V_k(z)\} \in \mathbb{R}^{n_b \times m}[z]$ defined by

$$V_k(z) = V_1(z) G_b^{k-1}(z) \tag{10.21}$$

$$V_1(z) = [zI - A_b]^{-1} B_b \tag{10.22}$$

constitutes a basis for $\mathcal{H}_2^m$ in the sense that any function $G(z) \in \mathcal{H}_2^{1 \times m}$ has a unique expansion

$$G(z) = \sum_{k=1}^{\infty} L_k^T V_k(z), \qquad L_k \in \mathbb{R}^{n_b}, \tag{10.23}$$

or in an extended form a basis for $\mathcal{H}_2^{p \times m}$, such that any any function $H(z) \in \mathcal{H}_2^{p \times m}$ has a unique expansion

$$H(z) = \sum_{k=1}^{\infty} L_k^T V_k(z) \qquad L_k \in \mathbb{R}^{n_b \times p}. \tag{10.24}$$

See [126, 132] for a detailed discussion on these basis functions and a proof of the completeness.

An important issue here is the construction *cf.* parameterization of square all-pass functions. Where in the scalar case these functions can be written as Blaschke products and thus are completely determined by the poles, the multi-variable case is more involved and inhibits much more freedom. Here also the structure and/or dynamic directions are of importance, but unfortunately not yet completely understood. See [308, 309] for an application of multi-variable Hambo functions in identification.

A nice property of the expansion (10.24) is that the expansion coefficients $\{L_k\}$ can (as in the scalar case) be seen as Markov parameters of a system with known poles. In the scalar case ($p = m = 1$) these poles are given by (see Chapters 2 and 12):

$$\lambda_i = G_b(\mu_i^{-1})$$

where $\{\mu_i\}$ is the set of poles of $H(z)$. For the multi-variable case this can be extended, resulting in the following proposition from [126]:

Proposition 10.1. *Let $H \in \mathcal{H}_2^{p\times m}$, with $H(z) = \sum_{k=1}^{\infty} L_k^T V_k(z)$, where $\{V_k(z)\}$ is the set of basis function vectors created from an $(m \times m)$ inner function G_b. Let H have McMillan degree n_h and poles $\{\mu_1, \cdots, \mu_{n_h}\}$. Denote by $\mathcal{V}(L_k)$ the vector resulting from stacking the columns of L_k,* i.e.

$$\mathcal{V}(L_k) \doteq [(L_k)_{11} \cdots (L_k)_{p1} (L_k)_{12} \cdots (L_k)_{p2} (L_k)_{13} \cdots\cdots (L_k)_{pm}]^T$$

Then there exist matrices $R \in \mathbb{R}^{n_h \cdot m \times n_h \cdot m}, S \in \mathbb{R}^{p \cdot n_b \times n_h \cdot m}, T \in \mathbb{R}^{n_h \cdot m \times 1}$ such that

1. *$\{\mathcal{V}(L_k)\}$ can be interpreted as the Markov parameters of the system (R, T, S):*
$$\mathcal{V}(L_k) = SR^{k-1}T$$
2. *The set of eigenvalues of R is the union of the sets of eigenvalues of $G_b(\mu_i^{-1})$:*
$$\sigma(R) = \bigcup_{i=1}^{n_h} \sigma\left(G_b(\mu_i^{-1})\right)$$
3. *The pair (R, T) is controllable.*

This proposition shows that there is a MIMO equivalent of the Hambo signal transform concept (see Chapters 2 and 12), which is defined by

$$\breve{H}(\lambda) = \sum_{k=1}^{\infty} \mathcal{V}(L_k)\lambda^{-k} = \sum_{k=1}^{\infty} SR^{k-1}T\lambda^{-k}.$$

This transform is always stable because the eigenvalues of $G_b(\mu_i^{-1})$ are inside the unit disk for $|\mu_i| < 1$.

Note that there is no assertion about minimality in Proposition 10.1. It depends on the system $H(z)$ whether the pair (R, S) is observable. For instance, if $H(z) = C[zI - A_b]^{-1}B_b = CV_1(z)$, it follows that $L_1 = C^T$ and $L_k \equiv 0, k > 1$. Hence

$$\breve{H}(\lambda) = \mathcal{V}(C^T)\lambda^{-1},$$

which is a first-order system, where $R \in \mathbb{R}^{n_b \cdot m \times n_b \cdot m}$.

Choice of Basis Functions

For identification purposes, these infinite expansions (10.24) will again be approximated by a finite expansion. An obvious problem in this setup is the choice of the inner function G_b, which defines the basis functions. In this respect it should be noted that this choice imposes structural information on the resulting model, as the resulting model can be written as

$$\widehat{H}(z) = \sum_{k=1}^{n} \widehat{L}_k^T V_k(z = C_e \left[zI - A_e\right]^{-1} B_e, \tag{10.25}$$

where (A_e, B_e) are matrices, fixed by the choice of basis functions.

A common approach for this choice is the use of an initial model, resulting from prior knowledge or prior identification experiments. Say that an initial model

$$\widehat{H}(z) = \hat{C}\left[zI - \hat{A}\right]^{-1}\hat{B} \tag{10.26}$$

has been derived. An inner function is created from this model by choosing matrices $(\bar{C}, \bar{D})$, such that $G_b(z) = \bar{D} + \bar{C}\left[zI - \hat{A}\right]^{-1}\hat{B}$ is a stable all-pass function. This is always possible [126, 308], using the following construction:

Assume that the realization $\left(\hat{A}, \hat{B}, \hat{C}\right)$, where $\hat{A} \in \mathbb{R}^{n \times n}$, $\hat{B} \in \mathbb{R}^{n \times m}$, is such that $\hat{A}\hat{A}^T + \hat{B}\hat{B}^T = I$. This can always be achieved using standard balancing techniques.
Let the singular value decomposition of the pair (A, B) be given by

$$\left(A\ B\right) = \left(U_1\ U_2\right)\begin{pmatrix}\Sigma_1 & 0\\ 0 & 0\end{pmatrix}\begin{pmatrix}V_1^T\\ V_2^T\end{pmatrix}$$

with $U_1 \in \mathbb{R}^{n \times n}$, $U_2 \in \mathbb{R}^{n \times m}$, $V_1 \in \mathbb{R}^{(n+m) \times n}$, $V_2 \in \mathbb{R}^{(n+m) \times m}$. Define the matrices $(\bar{C}, \bar{D})$ by

$$\left(\bar{C}\ \bar{D}\right) = V_2^T,$$

then it easily verified that

$$\begin{pmatrix} \hat{A} & \hat{B} \\ \bar{C} & \bar{D} \end{pmatrix} \begin{pmatrix} \hat{A} & \hat{B} \\ \bar{C} & \bar{D} \end{pmatrix}^T = I$$

and hence the system $G_b(z)$ with balanced realization $\left(\hat{A}, \hat{B}, \bar{C}, \bar{D}\right)$ is an inner function.

The (A, C) Variant

Up to now, the construction of basis functions in state-space terms was based on the so-called (A, B) variant. Starting from an inner function with balanced realization (A_b, B_b, C_b, D_b), the vector function V_1 is defined by

$$V_1(z) = [zI - A_b]^{-1} B_b.$$

For the Hambo functions, this construction is extended to

$$V_k(z) = V_1(z) G_b^{k-1}(z).$$

An analogue of this construction, called the (A, C) variant, starts with a function $V_1^{(t)}$, denoted by

$$V_1^{(t)}(z) = \left[zI - A_b^T\right]^{-1} C_b^T,$$

and is for the Hambo case extended with

$$V_k^{(t)}(z) = V_1^{(t)}(z)(G_b^T)^{k-1}(z).$$

This follows immediately from application of the (A, B) construction on the inner function G^T with balanced realization $\left(A_b^T, C_b^T, B_b^T, D_b^T\right)$.

For the *scalar* case (*i.e.* G_b scalar, so $G_b = G_b^T$), these two constructions are equivalent in the sense that there exists a unitary matrix T such that,

$$V_k(z) = T V_k^{(t)}(z), \quad k \in \mathbb{N}.$$

This is explained by the fact that $C_b[zI - A_b]^{-1} B_b = B_b^T [zI - A_b^T]^{-1} C_b^T$ and hence there must exist a similarity transformation matrix T, such that

$$T A_b^T T^{-1} = A_b, \qquad T C_b^T = B_b, \qquad B_b^T T^{-1} = C_b,$$

showing that

$$T (A_b^T)^k C_b^T = A_b^k B_b, \quad k \in \mathbb{N}.$$

The fact that T is unitary follows from the orthonormality of both V_1 and $V_1^{(t)}$.

From this observation it immediately follows that for the scalar case, there will be no difference (in input/output sense) if we use one of the following model structures, where we assume that (A_b, B_b, C_b, D_b) is a balanced realization of an inner function and θ_i is the vector with unknown parameters,

$$y(t) = \theta_1[qI - A_b]^{-1}B_b u(t) + v(t) \tag{10.27}$$
$$y(t) = C_b[qI - A_b]^{-1}\theta_2 u(t) + v(t), \tag{10.28}$$

and it follows from the above that $\theta_2^T = \theta_1 T$.
In fact, the construction used in Section 10.3.2 to include the estimation of initial conditions, makes use of this equivalence property.

For *multi-variable* inner functions, such an equivalence relation can only be given for $V_1, V_1^{(t)}$, but the relation is more involved. In fact, one can show that there exist matrices T and S (not necessarily orthogonal), such that

$$T(A_b^T)^k C_b^T S = A_b^k B_b,$$

or

$$V_1(z) = TV_1^{(t)}(z)S, \quad k \in \mathbb{N}.$$

This implies that for MIMO estimation problems – even in the case of symmetric inner functions – the equivalence in model structures as in Equations (10.27, 10.28) does *not* hold, and thus the resulting estimates will be essentially different.

Estimation

As stated before, the estimation of a GOBF model can be formulated as the estimation of a state-space model

$$\hat{H}(z) = C_e\,[zI - A_e]^{-1}\,B_e.$$

In the (A, B) variant, the pair (A_e, B_e) is normalized $(A_e A_e^T + B_e B_e^T = I)$ and fixed, and C_e contains the parameters to be estimated. In the (A, C) variant, the pair (A_e, C_e) is normalized $(A_e^T A_e + C_e^T C_e = I)$ and fixed, and B_e contains the parameters.

It is straightforward that the (A, C) variant is best suited for the case where initial conditions have to be estimated, as in that case no products of parameters will appear in the regression equation, see Section 10.3.2.

These two variants can be used to improve the quality of the estimated models, by using an iterative optimization scheme, not to be mistaken with the iterative scheme in the previous section, though it can be used as the estimation part of that iterative scheme. It is based on the idea that the estimation of a MIMO GOBF model can be interpreted as the estimation of a model (A_e, B_e, C_e), where A_e is fixed modulo similarity transformation and B_e, C_e have to be estimated. The following scheme approaches this bilinear estimation problem in consecutive linear steps.

1. Assume that an initial model guess is available, with minimal realization (A_0, B_0, C_0) .
2. Normalize the pair (A_0, C_0) to an output balanced pair (A_1, C_1), so $A_1^T A_1 + C_1^T C_1 = I$.
3. Estimate a model (A_1, B_1, C_1) using the (A, C) variant, *i.e.* estimate the matrix B_1, if desired including initial conditions.
4. Subtract the estimated initial condition effect from the data.
5. Normalize the pair (A_1, B_1) to (A_2, B_2), so $A_2 A_2^T + B_2 B_2^T = I$.
6. Estimate a model (A_2, B_2, C_2) using the (A, B) variant, *i.e.* estimate the matrix C_2, without initial conditions, from the revised data.
7. Normalize the pair (A_2, C_2) to (A_3, C_3).
8. Estimate a model (A_3, B_3, C_3) using the (A, C) variant, *i.e.* estimate the matrix B_3, if desired including initial conditions, using the original data.
9. Subtract the estimated initial condition effect from the original data.
10. *etc.*

For the case without initial conditions, this scheme will always converge, as the criterion function can only decrease. Experimental experience indicates that in general more than 95% of the improvement (in terms of decrease of the criterion function) is reached within one or two iteration steps.
An important feature of this iterative scheme is that it creates flexibility to correct errors in the imposed structure.

10.5.3 Discussion

In this section, two approaches to MIMO GOBF modelling were discussed. While the use of scalar basis functions is a straightforward extension of SISO GOBF modelling, the number of parameters $(n \times (m + p))$ can be much larger than in the application of MIMO orthogonal basis functions $(n \times min(m, p))$. The latter yields the problem of the yet unknown parameterization of multivariable inner functions and imposes a specific, possibly erroneous, structure on the model, where the scalar approach has the merit of being very flexible in imposing restrictions on the model structure.
It was shown that the expansion coefficients for the MIMO Hambo case can

be seen as the Markov parameters of a linear system, and the eigenvalues of this realization were given.
The literature on MIMO GOBF modelling is scarce, see [126, 132, 209, 211, 308, 309] and the references therein.

10.6 Summary

This chapter discussed a variety of practical issues that are of importance in the use of orthogonal basis model structures. It was pointed out that prior information about pole locations including uncertainty can be used to define well-suited basis functions. This will be elaborated upon in detail in the next chapter. Furthermore, it was emphasized that basis function structures are so-called IIR filters, where initial conditions may be of importance. It was shown that these effects can be included in the linear-in-the-parameters framework, as is the case with static-gain and time-delay information. Finally, two approaches towards multi-variable model structures were discussed, using either scalar or multi-variable basis functions.

11

Pole Selection in GOBF Models

Tomás Oliveira e Silva

University of Aveiro, Aveiro, Portugal

This chapter addresses the problem of the selection of a good set of poles for the generalized orthonormal basis function (GOBF) model. Two methodologies to select pole positions will be described. One (Section 11.2), based only on the knowledge of the poles of the stable system to be identified, yields asymptotically optimal pole locations. The other (Section 11.3), based only on the availability of a (long) record of input/output data, aims to minimize a suitable norm of the output error signal. Section 11.1 presents a summary of the theory behind GOBF models, described in more detail in the first two chapters of this book, and makes this chapter (almost) self-contained.

11.1 Introduction

The expansion of the unit pulse response $g_0(t)$ of a stable and causal linear system by a (complete) orthonormal series expansion has the general form

$$g_0(t) = \sum_{k=1}^{\infty} c_k\, f_k(t).$$

Keeping only the first n terms of this expansion gives rise to an approximation to $g_0(t)$ of order n, of the form

$$g(t;n) = \sum_{k=1}^{n} c_k f_k(t).$$

In order for this approximation to be useful, its error, given by

$$\varepsilon_g(t;n) = g(t;n) - g_0(t),$$

should be 'small' for all t. In order to optimize the approximation of $g_0(t)$ by $g(t;n)$ it is thus necessary to define a way to measure the 'size' of the error of the approximation, *i.e.* to measure the 'size' of $\varepsilon_g(t;n)$.

It is standard practice to measure the 'size' of a signal with a norm. The use of a norm induced by an inner product is very popular, because in that case the approximation problem has a simple solution. Such a norm is given by the square root of the inner product of $x(t)$ with itself, *i.e.* by

$$\|x\| = \sqrt{\langle x, x\rangle}.$$

(The notation $\langle\cdot,\cdot\rangle$ is used in this chapter to denote different inner products; the nature of its arguments will resolve any ambiguity that may arise because of this.) In the current case, the inner product between the real causal signals $x(t)$ and $y(t)$, both with finite norm, is given by

$$\langle x, y\rangle = \sum_{t=0}^{\infty} x(t)y(t).$$

When this inner product vanishes, $x(t)$ is said to be orthogonal to $y(t)$, and *vice versa.* When $\|x\| = 1$, $x(t)$ is said to be normal. A set $\{x_k(t)\}_{k=1}^{\infty}$ is said to be orthonormal if

$$\langle x_k, x_l\rangle = \begin{cases} 0, & \text{if } k \neq l \text{ (orthogonality)},\\ 1, & \text{if } k = l \text{ (normality)}.\end{cases}$$

The main advantage of using the orthonormal set $\{f_k(t)\}_{k=1}^{\infty}$ to build an approximation to $g_0(t)$ is that the coefficients c_k of the expansion are uncoupled; they can be obtained from the projection of $g_0(t)$ on $f_k(t)$, *i.e.*

$$c_k = \langle g_0, f_k\rangle = \sum_{t=0}^{\infty} g_0(t)f_k(t).$$

Another advantage of an orthonormal expansion is that the square of the norm of the approximation error can be computed easily, being given by

$$\left\|\varepsilon_g(\,\cdot\,;n)\right\|^2 = \|g_0\|^2 - \sum_{k=1}^{n} c_k^2.$$

The orthonormal set $\{f_k(t)\}_{k=1}^{\infty}$ is said to be complete (or closed) if and only if

$$\lim_{n\to\infty} \sum_{k=1}^{n} c_k^2 = \|g_0\|^2$$

for all $g_0(t)$ with finite norm. In this case, it is possible to make the approximation error as small as desired; to attain this goal, it is only necessary to use a sufficiently large number of terms in the approximation. However, in any practical application of this approximation method, the number of terms should not be very large. If that happens, then it is certainly better to use

another orthonormal set, namely, one better suitable to approximate $g_0(t)$ with only a few terms. It follows that a given orthonormal set is only suitable to approximate unit pulse responses belonging to a certain subset of the set of the unit pulse responses of all stable and causal systems (see Section 11.2 for further details).
With the help of Parseval's theorem, it is possible to compute norms and inner products using Fourier or Z transforms. In particular,

$$\langle x, y \rangle = \langle X, Y \rangle = \frac{1}{2\pi} \int_{-\pi}^{\pi} X(e^{i\omega}) Y^*(e^{i\omega})\, d\omega,$$

where $X(e^{i\omega})$ and $Y(e^{i\omega})$ are the discrete-time Fourier transforms of $x(t)$ and $y(t)$. (Note that the discrete-time Fourier transform is the Z transform evaluated on the unit circle $\mathbb{T}$.) This inner product can sometimes be evaluated using contour integration, via the formula

$$\langle x, y \rangle = \langle X, Y \rangle = \frac{1}{2\pi} \oint_{\mathbb{T}} X(z) Y^*(1/z^*)\, \frac{dz}{z}.$$

Here, $X(z)$ and $Y(z)$ are the Z transforms of $x(t)$ and $y(t)$ and are assumed to be analytic on $\mathbb{T}$.

In many practical situations, including system identification, the unit pulse response $g_0(t)$ is not available; indeed, an estimate of this unit pulse response, or of its corresponding transfer function $G_o(z)$ or frequency response $G_o(e^{i\omega})$, is precisely what is usually desired. Instead, what is often available is the input signal $u(t)$ of the system and its corresponding output signal $y(t) = G_0(q)u(t) + v(t)$, measured for $t = 1, \dots, T$. It is often assumed, as done here, that the output signal is corrupted by the colored noise $v(t) = H_0(q)e_0(t)$, where $e_0(t)$ is white noise. In this situation, it makes sense to attempt to estimate the coefficients c_k by matching as best as possible $y(t)$ by $y(t;n) = G(q;n)u(t)$, *i.e.* by making the output error as small as possible. This leads to the model structure of Figure 11.1. In general, the coefficients c_k are no longer uncoupled; they will change with n. Hence the change of notation from c_k to $c_{n,k}$.
The inner product between two real stationary stochastic processes $x(t)$ and $y(t)$, each with zero mean and finite variance, is given by

$$\langle x, y \rangle = E\big[x(t)y(t)\big].$$

It can be computed in the frequency domain using the formula

$$\langle x, y \rangle = \frac{1}{2\pi} \int_{-\pi}^{\pi} \Phi_{xy}(e^{i\omega})\, d\omega,$$

where $\Phi_{xy}(e^{i\omega})$ is the cross-spectral density of $x(t)$ and $y(t)$, *i.e.* it is the discrete-time Fourier transform of their cross-correlation function. As before,

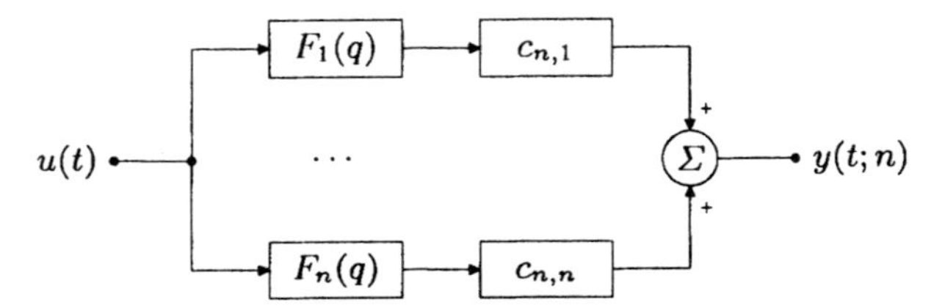

Fig. 11.1. The general linear-in-the-parameters model. The transfer function of the model is given by $G(z;n) = \sum_{k=1}^{n} c_{n,k} F_k(z)$, and its output signal is given by $y(t;n) = G(q;n)u(t)$. In certain cases, such as those described in this book, the parallel structure of the filter bank can be replaced by a series (cascade) structure.

the norm of a stationary stochastic process is given by the square-root of the inner product of itself with itself.

Assuming that the white noise process $e_0(t)$, with variance σ_0^2, is uncorrelated with the input signal $u(t)$, it can be verified that the square of the norm of the output prediction error signal $\varepsilon_y(t;n) = y(t;n) - y(t)$ is given by

$$\left\|\varepsilon_y(\cdot\,;n)\right\|^2 = \frac{1}{2\pi}\int_{-\pi}^{\pi} \left|G_0(e^{i\omega}) - G(e^{i\omega};n)\right|^2 \Phi_u(e^{i\omega})\,d\omega + \sigma_0^2\,\langle h_0, h_0\rangle.$$

Such an error function leads, for example, to asymptotically unbiased estimates of the coefficients $c_{n,k}$ as long as the power spectral density $\Phi_u(e^{i\omega})$ of the input signal obeys the condition $\Phi_u(e^{i\omega}) > \Phi^- > 0$ for all frequencies (this last condition is known as unconditional persistence of excitation).

Chapter 2 describes ways to construct sets of orthonormal sequences that have rational Z transforms. These are by far the easiest to use in practice, because they can be generated by finite-dimensional linear systems. These orthonormal sequences are, in general, the unit pulse responses from the input to the states of a balanced realization of an stable all-pass system. The orthonormal sequences generated by cascading the same balanced realization of an all-pass system, which is itself all-pass and balanced, are particularly interesting. The resulting model structure, depicted in Figure 11.2, besides having fewer free parameters (fewer distinct poles), has a certain regularity that facilitates the analysis of its asymptotic properties (when n goes to infinity). If $n_b = 1$ this model becomes the Laguerre model [313], and if $n_b = 2$ it becomes the two-parameter Kautz model [315]. The general case gives the GOBF model.

Let (A, B, C, D) be a balanced realization of the stable all-pass transfer function $G_b(z)$. The transfer function from the input of the system to the states of the kth all-pass section, is given by

$$V_k(z) = (zI - A)^{-1} B\, G_b^{k-1}(z) = G_b^{k-1}(z)\, V_1(z).$$

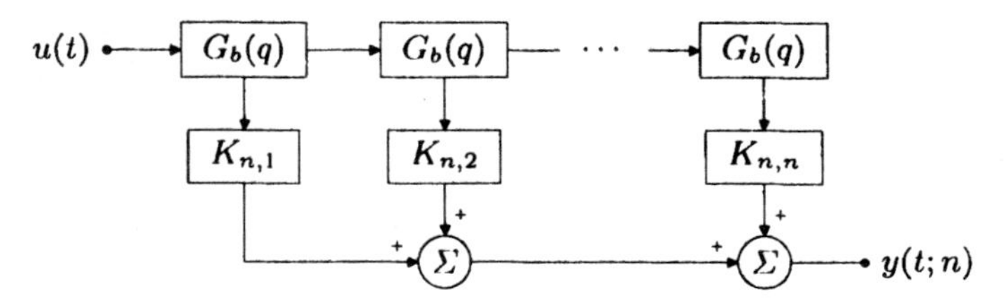

Fig. 11.2. The generalized orthonormal basis functions (GOBF) model. $G_b(z)$ is a stable all-pass transfer function of order n_b (implemented using a balanced state-space realization); its output is fed to the next transfer function in the cascade, and its states are fed to the gains $K_{n,k}$.

Using this notation, the transfer function of the GOBF model is given by

$$G(z;n) = \sum_{k=1}^{n} K_k^T V_k(z),$$

where K_k is a column vector with n_b elements. Any strictly causal unit pulse response $g_0(t)$ with finite norm can be expanded as follows

$$g_0(t) = \sum_{k=1}^{\infty} K_k^T v_k(t),$$

or, equivalently (using Z transforms),

$$G_0(z) = \sum_{k=1}^{\infty} K_k^T V_k(z), \tag{11.1}$$

where (for matrices, * denotes complex conjugate transposition)

$$K_k^T = [\![G_0, V_k]\!] = \frac{1}{2\pi}\int_{-\pi}^{\pi} G_0(e^{i\omega})\, V_k^*(e^{i\omega})\, d\omega.$$

(Recall that the notation $[\![\cdot,\cdot]\!]$, with column vector arguments, denotes the cross-Gramian between the two vectors, *i.e.* it is a matrix whose elements are all possible inner products between an element of the first vector and an element of the second vector.) Of course, an equivalent formula exists in the time domain.

11.1.1 Pole Parameterizations

In order to be able to adjust the poles of $G_b(z)$, it is necessary to find ways to parameterize them. The obvious way, which is to use the real and imaginary parts of the poles, as in

$$\xi_k = \alpha_k + i\beta_k,$$

has three disadvantages:

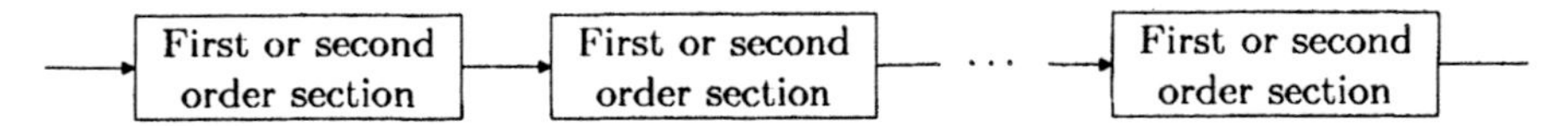

Fig. 11.3. Possible structure of $G_b(z)$ induced by the parameterization of its poles by their real and imaginary parts. The number of first-order sections is n_b and that of second-order sections with complex conjugate poles is n_c. Any permutation of the order of the sections is possible.

1. It is necessary to specify beforehand how many real poles will be used in the parameterization (because the two poles of a pair of complex conjugate poles share the same parameters);
2. The parameterization is not unique, as any other permutation of the poles will give essentially the same model;
3. The allowed range of each parameter may depend on the value of another parameter (the stability of $G_b(z)$ requires that $\alpha_k^2 + \beta_k^2 < 1$).

Nonetheless, there exist situations where this direct parameterization is useful. The determination of the stationarity conditions of the squared error of a GOBF model with respect to (w.r.t.) its pole positions is one of them (see Section 11.3). In order to describe a parameterization of this kind, it is necessary to assume that there are n_r real poles and n_c pairs of complex conjugate poles (with $n_b = n_r + 2n_c$). One possible parameterization for the poles of $G_b(z)$ uses $n_b + n_c$ parameters to specify the real parts of the poles and n_c parameters to specify their imaginary parts. In this case, a balanced realization of $G_b(z)$ can be implemented by cascading first and second-order balanced all-pass sections (*cf.* Figure 11.3), corresponding to real and pairs of complex conjugate poles.

Fortunately, there exists a simple parameterization of the poles of $G_b(z)$ that does not suffer from the shortcomings of the pole parameterization described in the previous paragraph. This quite useful parameterization is related to the Schur algorithm applied to $G_b(z)$ and is also related to the Schur-Cohn stability test applied to the denominator of $G_b(z)$. For stable and causal $G_b(z)$, with real and/or complex conjugate poles, the n_b real parameters of this parameterization, which will be denoted by γ_k, have modulus strictly smaller than one. These parameters are also connected to a specific balanced realization of $G_b(z)$, called the normalized ladder structure [107], depicted in Figures 11.4–11.6. This structure is a feedback connection of balanced realizations of 2-input 2-output all-pass transfer functions. Further details about this remarkable parameterization can be found in Section 2.6.

Using the recursions

$$\begin{bmatrix} N_k(z) \\ D_k(z) \end{bmatrix} = \begin{bmatrix} 1 & -\gamma_k \\ -\gamma_k & 1 \end{bmatrix} \begin{bmatrix} N_{k-1}(z) \\ z\,D_{k-1}(z) \end{bmatrix},$$

with $N_0(k) = D_0(k) = 1$, it can be verified that

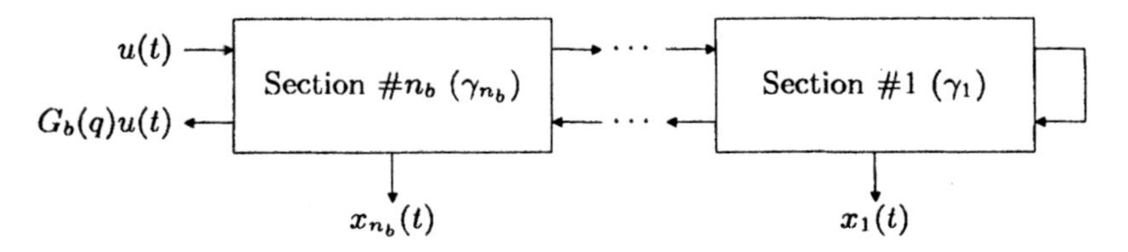

Fig. 11.4. The normalized ladder structure.

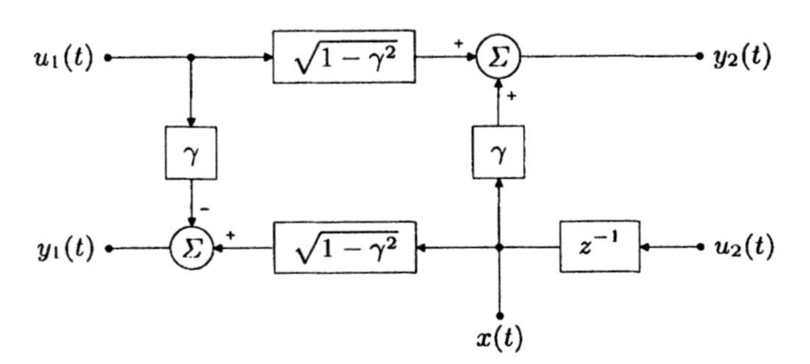

Fig. 11.5. A block (2-input 2-output all-pass section) of the normalized ladder structure; $x(t)$ is the state variable.

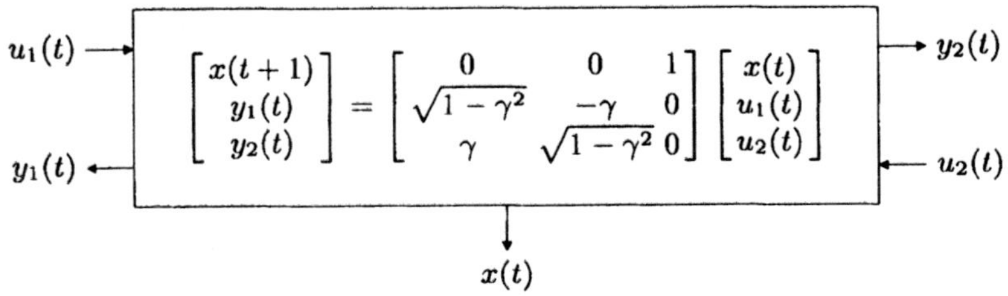

$u_1(t)$ → ; $y_1(t)$ ← ; → $y_2(t)$; ← $u_2(t)$

$$\begin{bmatrix} x(t+1) \\ y_1(t) \\ y_2(t) \end{bmatrix} = \begin{bmatrix} 0 & 0 & 1 \\ \sqrt{1-\gamma^2} & -\gamma & 0 \\ \gamma & \sqrt{1-\gamma^2} & 0 \end{bmatrix} \begin{bmatrix} x(t) \\ u_1(t) \\ u_2(t) \end{bmatrix}$$

↓ $x(t)$

Fig. 11.6. The state space equations of the 2-input 2-output all-pass sections of the normalized ladder structure.

$$\frac{X_k(z)}{U(z)} = \left[\prod_{l=k}^{n_b} \sqrt{1-\gamma_l^2}\right] \frac{N_{k-1}(z)}{D_{n_b}(z)}$$

and that

$$G_b(z) = \frac{N_{n_b}(z)}{D_{n_b}(z)}.$$

$N_k(z)$ and $D_k(z)$ are reciprocal polynomials, *i.e.* $N_k(z) = z^{-k} D_k(1/z)$. For future reference, using this parameterization for $n_b = 1$ yields

$$G_b(z) = \frac{1-\gamma_1 z}{z-\gamma_1},$$

and using it for $n_b = 2$ yields

Table 11.1. Parameters of the nominal test model ($n_0 = 5$)

j	p_j^0	a_j^0
1	$7/9$	$19/9$
2	$(1+i)/2$	$(4+7i)/5$
3	$(1-i)/2$	$(4-7i)/5$
4	$(4+7i)/13$	$(-8-10i)/13$
5	$(4-7i)/13$	$(-8+10i)/13$

$$G_b(z) = \frac{1 - \gamma_1(1-\gamma_2)z - \gamma_2 z^2}{z^2 - \gamma_1(1-\gamma_2)z - \gamma_2}.$$

The first parameterization is the usual one for Laguerre models [313], and the second one is the usual one for two-parameter Kautz models [315].

11.1.2 The Nominal Model Used in All Numerical Examples

All numerical examples presented in this chapter will attempt to approximate a nominal test model with transfer function [237]

$$G_0(z) = \sum_{j=1}^{n_0} \frac{a_j^0}{z - p_j^0}.$$

The parameters of this nominal test model are presented in Table 11.1. One interesting characteristic of this system is that its unit pulse response is always positive.

11.2 Asymptotically Optimal Pole Locations

Assume that for a large enough n it is true that

$$\big\|\varepsilon_g(\,\cdot\,;n+k)\big\| \approx \rho^k \big\|\varepsilon_g(\,\cdot\,;n)\big\|,$$

with $0 \le \rho < 1$. In this case, it is said that the norm of the error of the approximation has an exponential convergence factor of ρ. A small convergence factor corresponds to a fast convergence to zero of the error of the approximation. The (exponential) rate of decay is the inverse of the convergence factor.
In this section, the following problems will be solved:

1. Given the poles of a rational nominal model $G_0(z)$, find the poles of $G_b(z)$ that guarantee the best (largest) rate of decay of the norm of the approximation error of $G_0(z)$ by $G(z;n)$, when n goes to infinity;
2. Given the poles of *several* rational nominal models $G_{0,\iota}(z)$, find the poles of $G_b(z)$ that guarantee the best rate of decay of the norm of the approximation error of *any* of the $G_{0,\iota}(z)$ by $G(z;n)$, when n goes to infinity;

3. Given the (discretized) boundary of a region, or regions, of the complex plane where the poles of the rational nominal model $G_0(z)$ are guaranteed to lie, find the poles of $G_b(z)$ that guarantee the best rate of decay of the norm of the approximation error of any possible $G_0(z)$ by $G(z;n)$, when n goes to infinity;
4. For a given $G_b(z)$, characterize the class of systems for which the decay rate of the approximation error is at least a given number.

All these problems will be solved for a norm induced by an inner product, because that leads to optimization problems that can be easily solved.
The zeros of the nominal models do not appear in the formulation of the first three problems. It will be shown that asymptotically the position of the zeros of the nominal model(s) will not influence the best $G_b(z)$, nor will they influence the best decay rate of the norm of the error of the approximation that can be achieved. Of course, for small n large deviations from the asymptotically optimal $G_b(z)$ are to be expected. It will also be seen that the overall decay rate will be equal to the smallest decay rate for the individual modes of the nominal model(s). Thus, the first three problems posed above are essentially equivalent, as it is easy to solve the second problem by solving the first one for a system with the poles of all the nominal models, and it is easy to solve the third problem by solving the first one for a nominal model whose poles are the boundary points of the third problem. Moreover, the fourth problem also admits a simple solution: the decay rate for each one of the modes of the system has to be larger or equal to the specified limit.

11.2.1 The Decay Rates of $\|\varepsilon_y(\cdot\,;n)\|$ and $\|\varepsilon_g(\cdot\,;n)\|$

(Recall that ε_y is the output error and that ε_g is the impulse response error.) Suppose that it is necessary to find the best approximation $G(z;n)$ to $G_0(z)$ using the minimization of the value of $\|\varepsilon_y(\cdot\,;n)\|$ as the optimization criterion. It is possible to estimate $\|\varepsilon_g(\cdot\,;n)\|$ from $\|\varepsilon_y(\cdot\,;n)\|$, and *vice versa*, if it is known that

$$0 < \Phi^- = \min_\omega \Phi_u(e^{i\omega}) \le \Phi_u(e^{i\omega}) \le \max_\omega \Phi_u(e^{i\omega}) = \Phi^+ < \infty. \tag{11.2}$$

In this case, because

$$\|\varepsilon_y(\cdot\,;n)\|^2 = \frac{1}{2\pi}\int_{-\pi}^{\pi} \left|G_0(e^{i\omega}) - G(e^{i\omega};n)\right|^2 \Phi_u(e^{i\omega})\,d\omega + \sigma^2,$$

where $\sigma^2 = \sigma_0^2\,\langle h_0, h_0\rangle$ is the variance of the uncorrelated noise $v(t)$, it follows that

$$\Phi^-\|\varepsilon_g(\cdot\,;n)\|^2 \le \|\varepsilon_y(\cdot\,;n)\|^2 - \sigma^2 \le \Phi^+\|\varepsilon_g(\cdot\,;n)\|^2$$

and that

$$\frac{\|\varepsilon_y(\cdot\,;n)\|^2 - \sigma^2}{\Phi^+} \le \|\varepsilon_g(\cdot\,;n)\|^2 \le \frac{\|\varepsilon_y(\cdot\,;n)\|^2 - \sigma^2}{\Phi^-}.$$

These inequalities are valid for any choice of the coefficients $c_{n,k}$ (or $K_{n,k}$). The choice of these coefficients that minimizes $\|\varepsilon_y(\cdot\,;n)\|$ will not, in general, minimize at the same time $\|\varepsilon_g(\cdot\,;n)\|$, and *vice versa*. Thus, it follows that (in the next two formulas $c_n = [\, c_{n,0} \ \cdots \ c_{n,n} \,]$)

$$\Phi^- \min_{c_n} \|\varepsilon_g(\cdot\,;n)\|^2 \le \min_{c_n} \|\varepsilon_y(\cdot\,;n)\|^2 - \sigma^2 \le \Phi^+ \min_{c_n} \|\varepsilon_g(\cdot\,;n)\|^2,$$

and that

$$\frac{\min_{c_n} \|\varepsilon_y(\cdot\,;n)\|^2 - \sigma^2}{\Phi^+} \le \min_{c_n} \|\varepsilon_g(\cdot\,;n)\|^2 \le \frac{\min_{c_n} \|\varepsilon_y(\cdot\,;n)\|^2 - \sigma^2}{\Phi^-}.$$

The main conclusion that can be drawn from these inequalities is that the average rates of decay to zero of $\|\varepsilon_y(\cdot\,;n)\|^2 - \sigma^2$ and of $\|\varepsilon_g(\cdot\,;n)\|^2$ are the same. This argument also shows that provided (11.2) is satisfied and provided the orthonormal set $\{f_k(t)\}_{k=1}^{\infty}$ is complete, the convergence of $g(t;n)$, obtained from the minimization of $\|\varepsilon_y(\cdot\,;n)\|$, to $g_0(t)$ is assured when n goes to infinity.

◇ The average decay rates of $\|\varepsilon_g(\cdot\,;n)\|^2$ and $\|\varepsilon_y(\cdot\,;n)\|^2 - \sigma^2$ are the same.

11.2.2 The Decay Rate of $\|\varepsilon_g(\cdot\,;n)\|$ when $G_0(z)$ Is Rational

Because the average decay rates to zero of $\|\varepsilon_y(\cdot\,;n)\|^2 - \sigma^2$ and of $\|\varepsilon_g(\cdot\,;n)\|^2$ are the same, it is only necessary to study one of them. The one that is easier to analyse is $\|\varepsilon_g(\cdot\,;n)\|^2$, because it does not depend of the power spectral density of the input signal.
The decay rate of $\|\varepsilon_g(\cdot\,;n)\|$, which is the square root of the decay rate of $\|\varepsilon_g(\cdot\,;n)\|^2$, will only be computed here for nominal models with rational transfer functions. There are two reasons for this choice, namely,

1. For this kind of nominal models the rate of decay is easy to compute;
2. For this kind of nominal models $\|\varepsilon_g(\cdot\,;n)\|$ converges exponentially to zero (the use of a decay rate thus makes sense).

Let the transfer function of the (stable) nominal model be given by

$$G_0(z) = \sum_j \frac{a_j^0}{z - p_j^0}$$

(the number of poles is irrelevant to the present discussion). To simplify the following argument, it will be assumed that $G_0(z)$ has only simple poles; the results that will be obtained remain true if $G_0(z)$ has multiple poles, but their deduction becomes more involved. Assuming simple poles, the coefficients K_k

of the orthonormal expansion (11.1) can be easily evaluated using the residue theorem. The final result, when $g_0(t)$ is a real unit pulse response, is given by

$$K_k^T = \sum_j (a_j^0)^* V_k^*(1/p_j^0) = \sum_j (a_j^0)^* V_1^*(1/p_j^0) \left[G_b(1/p_j^0)\right]^{k-1}.$$

When n increases by one, the modulus of each one of the terms of this formula decreases by a factor of $|G_b(1/p_j^0)|$, which is a number smaller than 1 because the modulus of $G_b(z)$ is smaller than 1 for any z outside of the closed unit circle. Thus, for a large enough k the term, or terms, with the largest $|G_b(1/p_j^0)|$ will dominate the convergence factor (inverse of the decay rate) of K_k^T to the zero vector. The zeros of $G_0(z)$ influence only the residues a_j^0. Thus, they do not influence the asymptotic decay rate of K_k^T. With the definition

$$\rho = \max_j |G_b(1/p_j^0)|$$

(note that ρ is precisely the quantity that should be minimized to reduce the bias in the identification problem), it becomes possible to put K_k in the form

$$K_k^T = \rho^k \sum_j (a_j^0)^* V_1^*(1/p_j^0) \left[\frac{G_b(1/p_j^0)}{\rho}\right]^{k-1}. \tag{11.3}$$

Each number inside square brackets belongs to either $\mathbb{T}$ ($|z| = 1$) or to $\mathbb{D}$ ($|z| < 1$), with at least one belonging to $\mathbb{T}$. Thus, the summation present in the previous formula represents a function that is bounded for all k, and which does not converge to the zero vector; it is in general a pseudo-periodic function. As a result, for large values of k the average convergence factor of K_k is given by ρ.
The previous result can be used to infer the average convergence factor of the square of the norm of the approximation error, which is given by

$$\left\|\varepsilon_g(\cdot\,;n)\right\|^2 = \|g_0\|^2 - \sum_{k=1}^{n} K_k^T K_k = \sum_{k=n+1}^{\infty} K_k^T K_k.$$

From (11.3) it can be inferred that $K_k^T K_k$ behaves, for large k and ignoring pseudo-periodic fluctuations, as a geometric progression. Without entering into rigorous mathematical details, it can then be expected that the use of one more coefficient in the orthonormal expansion will decrease the square of the norm of the error by a factor of ρ^2 (on the average). This error decrease rate will be higher for some values of n and lower for others; this will depend on the pseudo-periodic function.

◇ For large n, $\left\|\varepsilon_g(\cdot\,;n)\right\|$ decreases on the average by a factor of ρ for each additional all-pass section used in the orthonormal expansion. The decay rate per each additional coefficient is thus $-(20\log_{10}\rho)/n_b$ dB.

The second and third problems posed at the beginning of Section 11.2 can be solved in exactly the same way as was done above for the first problem. Let

$$G_0(z;l) = \sum_j \frac{a_{lj}^0}{z - p_{lj}^0}$$

be the transfer function of the lth (stable) nominal model, and let

$$\rho(l) = \max_j \left| G_b(1/p_{lj}^0) \right|.$$

be its corresponding convergence factor. Finally, let

$$\rho = \max_l \rho(l) = \max_l \max_j \left| G_b(1/p_{l;j}^0) \right|$$

be the overall convergence factor. From the way ρ is computed it is clear that it is irrelevant if the poles p_{lj}^0 came from just one nominal model or from several nominal models. What matters is that ρ is equal to the largest convergence factor of all the individual modes (poles) of the nominal model(s).

This makes the first two problems equivalent. The same can be said about the third problem. Indeed, the maximum of $\left|G_b(1/z)\right|$ when z belongs to a closed region of $\mathbb{D}$ must occur in the boundary of that region (maximum modulus principle). If a discretized set of points of this boundary is used in the computation of ρ, the formulation of the third problem is exactly of the same form as that of the first problem, with the boundary points taking the role of the poles of the nominal model.

> ◇ Problems one, two, and three, as described in the beginning of Section 11.2, are equivalent.

11.2.3 Minimization of ρ for the Laguerre Model

(The continuous-time version of this problem was treated in [38] and in [265].) For the case $n_b = 1$ (Laguerre model), the second parameterization of the poles of an all-pass transfer function described in Section 11.1.1 gives

$$G_b(z) = \frac{N_1(z)}{D_1(z)} = \frac{1 - \gamma_1 z}{z - \gamma_1} = \frac{z^{-1} - \gamma_1}{1 - \gamma_1 z^{-1}}.$$

In this case, ρ is a function of γ_1 and is given by

$$\rho(\gamma_1) = \max_j \left| \frac{p_j^0 - \gamma_1}{1 - \gamma_1 p_j^0} \right|.$$

Because γ_1 is real, only poles of the model with non-negative imaginary parts need to be considered in this formula. The minimization of $\rho(\gamma_1)$ can be found using analytic means. To this end, let

$$J(\gamma_1) = \arg\max_j \left| \frac{p_j^0 - \gamma_1}{1 - \gamma_1 p_j^0} \right|$$

be the index, or the first of the indexes, for which $\left|(p_j^0 - \gamma_1)/(1 - \gamma_1 p_j^0)\right|$ attains its largest value for a given value of γ_1. It follows that

$$\rho(\gamma_1) = \left| \frac{p_{J(\gamma_1)}^0 - \gamma_1}{1 - \gamma_1 p_{J(\gamma_1)}^0} \right|.$$

In the interval $(-1, 1)$ the function $J(\gamma_1)$ is piecewise constant. Thus, to find the global minimum of $\rho(\gamma_1)$, it is only necessary to analyse the value of this function in two sets of points: the set of points where $J(\gamma_1)$ is discontinuous, and the set of points where each of the functions $\left|(p_j^0 - \gamma_1)/(1 - \gamma_1 p_j^0)\right|$ has a (global) minimum. The first of these two sets can be obtained by solving the equations

$$\left| \frac{p_j^0 - \gamma_1}{1 - \gamma_1 p_j^0} \right| = \left| \frac{p_k^0 - \gamma_1}{1 - \gamma_1 p_k^0} \right|$$

for $1 \leq j \leq n_0$ and $j < k \leq n_0$, and with $p_k^0 \neq (p_j^0)^*$. The second set can be obtained by solving the minimization problems

$$\arg\min_{\gamma_1} \left| \frac{p_j^0 - \gamma_1}{1 - \gamma_1 p_j^0} \right|$$

for $1 \leq j \leq n_0$. It is possible to verify that the solution of both types of problems involve the determination of the roots of second-order reciprocal polynomials with real coefficients, respectively

$$\left[|p_k^0|^2 - |p_j^0|^2\right]\gamma_1^2 + 2\left[\mathrm{Re}[p_j^0](1 - |p_k^0|^2) - \mathrm{Re}[p_k^0](1 - |p_j^0|^2)\right]\gamma_1 + \left[|p_k^0|^2 - |p_j^0|^2\right] = 0 \tag{11.4}$$

and

$$\mathrm{Re}[p_j^0]\gamma_1^2 - (1 + |p_j^0|^2)\gamma_1 + \mathrm{Re}[p_j^0] = 0. \tag{11.5}$$

Each problem will then have at most one real solution inside the interval $(-1, 1)$. In some cases no useful solution to (11.4) may exist.

◇ For the Laguerre model, the minimization of $\rho(\gamma_1)$ can be found analytically, and involves the analysis of a finite number of values of this function. The values of γ_1 that have to be analysed are the solutions of the equations (11.4) and (11.5) that belong to the interval $(-1, 1)$.

Table 11.2. Data required to determine the global minimum of $\rho(\gamma_1)$. Rows with a k value are solutions of (11.4), and those without one are solutions of (11.5)

j	k	γ_1	$\rho(\gamma_1)$	$\rho(\gamma_1)$ (in dB)
0		0.777778	0.823154	−1.690382
0	1	0.298649	0.624095	−4.094985
0	3	0.345294	0.619265	−4.162469
1		0.381966	0.618034	−4.179753
1	3	0.451416	0.622868	−4.112075
3		0.234436	0.664508	−3.549999

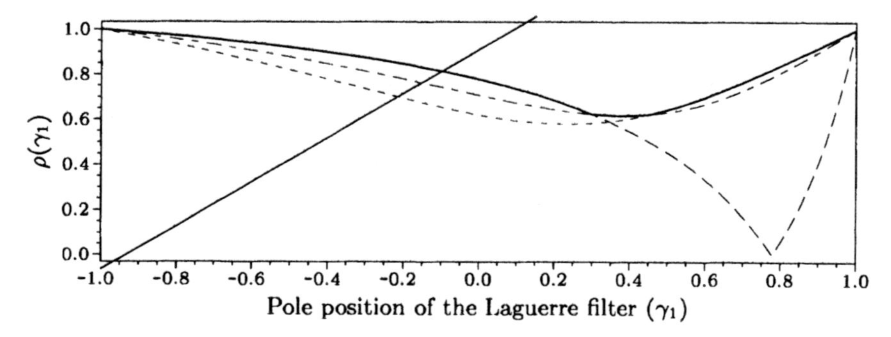

Fig. 11.7. The graph of $\rho(\gamma_1)$, drawn with a thick line, and the graphs of the functions $|G_b(1/p_j^0)|$, drawn with different dash patterns.

The functions

$$\left| \frac{p_j^0 - \gamma_1}{1 - \gamma_1 p_j^0} \right|$$

are unimodal, *i.e.* they have only one minimum in the interval $(-1, 1)$. As a consequence, $\rho(\gamma_1)$ will also be an unimodal function. It is thus possible to use a quasi-convex optimization procedure to find the best value of γ_1.

An Example

The data required to minimize $\rho(\gamma_1)$ for the nominal test model of Section 11.1.2 is presented in Table 11.2. A perusal of this table reveals that the global minimum of $\rho_1(\gamma_1)$ occurs for the data presented in the row with $j = 1$ without a value of k. Figure 11.7 presents a graph of $\rho(\gamma_1)$, together with those of the convergence factors of each mode of the nominal test model. It is instructive to compare the best convergence factor of the Laguerre model, which in this case is 0.618034, with the convergence factor of a FIR model (*i.e.* a Laguerre model with $\gamma_1 = 0$), which in this case is 0.777778. For the Laguerre model, the norm of the approximation error will then decrease to zero much faster that that of the FIR model. For large orders and the same approximation error, the Laguerre model will require a number of coefficients that is only $100 \log(0.777778)/\log(0.618034) \approx 52\%$ of the number of coefficients required for a comparable FIR model.

11.2.4 Minimization of ρ for the GOBF Model with $n_b > 1$

Let $\gamma = [\gamma_1 \; \cdots \; \gamma_{n_b}]$. The minimization of the continuous function

$$\rho(\gamma) = \max_j \left| \frac{N_{n_b}(1/p_j^0)}{D_{n_b}(1/p_j^0)} \right|$$

for a GOBF model with $n_b > 1$ is considerably more difficult than the minimization of $\rho(\gamma)$ for the Laguerre model. There are three main reasons for this, namely,

1. The minimization involves several independent variables;
2. $\rho(\gamma)$ may have local minima;
3. In some regions of the search space, including the neighborhood of local and/or global minima, $\rho(\gamma)$ is not differentiable.

Some minimization routines experience difficulties with this last property of $\rho(\gamma)$. The non-differentiability of $\rho(\gamma)$ at some points of the search space, which in the example given in the Laguerre case is already apparent in Figure 11.7, is a consequence of the fact that the function

$$J(\gamma) = \arg\max_j \left| \frac{N_{n_b}(1/p_j^0)}{D_{n_b}(1/p_j^0)} \right|$$

is piecewise constant in the unit hypercube $(-1, 1)^{n_b}$. Each minimum of $\rho(\gamma)$ occurs either in the interior of one of the regions where $J(\gamma)$ is constant, or it occurs in a point where $J(\gamma)$ is discontinuous. In the second case, the search for local minima is somewhat more difficult. For example, for $n_b = 2$ the optimization routine may be forced to descend some curvilinear v-shaped valleys (*cf.* Figure 11.8).

> ◇ For the GOBF model with $n_b > 1$, the minimization of $\rho(\gamma)$ must be done numerically. $\rho(\gamma)$ may have local minima. The minimization routine is likely to encounter points (quite probably even the global minimum) where $\rho(\gamma)$ is non-differentiable.

The poles of $G_b(z)$ found by the minimization of $\rho(\gamma)$ are asymptotically optimal, in the sense that when the order of the model increases, the optimal poles of $G_b(z)$ will converge to these values. Given the way $\rho(\gamma)$ is computed, it can be expected that the smallest value of this function will be closer to zero when the poles of the nominal model form at most n_b clusters. In this case, each pole of $G_b(z)$ can be used to reduce significantly the convergence factor of poles belonging to one cluster. The n_b poles of $G_b(z)$ can then cover adequately the n_b clusters. This observation may serve as a rough guideline to select a value of n_b that is smaller than the order of the nominal model.

In order to compare several values of n_b, the convergence factor per coefficient should be used. Otherwise the comparison will not be fair for the smaller values of n_b, which use fewer coefficients per all-pass section.

◇ To select the best value of n_b among several possible candidates, a fair figure of merit is $20\log\bigl(\min_\gamma \rho(\gamma)\bigr)/n_b$.

An Example

The results of the minimization of $\rho(\gamma)$ for $n_b = 1,\ldots,4$, and for the nominal test model of Section 11.1.2 are presented in Table 11.3. (The last one or two digits of the results for $n_b = 4$ may be inaccurate.) For $n_b = 5$, $\rho(\gamma)$ will vanish when the poles of $G_b(z)$ coincide with those of the nominal test model.
A perusal of this table reveals that when n_b increases, the best convergence factor can decrease significantly. However, the convergence factor per coefficient may not decrease. Thus, when n_b is smaller than the number of poles of the nominal model, it may happen that a larger value of n_b may be worse than some lower value. For example, for the nominal test model $n_b = 4$ is worse than $n_b = 3$. Of course, in this example $n_b = 5$ is the best possible choice. Note that when n_b increases the poles of the best $G_b(z)$ approach the poles of the nominal model.
To give an idea of the shape of $\rho(\gamma)$, in Figure 11.8 some contour lines of this function for the nominal test model and for $n_b = 2$ are presented. Observe

Table 11.3. The results of the minimization of $\rho(\gamma)$ for the nominal test model. The numbers inside parentheses in the third column are the convergence factors (in dB) divided by n_b

n_b	$\rho(\gamma)$	$\rho(\gamma)$ (in dB)	k	γ_k	ξ_k
FIR	0.777778	−2.182889	1	0.000000	0.000000
1	0.618034	−4.179753	1	0.381966	0.381966
2	0.340951	−9.346172	1	0.693791	$0.419460 + 0.182306i$
		(−4.673086)	2	−0.209182	$0.419460 - 0.182306i$
3	0.111915	−19.022242	1	0.836695	0.696383
		(−6.340747)	2	−0.592965	$0.404253 + 0.503254i$
			3	0.290172	$0.404253 - 0.503254i$
4	0.067604	−23.400551	1	0.882211	0.344364
		(−5.850138)	2	−0.695266	0.694331
			3	0.448079	$0.406609 + 0.502784i$
			4	−0.099974	$0.406609 - 0.502784i$

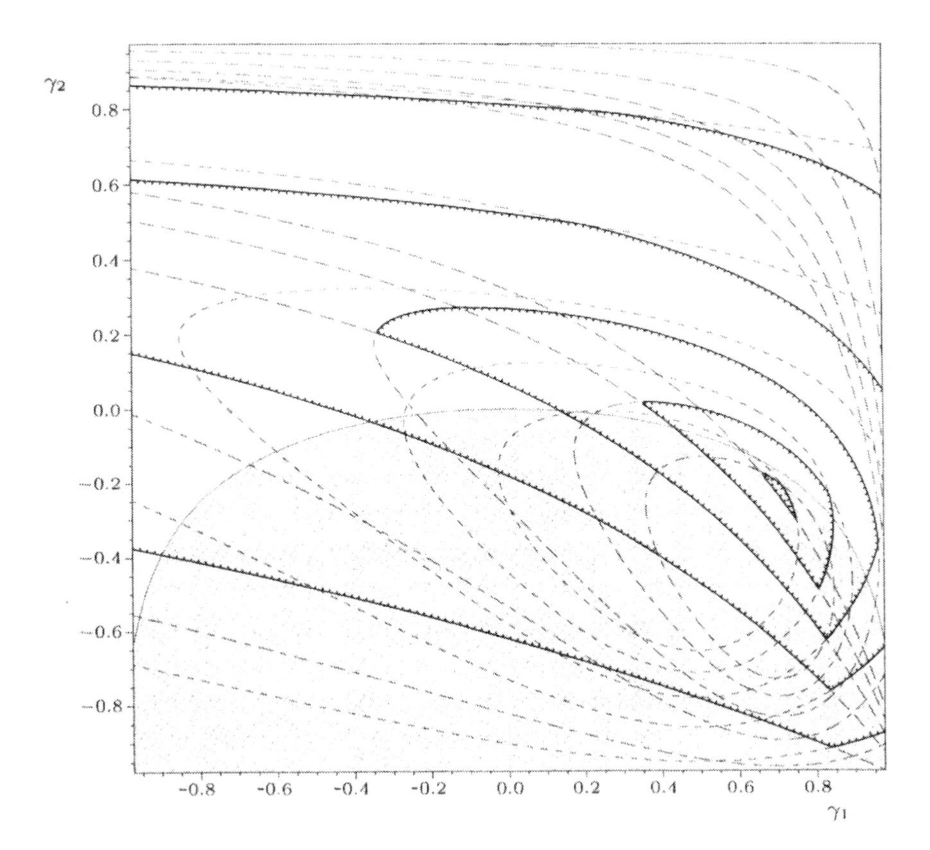

Fig. 11.8. Contour lines of the function $\rho(\gamma_1, \gamma_2)$ for -1, -3, -5, -7, and -9 dB, drawn in thick 'barbed' lines, and of the functions $|G_b(1/p_j^0)|$, drawn with different dash patterns. The shaded area corresponds to complex conjugate poles.

that the global minimum of $\rho(\gamma)$, which occurs for $\gamma_1 \approx 0.7$ and $\gamma_2 \approx -0.2$, lies in the intersection of three valleys (which manifest themselves as sudden changes in direction of the contour lines). This makes the convergence factors of all modes of the nominal test model equal at this global minimum. (A similar situation occurs for $n_b = 3$ and $n_b = 4$.) In these cases, $\rho(\gamma)$ is not differentiable at the global minimum, which makes the determination of its exact position somewhat more difficult.

11.2.5 A Kolmogorov n-Width Interpretation of $\rho(\gamma)$

This subsection will address the last problem presented in Section 11.2, which asks for a characterization of the class of systems for which $\rho(\gamma) \leq \rho_{\max} < 1$. From Section 11.2.2 it is known that

$$\rho(\gamma) = \max_j \left| \frac{N_{n_b}(1/p_j^0)}{D_{n_b}(1/p_j^0)} \right| = \max_j \left| \frac{D_{n_b}(p_j^0)}{N_{n_b}(p_j^0)} \right| .$$

It is then clear that $\rho(\gamma) \leq \rho_{\max}$ if and only if all the poles of the nominal model belong to the region

$$\mathcal{Z}(\gamma) = \left\{ z \in \mathbb{C} \mid \left| G_b(z^{-1}) \right| \leq \rho_{\max} \right\}$$

of the complex plane. This result admits a nice theoretical interpretation in terms of what is known as the Kolmogorov n-width [235, 251, 319].

To simplify the presentation, the Kolmogorov n-width will only be discussed in a Hilbert space setting; the concept can also be applied to Banach spaces. Let $\Phi = \{\phi_k\}_{k=1}^{\infty}$ be a complete set of linearly independent elements of a Hilbert space $\mathcal{H}$, and let Φ_n be the linear subspace of $\mathcal{H}$ spanned by the first n elements of Φ. The distance between $G \in \mathcal{H}$ and Φ_n is defined by

$$e_{\mathcal{H}}(G; \Phi_n) = \min_{\phi \in \Phi_n} \|G - \phi\|_{\mathcal{H}},$$

where $\|\cdot\|_{\mathcal{H}}$ denotes the norm of $\mathcal{H}$ (induced by its inner product). Because Φ is a complete set, $\min_{n\to\infty} e_{\mathcal{H}}(G; \Phi_n) = 0$ for all $G \in \mathcal{H}$. But, by choosing G properly, $e_{\mathcal{H}}(G; \Phi_n)$ can be made to converge to zero as slowly as desired. For this reason, in any practical situation each set Φ should be used only to approximate systems belonging to the subset of $\mathcal{H}$ for which the norm of the approximation error converges to zero 'quickly enough'.

Let $\mathcal{S}$ be a bounded set of $\mathcal{H}$, and let $M_n(\mathcal{H})$ be the collection of all n-dimensional subspaces of $\mathcal{H}$. The Kolmogorov n-width of $\mathcal{S}$ in $\mathcal{H}$, denoted by $d_n(\mathcal{S}; \mathcal{H})$, is defined by

$$d_n(\mathcal{S}; \mathcal{H}) = \inf_{\Phi_n \in M_n(\mathcal{H})} \sup_{G \in \mathcal{S}} e_{\mathcal{H}}(G; \Phi_n).$$

◇ The Kolmogorov n-width is the smallest possible norm of the error of approximation for the worst possible G of $\mathcal{S}$.

A subspace Ψ_n of dimension at most n satisfying

$$\sup_{G \in \mathcal{S}} e_{\mathcal{H}}(G; \Psi_n) = d_n(\mathcal{S}; \mathcal{H})$$

is called an optimal subspace for $d_n(\mathcal{S}; \mathcal{H})$. Let $\mathcal{S}(\gamma; M)$ be the set of all transfer functions with poles belonging to $\mathcal{Z}(\gamma)$, and for which the line integral of the square of their modulus in the boundary of $\mathcal{Z}(\gamma)$ is not larger than M. (This last technical condition is required to bound the norm of the elements of $\mathcal{S}(\gamma; M)$.) Then, it turns out [235] that the space spanned by $\left\{V_k(z)\right\}_{k=1}^{n}$

is an optimal subspace of $\mathcal{H}_2$ in the Kolmogorov $n_b n$-width sense for the set $\mathcal{S}(\gamma; M)$ of transfer functions.

◇ If the only available information about the nominal model is that all its poles belong to $\mathcal{Z}(\gamma)$ then the subspace spanned by $\{V_k(z)\}_{k=1}^{n}$ is the best one that can be used to approximate that nominal model. Best here means that the approximation will have the smallest possible norm for the worst possible nominal model with a given unit pulse response norm.

For example, if it is known that the nominal model has a certain degree of exponential stability, meaning that its poles belong to the disk $|z| < \rho_{\max}$, then the FIR model is the best model that can be used [178, 319]. If that region is a disk centered at a point of the real axis, then a Laguerre model is optimal [319]. In general, a GOBF will be optimal if all poles of the nominal model belong to $\mathcal{Z}(\gamma)$ [235].

An Example

Figure 11.9 presents the shape of the set $\mathcal{Z}(\gamma)$ for the data of Table 11.3. Observe that for $n_b = 3$, the set $\mathcal{Z}(\gamma)$ is disconnected. According to Table 11.3, it is for this case that the convergence factor per coefficient is the smallest (*i.e.* the decay rate is the largest); it is thus the most favorable case. This reflects the fact that in this case the poles of the nominal test model can be grouped in 3 'clusters'. For $2 \leq n_b \leq 4$ all the poles of the nominal test model are in the boundary of $\mathcal{Z}(\gamma)$, confirming that the convergence factors are equal for all of them.

11.3 Optimal Pole Location Conditions

Let $\|\cdot\|_p$ denote a p-norm, with $1 < p < \infty$, defined either in the time or in the frequency domains. The pth power of the p-norm is given by

$$\|x\|_p^p = \sum_{t=0}^{\infty} |x(t)|^p$$

for time-domain causal signals, and by

$$\|X\|_p^p = \frac{1}{2\pi} \int_{-\pi}^{\pi} |X(e^{i\omega})|^p \, d\omega$$

for frequency-domain signals. These two norms give identical results when $p = 2$ (Parseval's theorem).

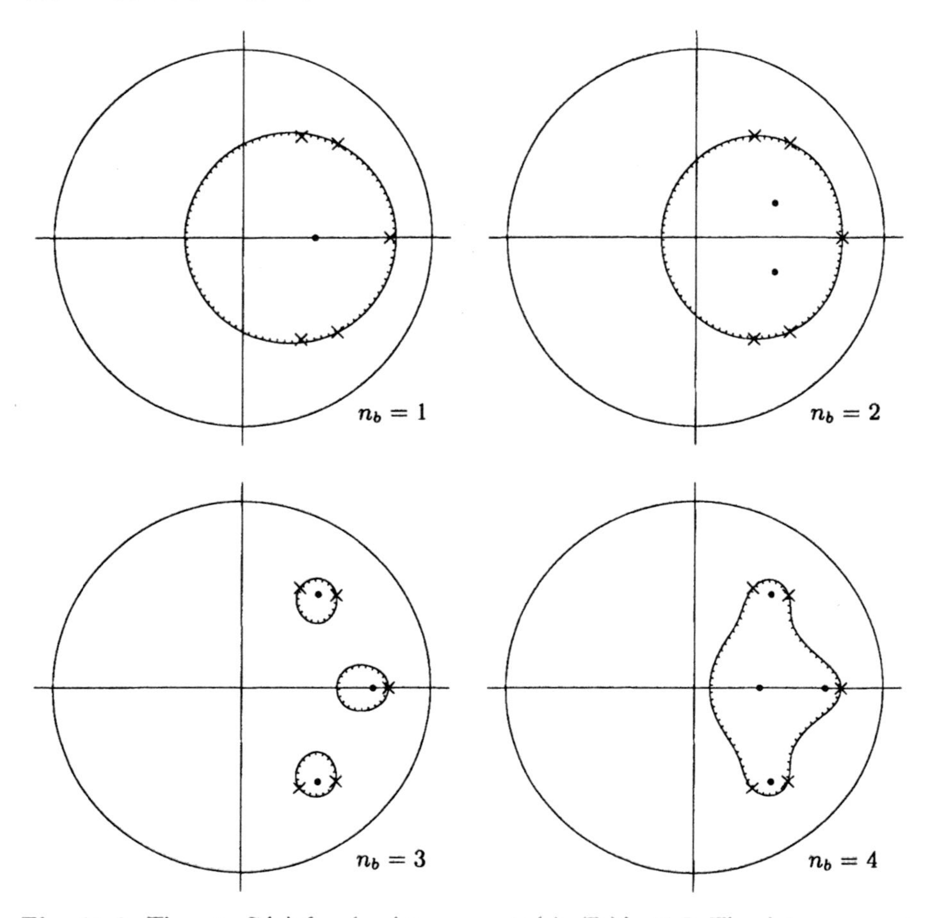

Fig. 11.9. The sets $\mathcal{Z}(\gamma)$ for the data presented in Table 11.3. The dots represent the poles of $G_b(z)$, and the crosses represent the poles of the nominal test model.

Other chapters of this book describe ways to estimate the coefficients K_k of the GOBF model, based on a possibly noise corrupted input/output data record of the system to be identified. In these procedures, the poles of the GOBF model are pre-specified. This section states the conditions that the poles of the GOBF model must satisfy in order for them to be optimal. More precisely, this section presents a detailed description of the solutions to the following two problems:

1. Find the conditions that the poles of the Laguerre or two-parameter Kautz model must satisfy in the global minimum of the time-domain or frequency-domain p-norm of the output error signal $\varepsilon_y(t;n)$, when the model's impulse response or frequency response is constrained to satisfy zero or more interpolation conditions;

2. Find the conditions that the poles of the GOBF model must satisfy in the global minimum of $\|\varepsilon_y(\cdot\,;n)\|_2$.

These are non-linear optimization problems, and so local minima may be encountered; the (stationarity) conditions that the optimal poles must satisfy are also satisfied at local minima, maxima, and saddle points of the optimization criterion.
The form of the solution of the first problem is the same, irrespective of the value of p (as long as $1 < p < \infty$), and irrespective of the interpolation constraints being forced (assuming that the number of coefficients is larger than the number of constraints). The interpolation constraints can be used to include some *a priori* information about the system to be identified in the model, such as, for example,

1. A given DC gain;
2. A given relative order;
3. Zeros of the frequency response at given frequencies;
4. Given gain and/or phase margins at given frequencies.

The second problem addresses only the case $p = 2$. Presumably, the form of the solution to this problem remains the same when other values of p are used. Let $K_{n,k}$ be the coefficients of the GOBF model with n all-pass sections. The solutions to the first problem have the simple form $K_{n,n} = 0$ or $K_{n+1,n+1} = 0$; the first of these conditions is usually associated with maxima, and the second to minima. These are also solutions to the second problem, but, unfortunately, in this case solutions with other forms are also possible.

11.3.1 Optimal Pole Conditions for the Laguerre Model

Consider first that $p = 2$ and that the input signal is white noise with unit variance. In this case $\Phi_u(e^{i\omega}) = 1$, and so $\|\varepsilon_y(\cdot\,;n)\| = \|\varepsilon_g(\cdot\,;n)\|$. Because $p = 2$, it is necessary to analyze the local minima of the truncated Laguerre series expansion of $G_0(z)$. Let

$$G_0(z) = \sum_{k=1}^{\infty} c_k(\gamma_1)\, L_k(z;\gamma_1)$$

be the Laguerre series expansion of a strictly proper $G_0(z)$. In this expansion

$$L_k(z;\gamma_1) = \sqrt{1-\gamma_1^2}\,\frac{(1-\gamma_1 z)^{k-1}}{(z-\gamma_1)^k}$$

are the Z transform of the Laguerre sequences $l_k(t;\gamma_1)$, and

$$c_k(\gamma_1) = \langle G_0, L_k(\cdot\,;\gamma_1)\rangle = \langle g_0, l_k(\cdot\,;\gamma_1)\rangle.$$

Note the dependence on γ_1, which is the pole position of the Laguerre model, of the coefficients of the expansion. Let

$$\varepsilon_g(t;n,\gamma_1) = g_0(t) - \sum_{k=1}^{n} c_k(\gamma_1)\, l_k(t;\gamma_1)$$

be the error of the approximation of $g_0(t)$ by the first n terms of the Laguerre expansion. It is known that

$$\left\|\varepsilon_g(\,\cdot\,;n,\gamma_1)\right\|^2 = \|g_0\|^2 - \sum_{k=1}^{n} c_k^2(\gamma_1).$$

The determination of the first derivative of $\left\|\varepsilon_g(\,\cdot\,;n,\gamma_1)\right\|^2$ w.r.t. γ_1 was addressed in [56, 147] (continuous-time case) and in [182, 229] (discrete-time case). This derivative is given by

$$\frac{\partial \left\|\varepsilon_g(\,\cdot\,;n,\gamma_1)\right\|^2}{\partial \gamma_1} = -2\sum_{k=1}^{n} c_k(\gamma_1)\,\frac{\partial\, c_k(\gamma_1)}{\partial \gamma_1}. \tag{11.6}$$

Because

$$\frac{\partial\, L_k(z;\gamma_1)}{\partial \gamma_1} = \frac{k\, L_{k+1}(z;\gamma_1) - (k-1)\, L_{k-1}(z;\gamma_1)}{1-\gamma_1^2}, \tag{11.7}$$

and because the inner product is linear (for real signals) in its second argument, it follows that

$$\frac{\partial\, c_k(\gamma_1)}{\partial \gamma_1} = \frac{k\, c_{k+1}(\gamma_1) - (k-1)\, c_{k-1}(\gamma_1)}{1-\gamma_1^2}. \tag{11.8}$$

Using this formula in (11.6) yields

$$\frac{\partial \left\|\varepsilon_g(\,\cdot\,;n,\gamma_1)\right\|^2}{\partial \gamma_1} = -\frac{2n}{1-\gamma_1^2}\, c_n(\gamma_1)\, c_{n+1}(\gamma_1).$$

Equating this derivative to zero gives the condition

$$c_n(\gamma_1)\, c_{n+1}(\gamma_1) = 0, \tag{11.9}$$

i.e. either $c_n(\gamma_1) = 0$ or $c_{n+1}(\gamma_1) = 0$ (or both). It makes sense that at local minima of $\left\|\varepsilon_g(\,\cdot\,;n,\gamma_1)\right\|$ the condition $c_{n+1}(\gamma_1) = 0$ be the one satisfied, as this coefficient is the first one not used in the expansion; in the other case, the order n model will be equal to the order $n-1$ model, as the last coefficient used in the expansion will be zero. Although the condition $c_{n+1}(\gamma_1) = 0$ is the one almost always satisfied at local minima, it may happen that a local minimum, and even the global minimum, may satisfy $c_n(\gamma_1) = 0$.

◇ The stationary points of the 2-norm of the error of a truncated Laguerre expansion with n terms satisfy the condition $c_n(\gamma_1)\, c_{n+1}(\gamma_1) = 0$. At local minima of $\left\|\varepsilon_g(\,\cdot\,;n,\gamma_1)\right\|$ usually $c_{n+1}(\gamma_1) = 0$, and at local maxima usually $c_n(\gamma_1) = 0$.

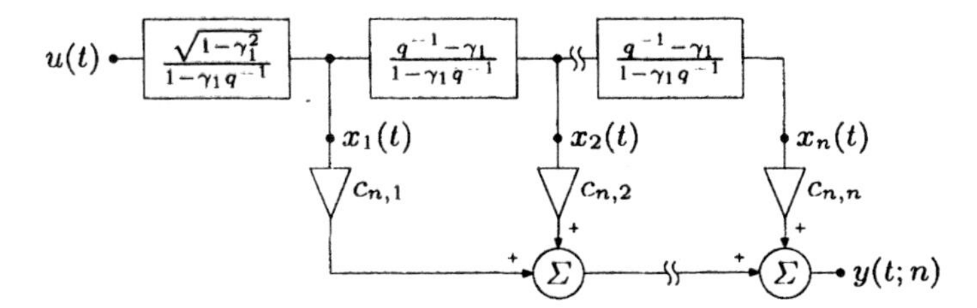

Fig. 11.10. The Laguerre model in transversal form.

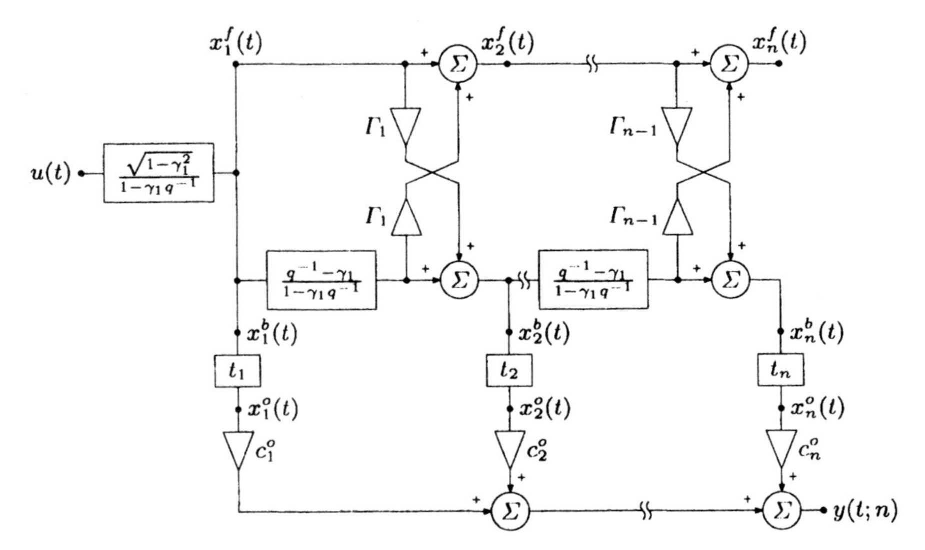

Fig. 11.11. The Laguerre model in lattice/ladder form.

Consider now the case of a general input signal (but still with $p = 2$). It turns out that the analysis of the problem is considerably simplified if the signals (states) of the Laguerre model, which is depicted in Figure 11.10, are orthogonalized (using, for example, the Gram-Schmidt procedure). The details of the orthogonalization of these signals [232], which relies on the Toeplitz structure of the correlation matrix of the states of the Laguerre model [313], will not be described here (*cf.* Section 11.6). The end result of this orthogonalization is the Laguerre model in lattice/ladder form, which is depicted in Figure 11.11. Referring to Figures 11.10 and 11.11, the signals $x_k^b(t)$ of the lattice/ladder model are the orthogonalized versions of the signals $x_k(t)$ of the transversal model. The signals $x_k^o(t)$ are the normalized versions of the signals $x_k^b(t)$, *i.e.* $t_k = 1/\|x_k^b\|$. They hence form an orthonormal set (it is assumed here that the signals $x_k(t)$ of the Laguerre model are linearly independent). The so-called reflection coefficients Γ_k of the lattice structure can be computed from the requirement that $\langle x_k^b, x_{k+1}^b \rangle = 0$. It is possible to verify that

$$\|x_{k+1}^b\|^2 = (1 - \Gamma_k^2)\,\|x_k^b\|^2,$$

which implies that $|\Gamma_k| \le 1$ (the equality sign can only occur if the first $k+1$ signals of the Laguerre model are linearly dependent).
The output signal of the Laguerre lattice/ladder model is given by

$$y(t;n) = \sum_{k=1}^{n} c_k^o\, x_k^o(t),$$

where

$$c_k^o = \langle g_0, x_k^o \rangle.$$

Because the signals $x_k^o(t)$ form an orthonormal set, it follows that

$$\big\|\varepsilon_y(\,\cdot\,;n)\big\|^2 = \|g_0\|^2 - \sum_{k=1}^{n} (c_k^o)^2.$$

To determine the derivative of $\big\|\varepsilon_y(\,\cdot\,;n)\big\|$ w.r.t. γ_1, it is therefore necessary to compute the derivative of the coefficients c_k^o w.r.t. γ_1. Fortunately, there exists in this case a counterpart to formula (11.8), which was fundamental in the Laguerre series case, namely [230, 232]

$$\frac{\partial c_k^o}{\partial \gamma_1} = \frac{1}{1-\gamma_1^2}\left[\frac{k\,t_k}{t_{k+1}}\,c_{k+1}^o - \frac{(k-1)\,t_{k-1}}{t_k}\,c_{k-1}^o\right].$$

Using this formula in the computation of the derivative of $\big\|\varepsilon_y(\,\cdot\,;n)\big\|^2$ yields

$$\frac{\partial\,\big\|\varepsilon_y(\,\cdot\,;n)\big\|^2}{\partial \gamma_1} = -\frac{2n}{1-\gamma_1^2}\,\frac{t_{n-1}}{t_n}\,c_n^o\,c_{n+1}^o,$$

which is the desired result. Equating this derivative to zero, and making explicit the dependence on γ_1 of the coefficients c_k^o, gives the condition

$$c_n^o(\gamma_1)\,c_{n+1}^o(\gamma_1) = 0, \tag{11.10}$$

i.e. either $c_n^o(\gamma_1) = 0$ or $c_{n+1}^o(\gamma_1) = 0$ (or both). It is possible to prove that the optimal values of $c_{n,n}$ and c_n^o are related by $c_{n,n} = t_n c_n^o$. If $t_k > 0$ for $1 \le k \le n+1$, which happens when the signals $x_k(t)$ of the Laguerre model of order $n+1$ are linearly independent, it is possible to put (11.10) in the form

$$c_{n,n}(\gamma_1)\,c_{n+1,n+1}(\gamma_1) = 0.$$

Like the truncated Laguerre series case, the condition $c_{n+1,n+1}(\gamma_1) = 0$ is usually associated with local minima of $\big\|\varepsilon_y(\,\cdot\,;n)\big\|$, and the condition $c_{n,n}(\gamma_1) = 0$ is usually associated with local maxima of $\big\|\varepsilon_y(\,\cdot\,;n)\big\|$.

◇ The stationary points of the 2-norm of the error signal of a Laguerre model with n all-pass sections satisfy the condition $c_{n,n}(\gamma_1)\,c_{n+1,n+1}(\gamma_1) = 0$. The corresponding condition for the Laguerre lattice/ladder model is $c_n^o(\gamma_1)\,c_{n+1}^o(\gamma_1) = 0$.

The case $1 < p < \infty$ can be analyzed in essentially the same way as the transversal Laguerre model case for $p = 2$ (some of the useful properties of the Laguerre lattice/ladder model hold only for $p = 2$). In this case the use of a time-domain p-norm gives optimal coefficients different from those obtained when a frequency-domain p-norm is used. Because the analysis of the two cases is similar, only the time-domain case will be treated here. Assuming that θ is a real parameter, it is known that

$$\begin{aligned}\frac{\partial |x|^p}{\partial\theta} &= p\,|x|^{p-2}\left(\mathrm{Re}[x]\frac{\partial\,\mathrm{Re}[x]}{\partial\theta} + \mathrm{Im}[x]\frac{\partial\,\mathrm{Im}[x]}{\partial\theta}\right)\\ &= p\,|x|^{p-2}\,\mathrm{Re}\left[x\left(\frac{\partial x}{\partial\theta}\right)^*\right]\end{aligned}$$

(as $p > 1$ this last formula is well defined when $x = 0$). The derivative of the time-domain p-norm w.r.t. θ is then given by

$$\frac{\partial \|x\|_p^p}{\partial\theta} = p\sum_{t=-\infty}^{\infty} |x(t)|^{p-2}\,\mathrm{Re}\left[x(t)\left(\frac{\partial x(t)}{\partial\theta}\right)^*\right].$$

Assuming real signals, the optimal coefficients of the Laguerre model satisfy the following set of 'normal' equations

$$\sum_{t=-\infty}^{\infty} |\varepsilon_y(t;n)|^{p-2}\,\mathrm{Re}\left[\varepsilon_y(t;n)\,l_k^*(t)\right] = 0, \qquad k = 1,\ldots,n, \tag{11.11}$$

obtained by differentiating $\|\varepsilon_y(\,\cdot\,;n)\|_p^p$ with respect to $c_{n,k}$ and equating the result to zero. The extraction of the real part and the complex conjugation where kept in this formula, despite the fact that the signals are real, to make the formula similar to the one that would be obtained if a frequency-domain p-norm had been used. For $p = 2$ this formula reduces to the classical normal equations. Because the Banach space associated with the p-norm used here is strictly convex [2], the optimal coefficients of the Laguerre model are uniquely defined, and can be found using a convex optimization procedure.
Differentiating $\|\varepsilon_y(\,\cdot\,;n)\|_p^p$ also with respect to γ_1, equating the result to zero, and using the time-domain version of (11.7) and the 'normal equations' (11.11) to simplify the result, yields the conditions $c_{n,n} = 0$ and/or

$$\sum_{t=-\infty}^{\infty} |\varepsilon_y(t;n)|^{p-2}\,\mathrm{Re}\left[\varepsilon_y(t;n)\,l_{n+1}^*(t)\right] = 0.$$

This last condition can be seen to be one of the 'normal equations' for a Laguerre model of order $n+1$, in which the last coefficient is zero, *i.e.* in which $c_{n+1,n+1} = 0$. Thus, the stationarity conditions take the same form as those found before for the case $p = 2$.

> ◇ The stationary points of the p-norm of the error signal of a Laguerre model with n all-pass sections satisfy the condition $c_{n,n}(\gamma_1)\, c_{n+1,n+1}(\gamma_1) = 0$.

It is now time to deal with the introduction of interpolation constraints. To avoid the introduction of too much extra notation, only frequency-domain interpolation conditions will be considered. It is straightforward to modify the argument presented here to time-domain interpolation constraints or to a mixture of both. Let the M interpolation conditions be given by

$$Y(e^{i\omega_m}; n) = \sum_{k=1}^{n} c_{n,k}\, L_k(e^{i\omega_m}) = G_{0m}, \qquad m = 1, \ldots, M.$$

Let the optimization criterion be the minimization of

$$\big\|\varepsilon_y(\,\cdot\,; n)\big\|_p^p + \mu \sum_{m=1}^{M} \big|Y(e^{i\omega_m}; n) - G_{0m}\big|^p \tag{11.12}$$

for a given positive value of the regularization parameter μ. The case $\mu = 0$ corresponds to the unconstrained case solved previously, and the case $\mu \to +\infty$ corresponds to the case in which the constraints are enforced. Values of μ between these two extreme values are also possible; the value of μ reflects the degree of confidence one has on the interpolation constraints. It is even possible to associate to each interpolation constraint a different regularization parameter (say, μ_m). This is not done here to simplify the presentation.
The differentiation of (11.12) w.r.t. $c_{n,k}$ and with respect to γ_1 can proceed exactly as in the previous case, yielding 'normal equations' with a form similar to (11.11), namely,

$$\mu \sum_{k=1}^{M} \big|Y(e^{i\omega_m}; n) - G_{0m}\big|^{p-2}\, \mathrm{Re}\left[\big(Y(e^{i\omega_m}; n) - G_{0m}\big)\, L_k^*(e^{i\omega_m})\right] +$$
$$\sum_{t=-\infty}^{\infty} \big|\varepsilon_y(t; n)\big|^{p-2}\, \mathrm{Re}\left[\varepsilon_y(t; n)\, l_k^*(t)\right] = 0, \qquad k = 1, \ldots, n,$$

Equating the derivative of (11.12) w.r.t. γ_1 to zero yields, after some algebra, an equation similar to the above, with $k = n+1$, and with the left-hand side multiplied by $c_{n,n}$. It follows, as before, that the stationarity conditions, assuming optimal coefficients, are given by $c_{n,n} = 0$ and/or $c_{n+1,n+1} = 0$. Because the optimal coefficients are continuous functions of μ, this result also holds when $\mu \to +\infty$.

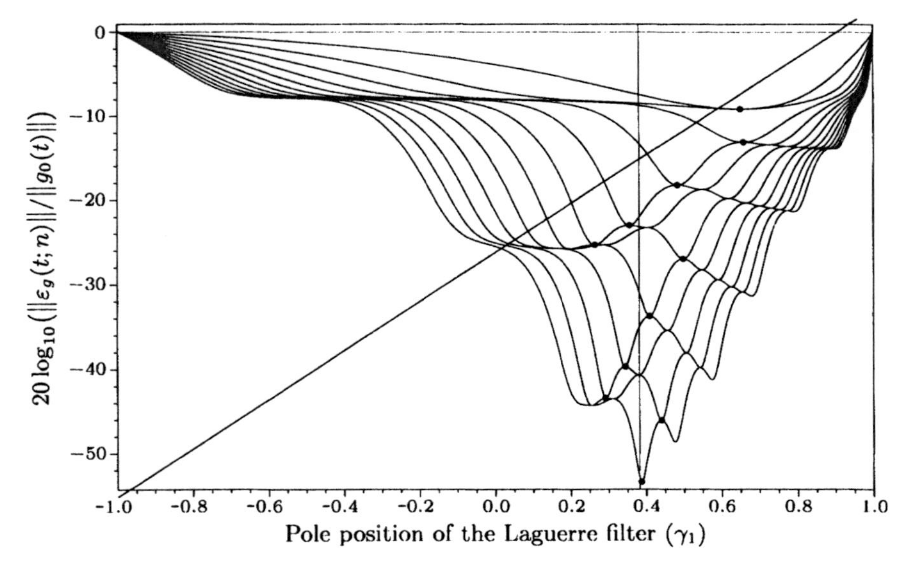

Fig. 11.12. Normalized squared error (in dB) of the approximation of the nominal test model by Laguerre models with orders up to 12. The thin vertical line marks the asymptotically best value of γ_1, and the small dots mark the location of the global minimum of each curve.

> ◇ The optimal pole position of a Laguerre model with n all-pass sections, when the optimization criterion of the model is the minimization of a function of the form (11.12), possibly with the inclusion of similar time-domain interpolation constraints, with $1 < p < \infty$ and $0 \leq \mu \leq \infty$, satisfies the condition $c_{n,n}(\gamma_1)\, c_{n+1,n+1}(\gamma_1) = 0$, where the coefficients of the model are the best possible for their respective model order.

An Example

Figure 11.12 presents the normalized squared errors of the approximation of the nominal test model of Section 11.1.2 by several truncated Laguerre series, with a number of terms used in the approximation ranging from 1 to 12. Observe that at each minima and maxima of these curves (with the exception of minima of the lower curve) there are two curves that touch. This is a consequence of the condition (11.9), as a truncated Laguerre series with k terms is equal to a truncated Laguerre series with $k-1$ terms when $c_k = 0$. The second curve, counting from the top, of Figure 11.12 illustrates a rare situation alluded to after (11.9). The global minimum of that curve occurs for a pole position for which the last coefficient used in the truncated series expansion vanishes (the first and second curves touch at that point). This situation occurs again for the penultimate curve, this time for a local minimum.

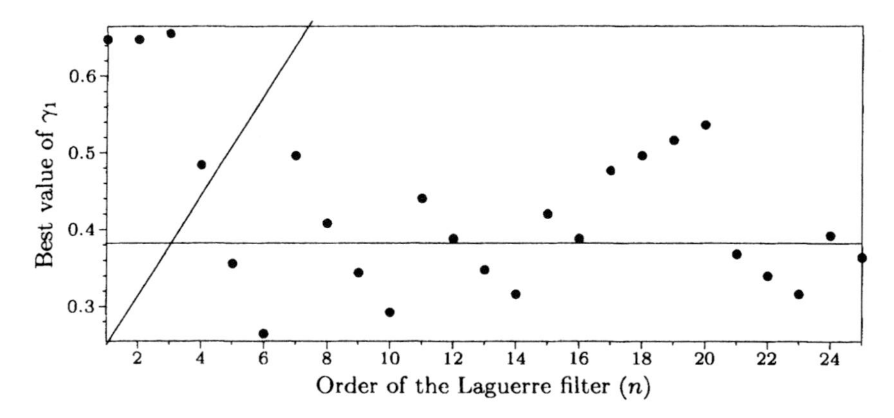

Fig. 11.13. The best pole position of the Laguerre models of orders up to 25. The horizontal line marks the asymptotically best value of γ_1.

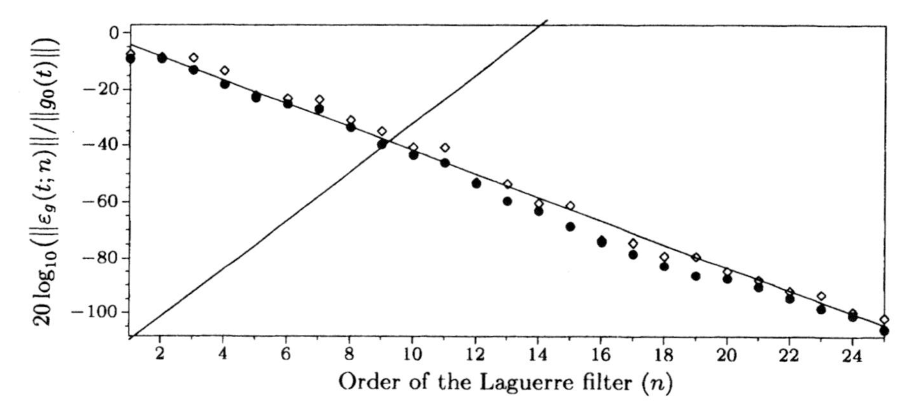

Fig. 11.14. The evolution of the normalized squared error (NSE) for Laguerre models of orders up to 25. Each • represents a NSE global minimum, each ◇ represents the NSE for the asymptotically optimal value of γ_1, and the straight line represents the expected NSE for this value of γ_1, given by $20n\log_{10}\rho(\gamma_1)$.

The global minima of the squared error of the approximations of the various model orders appear to be located near the asymptotically optimal value of γ_1 (the one that minimizes ρ, *cf.* Section 11.2.3). This is illustrated in Figure 11.13. In this specific case, the convergence of the best γ_1 to its asymptotically optimal value is quite slow.
Figure 11.14 presents the smallest value of the normalized squared error of the approximation as a function of the Laguerre model order, together with the normalized squared error for the asymptotically optimal value of γ_1 (the one that maximizes the decay rate). These two numbers, which are never far apart, should be compared to the estimate of the normalized squared error

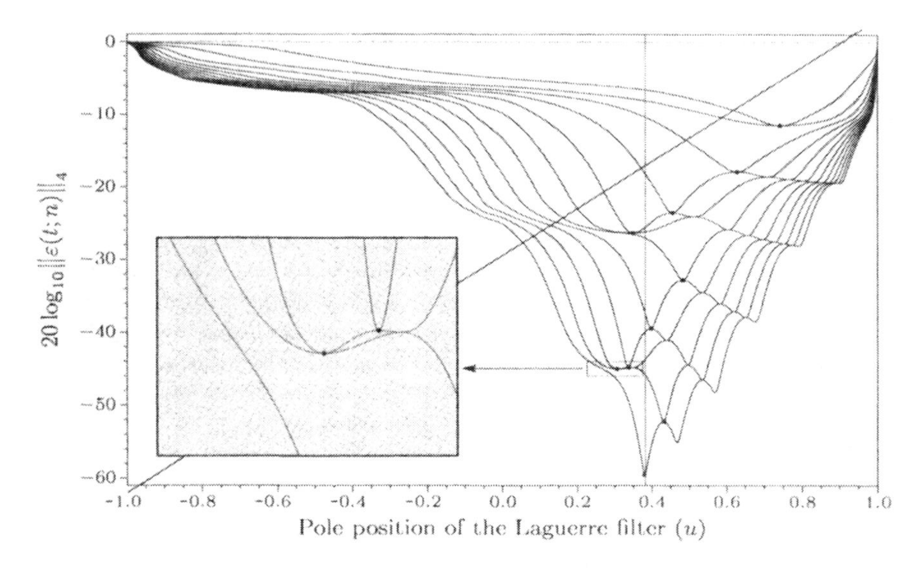

Fig. 11.15. Normalized 4-norm (in dB) of the error of the approximation of the nominal test model by Laguerre models with orders up to 12. The thin vertical line marks the asymptotically best value of γ_1, and the small dots mark the location of the global minimum of each curve.

obtained from the asymptotically best value of γ_1 (shown in the figure as a straight line).

Figure 11.15 presents the 4-norms of the errors of the approximation of the nominal test model of Section 11.1.2 by several Laguerre models with a number of terms used in the approximation ranging from 1 to 12, excited by a signal whose power spectral density is given by

$$\Phi_u(e^{i\omega}) = \frac{A^2}{|e^{i\omega} - 0.6|^2};$$

the constant A is such that the 4-norm of the output signal of the nominal test model is equal to one. The optimality conditions are again clearly visible.

11.3.2 Optimal Pole Conditions for the Two-parameter Kautz Model

It is possible to adapt the results of the Laguerre case to the two-parameter Kautz model case [233], which is a GOBF model with $n_b = 2$. To conserve space, the details of this adaptation will be omitted (see next subsection for a summary of this and other cases). Quite remarkably, the stationarity conditions of the p-norm of the output error of a two-parameter Kautz model with complex conjugate poles is similar to those for the Laguerre case, namely,

$$K_{n,n} = 0 \qquad \text{and/or} \qquad K_{n+1,n+1} = 0.$$

Note that $K_{k,k}$ is a vector with 2 elements. The condition $K_{n,n} = 0$ is usually associated with local maxima of $\|\varepsilon_y(\,\cdot\,;n)\|$, as in this case the approximation is actually of order $n-1$, and the condition $K_{n+1,n+1} = 0$ is usually associated with local minima. An example will be presented in the next subsection.

◇ The optimal complex conjugate pole positions of a two-parameter Kautz model with n all-pass sections, when the optimization criterion of the model is the minimization of a function of the form (11.12), possibly with the inclusion of similar time-domain interpolation constraints, with $1 < p < \infty$ and $0 \leq \mu \leq \infty$, satisfies the condition $K_{n,n}(\gamma_1)\,K_{n+1,n+1}(\gamma_1) = 0$, where the coefficients of the model are the best possible for their respective model order.

11.3.3 Optimal Pole Conditions for the General GOBF Model

The determination of the conditions that the optimal pole positions must satisfy for the GOBF model is considerably more difficult than the case for which $n_b = 1$ (Laguerre case) or for which $n_b = 2$ and the poles are complex (two-parameter Kautz case with complex conjugate poles). The main steps of the determination of the partial derivatives of the squared error of the approximation w.r.t. any parameter that affects $V_k(z)$ are the same as those used in the Laguerre case.

It is possible to block-orthogonalize (*cf.* Section 11.6 or [237]) the signals $x_k(t)$ of the GOBF model, shown in Figure 11.16, giving rise to the signals $x_k^b(t)$ of the GOBF model in lattice/ladder form, shown in Figure 11.17. These signals can in turn be normalized, an operation performed by the lower-triangular matrices T_k, giving rise to the orthonormal signals $x_k^o(t)$.

The output signal of the GOBF lattice/ladder model is given by

$$y(t;n) = \sum_{k=1}^{n} (K_k^o)^T\, x_k^o(t),$$

where

$$(K_k^o)^T = [\![g_0, x_k^o]\!].$$

Because the elements of the vectors $x_k^o(t)$ form an orthonormal set, it follows that

$$\|\varepsilon_y(\,\cdot\,;n)\|^2 = \|g_0\|^2 - \sum_{k=1}^{n} (K_k^o)^T\, K_k^o. \tag{11.13}$$

Quite remarkably, it is possible to prove that [237]

$$\frac{\partial\, x_k^o(t)}{\partial\,\theta} = T_{k,k-1}^{[\theta]}\, x_{k-1}^o(t) + T_{k,k}^{[\theta]}\, x_k^o(t) + T_{k,k+1}^{[\theta]}\, x_{k+1}^o(t),$$

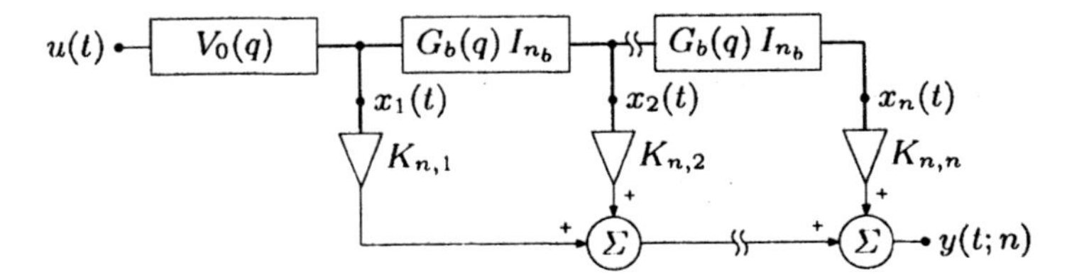

Fig. 11.16. The GOBF model in transversal form.

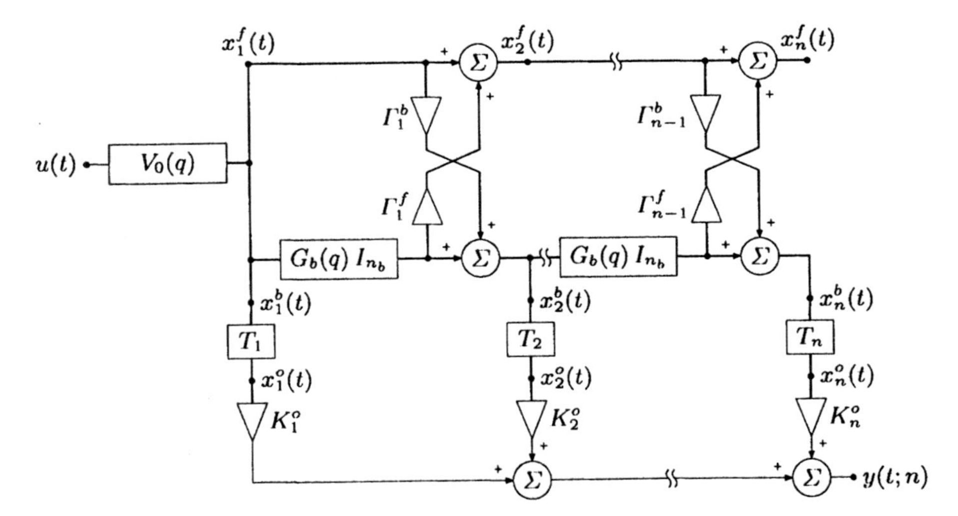

Fig. 11.17. The GOBF model in lattice/ladder form.

where the square matrices $T_{k,l}^{|\theta|}$ satisfy the conditions

$$T_{k,l}^{|\theta|} + T_{l,k}^{|\theta|,T} = 0$$

for all $k, l \geq 0$. Here θ is any real parameter that affects $V_k(z)$. The superscript in $T_{k,l}^{|\theta|}$ means that this square matrix depends of the parameter θ chosen. Obviously, a similar three-term formula exists for the partial derivative of K_k. Using that formula in (11.13) yields the important result

$$\frac{\partial \left\|\varepsilon_y(t;n)\right\|^2}{\partial \theta} = -2K_n^{o,T}\, T_{n,n+1}^{|\theta|} K_{n+1}^{o},$$

which is valid for any parameter θ. This formula suggests immediately two possible stationarity conditions:

$$K_n^o = 0 \qquad \text{and/or} \qquad K_{n+1}^o = 0. \tag{11.14}$$

Unfortunately, in general these are not the only two possible cases.

It is possible to prove [237] that for the parameterization of the poles of the GOBF model by their real and imaginary parts (the first parameterization discussed in Section 11.1.1), the stationarity conditions of the GOBF model take the following 'simple' form: a stationary point will exist if the element-wise product of $K_{n,n}$ and $K_{n+1,n+1}$ vanishes, *i.e.* the elements of the diagonal of $K_{n,n}K_{n+1,n+1}^T$ are all zero. It is possible to put this result in the following form:

◇ If for some permutation of the poles and for some $0 \leq j \leq n_b$ the last j elements of $K_{n,n}$ vanish and the first $n_b - j$ elements of $K_{n+1,n+1}$ also vanish, then $\|\varepsilon_y(\,\cdot\,;n)\|$ will have a stationary point.

Actually, the stationarity conditions are somewhat more involved in the case of confluent poles, *i.e.* when two or more poles of $G_b(z)$ have the same value (these cases are degenerate). It can be shown that there are only two cases for which the stationarity conditions reduce to (11.14). These are the Laguerre model case (obviously), and the two-parameter Kautz model case with complex conjugate poles. For all other cases, including the two-parameter Kautz model with real poles, the stationarity conditions have the general form described above. Nonetheless, it is often the case that the stationarity condition that holds for the global minimum of $\|\varepsilon_y(\,\cdot\,;n)\|$ is $K_{n+1,n+1} = 0$.

An Example

All the following results of the approximation of the output of the nominal test model of Section 11.1.2 by the output of a GOBF model were obtained for the special case of an white noise input signal (with unit variance). This was done only as a matter of convenience; in this case it is possible to compute the coefficients K_k analytically (*cf.* Section 11.2.2).

The first approximation example uses a two-parameter Kautz model ($n_b = 2$). Table 11.4 presents some interesting results of many minimization attempts of $\|\varepsilon_g(\,\cdot\,;4)\|$ (note the use of $n = 4$ all-pass sections). In this and in the following tables, the stationarity conditions are represented by lowercase letters after the coefficients. At least one of the two coefficients (of the same entry) with the same letter must vanish. Also, J_n will denote the normalized squared error of the approximation, *i.e.* $J_n = \|\varepsilon_g(\,\cdot\,;n)\|^2/\|g_0\|^2$.

The data of Table 11.4 demonstrate some of the possibilities for the stationarity conditions. The first entry, which has a double pole, has a stationarity condition similar to that of a Laguerre model; namely, the last scalar coefficient used in the approximation must vanish and/or the first one not used must vanish. Note the unusual presence of a local minimum when the last (scalar) coefficient used in the approximation vanishes. Two other entries,

Table 11.4. Some interesting results of the minimization of $\|\varepsilon_g(\cdot\,;4)\|$. All entries correspond to local minima. K_4 is the last (vector) coefficient used in the approximation. This table is incomplete; more (uninteresting) stationarity points were found during the many minimization attempts that were made when this chapter was being written. The stationarity conditions state that the product of the elements of K_4 and K_5 associated with the same letter must vanish. In degenerate cases (*e.g.* first row), the number of conditions is smaller than n_b

ξ_k	K_4	K_5	J_4	J_5
0.199337	0.284976	0.025037 b	−25.784881	−28.597806
0.199337	0.000000 b	0.111979		
$0.692323 + 0.343864i$	0.154462 a	0.000000 a	−25.804118	−25.804118
$0.692323 - 0.343864i$	0.203762 b	0.000000 b		
0.067580	0.243139 a	0.000000 a	−25.862876	−28.988657
0.323944	0.000000 b	0.117976 b		
−0.121513	0.569957 a	0.000000 a	−25.987832	−25.987832
0.299405	0.382641 b	0.000000 b		
$0.440194 + 0.220106i$	0.230393 a	0.000000 a	−39.522022	−39.522022
$0.440194 - 0.220106i$	0.019622 b	0.000000 b		

Table 11.5. Global minimum data when $n_b = 3$ and $n = 2$

ξ_k	K_2	K_3	J_2	J_3
$0.414415 + 0.493217i$	0.599432 a	0.000000 a	−49.196847	−52.875554
$0.414415 - 0.493217i$	−0.250482 b	0.000000 b		
0.772672	0.000000 c	0.008481 c		

with complex conjugate poles, display the typical behavior of the stationarity conditions; namely, the first (vector) coefficient not used in the approximation vanishes. The remaining two entries, with different real poles, illustrate the stationarity conditions for that case. As an aside, note that the normalized squared error, which in this case is −39.522 dB, can be estimated from the data of Table 11.3 by the formula $-4 \cdot 9.346172 \approx -37.385$ dB.

Figure 11.18 presents some contour lines of $J_4(\gamma_1, \gamma_2)$ for the approximation problem described above. The lines where one of the two elements of K_5 vanishes are also presented in this figure. The three minima of Table 11.4 for which $K_5 = 0$ can be readily located in this figure as the points where these lines intersect. A fourth stationary point, not presented in the table, is also visible on the right slightly above the middle of the figure.

The second (and last) approximation example uses a GOBF model with $n = 2$ all-pass sections, each of which with $n_b = 3$ poles. Table 11.5 presents the global minimum data for this case. As can be seen in this table, for the global minimum K_2 does not vanish (although most of its elements do). This is again a bit unusual.

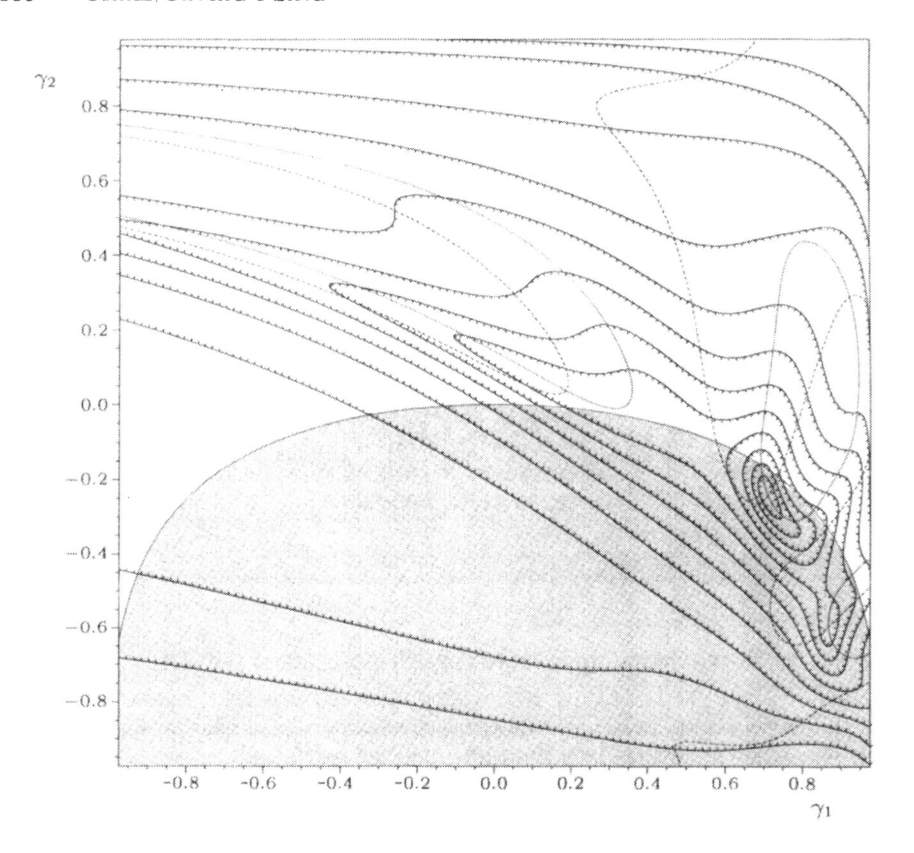

Fig. 11.18. Contour lines, in intervals of 3 dB, of the normalized squared error of the approximation of the nominal test model by a GOBF model with $n_b = 2$ and $n = 4$, drawn in thick 'barbed' lines. Also shown are the contour lines, with a level of zero, of the two coefficients of the first unused all-pass section. The shaded area corresponds to complex conjugate poles.

11.4 Notes

This chapter dealt always with discrete-time signals and systems. All the results that were obtained for 2-norms can be easily adapted to the case of continuous-time signals and systems (with respect to this see [281]). Furthermore, similar results also hold for the p-norms.
Based on the fundamental formula (11.7), it is possible to compute analytically the first l derivatives of $c_{n,n}(\gamma_1)$ for the Laguerre model. That computation requires only the knowledge of the coefficients of all Laguerre models up to order $n + l$. With these derivatives, it is possible to compute a truncated Taylor series or, better yet, a Padé approximant of $c_{n,n}(\gamma_1)$, which in turn can be used to estimate accurately the zeros of this coefficient near a given

value of γ_1. This approach to the determination of the stationary points of $\|\varepsilon_y(\cdot\,;n)\|$ for Laguerre models is described in detail in [231].
Using the Laguerre lattice/ladder model, it is possible to devise algorithms that adapt not only the coefficients and reflection coefficients of the model, but also its pole position. Some preliminary results in that direction can be found in [236].
The stationary conditions for the squared error of truncated two-parameter Kautz series expansions can be found in [73]. These results were later generalized to any input signal [233], and to general GOBF models [237].
The nominal test model that was used in all our examples produces GOBF models with squared error functions that, in some cases, have interesting (unusual) stationary points. It is not known why this has happened. Maybe this phenomenon is related to the fact that its unit pulse response is positive for all time instants (positive system).
All iterative algorithms that seek a minimum of a function require at least an initial value for the function's arguments. In order to minimize the squared error of the GOBF model, one possibility is to begin with a random value of $\gamma \in (-1,1)^{n_b}$. Another possibility is to estimate from the input/output data an ARMA model with n_b poles, and then use the poles of this model as the initial poles of the GOBF model. This approach can be extended to the estimation of a higher order model, followed by a suitable model order reduction step. A further extension may involve a bootstrap process, using earlier GOBF models and a model order reduction to update the position of the poles of the GOBF model [125].
It is possible to estimate a reasonably good pole position for a Laguerre model, or for a two-parameter Kautz model, based on 'simple' measurements of the unit pulse response of the system to be identified. This has been done in [97,244,297,298,300,322] for a Laguerre model and in [199,299] for a two-parameter Kautz model. Other attempts to estimate a good pole position of a Laguerre filter are given in [47,181,260].

11.5 Summary

The main task of this chapter was to describe some ways of selecting the poles of the GOBF model. Two different scenarios were analysed:

1. In the first, one some *a priori* information is available regarding the system to be identified; namely, the position of its poles. In this respect, it was argued that the minimization of

$$\rho = \max_j \left| G_b(1/p_j^0) \right|$$

is desirable, as it makes the coefficients of the GOBF model to go (asymptotically) to zero as fast as possible (for rational systems).

2. In the second one, a way to estimate the coefficients of the GOBF model is assumed to be available. It was found that the stationarity conditions of a p-norm of the output error, with respect to the pole positions of the GOBF model, only involve the coefficients $K_{n,n}$ and $K_{n+1,n+1}$, of optimal (w.r.t. the coefficients) GOBF models with n and $n+1$ all-pass sections. In particular, for Laguerre and two-parameter Kautz models with complex conjugate poles these conditions take the simple form

$$K_{n,n} = 0 \qquad \text{and/or} \qquad K_{n+1,n+1} = 0,$$

even when some interpolation constraints that imposed on the model's unit pulse response and/or on its transfer function. So, instead of minimizing the p-norm directly, it is possible to look for solutions of these stationarity conditions. Usually, but not always, the global minimum satisfies the condition $K_{n+1,n+1} = 0$.

11.6 Appendix

This appendix presents a detailed constructive proof of the existence of the lattice/ladder GOBF model when the input signal is a stationary stochastic process. The main purpose of this model structure is to orthogonalize the state signals of the GOBF model of Figure 11.16. The special case of the Laguerre lattice/ladder case was presented in [232] and, in greater detail, in [230]. An RLS algorithm to adapt the coefficients of the Laguerre lattice/ladder filter (but not to adapt its pole position), was given in [191] (see also [187]). A stochastic gradient algorithm to adapt both the coefficients of the Laguerre lattice/ladder filter and its pole position was given in [236], and one to adapt only the filter coefficients was given in [87]. The general GOBF case was presented in [237]. This presentation will follow closely this last paper (see also [192] and [330]). Due to the extensive use of several index variables, in this appendix, i will denote one of these indices and not the imaginary unit, as done in the rest of this book.
Let $x_i(t)$, $i = 1, \ldots, n$, be the state vectors of each all-pass section of the GOBF model of Figure 11.16. The (block) Gram-Schmidt procedure used to orthogonalize $x_i(t)$ with respect to $x_1(t), \ldots, x_{i-1}(t)$ amounts to find the linear combination of the form

$$x_i^b(t) = \sum_{j=1}^{i} K_{i,i+1-j}^b x_j(t), \qquad \text{with} \qquad K_{i,1}^b = I, \tag{11.15}$$

for which the following orthogonality conditions hold:

$$[\![x_i^b, x_j]\!] = 0, \qquad j = 1, \cdots, i-1. \tag{11.16}$$

Recall that if $u = \begin{bmatrix} u_1 & \cdots & u_m \end{bmatrix}^T$ and if $v = \begin{bmatrix} v_1 & \cdots & v_n \end{bmatrix}^T$ then

$$[\![u, v]\!] = \begin{bmatrix} \langle u_1, v_1 \rangle & \cdots & \langle u_1, v_n \rangle \\ \vdots & \ddots & \vdots \\ \langle u_m, v_1 \rangle & \cdots & \langle u_m, v_n \rangle \end{bmatrix}.$$

Because $[\![x_i, x_j]\!]$ is the block at position (i, j) of the correlation matrix of the states of the GOBF model, and because this correlation matrix is block Toeplitz, it follows that $[\![x_i, x_j]\!]$ can be identified with a square matrix which depends only on the difference between i and j, *i.e.*

$$[\![x_i, x_j]\!] = R_{i-j}.$$

Because the block Toeplitz matrix is symmetric, $R_{-k} = R_k^T$. Explicit formulas for R_{i-j} can be found in other parts of this book and in [237, 238]. Using this formula in (11.16) yields

$$\sum_{k=1}^{i} K_{i,i+1-k}^b R_{k-j} = 0, \qquad j = 1, \cdots, i-1.$$

Using also the orthogonality conditions it can be verified that the square of the norms of the elements of $x_i^b(t)$ are given by the diagonal elements of the matrices

$$\Sigma_i^b = [\![x_i^b, x_i^b]\!] = [\![x_i^b, x_i]\!] = \sum_{k=1}^{i} K_{i,i+1-k}^b R_{k-j} = 0, \qquad j = i.$$

For reasons that will become clear in a moment, it is also useful to consider the signals

$$x_i^f(t) = \sum_{j=1}^{i} K_{i,j}^f x_j(t), \qquad \text{with} \qquad K_{i,1}^f = I, \tag{11.17}$$

for which the following orthogonality conditions hold:

$$[\![x_i^f, x_j]\!] = 0, \qquad j = 2, \cdots, i.$$

These last conditions can be put in the form

$$\sum_{k=1}^{i} K_{i,k}^f R_{k-j} = 0, \qquad j = 2, \cdots, i.$$

The square of the norms of the elements of $x_i^f(t)$ are given by the diagonal elements of the matrices

$$\Sigma_i^f = [\![x_i^f, x_i^f]\!] = [\![x_i^f, x_1]\!] = \sum_{k=1}^{i} K_{i,k}^f R_{k-j} = 0, \qquad j = 1.$$

The signal $x_i^b(t)$ can be considered to be the error of the best linear prediction of $x_i(t)$ by a linear combination of $x_1(t), \ldots, x_{i-1}(t)$. The signal $x_i^f(t)$ can be considered to be the error of the best linear prediction of $x_1(t)$ by a linear combination of $x_2(t), \ldots, x_i(t)$. For the classical transversal filter, which has $x_i(t) = x(t+1-i)$, $x_i^b(t)$ is the prediction of a past sample based on more recent ones (a backward prediction), and $x_i^f(t)$ is the prediction of the most recent sample based on past ones (a forward prediction).
Collecting the above equations in a single (large) matricial equation yields the upper part of

$$\left[\begin{array}{cccc} I & K_{i,2}^f & \cdots & K_{i,i-1}^f & K_{i,i}^f \\ K_{i,i}^b & K_{i,i-1}^b & \cdots & K_{i,2}^b & I \\ \hline I & K_{i-1,2}^f & \cdots & K_{i-1,i-1}^f & 0 \\ 0 & K_{i-1,i-1}^b & \cdots & K_{i-1,2}^f & I \end{array}\right] \left[\begin{array}{ccccc} R_0 & R_{-1} & \cdots & R_{2-i} & R_{1-i} \\ R_1 & R_0 & \cdots & R_{3-i} & R_{2-i} \\ \vdots & \vdots & \ddots & \vdots & \vdots \\ R_{i-2} & R_{i-3} & \cdots & R_0 & R_{-1} \\ R_{i-1} & R_{i-2} & \cdots & R_1 & R_0 \end{array}\right] = \left[\begin{array}{ccccc} \Sigma_i^f & 0 & \cdots & 0 & 0 \\ 0 & 0 & \cdots & 0 & \Sigma_i^b \\ \hline \Sigma_{i-1}^f & 0 & \cdots & 0 & \Delta_{i-1}^f \\ \Delta_{i-1}^b & 0 & \cdots & 0 & \Sigma_{i-1}^b \end{array}\right]. \quad (11.18)$$

The lower part of this equation collects the equations for the next smaller order. Because of the way these smaller order equations are embedded into the larger equations, it is necessary to introduce two extra matrices, Δ_{i-1}^f and Δ_{i-1}^b.
Because $R_{-k} = R_k^T$, the equation formed by the first row of the matrix on the left of (11.18) and the column number k of the R matrix (the matrix in the middle) is, apart from the replacement of $K_{i,j}^f$ by $K_{i,j}^b$, the transpose of the equation formed by the second row of the matrix on the left by the column number $i+1-k$ of the R matrix. Because this holds for $k = 1, \ldots, i$, it follows that $K_{i,k}^b = (K_{i,k}^f)^T$ for $k = 1, \ldots, i$. The solution of the 'forward prediction problem' is then essentially the same as that of the 'backward prediction problem'.
The order $i-1$ equations suggest a way to compute the solution for the order i equations: a special linear combination of the forward order $i-1$ solution and of the 'shifted' backward order $i-1$ solution gives the solution to the order i equations. More precisely, if the square matrices Γ_i^f and Γ_i^b satisfy the following equation (this equation amounts to use linear combinations of the bottom two rows of the matrix on the right-hand side of (11.18) to give rise to the top two rows of the same matrix)

$$\begin{bmatrix} \Sigma_i^f & 0 \\ 0 & \Sigma_i^b \end{bmatrix} = \begin{bmatrix} I & \Gamma_i^f \\ \Gamma_i^b & I \end{bmatrix} \begin{bmatrix} \Sigma_{i-1}^f & \Delta_{i-1}^f \\ \Delta_{i-1}^b & \Sigma_{i-1}^b \end{bmatrix}$$

then the matrices $K_{i,j}^f$ and $K_{i,j}^b$ will be given by

$$\begin{bmatrix} K^f_{i,j} \\ K^b_{i,i+1-j} \end{bmatrix} = \begin{bmatrix} I & \Gamma^f_i \\ \Gamma^b_i & I \end{bmatrix} \begin{bmatrix} K^f_{i-1,j} \\ K^b_{i-1,i+1-j} \end{bmatrix}, \qquad j = 1, \cdots, i \tag{11.19}$$

with $K^f_{i-1,i} = 0$ and $K^b_{i-1,i} = 0$. Note that

$$\Gamma^f_i = -\Delta^f_{i-1}(\Sigma^b_{i-1})^{-1} \qquad \text{and} \qquad \Gamma^b_i = -\Delta^b_{i-1}(\Sigma^f_{i-1})^{-1},$$

and that

$$\Sigma^f_i = (I - \Gamma^f_i \Gamma^b_i)\Sigma^f_{i-1} \qquad \text{and} \qquad \Sigma^b_i = (I - \Gamma^b_i \Gamma^f_i)\Sigma^b_{i-1}.$$

Using (11.19) in (11.17) and (11.15), and because $x_{j+1}(t) = G_b(q)x_j(t)$, the order update formulas for the signals $x^f_i(t)$ and $x^b_i(t)$ are given by

$$\begin{bmatrix} x^f_i(t) \\ x^b_i(t) \end{bmatrix} = \begin{bmatrix} I & \Gamma^f_i \\ \Gamma^b_i & I \end{bmatrix} \begin{bmatrix} x^f_{i-1}(t) \\ G_b(q)x^b_{i-1}(t) \end{bmatrix},$$

with initial conditions $x^f_0(t) = x^b_0(t) = x_0(t)$. These recursion formulas define the lattice part of the GOBF lattice/ladder filter, shown in the top part of Figure 11.17.
Besides the above order-recursive formulas, it is useful to have explicit formulas for the 'reflection coefficients' Γ^f_i and Γ^b_i that involve only local signals (to these coefficients). This can be done using the fact that $x^f_i(t)$ is orthogonal to $x_j(t)$, for $j = 2, \cdots, i$; in particular, $x^f_i(t)$ is orthogonal to $G_b(q)x^b_{i-1}(t)$. It follows that

$$\Gamma^f_i = -\langle x^f_{i-1}(t), G_b(q)x^b_{i-1}(t)\rangle(\Sigma^b_{i-1})^{-1}.$$

In a similar manner, it is possible to show that

$$\Gamma^b_i = -\langle G_b(q)x^b_{i-1}(t), x^f_{i-1}(t)\rangle(\Sigma^f_{i-1})^{-1}.$$

The signals in $x^b_i(t)$ are orthogonal to the signals in $x^b_j(t)$ if $j \neq i$. However, they are not orthogonal to the other signals of the same vector. Performing a Gram-Schmidt orthogonalization with normalization on these signals gives rise to the orthonormal signals

$$x^o_i(t) = T_i x^b_i(t),$$

where T_i is a square root of $(\Sigma^b_i)^{-1}$; the inverse of one of the Cholesky factors of Σ_i gives one possible T_i, with a lower or upper triangular structure. The elements of the signals $x^o_i(t)$, $i = 1, \ldots, n$ will then form an orthonormal set, *i.e.*

$$[\![x^o_i, x^o_j]\!] = \delta_{ij} I.$$

(It is assumed that the elements of the signals $x_i(t)$, $i = 1, \ldots, n$, are linearly independent!) Because the signals $x^o_i(t)$, $i = 1, \ldots, n$ are linear combinations of the signals $x_i(t)$, $i = 1, \ldots, n$, the model's output signal is given by

$$y(t;n) = \sum_{i=1}^{n} K_i^T x_i(t) = \sum_{i=1}^{n} (K_i^o)^T x_i^o(t). \tag{11.20}$$

This formula defines the remaining part (the ladder part) of the GOBF lattice/ladder filter, shown in the bottom part of Figure 11.17.
The coefficients K_i^o of (11.20) that make $y(t;n)$ the best approximation to a given desired output signal $y(t)$ are given by

$$K_i^o = [\![x_i^o, y]\!],$$

i.e. the computation of each weight of the ladder part of the GOBF lattice/ladder filter is independent of the computation of the others. That's the main advantage of an orthonormal (or orthogonal) expansion, and that's the main advantage of the lattice/ladder GOBF model. In particular, adding one more section requires only the computation of two reflection coefficients, Γ_{n+1}^f and Γ_{n+1}^b, and the computation of one output weight, K_{n+1}^o; these computations do not affect the previously computed reflection coefficients and output weights.

12

Transformation Theory

Peter Heuberger[1] and Thomas de Hoog[2]

[1] Delft University of Technology, Delft, The Netherlands
[2] Philips Research Laboratories, Eindhoven, The Netherlands

12.1 Introduction

In the previous chapters, it was shown how the classes of pulse, Laguerre, and Kautz functions can be generalized to the family of Takenaka-Malmquist functions. These functions constitute a basis for the Hardy space of strictly proper stable linear dynamical systems. Furthermore, the Hambo functions were introduced, a subset of the Takenaka-Malmquist functions that result when the generating poles of the basis functions are taken in a repetitive manner from a finite set of poles. In this chapter, it is shown how series expansions of systems and signals in these orthonormal basis functions yield alternative descriptions of these systems and signals, both in frequency and time domain. These alternative descriptions can be viewed as generalizations of the Z-transform and the Laguerre transform. In this, chapter we will give a detailed analysis of the transform that arises from the Hambo basis. An equivalent, but more involved, theory for the more general case of Takenaka-Malmquist functions is presented in [64].
The focus of the chapter will be on presenting the main concepts of the theory. For a detailed account see [64, 128]. The theory developed here will be the keystone of the realization theory described in Chapter 13.
Preliminary results on this transform theory have appeared earlier in the analysis of system identification algorithms [306], in system approximation [126, 129, 317] and in minimal partial realization [68, 291]. The theory can be viewed as a generalization of the Laguerre transform theory for signals and systems that was developed and applied in [149, 173, 225, 226, 242].

12.2 Preliminaries

We first introduce the notation $\mathcal{H}_{2-}$ for the Hardy space, which consists of all functions in $\mathcal{H}_2(\mathbb{E})$ that are zero at infinity, such as strictly proper stable systems. We furthermore will use its L_2 complement $(\mathcal{H}_{2-})^{\perp}$. The symbols

$\mathcal{RH}_2, \mathcal{RH}_{2-}$ denote the subsets of rational functions with real-valued coefficients, *i.e.* the sets of stable transfer functions with real-valued impulse responses. It was shown in Chapter 2 that the Takenaka-Malmquist functions and Hambo functions constitute an orthonormal basis for $\mathcal{H}_{2-}$.
In this chapter, we will consistently use the notation G_b for the basis generating inner function with McMillan degree n_b and minimal balanced state-space realization (A_b, B_b, C_b, D_b). As in the previous chapters we will use the symbols $v_k(t)$ and $V_k(z)$ for the vectorized Hambo functions, in the time and frequency domain. It has been shown that any rational system $H(z) \in \mathcal{H}_{2-}$, or equivalently any signal $y(t) \in \ell_2(\mathbb{N})$, can be written in a unique series expansion,

$$H(z) = \sum_{k=1}^{\infty} \breve{h}^T(k) V_k(z), \quad \breve{h}(k) = [\![V_k, H]\!], \tag{12.1}$$

$$y(t) = \sum_{k=1}^{\infty} \breve{y}^T(k) v_k(t), \quad \breve{y}(k) = [\![v_k, y]\!]. \tag{12.2}$$

The vector coefficient sequences $\{\breve{h}(k)\}_{k\in\mathbb{N}}$ and $\{\breve{y}(k)\}_{k\in\mathbb{N}}$, defined in Equations (12.1) and (12.2) will be essential in the definitions and theory on Hambo transforms.

12.2.1 Dual Bases

It is straightforward to show that these Hambo bases in time and frequency domain give rise to dual bases for $\ell_2^{n_b}(\mathbb{N})$ and $\mathcal{H}_{2-}^{n_b}$. For the time domain, this dual basis consists of vector functions $w_k(t)$, defined by

$$w_k(t) = v_t(k).$$

An interesting property of these functions is that they also are connected by a shift structure. We recollect that the functions $v_k(t)$ and $V_k(z)$, induced by an inner function $G_b(z)$ with minimal balanced realization (A_b, B_b, C_b, D_b), are related by

$$\begin{aligned} v_1(t) &= A_b^{t-1} B_b \\ v_{k+1}(t) &= G_b(q) v_k(t) \end{aligned}$$

and

$$\begin{aligned} V_1(z)) &= [zI - A_b]^{-1} B_b \\ V_{k+1}(z) &= V_1(z) G_b^{k-1}(z). \end{aligned}$$

It turns out that a similar relation holds for $\{w_k(t)\}$. Let $N(z)$ be the matrix valued inner function (see also Equation (3.4) in Chapter 3), with minimal balanced realization (D_b, C_b, B_b, A_b),

$$N(z) = A_b + B_b[zI - D_b]^{-1}C_b, \tag{12.3}$$

then it holds that

$$w_1(t) = B_b D_b^{t-1}$$
$$w_{k+1}(t) = N(q)w_k(t).$$

If we denote the Z-transform of these functions by $W_k(z)$, then it follows that

$$W_1(z) = B_b[zI - D_b]^{-1}$$
$$W_{k+1}(z) = N(z)W_k(z).$$

Given these new bases for $\mathcal{H}_{2-}^{n_b}$ and $\ell_2^{n_b}(\mathbb{N})$, it follows that for any strictly proper system $\breve{H}(z) \in \mathcal{H}_{2-}^{n_b}$ or signal $\breve{y}(t) \in \ell_2^{n_b}(\mathbb{N})$, there exists a unique series expansion

$$\breve{H}(z) = \sum_{k=1}^{\infty} h(k)W_k(z), \quad h(k) = \langle \breve{H}, W_k \rangle, \tag{12.4}$$

$$\breve{y}(t) = \sum_{k=1}^{\infty} y(k)w_k(t), \quad y(k) = \langle \breve{y}, w_k \rangle. \tag{12.5}$$

In fact, these are exactly the counterparts of the expansions given by Equations (12.1) and (12.2), in the sense that $\breve{H}(z) = \sum_{k=1}^{\infty} \breve{h}(k)z^{-k}$.

Example 12.1. A simple but non-trivial inner function is $G_b(z) = z^{-2}$, with balanced realization

$$(A_b, B_b, C_b, D_b) = (\begin{bmatrix} 0 & 0 \\ 1 & 0 \end{bmatrix}, \begin{bmatrix} 1 \\ 0 \end{bmatrix}, \begin{bmatrix} 0 & 1 \end{bmatrix}, 0).$$

The corresponding basis functions are given by

$$V_1(z) = \begin{bmatrix} z^{-1} \\ z^{-2} \end{bmatrix}, \qquad V_2(z) = \begin{bmatrix} z^{-3} \\ z^{-4} \end{bmatrix}, \qquad \cdots$$

It is immediate that the (scalar) elements of these basis functions are the classical pulse functions $\{z^{-k}\}$.
For the basis functions $W_k(z)$, we consider the multi-variable inner function

$$N(z) = A_b + B_b[zI - D_b]^{-1}C_b = \begin{bmatrix} 0 & z^{-1} \\ 1 & 0 \end{bmatrix}.$$

It follows that

$$W_1(z) = \begin{bmatrix} z^{-1} \\ 0 \end{bmatrix}, W_2(z) = \begin{bmatrix} 0 \\ z^{-1} \end{bmatrix}, W_3(z) = \begin{bmatrix} z^{-2} \\ 0 \end{bmatrix}, W_4(z) = \begin{bmatrix} 0 \\ z^{-2} \end{bmatrix} \cdots,$$

which clearly constitutes a basis for the space $\mathcal{H}_{2-}^2$.

12.2.2 Extension to L_2

A second extension of the Takenaka-Malmquist or Hambo bases arises from the observation that any basis for $\mathcal{H}_{2-}$ also induces a basis the complementary space $(\mathcal{H}_{2-})^{\perp}$. Let $\{P_k(z), k \in \mathbb{N}\}$ be a basis for $\mathcal{H}_{2-}$, then it is easy to see that the set of functions $\{\frac{1}{z}P_k(\frac{1}{z})\}$ constitutes a basis for $(\mathcal{H}_{2-})^{\perp}$.

It follows that – given two bases $\{P_k(z)\}$ and $\{Q_k(z)\}$ for $\mathcal{H}_{2-}$ – the set of functions

$$\{P_k(z), \frac{1}{z}Q_k(\frac{1}{z})\}$$

constitutes a basis for $L_2 = \mathcal{H}_{2-} \cup (\mathcal{H}_{2-})^{\perp}$.

Now note that if (A_b, B_b, C_b, D_b) is a balanced realization for the scalar inner function G_b, then the same statement holds for the 'transformed' realization $(A_b^T, C_b^T, B_b^T, D_b)$. If we use the latter realization to derive basis functions, then this results in functions

$$V_k^{(t)}(z) = [zI - A^T]^{-1} C_b^T G_b^{k-1}(z),$$

where we use the symbol $V_k^{(t)}$ to distinguish from the 'standard' functions V_k. Using the intrinsic structure of balanced realizations of inner functions, it can be shown that

$$\frac{1}{z}V_1^{(t)}(\frac{1}{z}) = V_1(z)G_b(\frac{1}{z}) = V_1(z)G_b^{-1}(z),$$

which can easily be extended to

$$\frac{1}{z}V_k^{(t)}(\frac{1}{z}) = V_1(z)G_b^{-k}(z).$$

If we extend the definition of the functions $V_k = V_1 G_b^{k-1}$ to $k \in \mathbb{Z}$, then this property yields the following extensions of the Hambo basis:

Proposition 12.1 ([128]). *The set $\{V_k(z), k \in J\}$ constitutes a basis for:*

$$\begin{aligned} \mathcal{H}_{2-} \quad & if\ J = \mathbb{N}, \\ (\mathcal{H}_{2-})^{\perp} \quad & if\ J = \mathbb{Z} \setminus \mathbb{N}, \\ L_2 \quad & if\ J = \mathbb{Z}. \end{aligned}$$

Analogously we can extend the dual bases for $\ell_2^{n_b}(\mathbb{N})$ and $\mathcal{H}_{2-}^{n_b}$ to $\ell_2^{n_b}(\mathbb{Z})$ respectively $L_2^{n_b}$.

Example 12.2. Using the fact that both $G_b(z)$ and $N(z)$ are inner functions, together with the properties of balanced realizations, it is straightforward to derive the following expressions for the functions $V_0(z)$ and $W_0(z)$:

$$V_0(z) = V_1(z)G_b^{-1}(z) = (I - zA_b^T)^{-1}C_b^T \tag{12.6}$$
$$W_0(z) = N^{-1}(z)W_1(z) = C_b^T(I - zD_b^T)^{-1}. \tag{12.7}$$

For instance, the expression for V_0 can be explained by evaluating $C_b^T G_b(z)$, using the balancing properties $C_b^T D_b = -A_b^T B_b$ and $C_b^T C_b = I - A_b^T A_b$.

12.2.3 Hankel Framework

We recollect the representation of the Hankel operator associated with a finite dimensional linear system $G(z) = \sum_{k=1}^{\infty} g_k z^{-k}$. For scalar systems, this operator is a mapping from past input signals $u \in \ell_2(\mathbb{Z} \setminus \mathbb{N})$ to future output signals $y \in \ell_2(\mathbb{N})$. When employing the canonical bases for the input and output space, the Hankel operator can be represented by an infinite Hankel matrix $\mathbf{H}$ that operates on the infinite vectors $\mathbf{u}$ and $\mathbf{y}$, as in:

$$\mathbf{y} = \begin{bmatrix} y(1) \\ y(2) \\ y(3) \\ \vdots \end{bmatrix} = \begin{bmatrix} g_1 & g_2 & g_3 & \cdots \\ g_2 & g_3 & g_4 & \\ g_3 & g_4 & g_5 & \\ \vdots & & & \ddots \end{bmatrix} \begin{bmatrix} u(0) \\ u(-1) \\ u(-2) \\ \vdots \end{bmatrix} = \mathbf{H}\mathbf{u} \tag{12.8}$$

Later on in this chapter, we will show how alternative representations of the Hankel operator emerge that result from non-canonical orthonormal bases for the input and output space.

A well-known fact is that the rank of the Hankel matrix $\mathbf{H}$ in (12.8) is equal to the McMillan degree n of the system G. Let (A, B, C, D) be a minimal state-space realization for G, then the Markov parameters g_k can be written as $g_k = CA^{k-1}B$, $k \in \mathbb{N}$, which leads to the following full rank decomposition of $\mathbf{H}$:

$$\mathbf{H} = \mathbf{\Gamma} \cdot \mathbf{\Delta} = \begin{bmatrix} C \\ CA \\ CA^2 \\ \vdots \end{bmatrix} \cdot \begin{bmatrix} B & AB & A^2B & \cdots \end{bmatrix}. \tag{12.9}$$

The (rank n) matrices $\mathbf{\Gamma}$ and $\mathbf{\Delta}$ are known as the observability, respectively controllability matrix of the state-space realization and it follows that $\mathbf{H}$ must have rank n.

12.2.4 Products of Basis Functions

In this chapter, as well as in Chapter 13 on realization theory, we will encounter the product of elements of $V_1(z)$. For ease of notation we denote

$$\phi_i(z) = (V_1(z))_i, \quad i = 1, \cdots, n_b,$$

and we are interested in expressions of products of the form

$$V_1(z)\phi_i(z).$$

This issue was first addressed in [287] and closed form expressions were given in [68, 291] as follows

$$V_1(z)\phi_i(z) = P_i V_1(z) + Q_i V_2(z), \tag{12.10}$$

where the matrices P_i, Q_i result from the following Sylvester equations (with e_i the ith canonical unit vector):

$$P_i = A_b P_i A_b^T + B_b e_i^T A_b \tag{12.11}$$
$$Q_i = A_b^T Q_i A_b + C_b^T e_i^T \tag{12.12}$$

Note that the existence of the matrices P_i and Q_i is guaranteed by the stability of the matrix A_b. See Appendix 5A2 in [64] for more properties of these matrices.

12.3 Signal and Operator Transforms

In this section, we will introduce three transform concepts:

1. The Hambo signal transform, defined for arbitrary ℓ_2 signals,
2. The Hambo signal transform of a system, defined on the impulse response of a system,
3. The Hambo operator or system transform, defined on a system in $\mathcal{H}_2(\mathbb{E})$ that operates on ℓ_2-signals.

In Equation (12.2), it was shown that any signal $y \in \ell_2(\mathbb{N})$ can be written in a series expansion in term of Hambo functions $\{v_k\}_{k\in\mathbb{N}}$, with expansion coefficients $\breve{y}(k) = [\![v_k, y]\!] \in \mathbb{R}$. We will refer to the signal $\breve{y} \in \ell_2^{n_b}(\mathbb{N})$ as the *Hambo signal transform* of the signal $y \in \ell_2(\mathbb{N})$.
This concept can easily be extended to signals in $\ell_2(\mathbb{Z} \setminus \mathbb{N})$ or $\ell_2(\mathbb{Z})$ using the extended Hambo basis as discussed in the previous section. Furthermore, a similar definition can be given for the Hambo transform of vector signals $x \in \ell_2^n(\mathbb{N})$.

Definition 12.1. *(Hambo Signal Transform). Given a signal $x \in \ell_2^n(J)$, where $J = \mathbb{N}, \mathbb{Z}$, or $\mathbb{Z} \setminus \mathbb{N}$, the Hambo signal transform of x is the sequence $\{\breve{x}(k)\}_{k \in J}$, with $\breve{x}(k) \in \mathbb{R}^{n_b \times n}$ given by*

$$\breve{x}(k) = [\![v_k, x]\!] \,. \tag{12.13}$$

Furthermore, we define the λ-domain representation of the Hambo signal transform as

$$\breve{X}(\lambda) = \sum_{k \in j} \breve{x}(k)\lambda^{-k}$$

Here, $\breve{X}(\lambda)$ is simply the Z-transform with λ instead of z to avoid any confusion. Because this is just an alternative representation of the Hambo signal transform, it will commonly also be addressed as the Hambo signal transform. In order to define the *Hambo signal transform of a system*, we note that any L_2 system G is uniquely represented by its impulse response $\{g(k)\} \in \ell_2^n(\mathbb{Z})$.

Definition 12.2. *Consider a system $G \in L_2^n$ and a Hambo basis $\{V_k\}_{k \in \mathbb{Z}}$. The Hambo signal transform of $G(z)$, denoted as $\breve{G}(\lambda)$, is defined as*

$$\breve{G}(\lambda) = \sum_{k=-\infty}^{\infty} \breve{g}(k)\lambda^{-k}, \quad \text{where} \quad \breve{g}(k) = [\![V_k, G]\!]$$

Thus, the Hambo signal transform of a system is the same as the Hambo signal transform of its impulse response.

Example 12.3. To illustrate the signal transform of a system, we consider some special cases:

1. $G(z) = V_1(z)$.
   In this case, the expansion coefficients are given by $\breve{g}_k = \delta_k I_{n_b}$. Hence $\breve{G}(\lambda) = \lambda^{-1} I_{n_b}$. This can easily be extended to show that the Hambo signal transform of $V_k(z)$ is given by

$$\breve{V}_k(\lambda) = \lambda^{-k} I_{n_b}.$$

2. $G(z) = 1$.
   Now $\breve{g}(k) = [\![V_k, G]\!]$ and because $V_k(z)$ is strictly proper for $k \in \mathbb{N}$, it follows that $\breve{g}(k) = 0, k \in \mathbb{N}$. For $k = 0$ we have $\breve{g}(0) = [\![V_1(z) G_b^{-1}(z), 1]\!] = [\![V_1(z), G_b(z)]\!] = [\![V_1(z), D_b + C_b V_1(z)]\!] = C_b^T$. It is straightforward to extend this to $\breve{g}(k) = C_b^T D_b^{-k}$ for $k \in \mathbb{Z} \setminus \mathbb{N}$. Hence it follows that $\breve{G}(\lambda) = C_b^T \sum_{k=0}^{\infty} D_b^k \lambda^k = C_b^T (1 - D_b \lambda)^{-1}$.

3. $G(z) = \bar{D} + \bar{C}V_1(z)$.
   This is the case where G has the same 'pole structure' as G_b (the matrices $\bar{D}, \bar{C}$ are arbitrary). A simple exercise yields

$$\breve{G}(\lambda) = C_b^T \bar{D}(1 - D_b\lambda)^{-1} + \bar{C}^T \lambda^{-1}, \tag{12.14}$$

   which shows that for the special case $G = G_b$, we get

$$\breve{G}_b(\lambda) = C_b^T D_b(1 - D_b\lambda)^{-1} + C_b^T \lambda^{-1} = C_b^T \lambda^{-1}(1 - D_b\lambda)^{-1}.$$

The third example illustrates a property that we will elaborate upon later in this chapter, the effect that the Hambo transform can be used to reduce the complexity in terms of McMillan degree. In the example, we find for instance that for systems in $\mathcal{RH}_{2-}$ (*i.e.* $\bar{D} = 0$), the McMillan degree of the Hambo transform can even be reduced to 1, as in that case (12.14) reduces to $\breve{G}(\lambda) = \bar{C}^T \lambda^{-1}$. Note however that this reduction of complexity also includes a rise in input/output dimension.

The last transform concept we will introduce is the *Hambo operator transform*, also referred to as the *Hambo system transform*, which emerges when we consider the relation between the Hambo signal transforms of the input and output signals of a system in $\mathcal{H}_2(\mathbb{E})$. It turns out that this relation is again given by a stable rational function in $\mathcal{H}_2^{n_b \times n_b}$. The operator transform is formally defined as follows:

Definition 12.3. *(Hambo Operator Transform). Consider a system $G(z) \in \mathcal{H}_2(\mathbb{E})$ and a Hambo basis $\{V_k(z)\}_{k \in \mathbb{N}}$, associated with the inner function $G_b(z)$. We define the* Hambo operator transform *of $G(z)$, denoted by $\widetilde{G}(\lambda)$, as*

$$\widetilde{G}(\lambda) = \sum_{\tau=0}^{\infty} M_\tau \lambda^{-\tau}, \tag{12.15}$$

$$\textit{where} \quad M_\tau = [\![V_1(z) G_b^\tau(z), V_1(z) G(z)]\!] . \tag{12.16}$$

This expression for $\widetilde{G}$ can be seen as a Laurent expansion around $\lambda = \infty$, and $\{M_\tau\}$ can be interpreted as the Markov parameters of $\widetilde{G}$.
This series expansion can be rewritten, by noting that the matrix 'outer' products are evaluated on the unit circle, *i.e.* $z = e^{i\omega}$, and realizing that $|G_b(e^{i\omega})| = 1$, because G_b is inner. This implies that for $|\lambda| > 1$ we have $|G_b(e^{i\omega})\lambda^{-1}| < 1$. Now we can apply the property that $\sum_{k=0}^{\infty} R^k = (I - R)^{-1}$ for all R with $\rho(R) < 1$, and derive that $\widetilde{G}$ obeys,

$$\widetilde{G}(\lambda) = [\![V_1(z)\lambda(\lambda - G_b(z))^{-1}, V_1(z)G(z)]\!] . \tag{12.17}$$

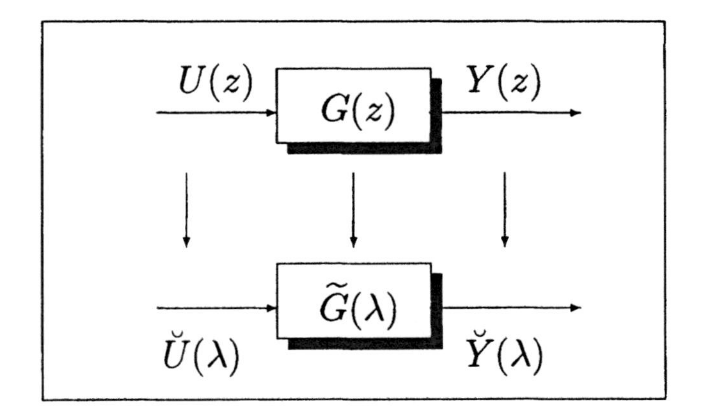

Fig. 12.1. Schematic overview of the Hambo signal and system transform.

Note furthermore that for $|\lambda| > 1$, the function $(\lambda - G_b(z))^{-1}$ has no poles inside the unit disc ($|G_b(z)| \leq 1$, for $|z| \geq 1$), which implies that the inner product above is finite. Hence $\widetilde{G}$ has no poles $|\lambda| > 1$ and is therefore a stable function.

The following proposition formalizes the aforementioned relation between the signal transforms of the input and output signals of a system and the operator transform of that system, also depicted in Figure 12.1.

Proposition 12.2 ([64, 128]). *Consider signals* $u, y \in \ell_2(\mathbb{N})$ *and a system* $G \in \mathcal{H}_2(\mathbb{E})$ *such that* $y(t) = G(q)u(t)$. *With* $\widetilde{G}$ *the Hambo operator transform of* G, *it holds that* $\breve{Y}(\lambda) = \widetilde{G}(\lambda)\breve{U}(\lambda)$.

Special Cases

1. Laguerre basis.
   It is straightforward to calculate a closed expression for $\widetilde{G}$ for the Laguerre case, where $G_b(z) = \frac{1-za}{z-a}$ and $V_1(z) = \frac{\sqrt{(1-a^2)}}{z-a}$, using Equation (12.17). This results in the well-known Laguerre system transform ([173,225,226]),
   $$\widetilde{G}(\lambda) = G\left(\frac{a+\lambda}{1+a\lambda}\right).$$
2. $G(z) = G_b(z)$.
   In this case, the system transform has a very simple form. Substitute G_b for G in Equation (12.16): $M_\tau = [\![V_1(z)G_b^\tau(z), V_1(z)G_b(z)]\!] = \delta_{\tau-1}$.

Consequently, it follows that a multiplication in the Z-domain with $G_b(z)$ is equivalent with a shift operator in the λ-domain,

$$\widetilde{G}_b(\lambda) = \lambda^{-1} I_{n_b}. \tag{12.18}$$

3. $G(z) = 1$.
   This leads to $M_\tau = [\![V_1(z)G_b^\tau(z), V_1(z)]\!] = \delta_\tau I$, and hence

$$\widetilde{G}(\lambda) = I_{n_b}.$$

4. $G(z) = z^{-1}$.
   First we recollect that $zV_1(z) = z(zI - A_b)^{-1}B_b = B_b + A_bV_1(z)$ and that $G_b(z) = D_b + C_bV_1(z)$. Now use the definition of M_τ in Definition 12.3 to calculate

$$\begin{aligned} M_\tau &= [\![V_1(z)G_b^\tau(z), V_1(z)z^{-1}]\!] \\ &= [\![V_1(z)zG_b^\tau(z), V_1(z)]\!] \\ &= [\![(B_b + A_bV_1(z))G_b^\tau(z), V_1(z)]\!] \\ &= A_b\delta_\tau + [\![B_b(D_b + C_bV_1(z))G_b^{\tau-1}(z), V_1(z)]\!] \\ &= A_b\delta_\tau + B_bC_b\delta_{\tau-1} + B_bD_bC_b\delta_{\tau-2}\cdots \end{aligned}$$

Hence
$$\widetilde{G}(\lambda) = A_b + \sum_{k=1}^{\infty} B_b D_b^{k-1} C_b \lambda^{-k} = N(\lambda), \tag{12.19}$$

with N defined by Equation (12.3). Thus, a shift in the Z-domain corresponds with a multiplication with the multi-variable inner function $N(\lambda)$ in the λ-domain.

5. $G(z) = \phi_i(z) = (V_1(z))_i$.
   To calculate the Hambo transform of this function, we first recall from (12.10) that $V_1(z)\phi_i(z) = P_iV_1(z) + Q_iV_2(z)$. It now follows from (12.16) that $M_\tau = [\![V_1(z)G_b^\tau(z), V_1(z)(P_iV_1(z) + Q_iV_2(z))]\!] = P_i^T\delta_\tau + Q_i^T\delta_{\tau-1}$. And hence we have

$$\widetilde{(V_1)}_i(\lambda) = P_i^T + Q_i^T\lambda^{-1} \tag{12.20}$$

6. $G(z) = \bar{D} + \bar{C}V_1(z)$.
   So G has the same (A, B) matrices as G_b. It follows (using Equation (12.10)) that

$$\begin{aligned} M_\tau &= [\![V_1(z)G_b^\tau(z), V_1(z)(\bar{D} + \bar{C}V_1(z))]\!] \\ &= \bar{D}I_{n_b}\delta_\tau + [\![V_1(z)G_b^\tau(z), V_1(z)V_1^T(z)\bar{C}^T]\!] \\ &= \bar{D}I_{n_b}\delta_\tau + \left[\!\!\left[V_1(z)G_b^\tau(z), \sum_{k=1}^{n_b}(\bar{C})_k(P_kV_1(z) + Q_kV_2(z))\right]\!\!\right] \\ &= \left(\bar{D}I_{n_b} + \sum_{k=1}^{n_b}(\bar{C})_kP_k\right)\delta_\tau + \left(\sum_{k=1}^{n_b}(\bar{C})_kQ_k\right)\delta_{\tau-1} \end{aligned}$$

Hence it follows that

$$\widetilde{G}(\lambda) = M_0 + M_1\lambda^{-1}$$

This last special case is often referred to as the *2 Markov parameter property*, which can be formulated as

> For any finite dimensional system $G \in \mathcal{RH}_2$, there exists a Hambo basis such that only the first two Markov parameters of the Hambo operator transform are non-zero.

An optimal basis in this sense is one where the basis poles are exactly the poles of the system G.
In the next section, we will elaborate on expressions for the Hambo system transform for arbitrary systems.
An important feature of the Hambo system transform is that it is a multiplicative and commutative mapping:

Proposition 12.3 ([64, 128]). *Let $G_1, G_2 \in \mathcal{H}_2(\mathbb{E})$ be given. It holds that*

$$(\widetilde{G_1 G_2})(\lambda) = \widetilde{G_1}(\lambda)\widetilde{G_2}(\lambda) = \widetilde{G_2}(\lambda)\widetilde{G_1}(\lambda) \tag{12.21}$$

We conclude this section with a proposition that shows the isomorphic relation between signals, systems, and their Hambo transforms.

Proposition 12.4 ([64, 128]). *(Hambo Transform Isomorphisms).*
With $X \in L_2^{n_x}$, $Y \in L_2^{n_y}$ and $G_1, G_2 \in \mathcal{H}_2(\mathbb{E})$, it holds that

a. $[\![X, Y]\!] = \left[\!\left[\breve{X}^T, \breve{Y}^T\right]\!\right]$.
b. $[\![V_k G_1, V_k G_2]\!] = \left[\!\left[\widetilde{G_1}^T, \widetilde{G_2}^T\right]\!\right]$.
c. $\langle G_1, G_2\rangle = \left\langle \widetilde{G_1} W_k, \widetilde{G_2} W_k \right\rangle$.

12.4 Hambo Transform Expressions

In the previous section, it was shown that the Hambo system transform of a scalar, causal, stable, linear, time-invariant system is again a causal, stable, linear, time-invariant, but multi-variable, system and we derived simple expressions for a number of special cases. In this section, we will show how for arbitrary systems, the system transform can actually be calculated with a well defined numerical procedure.
In the past, various approaches have been proposed for the calculation of the Hambo transform. This development started with non-minimal state-space

realizations in [126]. Variable substitution methods have been reported in [126, 306] and finally robust algorithms yielding minimal state-space realizations were reported in [64,68,128,291].
An intriguing form is based on a variable substitution in the Laurent expansion of a given system $G \in \mathcal{H}_2(\mathbb{E})$, given by

$$G(z) = \sum_{k=0}^{\infty} g_k z^{-k}$$

The Hambo transform of this system can be found by substituting the shift operator z^{-1} with the stable system $N(\lambda)$, as defined in Equation (12.3), extending the result presented in Equation (12.19) combined with the multiplicative property of Proposition 12.3:

Proposition 12.5 ([306]). *(Variable Substitution Property).*
Let $N(\lambda) = A_b + B_b(\lambda I - D_b)^{-1}C_b$. Then the Hambo operator transform $\widetilde{G}$ of a given system $G \in \mathcal{H}_2(\mathbb{E})$ is equal to

$$\widetilde{G}(\lambda) = \sum_{k=0}^{\infty} g(k) N^k(\lambda). \tag{12.22}$$

With abuse of notation, this property is sometimes also stated as

$$\widetilde{G}(\lambda) = G(z)|_{z^{-1}=N(\lambda)} \,.$$

Note that an immediate result of this proposition is the commutation property

$$N(\lambda)\widetilde{G}(\lambda) = \widetilde{G}(\lambda)N(\lambda).$$

A direct consequence of Proposition 12.5 and the use of the dual basis as in Equation (12.4), *i.e.* $\breve{G}(\lambda) = \sum_{t=1}^{\infty} g(t)W_t(\lambda)$, is the following relation between the Hambo operator respectively signal transform of a strictly proper system:

Proposition 12.6 ([128, 129]). *Let $G \in \mathcal{H}_{2-}$, then*

$$\breve{G}(\lambda) = \widetilde{G}(\lambda)W_0(\lambda), \tag{12.23}$$

where W_0 is defined by Equation (12.7).

Later in this chapter, we will also discuss the inverse Hambo operator transform, *i.e.* how to retrieve G from $\widetilde{G}$. The first account of this inverse was given in [129], where it was shown that G can be obtained from $\widetilde{G}$ using variable substitution as follows

$$G(z) = zV_1^T(z)\widetilde{G}(\lambda)W_1(\lambda)\lambda\Big|_{\lambda^{-1}=G_b(z)},$$

which can also be formulated as:

$$G(z) = V_0^T(z)\widetilde{G}(\lambda)W_0(\lambda)\Big|_{\lambda^{-1}=G_b(z)}.$$

This results directly in a variable substitution method to retrieve (a strictly proper) $G(z)$ from the signal transform of a system:

$$G(z) = V_0^T(z)\breve{G}(\lambda)\Big|_{\lambda^{-1}=G_b(z)}. \tag{12.24}$$

Remark 12.1. Denote the McMillan degree of G and $\breve{G}$ by n_G and $n_{\breve{G}}$. A direct consequence of Equation (12.24) is that

$$n_G \leq n_b \times n_{\breve{G}}. \tag{12.25}$$

We outline the proof for the case that G en $\breve{G}$ are stable. Because G_b is inner, it follows that $\breve{G}\left(G_b^{-1}(z)\right)$ is again a rational stable transfer function with degree smaller or equal to $n_b \times n_{\breve{G}}$. Multiplication with $V_0^T(z)$, which is strictly unstable, cannot increase the degree, as this would imply that G has unstable modes.

12.4.1 Hankel Operator Expressions

In Section 12.2, we briefly discussed the representation of the Hankel operator associated with a system $G \in \mathcal{H}_{2-}$ in terms of the standard canonical basis. An alternative representation arises if we use the Hambo basis $\{V_k(z), k \in J\}$ for the signal space $\ell_2(J)$, where $J = \mathbb{N}, \mathbb{Z}\setminus\mathbb{N}$. Expand the input signal $u \in \ell_2(\mathbb{Z}\setminus\mathbb{N})$ and the output signal $y \in \ell_2(\mathbb{N})$ in the corresponding Hambo bases,

$$u(t) = \sum_{k=0}^{\infty} \breve{u}(-k)v_{-k}(t)$$

$$y(t) = \sum_{k=1}^{\infty} \breve{y}(k)v_k(t)$$

It follows that we have

$$\begin{bmatrix} \breve{u}(0) \\ \breve{u}(-1) \\ \breve{u}(-2) \\ \vdots \end{bmatrix} = \begin{bmatrix} v_0(0) & v_0(-1) & v_0(-2) & \cdots \\ v_{-1}(0) & v_{-1}(-1) & v_{-1}(-2) & \cdots \\ v_{-2}(0) & v_{-2}(-1) & v_{-2}(-2) & \cdots \\ \vdots & \vdots & \vdots & \vdots \end{bmatrix} \begin{bmatrix} u(0) \\ u(-1) \\ u(-2) \\ \vdots \end{bmatrix}$$

which we write in short as $\breve{\mathbf{u}} = \mathbf{V_p}\mathbf{u}$. Analogously, we can write $\breve{\mathbf{y}} = \mathbf{V_f}\mathbf{y}$, where $\mathbf{V_f}$ is given by $(\mathbf{V_f})_{ij} = v_i(j)$. Note that $\mathbf{V_f}, \mathbf{V_p}$ are unitary matrices, and therefore we can rewrite Equation (12.8) as $\mathbf{V_f}^T\breve{\mathbf{y}} = \mathbf{H}\mathbf{V_p}^T\breve{\mathbf{u}}$ leading to

$$\breve{\mathbf{y}} = \mathbf{V_f}\mathbf{H}\mathbf{V_p}^T\breve{\mathbf{u}}.$$

We conclude that the matrix $\widetilde{\mathbf{H}}$, defined by

$$\widetilde{\mathbf{H}} = \mathbf{V_f}\mathbf{H}\mathbf{V_p}^T \tag{12.26}$$

is an alternative representation of the Hankel operator of the system G. Now we recall Proposition 12.2, which shows the relation between $\widetilde{\mathbf{H}}$ and $\widetilde{G}$.

Proposition 12.7 ([68,128,291]). *The matrix $\widetilde{\mathbf{H}}$ is the block Hankel matrix, associated with the system $\widetilde{G}$.*

Hence we have that – with M_τ defined in Proposition 12.3 – the matrix $\widetilde{\mathbf{H}}$ obeys

$$\widetilde{\mathbf{H}} = \begin{bmatrix} M_1 & M_2 & \cdots \\ M_2 & M_3 & \cdots \\ \vdots & \vdots & \vdots \end{bmatrix}$$

An immediate consequence is that the McMillan degree of $\widetilde{G}$, or equivalently the rank of $\widetilde{\mathbf{H}}$ is equal to the rank of the Hankel matrix $\mathbf{H}$, which is the same as the McMillan degree of G. Even stronger is the conclusion that G and $\widetilde{G}$ have the same Hankel singular values, which follows from Equation (12.26) with the fact that $\mathbf{V_p}$ and $\mathbf{V_f}$ are unitary matrices. So we conclude that

Hankel singular values and McMillan degree are invariant under the Hambo operator transformation.

12.4.2 Minimal State Space Expressions

For a thorough derivation of the algorithms in this paragraph, we refer to [64,128]. The basis of the theory lies in realization theory and in fact the first algorithms derived in this context were a spin-off of the work on realization theory in terms of orthogonal basis functions [68,291]. This theory is the subject of Chapter 13. Here we will outline the main concept in transform terms.

Assume we have an LTI system $G \in \mathcal{RH}_2$ with minimal realization (A, B, C, D) and McMillan degree n. Now we use Equation (12.9) to associate the derived functions $\varGamma \in \mathcal{H}_{2-}^n$ and $\varDelta \in (\mathcal{H}_{2-}^n)^\perp$ with the observability and controllability matrices $\mathbf{\Gamma}$ and $\mathbf{\Delta}$ as

$$\varGamma(z) = \sum_{k=1}^{\infty} CA^{k-1}z^{-k} = C(zI - A)^{-1}$$

$$\varDelta(z) = \sum_{k=0}^{\infty} A^k B z^k = (I - zA)^{-1}B.$$

The basic ingredient of the theory is that the corresponding Hambo operator transform $\widetilde{G}$ has a minimal state space realization $\left(\widetilde{A},\widetilde{B},\widetilde{C},\widetilde{D}\right)$ such that

$$\widetilde{C}(\lambda I-\widetilde{A})^{-1}=\breve{\Gamma}^T(\lambda) \tag{12.27}$$

$$(I-\lambda\widetilde{A})^{-1}\widetilde{B}=\breve{\Delta}^T(\lambda). \tag{12.28}$$

Even stronger, this property works both ways. The property has the direct consequence that the realizations for G and $\widetilde{G}$, which are linked as described above, must have the same controllability Gramian X_c and observability Gramian X_o. Consider for instance the observability Gramian. then it holds that

$$X_o=\left[\!\left[\Gamma^T(z),\Gamma^T(z)\right]\!\right]=\left[\!\left[\breve{\Gamma}^T(\lambda),\breve{\Gamma}^T(\lambda)\right]\!\right]=\widetilde{X}_o,$$

because of the isomorphic relations in Proposition 12.4.

These properties – combined with tedious manipulations on the various Hambo transforms – are used in [64,128] to derive the following algorithms in terms of Sylvester equations:

Proposition 12.8 ([64,128]). *(Forward Hambo Operator Transform). Consider a system $G\in\mathcal{RH}_2$, with minimal realization (A,B,C,D), controllability Gramian X_c, and observability Gramian X_o. Then the Hambo system transform $\widetilde{G}$ has minimal realizations $\left(\widetilde{A_i},\widetilde{B_i},\widetilde{C_i},\widetilde{D_i}\right)$ $(i=1,2)$ that satisfy the following Sylvester equations.*

$$\begin{bmatrix}A & 0\\ C_b^T & A_b^T\end{bmatrix}\begin{bmatrix}\widetilde{A}_1X_c & \widetilde{B}_1\\ \widetilde{C}_1X_c & \widetilde{D}_1\end{bmatrix}\begin{bmatrix}A^T & 0\\ B_bB^T & A_b\end{bmatrix}+\begin{bmatrix}B\\ C_b^TD\end{bmatrix}\begin{bmatrix}D_bB^T & C_b\end{bmatrix}=\begin{bmatrix}\widetilde{A}_1X_c & \widetilde{B}_1\\ \widetilde{C}_1X_c & \widetilde{D}_1\end{bmatrix}$$

$$\begin{bmatrix}A^T & C^TC_b\\ 0 & A_b\end{bmatrix}\begin{bmatrix}X_o\widetilde{A}_2 & X_o\widetilde{B}_2\\ \widetilde{C}_2 & \widetilde{D}_2\end{bmatrix}\begin{bmatrix}A & BB_b^T\\ 0 & A_b^T\end{bmatrix}+\begin{bmatrix}C^TD_b\\ B_b\end{bmatrix}\begin{bmatrix}C & DB_b^T\end{bmatrix}=\begin{bmatrix}X_o\widetilde{A}_2 & X_o\widetilde{B}_2\\ \widetilde{C}_2 & \widetilde{D}_2\end{bmatrix}.$$

Proposition 12.9 ([64,128]). *(Inverse Hambo Operator Transform). Consider a Hambo system transform $\widetilde{G}$ of a system $G\in\mathcal{RH}_2$, with minimal state-space realization $(\widetilde{A},\widetilde{B},\widetilde{C},\widetilde{D})$, controllability Gramian $\widetilde{X}_o$, and observability Gramian $\widetilde{X}_o$. Then the system G has minimal state-space realizations (A_i,B_i,C_i,D_i) $(i=1,2)$ that satisfy the following Sylvester equations.*

$$\begin{bmatrix}\widetilde{A} & 0\\ C_b\widetilde{C} & D_b\end{bmatrix}\begin{bmatrix}A_1\widetilde{X}_c & B_1\\ C_1\widetilde{X}_c & D_1\end{bmatrix}\begin{bmatrix}\widetilde{A}^T & 0\\ B_b^T\widetilde{B}^T & D_b^T\end{bmatrix}+\begin{bmatrix}\widetilde{B}\\ C_b\widetilde{D}\end{bmatrix}\begin{bmatrix}A_b^T\widetilde{B}^T & C_b^T\end{bmatrix}=\begin{bmatrix}A_1\widetilde{X}_c & B_1\\ C_1\widetilde{X}_c & D_1\end{bmatrix}$$

$$\begin{bmatrix}\widetilde{A}^T & \widetilde{C}^TC_b^T\\ 0 & D_b^T\end{bmatrix}\begin{bmatrix}\widetilde{X}_oA_2 & \widetilde{X}_oB_2\\ C_2 & D_2\end{bmatrix}\begin{bmatrix}\widetilde{A} & \widetilde{B}B_b\\ 0 & D_b\end{bmatrix}+\begin{bmatrix}\widetilde{C}^TA_b^T\\ B_b^T\end{bmatrix}\begin{bmatrix}\widetilde{C} & \widetilde{D}B_b\end{bmatrix}=\begin{bmatrix}\widetilde{X}_oA_2 & \widetilde{X}_oB_2\\ C_2 & D_2\end{bmatrix}.$$

All these Sylvester equations are in fact alternative formulations of matrix 'inner products', whose existence is guaranteed because the inner product terms are stable systems [64,128]. Because the systems G in Proposition 12.8 and $\widetilde{G}$ in Proposition 12.9 are assumed to be minimal, the corresponding controllability and observability matrices are full rank, which implies that the realizations $\left(\widetilde{A_i}, \widetilde{B_i}, \widetilde{C_i}, \widetilde{D_i}\right)$ respectively (A_i, B_i, C_i, D_i) $(i = 1, 2)$ can be deduced directly from the solutions of the Sylvester equations.
Note that the equations in Propositions 12.8 and 12.9 can be simplified by using input or output balanced realizations, which have the property that the controllability or observability matrix respectively, is equal to the identity matrix.
The first account of these methods to calculate minimal state space realizations were given in the context of realization theory [68, 291] and resulted in slightly different, but numerically 'cheaper' algorithms:

Proposition 12.10 ([68]). *(Forward Operator Transform - Reduced Form). Consider the situation as in Proposition 12.8. Then $\widetilde{G}$ has minimal realizations $\left(\widetilde{A_i}, \widetilde{B}, \widetilde{C}, \widetilde{D}\right)$ $(i = 1, 2)$ that satisfy the following Sylvester equations.*

$$
\begin{aligned}
\widetilde{B} &= A\widetilde{B}A_b + BC_b \\
\widetilde{C} &= A_b\widetilde{C}A + B_bC \\
\widetilde{D} &= A_b\widetilde{D}A_b^T + \left(B_bD + A_b\widetilde{C}B\right)B_b^T \\
X_0\widetilde{A_1} &= A^T\left(X_0\widetilde{A_1}\right)A + +C^T\left(D_bC + C_b\widetilde{C}A\right) \\
\widetilde{A_2}X_c &= A\left(\widetilde{A_2}X_c\right)A^T + +\left(BD_b + A\widetilde{B}B_b\right)B^T
\end{aligned}
$$

Proposition 12.11 ([68]). *(Inverse Operator Transform - Reduced Form). Consider the situation as in Proposition 12.9. Then G has minimal realizations (A_i, B, C, D) $(i = 1, 2)$ that satisfy the following Sylvester equations.*

$$
\begin{aligned}
B &= \widetilde{A}BD_b^T + \widetilde{B}C_b^T \\
C &= D_b^TC\widetilde{A} + B_b^T\widetilde{C} \\
D &= D_bDD_b^T + C_b\left(\widetilde{D}C_b^T + \widetilde{C}BD_b^T\right) \\
\widetilde{X}_oA_1 &= \widetilde{A}^T\left(\widetilde{X}_oA_1\right)\widetilde{A} + \widetilde{C}^T\left(A_b^T\widetilde{C} + C_b^TC\widetilde{A}\right) \\
A_2\widetilde{X}_c &= \widetilde{A}\left(A_2\widetilde{X}_c\right)\widetilde{A}^T + \left(\widetilde{B}A_b^T + \widetilde{A}BB_b^T\right)\widetilde{B}^T
\end{aligned}
$$

Note that D_b is a scalar, which shows that the expressions for B, C, D in Proposition 12.11 can be reduced; for instance, the first equation for calculating B can be rewritten as

$$B = (I - D_b \cdot \widetilde{A})^{-1} \widetilde{B} C_b^T. \tag{12.29}$$

This last relation, Equation (12.29), can be applied to find a realization of the Hambo signal transform $\breve{G}$ of a system $G \in \mathcal{H}_{2-}$. From Proposition 12.6 we know that $\breve{G}(\lambda) = \widetilde{G}(\lambda) W_0(\lambda)$, so we can write

$$\begin{aligned}
\breve{G}(\lambda) &= \left(\widetilde{D} + \sum_{k=1}^{\infty} \widetilde{C} \widetilde{A}^{k-1} \widetilde{B} \lambda^{-k} \right) C_b^T (1 - D_b \lambda)^{-1} \\
&= \left(\widetilde{D} C_b^T + \sum_{k=1}^{\infty} \widetilde{C} \widetilde{A}^{k-1} \widetilde{B} C_b^T \lambda^{-k} \right) (1 - D_b \lambda)^{-1} \\
&= \left(\widetilde{D} C_b^T + \sum_{k=1}^{\infty} \widetilde{C} \widetilde{A}^{k-1} (I - D_b \cdot \widetilde{A}) B \lambda^{-k} \right) (1 - D_b \lambda)^{-1} \\
&= \left(\widetilde{D} C_b^T + \widetilde{C} B D_b + \sum_{k=1}^{\infty} \widetilde{C} \widetilde{A}^{k-1} B (1 - D_b \lambda) \lambda^{-k} \right) (1 - D_b \lambda)^{-1} \\
&= \left(\widetilde{D} C_b^T + \widetilde{C} B D_b \right) (1 - D_b \lambda)^{-1} + \sum_{k=1}^{\infty} \widetilde{C} \widetilde{A}^{k-1} B \lambda^{-k}
\end{aligned}$$

Now we know that $\breve{G} \in \mathcal{H}_{2-}^{n_b}$ and because the first part of the right-hand side is strictly unstable, it follows that this must be zero. Hence we conclude

Corollary 12.1 ([64, 68]). *$\breve{G}$ has a state-space realization* $(\widetilde{A}, B, \widetilde{C}, 0)$.

Note that this realization is not necessary minimal and that $\widetilde{G}$ and $\breve{G}$ can have different McMillan degrees. Consider for example $G \in \mathcal{H}_{2-}$ given by $G(z) = G_b(z) - D_b = C_b V_1(z)$. Then $\widetilde{G}(\lambda) = (\lambda^{-1} - D_b) I_{n_b}$ and $\breve{G}(\lambda) = C_b^T \lambda^{-1}$, with McMillan degrees of n_b and 1 respectively.

We conclude this section with two very intriguing expressions for $\widetilde{D}$ and $\widetilde{A}_i$ as given in Propositions 12.8 and 12.10, where we assume $G(z) = \sum_{k=0}^{\infty} g_k z^{-k}$ and $G_b(z) = \sum_{k=0}^{\infty} (g_b)_k z^{-k}$ (see [68, 128] for details):

$$\widetilde{D} = \sum_{k=0}^{\infty} g_k A_b^k$$

$$\widetilde{A}_i = \sum_{k=0}^{\infty} (g_b)_k A^k$$

A direct consequence of the last expression is that $\widetilde{A}_1 = \widetilde{A}_2$! For practical application, it is therefore recommended to check the numerical condition of the controllability c.q. observability matrix and to choose on the basis of this information which algorithm to use, as this involves the inverse of one of these matrices.

12.5 Hambo Transform Properties

In Proposition 12.4, it was already shown that the Hambo operator transform is a multiplicative mapping. This 'calculation rule' can be extended with linearity and inverse commutation:

$$\widetilde{(\alpha G_1 + \beta G_2)} = \alpha \widetilde{G_1} + \beta \widetilde{G_2} \tag{12.30}$$

$$\widetilde{G^{-1}} = \left(\widetilde{G}\right)^{-1}, \tag{12.31}$$

under the assumption that $G_1, G_2, G, G^{-1} \in \mathcal{H}_2(\mathbb{E})$ and $\alpha, \beta \in \mathbb{R}$.
It can be concluded that parallel and series connections of systems remain unchanged under Hambo operator transformation. The same goes for feedback connections and linear fractional transformations, under the condition that the inverses involved are stable.

12.5.1 Poles and Zeros

As shown before, the Hambo operator transform preserves the McMillan degree, which shows that G and $\widetilde{G}$ have the same number of poles. It turns out that the location of these poles can be specified precisely, with interesting consequences:

> If G has a pole (zero) in $z = a$, then $\widetilde{G}$ has a pole (zero) in $\lambda = G_b\left(\frac{1}{a}\right) = G_b^{-1}(a)$.
> This implies that the operator transform preserves stability.
> Furthermore, for each pole of G that coincides with a pole of G_b, the Hambo operator transform $\widetilde{G}$ has a pole in 0.

The first statement is due to [126]. Now we point out that because G_b is an inner function, it holds that $|G_b(z)| < 1$ for all z with $|z| > 1$. This shows that if $G \in \mathcal{H}_2(\mathbb{E})$, also $\widetilde{G} \in \mathcal{H}_2(\mathbb{E})$. The last statement follows from the fact that for each pole μ of G_b it holds that $G_b\left(\frac{1}{\mu}\right) = 0$.

12.5.2 Norms

It was shown that the Hambo operator transform preserves stability, Hankel singular values, McMillan degree and ℓ_2-gain (Proposition 12.4). These properties lead to the following observation

> For any $G \in \mathcal{H}_\infty$ the Hambo operator transform preserves the Hankel norm and the $\mathcal{H}_\infty$ norm.

Remark 12.2. This statement cannot be made for the $\mathcal{H}_2$ norm. We can however conclude from Proposition 12.4 (using $G_1 = G_2$) that for any $G \in \mathcal{H}_2(\mathbb{E})$, it holds that $\|\widetilde{G}\|_2 = \|V_k G\|_2, \;\; \forall k \in \mathbb{Z}$.

12.6 Extension to Unstable Systems

Until now, we merely considered Hambo operator transforms of stable systems. It is however straightforward – using the extended Hambo bases defined in Proposition 12.1 – to extend the transform theory to systems with unstable poles as well. For a strictly unstable system G, we thereto consider the derived system H, defined by $H(z) = G(\frac{1}{z})$, which is again stable and therefore has a Hambo transform $\widetilde{H}$. It can be shown that $\widetilde{G}(\lambda) = \widetilde{H}(\frac{1}{\lambda})$. Systems that have both stable and unstable poles can be dealt with by splitting into a stable and unstable part.
A nice feature in this respect is that the transformation formulas of Propositions 12.8 and 12.10 can still be applied for all systems $G \in L_2$, except for two classes of systems.

12.6.1 Problem Systems

These two classes of systems are:

1. Systems that have poles both in μ and in $\frac{1}{\mu}$.
2. Systems that have unstable poles that are reciprocals of basis poles.

The reason for these exceptions is that a Sylvester equation of the form

$$X = RXS + T$$

has no unique solution if the matrices R and S have reciproke eigenvalues. A close look at for instance the expressions of Proposition 12.10 then reveals 2 situations with such problems, *i.e.* the cases where A, respectively A and A_b have reciprocal eigenvalues.

For the first case, *i.e.* $\exists \mu$ such that $\mu, \frac{1}{\mu} \in \sigma(A)$, a simple solution exists. It consists of applying the algorithm with A replaced by kA, where $k \in \mathbb{R}, |k| < 1$, such that kA does not have reciprocal eigenvalues. It can be shown that the resulting transformed system is indeed correct, if the matrices X_o and/or X_c are calculated with kA as well. The effect of the multiplication factor k is cancelled out by multiplication with the inverse of X_o respectively X_c.

For the second case, *i.e.* $\exists \mu$ such that $\mu \in \sigma(A_b), \frac{1}{\mu} \in \sigma(A)$, there is unfortunately no generally applicable work around. Note that the Hambo operator transform is in this case still well defined. The problem only lies with the algorithm. The root cause here is that the resulting transformed system may be a non-causal system, which cannot be described with a standard state-space model. To illustrate this phenomenon we consider the case of Laguerre functions, with pole a ($|a| < 1$), so

$$G_b(z) = \frac{1 - za}{z - a} \qquad \text{and} \qquad N(\lambda) = \frac{1 + \lambda a}{\lambda + a}.$$

Let G be given by:

$$G(z) = \frac{z}{1 - za},$$

and calculate $\widetilde{G}$ with the variable substitution method (Proposition 12.5):

$$\widetilde{G}(\lambda) = G(z)|_{z^{-1} = N(\lambda)} = \frac{1}{N(\lambda) - a} = \frac{\lambda + a}{1 - a^2},$$

which is non-causal and cannot be represented by a standard state-space model.

Note that G has a pole in $z_1 = \frac{1}{a}$ and a zero in $z_2 = 0$, while $\widetilde{G}$ has a pole in $\lambda = \infty$ and a zero in $\lambda = -a$, which illustrates again the mapping of poles and zeros, as $G_b(\frac{1}{z_1}) = G_b(a) = \infty$ and $G_b(\frac{1}{z_2}) = G_b(\infty) = -a$.

12.7 Transformations Based on Takenaka-Malmquist Functions

As indicated in the introduction to this chapter, a more involved transform theory for the class of Takenaka-Malmquist functions exists and is reported in detail in [64]. The most important feature of this theory is that it results in an operator transform that transforms scalar time-invariant linear systems into scalar time-varying linear systems. When applied to the special case of repetitive poles, such as in the Hambo case, the resulting operator turns out to be a *periodic* time-varying operator, with period n_b. So we can write this operator in state-space form as

$$\begin{aligned} x(k+1) &= \mathcal{A}(k)x(k) + \mathcal{B}(k)u(k) \\ y(k) &= \mathcal{C}(k)x(k) + \mathcal{D}(k)u(k), \end{aligned}$$

where $\mathcal{A}(k + n_b) = \mathcal{A}(k)$ *etc.* Now it is well-known that such periodic time varying systems can be reformulated as multi-variable time-invariant systems using *lifting* [95]. This is exactly the procedure that yields the Hambo operator transform.

12.8 Transformations of Multi-variable Systems

In this chapter, we considered only the operator transform of scalar systems, which resulted in multi-variable systems of dimension $n_b \times n_b$. For the multi-variable case there are two approaches that can be followed (see also Section 10.5).

12.8.1 Approach 1: Orthonormal Basis Functions Using a Scalar Inner Function

This is the situation that has been considered in this chapter. It is straightforward that for a system $G \in \mathcal{H}_2^{p\times m}$, characterized by

$$\begin{bmatrix} Y_1(z) \\ Y_2(z) \\ \vdots \\ Y_p(z) \end{bmatrix} = \begin{bmatrix} G_{11}(z) & G_{12}(z) & \cdots & G_{1m}(z) \\ G_{21}(z) & G_{22}(z) & \cdots & G_{2m}(z) \\ \vdots & \vdots & \cdots & \vdots \\ G_{p1}(z) & G_{p2}(z) & \cdots & G_{pm}(z) \end{bmatrix} \begin{bmatrix} U_1(z) \\ U_2(z) \\ \vdots \\ U_p(z) \end{bmatrix}$$

we can expand the theory in this chapter by taking the Hambo signal transform of the scalar signals Y_i and U_i, resulting in a multi-variable formulation (see also [64])

$$\begin{bmatrix} \breve{Y}_1(\lambda) \\ \breve{Y}_2(\lambda) \\ \vdots \\ \breve{Y}_p(\lambda) \end{bmatrix} = \begin{bmatrix} \widetilde{G}_{11}(\lambda) & \widetilde{G}_{12}(\lambda) & \cdots & \widetilde{G}_{1m}(\lambda) \\ \widetilde{G}_{21}(\lambda) & \widetilde{G}_{22}(\lambda) & \cdots & \widetilde{G}_{2m}(\lambda) \\ \vdots & \vdots & \cdots & \vdots \\ \widetilde{G}_{p1}(\lambda) & \widetilde{G}_{p2}(\lambda) & \cdots & \widetilde{G}_{pm}(\lambda) \end{bmatrix} \begin{bmatrix} \breve{U}_1(\lambda) \\ \breve{U}_2(\lambda) \\ \vdots \\ \breve{U}_p(\lambda) \end{bmatrix}$$

Note that this results in a system $\widetilde{G} \in \mathcal{H}_2^{pn_b\times mn_b}$, which can become quite large and difficult to handle. The invariance of McMillan degree is still valid for the multi-variable case.

12.8.2 Approach 2: Orthonormal Basis Functions Using a Multi-variable Inner Function

In [126] it was shown that the definition of Hambo functions can be extended to define basis functions for the space $\mathcal{H}_{2-}^{m}$. Thereto we consider a square multi-variable inner function $G_b \in \mathcal{H}_2^{m\times m}$. Let as before (A_b, B_b, C_b, D_b) be a balanced realization for G_b and define

$$V_1(z) = (zI - A_b)^{-1}B_b \in \mathcal{H}_{2-}^{n_b\times m}.$$

It can be shown that the the functions

$$\{e_i^T V_1(z) G_b^{k-1}(z) \in \mathcal{H}_{2-}^{1\times m}\}$$

constitute a basis for this space $\mathcal{H}_{2-}^{1\times m}$ of m-dimensional signals, such that for each $X(z) \in \mathcal{H}_{2-}^{m}$, there exist unique coefficient vectors $\breve{X}_k \in \mathbb{R}^{n_b}$ such that

$$X^T(z) = \sum_{k=0}^{\infty} \breve{X}_k^T V_1(z) G_b^{k-1}(z).$$

As shown before, this can be used for the identification of multi-variable $p\times m$ systems, estimating a finite number parameters $L_k \in \mathbb{R}^{n_b\times p}$ in

$$Y(z) = \sum_{k=1}^{N} L_k^T V_1(z) G_b^{k-1}(z) U(z) + E(z)$$

This angle, using multi-variable inner functions, has to our knowledge never been used in the context of transformation, except for a theorem in [126], closely related to Proposition 10.1, where it is shown that for a given system $G \in \mathcal{H}_2^{p \times m}$, with series expansion,

$$G(z) = \sum_{k=1}^{\infty} L_k^T V_1(z) G_b^{k-1}(z)$$

the associated system $\breve{G} \in \mathcal{H}_2^{n_b \times p}$,

$$\breve{G}(\lambda) = \sum_{k=1}^{\infty} L_k \lambda^{-k},$$

has maximally McMillan degree $n \cdot m$, where n is the McMillan degree of G. This result is consistent with Corollary 12.1. It is our expectation that many other results, presented in this chapter for scalar basis functions, can be carried over to the multi-variable function case.

12.9 Summary

This chapter presented a detailed overview of the transform theory of signals and systems that is induced by the use of rational orthonormal bases, specifically Hambo bases. It was shown that if the input and output signals of a linear time-invariant (LTI) system are expanded in terms of these bases, then the corresponding expansion sequences, called the signal transforms, are mapped to each other via an LTI system as well, which is called the operator transform of the system. This operator transform leaves a number of interesting system properties invariant, such as Hankel singular values and $\mathcal{H}_\infty$-norm. Central results are the transform formulas that enable the calculation of the operator transform on state-space level.

13

Realization Theory

Peter Heuberger[1] and Thomas de Hoog[2]

[1] Delft University of Technology, Delft, The Netherlands
[2] Philips Research Laboratories, Eindhoven, The Netherlands

13.1 Introduction

This chapter will discuss how the concepts of classical realization theory can be generalized and embedded in the orthonormal basis framework, as described in the previous chapters. In particular, it will be shown how series expansions in terms of a Hambo basis can be related to the underlying system and how minimal (state-space) realizations can be derived from such expansions. Both the problems of exact realization, based on an infinite length expansion, and the problem of approximate realization, based on finite expansions, will be discussed.
Hence, the problem considered is as follows: given a generalized orthonormal basis function (GOBF) basis $\{F_k(z)\}$ and a series of expansion coefficients $\{\tilde{c}_k\}_{k=1,\cdots N}$, find a minimal state-space realization (A, B, C, D) of a system $G(z) = D + C(zI - A)^{-1}B$ of smallest order such that

$$G(z) = \sum_{k=1}^{\infty} c_k F_k(z)$$
$$c_k = \tilde{c}_k, \qquad k = 1, \cdots N.$$

When $N = \infty$, this is known as the exact minimal realization problem, and for $N < \infty$ this is referred to as the minimal partial realization problem.

This problem can be viewed as a generalization of the classical minimal (partial) realization problem that was solved in [134] and [301]. In the classical realization framework, one attempts to find a minimal state-space realization on the basis of full or partial knowledge of the pulse response parameters of a, possibly multi-variable, system. A solution for the case in which one has knowledge of all the coefficients was given by [134] and is known as the *Ho-Kalman algorithm*. It is based on full-rank decompositions of the (block) Hankel matrix built from the pulse response coefficients of the system.

It was shown in [301] that the Ho-Kalman algorithm can still be applied to find a minimal realization in the partial information case, provided that certain rank conditions on a finite Hankel matrix are satisfied.

In the case of Laguerre function expansions, the problem was solved in [225]. For the Hambo basis, case the realization problem was first considered in [287] and [291], where the problem was solved for the full-information case ($N \to \infty$). The minimal partial realization case has been solved in [68], by exploiting the Hambo transform theory, as discussed in the previous chapter, see also [64].
Besides being interesting from a system theoretic point of view, the generalized realization theory finds its motivation in the application of these bases in system identification. As discussed before in *e.g.* Chapter 4, because of the linear parameterization (c_k appears linearly in $G(z)$) attractive computational properties result, *e.g.* in least squares algorithms, which enables the handling of large-scale problems.
When using orthonormal basis functions in an approximation or identification setting, one will typically end up with a finite number of expansion coefficients, leading to a model

$$\hat{G}(z) = D + \sum_{k=1}^{n} c_k F_k(z).$$

It is straightforward to derive a state-space representation for this model (see also Chapters 2 and 10). For practical purposes, such as control design, this model will often be of a too high order. Hence the user will need some model reduction procedure, such as for instance balanced reduction or optimal Hankel norm reduction. The realization algorithms discussed in this chapter serve as an alternative to model reduction, motivated by the observation that the expansion coefficients carry more information than is reflected in the full-order state-space model.

Consider for example the expansion of a system $G(z)$ in the standard basis

$$G(z) = D + \sum_{k=1}^{\infty} c_k z^{-k}$$

and assume that we have exact knowledge of the first n coefficients, in this case Markov parameters. It is well-known that (for scalar systems) the Ho-Kalman realization algorithm will render the true system if $n \geq 2n_g$, where n_g is the McMillan degree of $G(z)$ [134,301]. A model reduction method applied to the full-order state-space realization of

$$\hat{G}(z) = D + \sum_{k=1}^{n} c_k z^{-k}$$

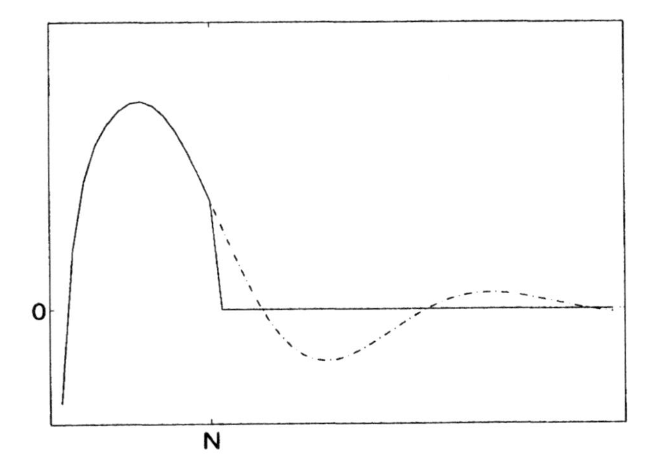

Fig. 13.1. Impulse response of true system (dash-dotted) and of FIR model (solid).

will often only render the true system if n is very high. This results from the fact that this model assumes that $c_k \equiv 0$, for $k > n$. See Figure 13.1 for an example of the phenomenon. Hence in this case it pays off to explore all the information contained in the expansion coefficients. A similar effect is present in the generalized basis expansions.
The major subject of this chapter is the problem of how we can, once a (finite) number of expansion coefficients is known, realize a minimal state-space model of which the corresponding expansion coefficients match the given ones, where minimality should be interpreted in the sense that there are no models of lower McMillan degree that solve the problem. It turns out that for the class of *Hambo* orthonormal basis functions, an elegant solution to this problem can be formulated, making use of the Hambo system transform defined in Chapter 12. The basic ingredient of this solution is the (generalized) shift structure that is present in the Hambo functions. This shift structure implies that there exists a one-to-one relationship between the expansion coefficients and the Markov parameters of the Hambo system transform of the system. The theory that is analysed in this chapter, relies heavily on the transformation results of Chapter 12.

In this chapter, we will consistently use the notation G_b for the basis generating inner function with McMillan degree n_b and minimal balanced state-space realization (A_b, B_b, C_b, D_b). As in the previous chapters, we will use the symbols $v_k(t)$ and $V_k(z)$ for the vectorized Hambo functions in the time and the frequency domain. Throughout the chapter we will restrict attention to the case of scalar transfer functions. Also, it is assumed that the basis functions are real rational implying that the associated coefficients are real valued. As in Chapter 12, we will use the symbol $\mathcal{H}_{2-}$ for the Hardy space, which consists of all functions in $\mathcal{H}_2(\mathbb{E})$ that are zero at infinity, such as strictly proper stable systems. The symbol $\mathcal{RH}_{2-}$ denotes the subset of rational functions in $\mathcal{H}_{2-}$ with real-valued coefficients.

13.2 From Expansion Coefficients to Markov Parameters

In this section, it will be explained how, given a system $G \in \mathcal{RH}_{2-}$, the Markov parameters of the Hambo operator transform $\widetilde{G}$, as defined in Definition 12.3, can be derived from the expansion coefficients pertaining to an orthonormal series expansion for G.
It was shown in Chapter 2 that – given an Hambo basis $\{V_k\}_{k=1}^{\infty}$ – any system $G \in \mathcal{H}_{2-}$ can be written in a unique series expansion,

$$G(z) = \sum_{k=1}^{\infty} \breve{g}^T(k) V_k(z), \quad \breve{g}(k) = [\![V_k, G]\!] .$$

Furthermore it was shown in Chapter 12 that the Hambo operator transform of $G \in \mathcal{RH}_{2-}$, denoted by $\widetilde{G}(\lambda) \in \mathcal{RH}_2^{n_b \times n_b}$ has the same McMillan degree as G and efficient algorithms have been presented to calculate minimal state-space realizations, as well for the forward as inverse transformation.

In this chapter, we deal with the situation where only expansion coefficients are given and we will use a route through the Hambo transform to calculate state-space realizations for the underlying system. We therefore start by showing how these expansion coefficients can be used to calculate the Markov parameters of the Hambo operator transform.

We recall from Equation (12.10) the relation between products of the elements of $V_1(z)$, where we denote $\phi_i(z) = (V_1(z))_i$,

$$\begin{aligned} V_1(z)\phi_i(z) &= P_i V_1(z) + Q_i V_2(z) \\ P_i &= A_b P_i A_b^T + B_b e_i^T A_b^T \\ Q_i &= A_b^T Q_i A_b + C_b^T e_i^T \end{aligned}$$

This relation was used to show in Equation (12.20) that the Hambo operator transform of $\phi_i(z)$ obeys

$$\widetilde{\Phi}_i(\lambda) = P_i^T + Q_i^T \lambda^{-1}.$$

Furthermore, we recollect from Equation (12.18) that the inner function G_b, used to generate the Hambo basis, is transformed into a shift in the λ-domain,

$$\widetilde{G_b}(\lambda) = \lambda^{-1} I.$$

Now we consider a system $G \in \mathcal{RH}_{2-}$, with a series expansion in terms of the Hambo basis $\{V_k\}$, as in $G(z) = \sum_{k=1}^{\infty} \breve{g}^T(k)V_k(z)$, and we rewrite this summation in scalar functions

$$G(z) = \sum_{k=1}^{\infty}\sum_{i=1}^{n_b} \breve{g}_i(k)\phi_i(z)G_b^{k-1}(z).$$

Application of the multiplicativity and linearity of the Hambo operator transform (see Proposition 12.4 and Equation (12.30)) then yields

$$\begin{aligned}\widetilde{G}(\lambda) &= \sum_{k=1}^{\infty}\sum_{i=1}^{n_b} \breve{g}_i(k)\widetilde{\Phi}_i(\lambda)\lambda^{1-k}\\ &= \sum_{k=1}^{\infty}\sum_{i=1}^{n_b} \breve{g}_i(k)\left(P_i^T + Q_i^T\lambda^{-1}\right)\lambda^{1-k}\end{aligned}$$

Now we compare this expression with the Markov parameter description $\widetilde{G}(\lambda) = \sum_{k=0}^{\infty} M_k\lambda^{-k}$, and conclude that

$$M_0 = \sum_{i=1}^{n_b} \breve{g}_i(1)P_i^T \tag{13.1}$$

$$M_k = \sum_{i=1}^{n_b} \left(\breve{g}_i(k+1)P_i^T + \breve{g}_i(k)Q_i^T\right) \quad k \in \mathbb{N} \tag{13.2}$$

Hence, given a sequence of N expansion coefficient vectors $\{\breve{g}(k)\}_{k=1}^{N}$, the expressions above enable the calculation of N Markov parameters $\{M_k\}_{k=0}^{N-1}$ of $\widetilde{G}$. Note specifically that M_k only depends on $\breve{g}(k)$ and $\breve{g}(k+1)$.

13.3 Realization Algorithms

In this section, we will discuss three algorithms, starting with the full knowledge case for illustrative purposes. For the next two algorithms we will consider the partial knowledge case, *i.e.* the case where only a finite number of expansion coefficients is available. First, we study the situation where the McMillan degree of the underlying system is known and finally, we discuss the actual minimal partial realization problem. In the sequel, we will use the symbol N for the number of expansion coefficient vectors.

In this section we, will use the Hambo operator transform, *i.e.* system representations in the λ-domain. It is important to stress here that (for $n_b > 1$) not all systems in $\mathcal{H}_2^{n_b \times n_b}$ are the result of the operator transform. The set of systems that indeed are Hambo operator transforms of a system in $\mathcal{H}_2$ constitutes only a subset of the space $\mathcal{H}_2^{n_b \times n_b}$, which is isomorphic with $\mathcal{H}_2$. In fact, it is the set of systems that commute with $N(\lambda) = A_b + B_b(\lambda I - D_b)^{-1}C_b$, defined in Equation (12.3) (see [64,68]).

13.3.1 Algorithm 1: N = ∞

The solution to the classical minimal realization problem, due to Ho and Kalman [134] and known as the Ho-Kalman algorithm, is based on the full rank decomposition of the Hankel matrix, as given by Equation (12.9). Given a decomposition $\mathbf{H} = \mathbf{\Gamma\Delta}$, the B and C matrices are obtained as the first column(s) of $\mathbf{\Delta}$, respectively the first row(s) of $\mathbf{\Gamma}$. The A-matrix is obtained from the equation $\mathbf{H}^{\leftarrow} = \mathbf{\Gamma} A \mathbf{\Delta}$, where $\mathbf{H}^{\leftarrow}$ is the matrix, that results when removing the first (block) column of $\mathbf{H}$.

If we have expansion coefficients $\{\breve{g}(k)\}_{k=1}^{\infty}$, then the previous section enables the calculation of the Markov parameters $\{M_k\}_{k=0}^{\infty}$ of the system $\widetilde{G}$. With these Markov parameters, we can construct a Hankel matrix

$$\widetilde{\mathbf{H}} = \begin{bmatrix} M_1 & M_2 & M_3 & \cdots \\ M_2 & M_3 & M_4 & \\ M_3 & M_4 & M_5 & \\ \vdots & & & \ddots \end{bmatrix}$$

Now we can apply the standard Ho-Kalman algorithm to derive – under the assumption that the system at hand is finite dimensional – a minimal state-space realization $(\widetilde{A}, \widetilde{B}, \widetilde{C}, M_0)$, such that $M_k = \widetilde{C}\widetilde{A}^{k-1}\widetilde{B}, \quad k \in \mathbb{N}$. Application of the inverse Hambo operator transform, as in Proposition 12.9 or Proposition 12.11, then yields a minimal state-space realization for the underlying system G.

In this situation, it is obvious that the system in the λ-domain must be a valid Hambo operator transform.
Though this algorithm can only be seen as a thought experiment, as we cannot perform these infinite length calculations, it explains clearly the basic steps that need to be taken care of in the generalized realization problems.

13.3.2 Algorithm 2: N < ∞, Known McMillan Degree

It is well-known that for the standard basis, the Ho-Kalman algorithm can also be applied to a finite submatrix $\mathbf{H}_{N_1,N_2}$ of the full Hankel matrix $\mathbf{H}$, where $N = N_1 + N_2$ is the number of available Markov parameters. If N is sufficiently large, the algorithm will lead again to an exact realization of the underlying system. The obvious problem here is of course that beforehand it is unknown how large 'sufficiently large' is. This is only known when the McMillan degree n of the underlying system is known, in which case the finite Hankel matrix must be so large that it has rank n. Now assume that n is known and we have expansion coefficients $\{\breve{g}(k)\}_{k=1}^{N}$, then again we can calculate the Markov parameters $\{M_k\}_{k=0}^{N-1}$. If N is sufficiently large, we can create a finite Hankel matrix $\widetilde{\mathbf{H}}_{N_1,N_2}$, with $N_1 + N_2 = N - 1$, that has rank n.

Application of the Ho-Kalman algorithm then again yields a minimal state-space realization $(\widetilde{A}, \widetilde{B}, \widetilde{C}, M_0)$ for $\widetilde{G}$, which can be transformed to a minimal state-space realization for G using the inverse Hambo operator transform, as in Proposition 12.9 or Proposition 12.11.
Whether the system in the λ-domain is a valid Hambo operator transform is now assured by the rank condition.

13.3.3 Algorithm 3: N < ∞, Unknown McMillan Degree

We now proceed with the actual minimal partial realization problem. This problem is stated as follows:

> Given a finite set of coefficient vectors $\{L_k\}_{k=1}^N, N < \infty$, find a minimal state-space realization $(A, B, C, 0)$ of a system $G(z) = C(zI - A)^{-1}B$ of smallest McMillan degree, such that $G(z) = \sum_{k=1}^{\infty} \breve{g}^T(k)V_k(z)$ and $L_k = \breve{g}(k)$, $k = 1 \cdots N$.

For the standard basis, it was shown in [301] that a unique solution (modulo similarity transformations) can be found using the Ho-Kalman algorithm, provided that a certain rank condition is satisfied by the sequence $\{L_k\}_{k=1}^N$, which in the standard case are the Markov parameters of the system.
A similar result can of course be derived for the Hankel matrix $\widetilde{\mathbf{H}}$, which can be created on the basis of the coefficient vectors and the procedure of the previous section for calculation of the generalized Markov parameters. Hence this would result in a realization $(\widetilde{A}, \widetilde{B}, \widetilde{C}, M_0)$ in the λ-domain. However *there is no guarantee that this system in $\mathcal{RH}_2^{n_b \times n_b}$ is a valid Hambo operator transform of a system $G \in \mathcal{RH}_2$.*

In order to solve the minimal partial realization problem, we have to consider a specific realizability condition. The major key to find this condition is provided by Corollary 12.1, which showed that for a system $G \in \mathcal{RH}_2$ the generalized Markov parameters and the coefficient vectors can be realized by state-space realizations with the same state transition matrix $\widetilde{A}$.
This motivates to consider the sequence of concatenated matrices

$$\mathcal{M}_k = \begin{bmatrix} M_k \; L_k \; L_{k+1} \end{bmatrix} \qquad k = 1 \cdots N-1,. \tag{13.3}$$

where L_{k+1} is included because M_k depends on both L_k and L_{k+1}.
The following lemma provides the conditions under which the minimal partial realization problem can be solved.

Lemma 13.1 ([68]). *Let $\{L_k\}_{k=1,..,N}$ be an arbitrary sequence of $n_b \times 1$ vectors and let M_k and $\mathcal{M}_k$ for $k = 1,..N-1$ be derived from L_k via relations (13.2) and (13.3). Then there exists a unique minimal realization (modulo* *similarity transformation)* $(\widetilde{A}, \widetilde{B}, \widetilde{C})$ *with McMillan degree* n*, and an* $n \times 1$ *vector* B *such that*

(a) $M_k = \widetilde{C}\widetilde{A}^{k-1}\widetilde{B}$ *for* $k = 1, .., N-1$*, and* $L_k = \widetilde{C}\widetilde{A}^{k-1}B$ *for* $k = 1, .., N$*,*
(b) the infinite sequences $[\,M_k\ L_k\,] := \widetilde{C}\widetilde{A}^{k-1}[\,\widetilde{B}\ B\,]$ *satisfy relation* (13.2) *for all* $k \in \mathbb{N}$*,*

if and only if there exist positive N_1, N_2 *such that* $N_1 + N_2 = N - 1$ *and*

$$\text{rank } \hat{\mathbf{H}}_{N_1,N_2} = \text{rank } \hat{\mathbf{H}}_{N_1+1,N_2} = \text{rank } \hat{\mathbf{H}}_{N_1,N_2+1} = n, \tag{13.4}$$

where $\hat{\mathbf{H}}_{i,j}$ *is the Hankel matrix built from the matrices* $\{\mathcal{M}_k\}$ *with block-dimensions* $i \times j$.

The condition formulated in (13.4) is the same as the one given in [301], applied to the sequence $\{\mathcal{M}_k\}_{k=1}^{N-1}$ and it follows immediately that this can only be verified if $N > 2$.
The actual algorithm now consists of the following steps:

1. Create $\{M_k\}_{k=1}^{N-1}$ and $\{\mathcal{M}_k\}_{k=1}^{N-1}$ using the relations (13.2) and (13.3).
2. Check whether positive N_1, N_2 exist such that $N_1 + N_2 = N - 1$ and Condition (13.4) holds
3. If so, apply the Ho-Kalman algorithm to the sequence $\{\mathcal{M}_k\}_{k=1}^{N-1}$ and derive a minimal realization $(\widetilde{A}, \left[\widetilde{B}\ X_2\ X_3\right], \widetilde{C}, \widetilde{D})$.
4. Apply the inverse Hambo operator transform (Propositions 12.9, and 12.11) to the realization $(\widetilde{A}, \widetilde{B}, \widetilde{C}, \widetilde{D})$ to derive the matrices A, B and C.

The special realizability condition, defined by Equation (13.4) ensures that the system with realization $(\widetilde{A}, \widetilde{B}, \widetilde{C}, \widetilde{D})$ is indeed a valid Hambo operator transform.

13.4 Approximate Realization

A well-known application of the classical partial realization algorithm is its use as a system identification method, as in *e.g.* [154, 337], where a Hankel matrix is created from (possibly noise corrupted) expansion coefficients and rank reduction is performed through singular value truncation. In that setting one first obtains estimates of the pulse response parameters of the system G on the basis of an identification experiment, *e.g.* by identifying a finite impulse response (FIR) model. The finite Hankel matrix $\mathbf{H}$ built from these parameter estimates can be used in the Ho-Kalman realization algorithm. In general, the Hankel matrix obtained in this manner will not be rank deficient, as required in Condition (13.4). In order to obtain a low-order state-space model, the realization problem has to be solved in some approximate sense. This is most commonly done by truncating the smaller singular values of the singular value decomposition of $\mathbf{H}$ [337]. This approach can similarly be applied to the generalized situation using Algorithm 2. Instead of estimating the FIR, coefficients one uses a finite number N of estimates of the expansion coefficients L_k in a Hambo function expansion. From these, one can then compute estimates of the first $N-1$ Markov parameters M_k. As in the classical situation, the finite block Hankel matrix $\widetilde{\mathbf{H}}$ that is constructed from these data will in general be full rank. A low-order (in terms of rank) approximation can then be obtained by applying the singular value decomposition (SVD) truncation approach.

An important issue that arises in this context is the fact that not every system in $\mathcal{H}_2^{n_b \times n_b}$ is the Hambo transform of a system in $\mathcal{H}_2$. A necessary and sufficient condition (see [64,68]) is that it commutes with the transfer matrix $N(\lambda)$, defined in Equation (12.3). Naturally, this commutative property of Hambo transforms is not true for general approximate realizations $\{\widetilde{A}, \widetilde{B}, \widetilde{C}\}$ obtained with Algorithm 2, respectively realizations $(\widetilde{A}, \left[X_1 | X_2 | X_3\right], \widetilde{C})$ obtained through Algorithm 3. Although the formulas for the inverse operator transform still can be applied, the resulting system will not have a one-to-one correspondence with the approximate realization. In the exact realization setting, this problem does not arise. The problem can be circumvented however by using the approximate realization $(\widetilde{A}, X_2, \widetilde{C})$ for $\{L_k\}$ resulting from Algorithm 3 and considering the approximation

$$\hat{G}(z) = \sum_{k=1}^{\infty} (\widetilde{C}\widetilde{A}^{k-1} X_2)^T V_k(z)$$

which (under the condition that $\widetilde{A}$ is stable) is finite dimensional with McMillan degree smaller or equal to $n_b \times \dim(\widetilde{A})$, as shown in Remark 12.1. A minimal realization for $\hat{G}(z)$ can subsequently be obtained by application of Algorithm 2.

Remark 13.1. The formulas for the inverse Hambo transformation make it possible to transfer not only the realization problem to the transform domain but also the whole identification procedure itself. This idea has been applied before for the Laguerre and two-parameter Kautz case in [79,92], where a subspace identification method is used for the extraction of the state-space matrices of the transformed system from expansions of the measured data in terms of these basis functions. A state-space representation in the time domain is then obtained by applying inversion formulas. This idea can straightforwardly be extended to the general case using the Hambo transform formulas. It should be noted however, as explained above, that in general the outcome of the identification procedure does not immediately result in a valid operator transform. Therefore, care should be taken when the inverse transformation formulas are applied in such a context.

13.5 Interpolation

It is well-known that the classical problem of minimal partial realization from the first N Markov parameters is equivalent to the problem of constructing a stable strictly-proper real-rational transfer function of minimal degree that interpolates to the first $N-1$ derivatives of $G(z)$ evaluated at $z=\infty$ [18]. Similarly, the least-squares approximation of a stable transfer function G in terms of a finite set of rational basis functions interpolates to the function $G(z)$ and/or its derivatives in the points $1/\xi_i$, with $\{\xi_i\}$ the poles of the basis functions involved [320].

In the basis construction considered in this book the error function of an Nth-order approximation $\hat{G}(z)=\sum_{k=1}^{N}\breve{g}(k)^T V_k(z)$ takes on the form

$$E(z)=\sum_{k=N+1}^{\infty}\breve{g}(k)^T V_k(z)=G_b^N(z)\sum_{k=1}^{\infty}\breve{g}(N+k)^T V_1(z)G_b(z)^{k-1}.$$

Due to the repetition of the all-pass function $G_b(z)$ in V_k, the error $E(z)$ will have as a factor the function $G_b^N(z)$. This means that $E(z)$ has zeros of order N at each of the points $1/\xi_i$ and subsequently $\hat{G}(z)$ interpolates to $\frac{d^{k-1}G}{dz^{k-1}}$ in $z=1/\xi_i$ for $k=1,..,N$ and $i=1,..,n_b$. This interpolation property in fact holds true for any model of which the first N expansion coefficient vectors match those of the system, in particular for a model found by solving the partial realization problem.

In view of the interpolating property of the basis function expansion, it is not surprising that there exists a one-to-one correspondence between the expansion coefficient vector sequence $\{\breve{g}(k)\}_{k=1,..,N}$ and the interpolation data $\{\frac{d^{k-1}G}{dz^{k-1}}(1/\xi_i)\}_{k=1,..,N}$. An explicit expression for this relation can be derived by exploiting the linear transformation that links the set of basis function vectors V_k and the set of vectors that consists of single-pole transfer functions as given by

$$S_k^T(z)=\left[\frac{1}{(z-\xi_1)^k},\cdots,\frac{1}{(z-\xi_{n_b})^k}\right]^T,$$

with $\{\xi_i\}$ the poles of the basis generating function G_b. See [64, 68] for a detailed account of these relations. One can hence solve the following interpolation problem, by means of Algorithm 3 in the previous section.

Problem 13.1 (Interpolation Problem). Given the interpolation conditions

$$\frac{d^{k-1}G}{dz^{k-1}}(1/\xi_i)=c_{i,k},\quad c_{i,k}\in\mathbb{C},$$

for $i=1,..,n_b$ and $k=1,..,N$, with $\xi_i\neq 0$ distinct points inside the unit disc, find the $\mathcal{H}_2$ rational transfer function of least possible degree that interpolates these points.

The problem is solved for $N > 2$ by constructing an inner function G_b with balanced realization (A_b, B_b, C_b, D_b) such that the eigenvalues of A_b are ξ_i. From this inner function, one can then obtain all the parameters that are necessary to compute the set of Markov parameters M_k that correspond to the interpolation data. The minimal partial realization algorithm then renders the desired transfer function, provided that the necessary conditions are satisfied. Note that the $\{\xi_i\}$ should come in conjugate pairs to ensure that the resulting transfer function has real-valued coefficients. See [64, 68] for details.

13.6 Examples

13.6.1 Exact Realization

To illustrate the power of the partial realization algorithm, we applied it to data from a sixth-order LTI system that was obtained through system identification of one of the input/output channels of a non-linear model of a fluidized catalytic cracking unit [309], given by

$$G(z) = 10^{-3} \times \frac{-0.5638z^5 + 43.93z^4 - 21.66z^3 - 1.041z^2 - 95.72z + 75.18}{z^6 - 3.3519z^5 + 4.8399z^4 - 4.4410z^3 + 3.1107z^2 - 1.4755z + 0.3180}.$$

An important feature of this process is that it incorporates a combination of fast and slow dynamics, illustrated by the step response of the system, as shown in Figure 13.2. The poles of the system are:

$$\big(-0.0036 \pm 0.7837i,\, 0.7446 \pm 0.2005i,\, 0.9941,\, 0.8758\big)\,.$$

We deliberately choose a 'bad' Hambo basis, generated by poles in $0.5 \pm 0.5i$. The motivation for doing so is that an expansion in this basis will need a considerable number of expansion coefficients in order to capture the system dynamics, whereas the partial realization algorithm can render the true system with much fewer coefficients. It turns out that the partial realization Algorithm 2 can exactly retain the true system from the first 8 expansion coefficients $\breve{g}(k)$.

On the other hand, the 16th-order model that incorporates these coefficients in

$$\hat{G}(z) = \sum_{k=1}^{8} \breve{g}(k) V_k(z)$$

is lacking information in the low frequency range. This is depicted in Figures 13.2 and 13.3, where the step responses and frequency responses of the example system and the 16th-order model are shown.

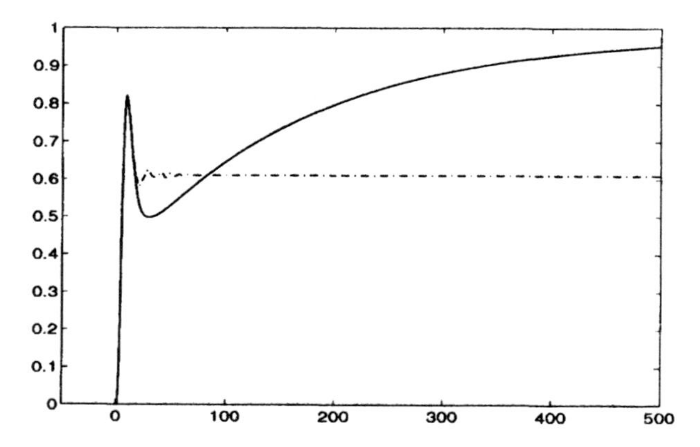

Fig. 13.2. Comparison of the step responses of the example system (solid) and the 16th-order model $\hat{G}(z)$ (dash-dotted).

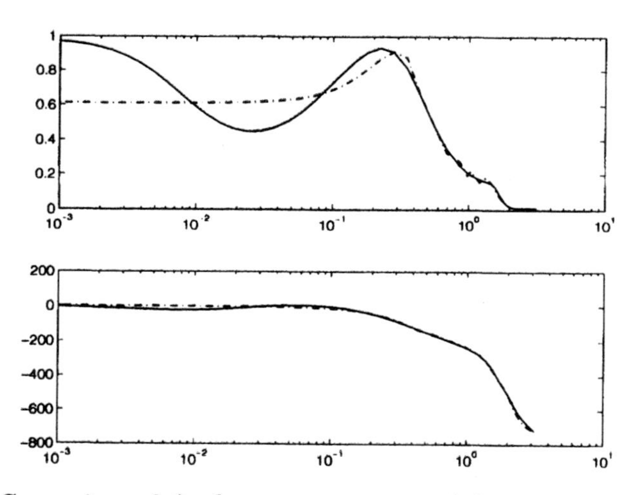

Fig. 13.3. Comparison of the frequency responses of the example system (solid) and the 16th-order model $\hat{G}(z)$ (dash-dotted). Upper: amplitude; lower: phase.

The coefficient vectors $\{\breve{g}(k)\}$ are depicted in Figure 13.4. In the upper part of the latter, the first 20 coefficient vectors are shown, while the lower part shows the coefficients $\breve{g}(21)$ to $\breve{g}(200)$. Although from the upper part of the figure it seems valid to say that the coefficients have converged, the lower part shows that the convergence is very slow. Hence the energy in the neglected tail is considerable, causing the bad quality of the approximation.
This example illustrates that the sequence of coefficients carries more information than is reflected by the full-order approximation and that partial realization can be a powerful tool.

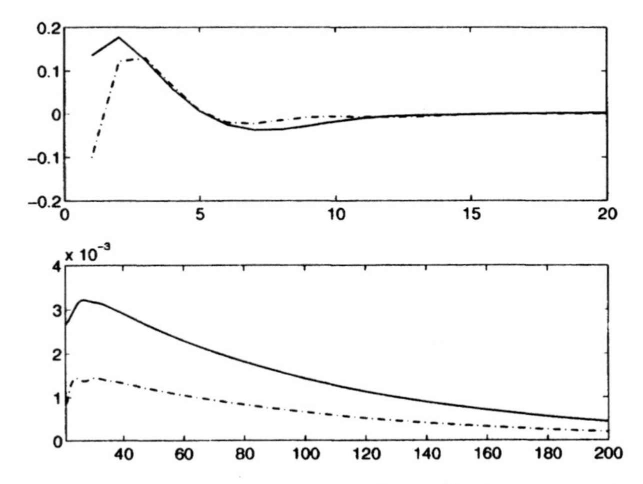

Fig. 13.4. Expansion coefficients vectors $\breve{g}(k) \in \mathbb{R}^2$ of the example system. Upper: $\breve{g}(1) - \breve{g}(20)$; lower: $\breve{g}(21) - \breve{g}(200)$

13.6.2 Approximate Realization, Using Algorithm 2

We use the same model as in the previous example to show the strength of the partial realization algorithm for the approximate case. For this purpose, 10 simulations are carried out, in which the response of the system $G(z)$ to a Gaussian white noise input with unit standard deviation is determined. An independent Gaussian noise disturbance with standard deviation 0.05 is added to the output. This amounts to a signal-to-noise ratio (in terms of RMS values) of about 17 dB. The length of the input and output data signals is taken to be 1000 samples. For each of the 10 data sets, two basis function models of the form

$$\hat{G}(z) = \sum_{k=1}^{N} \hat{\breve{g}}(k)^T V_k(z), \tag{13.5}$$

are estimated using the least-squares method described in Chapter 4. The first model is a 40th-order FIR model. Hence in this case, $V_k(z) = z^{-k}$ and $N = 40$. The second model uses a generalized basis that is generated by a second-order inner function with poles 0.5 and 0.9. For this model, 20 coefficient vectors are estimated. Hence the number of estimated coefficients is equal for both models.

We now apply the approximate realization method using the estimated expansion coefficients of both models, for all 10 simulations. In either case a sixth-order model is computed, through truncation of the SVD of the finite Hankel matrix, which in the classical setting is the algorithm proposed by Kung [154]. In Figure 13.5, step response plots of the resulting models are

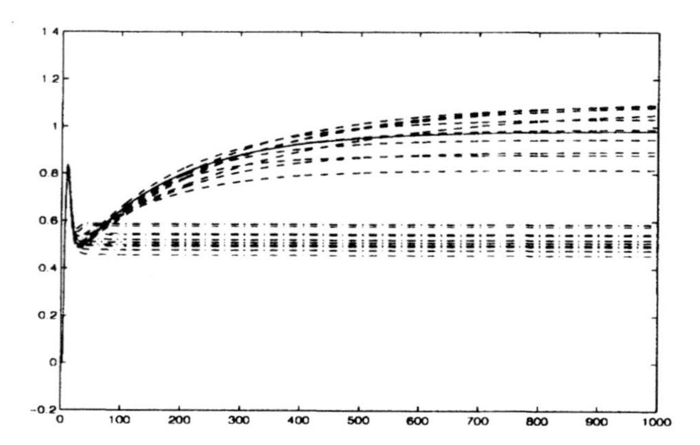

Fig. 13.5. Step response plots of the example system (solid) and the models obtained in 10 simulations with approximate realization using the standard basis (dash-dotted) and the generalized basis (dashed).

shown. It is seen that approximate realization using the standard basis results in a model that only fits the first samples of the response well. This is a known drawback of this method. Employing the generalized basis with poles 0.5 and 0.9 results in models that capture the transient behavior much better. Apparently, a sensible choice of basis can considerably improve the performance of the Kung algorithm.

13.6.3 Approximate Realization, Using Algorithm 3

To illustrate the fact that Algorithm 3 can lead to different results in an approximate setting, we consider an inner function G_b with poles $\{0.5 \pm 0.3i, 0.6, 0.7\}$. Let (A_b, B_b, C_b, D_b) be a balanced realization and V_k as before be given by $V_k(z) = [zI - A_b]^{-1}BG_b^{k-1}(z)$. We now create a system G as follows:

$$
\begin{aligned}
G(z) &= C_b V_1(z) + \frac{1}{2}C_b V_2(z) + \frac{1}{4}C_b V_3(z) + \cdots \\
&= (G_b(z) - D_b)\left(1 + \frac{1}{2}G_b(z) + \frac{1}{4}G_b^2(z) + \cdots\right) \\
&= \frac{G_b(z) - D_b}{1 - \frac{1}{2}G_b(z)}
\end{aligned}
$$

From the above, it follows that the expansion coefficients of G in terms of the Hambo basis V_k are given by $\breve{g}(k) = C_b 2^{k-1}$.
We now proceed at follows:

1. Add noise to the exact expansion coefficients, with a signal-to-noise ratio of about 17 dB (in terms of RMS values).

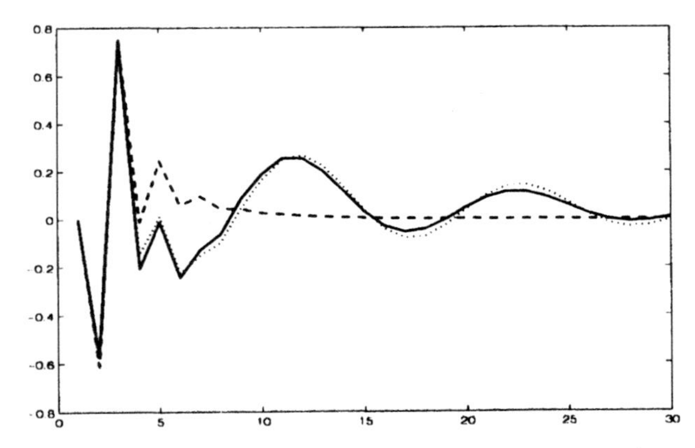

Fig. 13.6. Impulse response plots of the example system (solid) and the models obtained by Algorithm 3 (dashed) and by the application of Algorithm 3 followed by Algorithm 2 (dotted).

2. Use Algorithm 3 on the first 3 (noisey) expansion coefficients. Though the resulting Hankel matrix does not obey the generalized Tether condition, we let the algorithm create a second-order approximation.
3. Use this second-order realization to reconstruct the expansion coefficients for a large number (30 in this case).
4. Use the reconstructed expansion coefficients in conjunction with Algorithm 2. The resulting Hankel matrix has rank 4 and we create a fourth-order model.

Figure 13.6 shows the impulse responses of the resulting realizations. Clearly, the model that results from Algorithm 2 on the reconstructed expansion coefficients is superior to the other approximation. As it turns out, the second-order realization is too limited to represent the system behavior adequately, but still contains the information to reconstruct this behaviour.

13.7 Summary

In this chapter, it was shown how the classical realization theory can be expanded to a generalized theory in terms of orthonormal basis function expansions. It was shown that the exact as well as the partial realization problem in these terms can be solved by considering the expression for the Hankel operator in the Hambo operator transform domain. For the solution, efficient algorithms are given that can be used in an exact and in an approximative setting. Furthermore, the connection of the developed realization theory with interpolation theory was established.

References

1. S.S. Abeysekera and X. Yao. Design of narrow-band Laguerre filters using a min-max criterion. In *Proc. IEEE International Conf. on Acoustics, Speech, and Signal Processing Proceedings (ICASSP'00)*, volume 1, pages 137–140, Istanbul, Turkey, June 2000. IEEE.
2. N. I. Achieser. *Theory of Approximation.* Dover Publications, Inc., New York, 1992. Published for the first time by the Frederick Ungar Publishing Co. in 1956. Translated by Charles J. Hyman.
3. N.I. Achieser. *The Classical Moment Problem.* University Mathematical Monographs. Oliver and Boyd, Edinburgh, 1965.
4. H. Akçay and P.S.C. Heuberger. A frequency-domain iterative identification algorithm using general orthonormal basis functions. *Automatica*, 37(5):663–674, 2001.
5. H. Akçay. Orthonormal basis functions for modelling continuous-time systems. *Signal Processing*, 77:261–274, 1999.
6. H. Akçay. Continuous-time stable and unstable system modelling with orthonormal basis functions. *International Journal of Robust and Nonlinear Control*, 10:513–531, 2000.
7. H. Akçay. Discrete-time system modelling in L_p with orthonormal basis functions. *Systems & Control Letters*, 39:365–376, 2000.
8. H. Akçay. On the existence of a disk algebra basis. *Signal Processing*, 80:903–907, 2000.
9. H. Akçay. Paley inequalities for discrete-time rational orthonormal bases. *Signal Processing*, 80:2449–2455, 2000.
10. H. Akçay. General orthonormal bases or robust identification in $\mathcal{H}_\infty$. *SIAM Journal of Control and Optimization*, 40:947–968, 2001.
11. H. Akçay. Generalization of a standard inequality for Hardy space $\mathcal{H}_1$. *Automatica*, 37:1853–1857, 2001.
12. H. Akçay. On the uniform approximation of discrete-time systems by generalized Fourier series. *IEEE Transactions on Signal Processing*, 49:1461–1467, 2001.
13. H. Akçay. Synthesis of complete rational orthonormal bases with prescribed asymptotic order. *Automatica*, 37:559–564, 2001.
14. H. Akçay. A stochastic analysis of robust estimation algorithms in $\mathcal{H}_\infty$ with rational basis functions. *International Journal of Robust and Nonlinear Control*, 12:71–86, 2002.

15. H. Akçay and B. Ninness. Orthonormal basis functions for continuous-time systems and L_p convergence. *Mathematics of Control, Signals and Systems*, 12:295–305, 1999.
16. H. Akçay and B.M. Ninness. Rational basis functions for robust identification from frequency and time-domain measurements. *Automatica*, 34(9):1101–1117, 1998.
17. D. Alpay. *The Schur Algorithm, Reproducing Kernel Spaces and System Theory*. American Mathematical Society, Providence, RI, 1998.
18. B.D.O. Anderson and A.C. Antoulas. Rational interpolation and state-variable realizations. *Linear Algebra and its Applications*, 137/138:479–509, 1990.
19. N. Aronszajn. Theory of reproducing kernels. *Acta Mathematica*, 68:337–404, 1950.
20. K.J. Aström. Matching criteria for control and identification. In *Proc. of the 1993 European Control Conf. ECC93*, pages 248–251, Groningen, 1993. Foundation Systems and Control, Groningen, The Netherlands.
21. D.H. Baldelli, M.C. Mazzaro, and R.S. Sánchez Peña. Robust identification of lightly damped flexible structures by means of orthonormal bases. *IEEE Transactions on Control Systems Technology*, 9(5):696–707, 2001.
22. E.R. Barnes and A.J. Hoffman. On bounds for eigenvalues of real symmetric matrices. *Linear Algebra and its Applications*, 40:217–223, 1981.
23. D.S. Bayard. Multivariable frequency domain identification via 2-norm minimization. In *Proc. 1992 American Control Conf.*, pages 1253–1257, Chicago, IL, 1992. IEEE Press, Piscataway, NJ.
24. G.H. Blackwood, R.N. Jacques, and D.W. Miller. The MIT multipoint alignment testbed technology development for optical interferometry. In *SPIE Conf. on Active and Adaptive Optical Systems, San Diego, CA*. SPIE, Bellingham, WA, 1991.
25. P. Bodin, T. Oliveira e Silva, and B. Wahlberg. On the construction of orthonormal basis functions for system identification. In *Prepr. 13th IFAC World Congress*, volume I, pages 369–374, San Francisco, CA, 1996. Elsevier, Oxford, UK.
26. P. Bodin, L.F. Villemoes, and B. Wahlberg. Selection of best orthonormal rational basis. *SIAM Journal of Control and Optimization*, 38(4):995–1032, 2000.
27. P. Bodin and B. Wahlberg. Thresholding in high order transfer function estimation. In *Proc. 33rd IEEE Conf. on Decision and Control*, pages 3400–3405, Orlando, FL, December 1994. IEEE Press, Piscataway, NJ.
28. J. Bokor and M. Athans. Frequency domain identification of the MIT interferometer testbed in generalized orthogonal basis. In *Proc. 11th IFAC Symp. on System Identification (SYSID 1997)*, pages 1735–1739, Kitakyushu, Japan, 1997. Elsevier Science Ltd., Oxford, UK.
29. J. Bokor, L. Gianone, and Z. Szabó. Construction of generalised orthonormal bases in $\mathcal{H}_2$. Technical report, Computer and Automation Institute, Hungarian Academy of Sciences, 1995.
30. J. Bokor, P. Heuberger, B. Ninness, T. Oliveira e Silva, P. Van den Hof, and B. Wahlberg. Modelling and identification with orthogonal basis functions. In R. Smith, editor, *Proc. 12th IFAC Symp. on System Identification (SYSID 2000)*, Santa Barbara, CA, June 2000. Elsevier Science Ltd., Oxford, UK.

31. X. Bombois, B.D.O. Anderson, and M. Gevers. Mapping parametric confidence ellipsoids to Nyquist plane for linearly parametrized transfer functions. In G.C. Goodwin, editor, *Model Identification and Adaptive Control*, pages 53–71. Springer-Verlag, London, UK, 2001.
32. X. Bombois, M. Gevers, and G. Scorletti. A measure of robust stability for an identified set of parametrized transfer functions. *IEEE Transactions on Automatic Control*, 45:2141–2145, 2000.
33. A. Böttcher and B. Silbermann. *Invertibility and Asymptotics of Toeplitz Matrices*. Akademie-Verlag, Berlin, 1983.
34. G.E.P. Box and G.W. Jenkins. *Time Series Analysis; Forecasting and Control*. Holden-Day, San Francisco, CA, second edition, 1976.
35. J.W. Brewer. Kronecker products and matrix calculus in system theory. *IEEE Transactions on Circuits and Systems*, 25(9):772–781, 1978.
36. D.R. Brillinger. *Time Series: Data Analysis and Theory*. Holden-Day, 1981.
37. P.W. Broome. Discrete orthonormal sequences. *Journal of the Association for Computing Machinery*, 12(2):151–168, 1965.
38. C. Bruni. Analysis of approximation of linear and time-invariant systems pulse response by means of Laguerre finite term expansion. *IEEE Transactions on Automatic Control*, 9(4):580–581, 1964.
39. A. Bultheel and P. Carrette. Algebraic and spectral properties of general Toeplitz matrices. *SIAM Journal of Control and Optimization*, 41(5):1413–1439 (electronic), 2002.
40. A. Bultheel, P. González-Vera, E. Hendriksen, and O. Njåstad. *Orthogonal Rational Functions*. Cambridge Monographs on Applied and Computational Mathematics. Cambridge University Press, Cambridge, UK, 1999.
41. A. Bultheel, M. Van Barel, and P. Van gucht. Orthogonal bases in discrete least squares rational approximation. *Journal of Computational and Applied Mathematics*, 164-165:175–194, 2004.
42. P.L. Butzer and R.J. Nessel. *Fourier Analysis and Approximation*. Birkhäuser, New York, 1971.
43. F. Byron and R. Fuller. *Mathematics of Classical and Quantum Physics*, volume 1 and 2 of *Series in Advanced Physics*. Addison-Wesley, Reading, MA, 1969.
44. P.E. Caines. *Linear Stochastic Systems*. John Wiley and Sons, New York, 1988.
45. R.J.G.B. Campello, G. Favier, and W.C. do Amaral. Optimal expansions of discrete-time Volterra models using Laguerre functions. *Automatica*, 40:815–822, 2004.
46. M.C. Campi, R. Leonardi, and L. Rossi. Generalized super-exponential method for blind equalization using Kautz filters. In *Proc. of the IEEE Signal Processing Workhop on Higher Order Statistics*, pages 107–111, Caesarea, Italy, 1999. IEEE Computer Society, New York.
47. M. Casini, A. Garulli, and A. Vicino. On worst-case approximation of feasible system sets via orthonormal basis functions. *IEEE Transactions on Automatic Control*, 48(1):96–101, 2003.
48. J. Chen and G. Gu. *Control-Oriented System Identification. An $\mathcal{H}_\infty$ Approach*. John Wiley, New York, 2000.
49. J. Chen and C.A. Nett. The Carathéodory–Fejér problem and $\mathcal{H}_\infty/\ell_1$ identification: A time domain approach. *IEEE Transactions on Automatic Control*, AC-40:729–735, 1995.

50. J. Chen, C.A. Nett, and M.K.H. Fan. Optimal nonparametric identification from arbitrary corrupt finite time series. *IEEE Transactions on Automatic Control*, AC-40:769–776, 1995.
51. E.W. Cheney. *Introduction to Approximation Theory*. McGraw-Hill, New York, 1966.
52. G. Chesi, A. Garulli, A. Tesi, and A. Vicino. A convex approach to the characterization of the frequency response of ellipsoidal plants. *Automatica*, 38:249–259, 2002.
53. C.T. Chou, M. Verhaegen, and R. Johansson. Continuous-time identification of SISO systems using Laguerre functions. *IEEE Transactions on Signal Processing*, 47(2):349–362, 1999.
54. P.R. Clement. On completeness of basis functions used for signal analysis. *SIAM Review*, 5(2):131–139, 1963.
55. P.R. Clement. Laguerre functions in signal analysis and parameter identification. *Journal of the Franklin Institute*, 313(2):85–95, 1982.
56. G.J. Clowes. Choice of the time-scaling factor for linear system approximations using orthonormal Laguerre functions. *IEEE Transactions on Automatic Control*, 10(4):487–489, 1965.
57. W.R. Cluett and L. Wang. Some asymptotic results in recursive identification using Laguerre models. *International Journal of Adaptive Control and Signal Processing*, 5:313–333, 1991.
58. W.R. Cluett and L. Wang. Frequency smoothing using Laguerre model. *IEEE Proceedings-D*, 139(1):88–96, 1992.
59. J. Cousseau, P.S.R. Diniz, G. Sentoni, and O. Agamennoni. On orthogonal realizations for adaptive iir filters. *International Journal of Circuit Theory and Applications*, 28:481–500, 2000.
60. G.W. Davidson and D.D. Falconer. Reduced complexity echo cancellation using orthonormal functions. *IEEE Transactions on Circuits and Systems*, 38(1):20–28, 1991.
61. H.F. Davis. *Fourier Series and Orthogonal Functions*. Dover Publications, Inc., New York, 1989. Published for the first time by Allyn and Bacon, Inc., in 1963.
62. Ph.J. Davis. *Interpolation & Approximation*. Dover Publications, New York, 1975. Published for the first time by Blaisdell Publishing Company in 1963.
63. R.A. de Callafon and P.M.J. Van den Hof. Multivariable feedback relevant system identification of a wafer stepper system. *IEEE Transactions on Control Systems Technology*, 9(2):381–390, 2001.
64. T.J. de Hoog. *Rational Orthonormal Bases and Related Transforms in Linear System Modeling*. PhD thesis, Delft University of Technology, Delft, The Netherlands, 2001.
65. T.J. de Hoog, P.S.C. Heuberger, and P.M.J. Van den Hof. On partial realization and interpolation of models from orthogonal basis function expansions. In *Proc. 38th IEEE Conf. on Decision and Control*, pages 3212–3218, Phoenix, USA, December 1999. IEEE Press, Piscataway, NJ.
66. T.J. de Hoog, P.S.C. Heuberger, and P.M.J. Van den Hof. A general transform theory of rational orthonormal basis function expansions. In *Proc. 39th IEEE Conf. on Decision and Control*, pages 4649–4654, Sydney, Australia, December 2000. IEEE Press, Piscataway, NJ.

67. T.J. de Hoog, P.S.C. Heuberger, and P.M.J. Van den Hof. Partial realization in generalized bases: Algorithm and example. In R. Smith, editor, *Proc. 12th IFAC Symp. on System Identification (SYSID 2000)*, Santa Barbara, CA, June 2000. Elsevier Science Ltd., Oxford, UK.
68. T.J. de Hoog, Z. Szabó, P.S.C. Heuberger, P.M.J. Van den Hof, and J. Bokor. Minimal partial realization from generalized orthonormal basis function expansions. *Automatica*, 38(4):655–669, 2002.
69. D.K. de Vries. *Identification of Model Uncertainty for Control Design.* PhD thesis, Delft University of Technology, Delft, The Netherlands, 1994.
70. D.K. de Vries and P.M.J. Van den Hof. Quantification of uncertainty in transfer function estimation: a mixed-probabilistic - worst-case approach. *Automatica*, 31(4):543–557, 1995.
71. D.K. de Vries and P.M.J. Van den Hof. Frequency domain identification with generalized orthonormal basis functions. *IEEE Transactions on Automatic Control*, 43(3):656–669, 1998.
72. A.C. den Brinker. Laguerre-domain adaptive filters. *IEEE Transactions on Signal Processing*, 42(4):953–956, 1994.
73. A.C. den Brinker, F.P.A. Benders, and T.A.M. Oliveira e Silva. Optimality conditions for truncated Kautz series. *IEEE Transactions on Circuits and Systems II*, 43(2):117–122, 1996.
74. E. Deprettere and P. Dewilde. Orthogonal cascade realization of real multiport digital filters. *Circuit Theory and Applications*, 8:245–272, 1980.
75. R. Deutsch. *System Analysis Techniques.* Prentice Hall Inc., Englewood Cliffs N.J., 1969.
76. P. Dewilde and H. Dym. Schur recursions, error formulas, and convergence of rational estimators for stationary stochastic sequences. *IEEE Transactions on Information Theory*, IT-27:446–461, 1981.
77. P. Dewilde and H. Dym. Lossless inverse scattering, digital filters and estimation theory. *IEEE Transactions on Information Theory*, IT-30(4):664–662, 1984.
78. P. Dewilde, A. Vieira, and T. Kailath. On a generalised Szegö–Levinson realization algorithm for optimal linear predictors based on a network synthesis approach. *IEEE Transactions on Circuits and Systems*, CAS-25(9):663–675, 1978.
79. E.M. Diaz, B.R. Fischer, and A. Medvedev. Identification of a vibration process by means of Kautz functions. In *Proc. of the Fourth International Conf. on Vibration Control (MOVIC'98)*, pages 387–392, Zürich, Switzerland, August 1998.
80. M.M. Djrbashian. Orthogonal systems of rational functions on the unit circle. *Izvestiya Akademii Nauk Armyanskoi SSR, ser. Matematika*, 1, 2:3–24, 106–129, 1966. Translation by K. Müller and A. Bultheel, appeared as Report TW235, Dept. of Comp. Science, K.U. Leuven, Feb. 1997.
81. S.G. Douma. *From Data to Robust Control - System Identification Uncertainty and Robust Control Design.* PhD thesis, Delft University of Technology, Delft, The Netherlands, 2005.
82. G.A. Dumont, Y. Fu, and A.L. Elshafi. Orthonormal functions in identification and adaptive control. In *Proc. IFAC Intern. Symp. Intelligent Tuning and Adaptive Control*, Singapore, 1991. Elsevier Science Ltd., Oxford, UK.
83. P. L. Duren. *Theory of $\mathcal{H}_p$ spaces.* Academic Press, New York, 1970.

84. R.E. Edwards. *Fourier Series, a Modern Introduction.*, volume 2 of *Graduate texts in mathematics number 64*. Springer-Verlag, New York, 1979.
85. E.M. El Adel, M. Ouladsine, and L. Radouane. Predictive steering control using Laguerre series representation. In *Proc. 2003 American Control Conf.*, volume 1, pages 439–445, Denver, CO, June 2003. IEEE Press, Piscataway, NJ.
86. B. Epstein. *Orthogonal Families of Analytic Functions*. MacMillan, 1965.
87. Z. Fejzo and H. Lev-Ari. Adaptive Laguerre-lattice filters. *IEEE Transactions on Signal Processing*, SP-45(12):3006–3016, 1997.
88. A. Fettweiss. Factorisation of transfer matrices of lossless two-ports. *IEEE Transactions on Circuit Theory*, 17:86–94, 1970.
89. A. Fettweiss. Digital filter structures related to classical filter networks. *Archive für Elektronik und Übertragung*, 25:79–89, 1971.
90. A. Fettweiss and K. Meerkötter. Suppression of parasitic oscillations in wave digital filters. *IEEE Transactions on Circuits and Systems*, 22:239–246, 1975.
91. C. Finn, E. Ydstie, and B. Wahlberg. Contrained predictive control using orthogonal expansions. *AIChE Journal*, 39:1810–1826, 1993.
92. B.R. Fischer and A. Medvedev. Laguerre shift identification of a pressurized process. In *Proc. 1998 American Control Conf.*, pages 1933–1937, Philadelphia,PA, June 1998. IEEE Press, Piscataway, NJ.
93. B.R. Fischer and A. Medvedev. L_2 time delay estimation by means of Laguerre functions. In *Proc. 1999 American Control Conf.*, pages 455–259, San Diego, CA, June 1999. IEEE Press, Piscataway, NJ.
94. B.R. Fischer and A. Medvedev. On the concept of 'excitation order' in Laguerre domain identification. In *Proc. of the 1999 European Control Conf. ECC99 (on CD)*, Karlsruhe, September 1999. Rubicon-Agentur für digitale Medien, Aachen, Germany. paper ID F1064-3.
95. D.S. Flamm. A new shift-invariant representation of periodic linear systems. *Systems & Control Letters*, 17:9–14, 1991.
96. G. Freud. *Orthogonal Polynomials*. Pergamon Press, Oxford, 1971.
97. Y. Fu and G.A. Dumont. An optimum time scale for discrete Laguerre network. *IEEE Transactions on Automatic Control*, 38(6):934–938, 1993.
98. J.B Garnett. *Bounded Analytic Functions*. Academic Press, New York, 1981.
99. L.Y. Geronimus. *Orthogonal Polynomials*. Consultants Bureau, New York, 1961.
100. M. Gevers. Towards a joint design of identification and control. In H.L. Trentelman and J.C. Willems, editors, *Essays on Control: Perspectives in the Theory and its Applications*, pages 111–115. Birkhäuser, Boston, 1993.
101. L. Giarré, M. Milanese, and M. Taragna. $\mathcal{H}_\infty$ identification and model quality evaluation. *IEEE Transactions on Automatic Control*, AC-42(2):188–199, 1997.
102. K. Gilpin, M. Athans, and J. Bokor. Multivariable identification of the MIT interferometer testbed. In *Proc. 1992 American Control Conf.*, Chicago, IL, 1992. IEEE Press, Piscataway, NJ.
103. G.H. Golub and C.F. Van Loan. *Matrix Computations*. The John Hopkins University Press, Baltimore, 2nd edition, 1989.
104. J.C. Gomez and E. Baeyens. Identification of block-oriented nonlinear systems using orthonormal bases. *Journal of Process Control*, 14:685–697, 2004.
105. G.C. Goodwin, M. Gevers, and B. Ninness. Quantifying the error in estimated transfer functions with application to model order selection. *IEEE Transactions on Automatic Control*, 37(7):913–928, 1992.

106. G.C. Goodwin and R.L. Payne. *Dynamic System Identification.* Academic Press, New York, 1977.
107. A.H. Gray, Jr. and J.D. Markel. A normalized digital filter structure. *IEEE Transactions on Acoustics, Speech and Signal Processing*, 23(3):268–277, 1975.
108. U. Grenander and G. Szegö. *Toeplitz Forms and Their Applications.* Chelsea Publishing Company, New York, second edition, 1984. First edition published by University of California Press, Berkeley, 1958.
109. G. Gu and P. Khargonekar. A class of algorithms for identification in $\mathcal{H}_\infty$. *Automatica*, 28:192–312, 1992.
110. G. Gu and P. Khargonekar. Linear and nonlinear algorithms for identification in $\mathcal{H}_\infty$ with error bounds. *IEEE Transactions on Automatic Control*, AC-37:953–963, 1992.
111. G. Gu, P. Khargonekar, and E.B. Lee. Approximation of infinite dimensional systems. *IEEE Transactions on Automatic Control*, AC–34:610–618, 1989.
112. G. Gu, P. Khargonekar, and Y. Li. Robust convergence of two stage nonlinear algorithms for system identification in $\mathcal{H}_\infty$. *Systems & Control Letters*, 18:253–263, 1992.
113. L. Guo and L. Ljung. Performance analysis of general tracking algorithms. *IEEE Transactions on Automatic Control*, 40(8):1388–1402, 1995.
114. R.G. Hakvoort. Worst–case system identification in ℓ_1 error bounds. optimal models and model reduction. In *Proc. 31st IEEE Conf. on Decision and Control*, pages 499–504, Tucson, AZ, 1992. IEEE Press, Piscataway, NJ.
115. R.G. Hakvoort. A linear programming approach to the identification of frequency domain error bounds. In *Proc. 10th IFAC Symp. on System Identification (SYSID'94)*, volume 2, pages 195–200, Copenhagen, Denmark, 1994. Elsevier Science Ltd., Oxford, UK.
116. R.G. Hakvoort. *System Identification for Robust Process Control—Nominal Models and Error Bounds.* PhD thesis, Delft University of Technology, Delft, The Netherlands, November 1994.
117. R.G. Hakvoort and P.M.J. Van den Hof. Frequency domain curve fitting with maximum amplitude criterion and guaranteed stability. *International Journal of Control*, 60:809–825, 1994.
118. R.G. Hakvoort and P.M.J. Van den Hof. Consistent parameter bounding identification for linearly parametrized model sets. *Automatica*, 31(7):957–969, 1995.
119. R.G. Hakvoort and P.M.J. Van den Hof. Identification of probabilistic uncertainty regions by explicit evaluation of bias and variance errors. *IEEE Transactions on Automatic Control*, 42(11):1516–1528, 1997.
120. E.J. Hannan and D.F. Nicholls. The estimation of the prediction error variance. *Journal of the American Statistical Association*, 72(360, part 1):834–840, 1977.
121. E.J. Hannan and B. Wahlberg. Convergence rates for inverse Toeplitz matrix forms. *Journal Multivariate Analysis*, 31:127–135, 1989.
122. J.W. Head. Approximation to transient by means of Laguerre series. *Proc. Cambridge Philosophical Society*, 52:64–651, 1956.
123. A.J. Helmicki, C.A. Jacobson, and C.N. Nett. Identification in $\mathcal{H}_\infty$: linear algorithms. In *Proc. 1990 American Control Conf.*, pages 2418–2423, San Diego, CA, 1990. IEEE Press, Piscataway, NJ.
124. A.J. Helmicki, C.A. Jacobson, and C.N. Nett. Control oriented system identification: a worst case/deterministic approach in $\mathcal{H}_\infty$. *IEEE Transactions on Automatic Control*, AC–36:1163–1176, 1991.

125. P. S. C. Heuberger, P. M. J. Van den Hof, and O. H. Bosgra. A generalized orthonormal basis for linear dynamical systems. In *Proc. 32nd IEEE Conf. on Decision and Control*, volume 3, pages 2850–2855, San Antonio, TX, 1993. IEEE Press, Piscataway, NJ.
126. P.S.C. Heuberger. *On Approximate System Identification With System Based Orthonormal Functions*. PhD thesis, Delft University of Technology, Delft, The Netherlands, 1991.
127. P.S.C. Heuberger and T.J. de Hoog. Approximation and realization using generalized orthonormal bases. In *Proc. of the 1999 European Control Conf. ECC99 (on CD)*, Karlsruhe, Germany, 1999. Rubicon-Agentur für digitale Medien, Aachen, Germany. paper ID F1064-3.
128. P.S.C. Heuberger, T.J. de Hoog, P.M.J. Van den Hof, and B. Wahlberg. Orthonormal basis functions in time and frequency domain: Hambo transform theory. *SIAM Journal of Control and Optimization*, 42(4):1347–1373, 2003.
129. P.S.C. Heuberger and P.M.J. Van den Hof. The Hambo transform: A signal and system transform induced by generalized orthonormal basis functions. In *Preprints of the 13th IFAC World Congress*, volume I, pages 357–362, San Francisco, CA, 1996. Elsevier, Oxford, UK.
130. P.S.C. Heuberger, Z. Szabó, T.J. de Hoog, P.M.J. Van den Hof, and J. Bokor. Realization algorithms for expansions in generalized orthonormal basis functions. In *Proc. 14th IFAC World Congress*, volume H, pages 373–378, Beijing, 1999. Elsevier, Oxford, UK.
131. P.S.C. Heuberger and P.M.J. Van den Hof. The Hambo transform: A signal and system transform induced by generalized orthonormal basis functions. Technical Report N-512, Mechanical Engineering Systems and Control Group, Delft University of Technology, The Netherlands, April 1998.
132. P.S.C. Heuberger, P.M.J. Van den Hof, and O.H. Bosgra. A generalized orthonormal basis for linear dynamical systems. *IEEE Transactions on Automatic Control*, 40(3):451–465, 1995.
133. H. Hjalmarsson. From experiment design to closed loop control. *Automatica*, 41(3), 2005. To appear.
134. B.L. Ho and R.E. Kalman. Effective construction of linear state-variable models from input/output functions. *Regelungstechnik*, 14(12):545–592, 1966.
135. K. Hoffman. *Banach Spaces of Analytic Functions*. Prentice-Hall, Englewood Cliffs, NJ, 1962.
136. R.A. Horn and C.R. Johnson. *Matrix Analysis*. Cambridge University Press, Cambridge, 1985.
137. I.M. Horowitz. *Synthesis of Feedback Systems*. Academic Press, New York, 1963.
138. M. Huzmezan, G.A. Dumont, W.A. Gough, and S. Kovac. Adaptive control of delayed integrating systems: a pvc batch reactor. *IEEE Transactions on Control Systems Technology*, 11:390–398, 2003.
139. D. Jackson. *Fourier Series and Orthogonal Polynomials*, volume 6 of *The Carus Mathematical Monographs*. Mathematical Association of America, Oberlin, Ohio, 1941.
140. T. Kailath. Signal processing applications of some moment problems. In Henry J. Landau, editor, *Moments in Mathematics*, volume 37 of *AMS Short Course Lecture Notes. Proceedings of Symposia in Applied Mathematics*, pages 71–109. American Mathematical Society, Providence, RI, 1987.

141. T. Kailath, A. Vieira, and M. Morf. Inverses of Toeplitz operators, innovations, and orthogonal polynomials. *SIAM Review*, 20(1):106–119, 1978.
142. M. Karjalainen and T. Paatero. Frequency-dependent signal windowing. In *2001 IEEE Workshop on the Applications of Signal Processing to Audio and Acoustics*, pages 35–38, New Paltz, NY, October 2001. IEEE Press.
143. W.H. Kautz. Network synthesis for specified transient response. Technical Report 209, Massachusetts Institute of Technology, Research Laboratory of Electronics, April 1952.
144. W.H. Kautz. Transient synthesis in the time domain. *IRE Transactions on Circuit Theory*, 1:29–39, 1954.
145. L. Keviczky. Combined identification and control: another way. *Control Engineering Practice*, 4(5):685–698, 1996.
146. L. Keviczky and Cs. Bányász. On the dialectics of identification and control in iterative learning schemes. In *Valencia COSY Workshop ESF*, pages 1–6, Valencia, 1996. European Science Foundation.
147. J.J. King and T. O'Canainn. Optimum pole positions for Laguerre-function models. *Electronics Letters*, 5(23):601–602, 1969.
148. R.E. King and P.N. Paraskevopoulos. Digital Laguerre filters. *Circuit Theory and Applications*, 5:81–91, 1977.
149. R.E. King and P.N. Paraskevopoulos. Parametric identification of discrete time SISO systems. *International Journal of Control*, 30:1023–1029, 1979.
150. L. Knockaert and D. De Zutter. Laguerre-svd reduced-order modeling. *IEEE Transactions on Microwave Theory and Techniques*, 48:1469–1475, 2000.
151. L. Knockaert and D. De Zutter. Stable Laguerre-svd reduced-order modeling. *IEEE Transactions on Circuits and Systems I*, 50:576–579, 2003.
152. T.W. Körner. *Fourier Analysis*. Cambridge University Press, Cambridge, 1988.
153. E. Kreyszig. *Introductory Functional Analysis with Applications*. John Wiley & Sons, New York, 1989.
154. S. Kung. A new identification and model reduction algorithm via singular value decompositions. In *Proc. of the 12th Asilomar Conf. on Circuits, Systems and Computers*, pages 705–714, Pacific Grove, California, November 1978. IEEE Press, Piscataway, NJ.
155. A.J. Laub, M.T. Heath, C.C. Paige, and R.C. Ward. Computation of system balancing transformations and other applications of simultaneous diagonalization algorithms. *IEEE Transactions on Automatic Control*, AC-32(2):115–122, 1987.
156. A. Lecchini and M. Gevers. Explicit expression of the parameter bias in identification of Laguerre models from step responses. *Systems & Control Letters*, 52:149–165, 2004.
157. Y.W. Lee. Synthesis of electric networks by means of the Fourier transforms of Laguerre's functions. *Journal of Mathematics and Physics*, XI:83–113, 1933.
158. Y.W. Lee. *Statistical Theory of Communication*. John Wiley and Sons, New York, 1960.
159. E.C. Levy. Complex curve fitting. *IEEE Trans. Autom. Control*, 4:37–43, 1959.
160. P.L Lin and Y.C. Wu. Identification of multi–input multi–output linear systems from frequency response data. *J. Dyn. Syst. Meas. Contr.*, 4:37–43, 1982.
161. L. Ljung. Convergence analysis of parametric identification methods. *IEEE Transactions on Automatic Control*, AC-23(5):770–783, 1978.
162. L. Ljung. Asymptotic variance expressions for identified black box transfer function models. *IEEE Transactions on Automatic Control*, 30:834–844, 1985.

163. L. Ljung. Model validation and model error modeling. In B. Wittenmark and A. Rantzer, editors, *The Åström Symposium on Control*, pages 15–42, Lund, Sweden, August 1999. Studentlitteratur.
164. L. Ljung. *System Identification — Theory for the User.* Prentice-Hall, Upper Saddle River, NJ, second edition, 1999.
165. L. Ljung. *MATLAB System Identification Toolbox Users Guide, Version 6.* The Mathworks, 2003.
166. L. Ljung and P.E. Caines. Asymptotic normality of prediction error estimators for approximate system models. *Stochastics*, 3:29–46, 1979.
167. L. Ljung and B. Wahlberg. Asymptotic properties of the least squares method for estimating transfer functions and disturbance spectra. *Advances in Applied Probability*, 24:412–440, 1992.
168. L. Ljung and Z.D. Yuan. Asymptotic properties of black-box identification of transfer functions. *IEEE Transactions on Automatic Control*, 30:514–530, 1985.
169. G.G. Lorentz, M. v. Golitschek, and Y. Makovoz. *Constructive Approximation.* Springer-Verlag, Berlin, 1996.
170. L. Lublin and M. Athans. Application of $\mathcal{H}_2$ control to the MIT SERC interferometer testbed. In *Proc. 1993 American Control Conf.*, Boston, MA, 1993. IEEE Press, Piscataway, NJ.
171. L. Lublin and M. Athans. An experimental comparison of $\mathcal{H}_2$ and $\mathcal{H}_\infty$ design for an interferometer testbed. In *Lecture Notes in Control and Information Sciences: Feedback Control, Nonlinear Systems and Complexity.*, pages 150–172, Amsterdam, 1995. Francis B. and Tannenbaum A. Eds., Springer-Verlag, London.
172. M. Mackenzie and K. Tieu. Gaussian filters and filter synthesis using a Hermite/Laguerre neural network. *IEEE Transactions on Neural Networks*, 15:206–214, 2004.
173. B. Maione and B. Turchiano. Laguerre z-transfer function representation of linear discrete-time systems. *International Journal of Control*, 41(1):245–257, 1985.
174. G. Maione. Laguerre approximation of fractional systems. *Electronics Letters*, 38:1234–1236, 2002.
175. P.M. Mäkilä. Approximation of stable systems by Laguerre filters. *Automatica*, 26:333–345, 1990.
176. P.M. Mäkilä. Laguerre series approximation of infinite dimensional systems. *Automatica*, 26(6):985–995, 1990.
177. P.M. Mäkilä. Worst–case input–output identification. *International Journal of Control*, 56:673–689, 1992.
178. P.M. Mäkilä and J.R. Partington. Robust approximate modelling of stable linear systems. *International Journal of Control*, 58(3):665–683, 1993.
179. P.M. Mäkilä, P.R. Partington, and T.K. Gustafsson. Worst–case control relevant identification. *Automatica*, 32(12):1799–1819, 1995.
180. F. Malmquist. Sur la détermination d'une classe de fonctions analytiques par leurs valeurs dans un ensemble donné de points. In *Comptes Rendus du Sixième Congrès des mathématiciens scandinaves (1925)*, pages 253–259, Copenhagen, Denmark, 1926. Gjellerup, Copenhagen.
181. R. Malti, S. B. Ekongolo, and J. Ragot. Dynamic SISO and MISO system approximations based on optimal Laguerre models. *IEEE Transactions on Automatic Control*, 43(9):1317–1322, 1998.

182. M.A. Masnadi-Shirazi and N. Ahmed. Optimal Laguerre networks for a class of discrete-time systems. *IEEE Transactions on Signal Processing*, 39(9):2104-2108, 1991.
183. A. Mbarek, H. Messaoud, and G. Favier. Robust predictive control using Kautz model. In *Proc. 10th IEEE International Conf. Electronics, Circuits and Systems*, pages 184–187, Sharjah, United Arab Emirates, December 2003. IEEE, Piscataway, NJ.
184. T. McKelvey and H. Akcay. An efficient frequency domain state space identification algorithm: robustness and stochastic analysis. In *Proc. 33rd IEEE Conf. on Decision and Control*, pages 3348–3353, Lake Buena Vista, FL, 1994. IEEE Press, Piscataway, NJ.
185. T. McKelvey, H. Akcay, and L. Ljung. Subspace-based multivariable system identification from frequency response data. *IEEE Transactions on Automatic Control*, AC-41(7):960–979, 1996.
186. J.M. Mendel. A unified approach to the synthesis of orthonormal exponential functions useful in systems analysis. *IEEE Transactions on Systems Science and Cybernetics*, 2(1):54–62, 1966.
187. R. Merched. Extended RLS lattice adaptive filters. *IEEE Transactions on Signal Processing*, 51(8):2294–2309, 2003.
188. R. Merched and A.H. Sayed. Fast RLS Laguerre adaptive filtering. In *Proc. Allerton Conf. on Communication, Control and Computing*, pages 338–347, Allerton, IL, September 1999. UIUC Press, Champaign-Urbana, IL.
189. R. Merched and A.H. Sayed. Order-recursive RLS Laguerre adaptive filtering. *IEEE Transactions on Signal Processing*, 48(11):3000–3010, 2000.
190. R. Merched and A.H. Sayed. Extended fast fixed-order RLS adaptive filters. *IEEE Transactions on Signal Processing*, 49(12):3015–3031, 2001.
191. R. Merched and A.H. Sayed. RLS-Laguerre lattice adaptive filtering: Error-feedback, normalized, and array-based algorithms. *IEEE Transactions on Signal Processing*, 49(11):2565–2576, 2001.
192. D.G. Messerschmitt. A class of generalized lattice filters. *IEEE Transactions on Acoustics, Speech and Signal Processing*, ASSP-28(2):198–204, 1980.
193. Y. Meyer. *Ondelettes et Opérateurs I. Ondelettes*. Hermann, Paris, 1990. Also in English by Cambridge University Press.
194. M. Milanese, J. Norton, H. Piet-Lahanier, and E. Walter. *Bounding Approaches to System Identification*. Plenum Press, New York, 1996.
195. M. Milanese and R. Tempo. Optimal algorithms theory for robust estimation and prediction. *IEEE Transactions on Automatic Control*, AC-30:730–738, 1985.
196. M. Milanese and A. Vicinio. Information–based complexity and nonparametric worst–case system identification. *Journal of Complexity*, 9:427–446, 1993.
197. M. Milanese and A. Vicino. Optimal estimation theory for dynamic systems with set membership uncertainty: an overview. *Automatica*, 27(6):997–1009, 1991.
198. B.C. Moore. Principal component analysis in linear systems: controllability, observability, and model reduction. *IEEE Transactions on Automatic Control*, 26(1):17–32, 1981.
199. R. Morvan, N. Tanguy, P. Vilbé, and L. C. Calvez. Pertinent parameters for Kautz approximation. *Electronics Letters*, 36(8):769–771, 2000.

200. C.T. Mullis and R.A. Roberts. Roundoff noise in digital filters: Frequency transformations and invariants. *IEEE Transactions on Acoustics, Speech and Signal Processing*, 24(6):538–550, 1976.
201. B.Sz. Nagy. *Introduction to Real Functions and Orthogonal Expansions.* Oxford University Press, Oxford, New York, 1965.
202. V. Nalbantoglu, J. Bokor, G. Balas, and P. Gaspar. System identification with generalized orthonormal basis functions: an application to flexible structures. *Control Engineering Practice*, 11:245–259, 2003.
203. G.M. Natanson and V.V. Zuk. *Trigonometric Fourier Series and Approximation Theory.* Izdat. Leningrad Un–ta, 1983.
204. P. Nevai. *Orthogonal Polynomials*, volume 213 of *Memoirs of the American Philosophical Society.* The American Philosophical Society, Philadelphia, 1979.
205. L.S.H. Ngia. Recursive identification of acoustic echo systems using orthonormal basis functions. *IEEE Transactions on Speech and Audio Processing*, 11:278–293, 2003.
206. L.S.H. Ngia and F. Gustafsson. Using Kautz filter for adaptive acoustic echo cancellation. In *Conference Record of the Thirty-Third Asilomar Conference on Signals, Systems, and Computers*, volume 2, pages 1110–1114, Pacific Grove, CA, October 1999. IEEE Press, Piscataway, NJ.
207. B. Ninness and H. Hjalmarsson. Model structure and numerical properties of normal equations. In *Proc. 14th IFAC World Congress*, volume H, pages 271–276, Beijing, July 1999. Elsevier, Oxford, UK.
208. B. Ninness and H. Hjalmarsson. Variance error quantifications that are exact for finite model order. *IEEE Transactions on Automatic Control*, 49(12):12–26, 2004.
209. B.M. Ninness and J.C. Gómez. Asymptotic analysis of mimo systems estimates by the use of othonormal bases. In *Preprints of the 13th IFAC World Congress*, volume I, pages 363–368, San Francisco, CA, 1996. Elsevier, Oxford, UK.
210. B.M. Ninness and J.C. Gómez. Frequency domain analysis of tracking and noise performance of adaptive algorithms. *IEEE Transactions on Signal Processing*, 46(5):1314–1332, 1998.
211. B.M. Ninness, J.C. Gómez, and S. Weller. Mimo system identification using orthonormal basis functions. In *Proc. 34th IEEE Conf. on Decision and Control*, pages 703–708, New Orleans, December 1995. IEEE Press, Piscataway, NJ.
212. B.M. Ninness and G.C. Goodwin. Estimation of model quality. *Automatica*, 31(12):1771–1797, 1995.
213. B.M. Ninness and F. Gustafsson. A general construction of orthonormal bases for system identification. In *Proc. 33rd IEEE Conf. on Decision and Control*, pages 3388–3393, Orlando, FL, December 1994. IEEE Press, Piscataway, NJ.
214. B.M. Ninness and F. Gustafsson. A unifying construction of orthonormal bases for system identification. *IEEE Transactions on Automatic Control*, 42(4):515–521, 1997.
215. B.M. Ninness and F. Gustafsson. A unifying construction of orthonormal bases for system identification. Technical Report EE9432, Department of Electrical Engineering, University of Newcastle, Newcastle, NSA, Australia, 1994.
216. B.M. Ninness and H. Hjalmarsson. The analysis of variance error part I: Accurate quantifications. *Available at* `www.ee.newcastle.edu.au/brett`, 2001.
217. B.M. Ninness and H. Hjalmarsson. The analysis of variance error part II: Fundamental principles. *Available at* `www.ee.newcastle.edu.au/brett`, 2001.

218. B.M. Ninness and H. Hjalmarsson. Accurate quantification of variance error. In *Proc. 15th IFAC World Congress*, Barcelona, Spain, 2002. Elsevier, Oxford, UK.
219. B.M. Ninness and H. Hjalmarsson. Exact quantification of variance error. In *Proc. 15th IFAC World Congress*, Barcelona, Spain, 2002. Elsevier, Oxford, UK.
220. B.M. Ninness and H. Hjalmarsson. The effect of regularisation on variance error. *IEEE Transactions on Automatic Control*, 49(7):1142–1147, 2004.
221. B.M. Ninness and H. Hjalmarsson. On the frequency domain accuracy of closed loop estimates,. *To appear in Automatica*, 2005.
222. B.M. Ninness, H. Hjalmarsson, and F. Gustafsson. Generalised Fourier and Toeplitz results for rational orthonormal bases. Technical Report EE9740, Department of Electrical Engineering, University of Newcastle, Newcastle, NSA, Australia, 1997.
223. B.M. Ninness, H. Hjalmarsson, and F. Gustafsson. Generalised Fourier and Toeplitz results for rational orthonormal bases. *SIAM Journal of Control and Optimization*, 37(2):429–460, 1999.
224. B.M. Ninness, H. Hjalmarsson, and F. Gustafsson. On the fundamental role of orthonormal bases in system identification. *IEEE Transactions on Automatic Control*, 47(8):1384–1407, 1999.
225. Y. Nurges. Laguerre models in problems of approximation and identification of discrete systems. *Automation and Remote Control*, 48:346–352, 1987.
226. Y. Nurges and Y. Yaaksoo. Laguerre state equations for a multivariable discrete system. *Automation and Remote Control*, 42:1601–1603, 1981.
227. G.H.C. Oliveira, W.C. Amaral, and K. Latawiec. CRHPC using Volterra models and orthonormal basis functions: an application to CSTR plants. In *Proc. of the International IEEE Conf. on Control Applications*, pages 718–723, Istanbul, Turkey, June 2003.
228. G.H.C. Oliveira, R.J.G.B. Campello, and W.C. Amaral. Fuzzy models within orthonormal basis function framework. In *Proc. of the IEEE International Conf. on Fuzzy Systems (FUZZ-IEEE '99)*, volume 2, pages 957–962, Seoul, Korea, August 1999. IEEE.
229. T. Oliveira e Silva. Optimality conditions for truncated Laguerre networks. *IEEE Transactions on Signal Processing*, 42(9):2528–2530, 1994.
230. T. Oliveira e Silva. Laguerre filters – an introduction. *Revista do DETUA*, 1(3):237–248, 1995. Internal publication of the Electrical and Telecommunications Engineering Department of the University of Aveiro, Portugal. Available at `ftp://inesca.inesca.pt/pub/tos/English/revdetua.ps.gz`.
231. T. Oliveira e Silva. On the determination of the optimal pole position of Laguerre filters. *IEEE Transactions on Signal Processing*, 43(9):2079–2087, 1995.
232. T. Oliveira e Silva. Optimality conditions for Laguerre lattice filters. *IEEE Signal Processing Letters*, 2(5):97–98, 1995.
233. T. Oliveira e Silva. Optimality conditions for truncated Kautz networks with two periodically repeating complex conjugate poles. *IEEE Transactions on Automatic Control*, 40(2):342–346, 1995.
234. T. Oliveira e Silva. Rational orthonormal functions on the unit circle and on the imaginary axis, with applications in system identification. Available at `ftp://inesca.inesca.pt/pub/tos/English/rof.ps.gz`, October 1995.

235. T. Oliveira e Silva. A n-width result for the generalized orthonormal basis function model. In *Preprints of the 13th IFAC World Congress*, volume I, pages 375–380, San Francisco, CA, USA, 1996. Elsevier, Oxford, UK.
236. T. Oliveira e Silva. On the adaptation of the pole of Laguerre-lattice filters. In G. Ramponi, G. L. Sicuranza, S. Carrato, and S. Marsi, editors, *Signal Processing VIII: Theories and Applications (Proceedings of EUSIPCO-96, Trieste, Italy)*, volume II, pages 1239–1242. Edizioni LINT, Trieste, Italy, 1996.
237. T. Oliveira e Silva. Stationarity conditions for the L_2 error surface of the generalized orthonormal basis functions lattice filter. *Signal Processing*, 56(3):233–253, 1997.
238. T. Oliveira e Silva. On the asymptotic eigenvalue distribution of block-Toeplitz matrices related to the generalized orthonormal basis functions model. *Automatica*, 35(10):1653–1661, 1999.
239. T. Paatero, M. Karjalainen, and A. Harma. Modeling and equalization of audio systems using Kautz filters. In *Proc. IEEE International Conf. on Acoustics, Speech, and Signal Processing Proceedings (ICASSP'01)*, pages 3313–3316, Salt Lake City, Utah, May 2001. IEEE.
240. M. Padmanabhan and K. Martin. Feedback-based orthogonal digital filters. *IEEE Transactions on Circuits and Systems II*, 40(8):512–525, 1993.
241. M. Padmanabhan, K. Martin, and G. Peceli. *Feedback-Based Orthogonal Digital Filters: Theory, Applications, and Implementation.* Kluwer Academic Publishers, Boston-Dordrecht-London, 1996.
242. P.N. Paraskevopoulos. System analysis and synthesis via orthogonal polynomial series and Fourier series. *Mathematics and Computers in Simulation*, 27:453–469, 1985.
243. R.S. Parker. Nonlinear model predictive control of a continuous bioreactor using approximate data-driven models. In *Proc. 2002 American Control Conf.*, volume 4, pages 2885–2890, Anchorage, AK, May 2002. IEEE Press, Piscataway, NJ.
244. T. W. Parks. Choice of time scale in Laguerre approximations using signal measurements. *IEEE Transactions on Automatic Control*, AC-16(5):511–513, 1971.
245. J.R. Partington. Approximation of delay systems by Fourier-Laguerre series. *Automatica*, 27(3):569–572, 1991.
246. J.R. Partington. Robust identification and interpolation in $\mathcal{H}_\infty$. *International Journal of Control*, 54:1281–1290, 1991.
247. J.R. Partington. Robust identification in $\mathcal{H}_\infty$. *Journal of Mathematical Analysis and Applications*, 100:428–441, 1992.
248. Y.C. Pati and P.S. Krishnaprasad. Rational wavelets in model reduction and system identification. In *Proc. 33rd IEEE Conf. on Decision and Control*, pages 3394–3399, Lake Buena Vista, FL, 1994. IEEE Press, Piscataway, NJ.
249. H. Perez and S. Tsujii. A system identification algorithm using orthogonal functions. *IEEE Transactions on Signal Processing*, 38(3):752–755, 1991.
250. S. Pillai and T. Shim. *Spectrum Estimation and System Identification.* Springer-Verlag, New York, 1993.
251. A. Pinkus. *n-Widths in Approximation Theory.* Springer-Verlag, New York, 1985.
252. R. Pintelon, P. Guillaume, Y. Rolain, J. Schoukens, and H. Van Hamme. Parametric identification of transfer functions in the frequency domain-a survey. *IEEE Transactions on Automatic Control*, 39(11):2245–2260, 1994.

253. R. Pintelon and J. Schoukens. *System Identification - A Frequency Domain Approach.* IEEE Press, Piscataway, NJ, 2001.
254. P.A. Regalia. *Adaptive IIR Filtering in Signal Processing and Control*, volume 90 of *Electrical Engineering and Electronics.* Marcel Dekker, Inc., New York and Basel, 1995.
255. W. Reinelt, A. Garulli, and L. Ljung. Comparing different approaches to model error modelling in robust identification. *Automatica*, 38(5):787–803, 2002.
256. F. Riesz and B. Sz.-Nagy. *Functional Analysis.* Fredrick Ungar, New York, 1955.
257. R.A. Roberts and C.T. Mullis. *Digital Signal Processing.* Addison-Wesley Publishing Company, Reading, Massachusetts, 1987.
258. D.C. Ross. Orthonormal exponentials. *IEEE Transactions on Communication and Electronics*, 12(71):173–176, 1964.
259. W. Rudin. *Real and Complex Analysis.* McGraw-Hill Int. Ed., New York, third edition, 1987.
260. A.M. Sabatini. A hybrid genetic algorithm for estimating the optimal time scale of linear systems approximations using Laguerre models. *IEEE Transactions on Automatic Control*, 45(5):1007–1011, 2000.
261. G. Sansone. *Orthogonal Functions.* Dover Publications, Inc., New York, 1991. Published for the first time by Interscience Publishers in 1959.
262. D. Sarason. Generalized interpolation in $\mathcal{H}_\infty$. *Trans. Am. Math. Soc.*, 127:179–203, 1967.
263. B.E. Sarrouhk, A.C. den Brinker, and S.J.L. van Eijndhoven. Optimal parameters in modulated Laguerre series expansions. In *Proc. of the Fifth International Symp. on Signal Processing and Its Applications*, volume 1, pages 187–190, Brisbane, Australia, August 1999. IEEE Press, Piscataway, NJ.
264. B.E. Sarroukh, S.J.L. Van Eijndhoven, and A.C. Den Brinker. An iterative solution for the optimal poles in a Kautz series. In *Proc. IEEE International Conf. on Acoustics, Speech, and Signal Processing Proceedings (ICASSP'01)*, volume 6, pages 3949–3952, Salt Lake City, Utah, May 2001. IEEE.
265. M. Schetzen. Asymptotic optimum Laguerre series. *IEEE Transactions on Circuit Theory*, CT-18(5):493–500, 1971.
266. F. Schipp and J. Bokor. L_∞ system approximation algorithms generated by φ-summations. *Automatica*, 33:2019–2024, 1997.
267. F. Schipp and J. Bokor. Approximate identification in Laguerre and Kautz bases. *Automatica*, 34:463–468, 1998.
268. F. Schipp and J. Bokor. Rational bases generated by Blaschke-product systems. In *Proc. 13th IFAC Symp. on System Identification (SYSID 2003)*, Rotterdam, NL, 2003. Elsevier Science Ltd., Oxford, UK.
269. F. Schipp, J. Bokor, and L. Gianone. Approximate $\mathcal{H}_\infty$ identification using partial sum operators in disc algebra basis. In *Proc. 1995 American Control Conf.*, pages 1981–1985, Seattle, WA, 1995. IEEE Press, Piscataway, NJ.
270. F. Schipp, L. Gianone, J. Bokor, and Z. Szabó. Identification in generalized orthogonal basis - a frequency domain approach. In *Preprints of the 13th IFAC World Congress*, volume I, pages 387–392, San Francisco, CA, 1996. Elsevier, Oxford, UK.
271. J. Schoukens, T. Dobrowiecki, and R. Pintelon. Parametric and nonparametric identification of linear systems in the presence of nonlinear distortions a frequency domain approach. *IEEE Transactions on Automatic Control*, 43(2):176–190, 1998.

272. I. Schur. On power series which are bounded in the interior of the unit circle. I and II. In I. Gohberg, editor, *I. Schur Methods in Operator Theory and Signal Processing*, volume OT 18 of *Operator Theory: Advances and Applications*, pages 31–88. Birkhäuser Verlag, Boston, 1986. English translation of the well known 1917 and 1918 papers.
273. C.K. Shanatanan and J. Koerner. Transfer function synthesis as a ratio of two complex polynomials. *IEEE Transactions on Automatic Control*, 8:56–58, 1963.
274. M.D. Sidman, F.E. DeAngelis, and G.C. Verghese. Parametric system identification on logarithmic frequency response data. *IEEE Transactions on Automatic Control*, 36:1065–1070, 1991.
275. R.E. Skelton. Model error concepts in control design. *International Journal of Control*, 49(5):1725–1753, 1989.
276. R.S. Smith and J.C. Doyle. Towards a methodology for robust parameter identification. In *Proc. 1990 American Control Conf.*, San Diego, CA, 1990. IEEE Press, Piscataway, NJ.
277. R.S. Smith and J.C. Doyle. Model validation – a connection between robust control and identification. *IEEE Transactions on Automatic Control*, AC-36:942–952, 1992.
278. T. Söderström and P.G. Stoica. *System Identification*. Prentice-Hall International, Hemel Hempstead, U.K., 1989.
279. V. Solo and X. Kong. *Adaptive Signal Processing Algorithms*. Prentice-Hall, Englewood Cliffs, NJ, 1995.
280. J. T. Spanos. Algorithms for ℓ_2 and ℓ_∞ transfer function curve fitting. In *Proc. AIAA Guidance, Navigation and Control Conf.*, pages 1739–1747, New Orleans, LA, 1991. AIAA.
281. K. Steiglitz. The equivalence of digital and analog signal processing. *Information and Control*, 8(5):455–467, 1965.
282. K. Steiglitz and L.E. McBride. A technique for the identification of linear systems. *IEEE Transactions on Automatic Control*, AC-10:461–464, 1965.
283. P. Stoica and T. Söderström. On the parsimony principle. *International Journal of Control*, 36(3):409–418, 1982.
284. Z. Szabó. L_p norm convergence of rational operators on the unit circle. *Mathematica Pannonica*, 9(2):281–292, 1998.
285. Z. Szabó. *Rational Orthonormal Bases and Related Transforms in Linear System Modeling*. PhD thesis, Eötvös Loránd University, Budapest, Hungary, 2001. Available at `http://www.sztaki.hu/scl/mainth.pdf`.
286. Z. Szabó. Interpolation and quadrature formulae for rational systems on the unit circle. *Annales Univ. Sci. Budapestiensis Sectio Computatorica*, 21:41–56, 2002.
287. Z. Szabó and J. Bokor. Minimal state space realization for transfer functions represented by coefficients using generalized orthonormal basis. In *Proc. 36th IEEE Conf. on Decision and Control*, pages 169–174, San Diego, CA, December 1997. IEEE Press, Piscataway, NJ.
288. Z. Szabó and J. Bokor. L_p norm convergence of rational orthonormal basis function expansions. In *Proc. of the 38th IEEE Conf. on Decision and Control*, pages 3218–3223, Phoenix, AZ, December 1999. IEEE Press, Piscataway, NJ.
289. Z. Szabó, J. Bokor, and F. Schipp. Nonlinear rational approximation using generalized orthonormal basis. In *Proc. of the 1997 European Control*

Conf. ECC97, volume 6, Part B, pages 1–5, Brussels, Belgium, 1997. Ottignies, Louvain-la-Neuve, Belgium.

290. Z. Szabó, J. Bokor, and F. Schipp. Identification of rational approximate models in $\mathcal{H}_\infty$ using generalized orthonormal basis. *IEEE Transactions on Automatic Control*, 44(1):153–158, 1999.
291. Z. Szabó, P.S.C. Heuberger, J. Bokor, and P.M.J. Van den Hof. Extended Ho-Kalman algorithm for systems represented in generalized orthonormal bases. *Automatica*, 36(12):1809–1818, 2000.
292. O. Szász. On closed sets of rational functions. *Annali di Matematica Puru ed Applicata*, Serie Quarta – Tomo XXXXIV:195–218, 1953.
293. G. Szegö. *Orthogonal Polynomials*, volume 23 of *Colloquium Publications*. American Mathematical Society, New York, 1959.
294. G. Szegö. *Orthogonal Polynomials*. American Mathematical Society Colloquium Publications, Volume XXIII, Providence, RI, fourth edition, 1975.
295. L. Szili. On the summability of trigonometric interpolation process. *Acta Mathematica Hungarica.*, 91:131–158, 2001.
296. S. Takenaka. On the orthogonal functions and a new formula of interpolation. *Japanese Journal of Mathematics*, II:129–145, 1925.
297. N. Tanguy, R. Morvan, P. Vilbé, and L. C. Calvez. Improved method for optimum choice of free parameter in orthogonal approximations. *IEEE Transactions on Signal Processing*, 47(9):2576–2578, 1999.
298. N. Tanguy, R. Morvan, P. Vilbé, and L. C. Calvez. Online optimization of the time scale in adaptive Laguerre-based filters. *IEEE Transactions on Signal Processing*, 48(4):1184–1187, 2000.
299. N. Tanguy, R. Morvan, P. Vilbé, and L.C. Calvez. Pertinent choice of parameters for discrete Kautz approximation. *IEEE Transactions on Automatic Control*, 47(5):783–787, 2002.
300. N. Tanguy, P. Vilbé, and L. C. Calvez. Optimum choice of free parameter in orthonormal approximations. *IEEE Transactions on Automatic Control*, 40(10):1811–1813, 1995.
301. A.J. Tether. Construction of minimal linear state-variable models from finite input-output data. *IEEE Transactions on Automatic Control*, 15(4):427–436, 1970.
302. F. Tjärnström and L. Ljung. L_2 model reduction and variance reduction. *Automatica*, 38(9):1517–1530, 2002.
303. D.N.C. Tse, M.A. Dahleh, and J.N. Tsitsiklis. Optimal and robust identification in ℓ_1. In *Proc. 1993 American Control Conf.*, Boston, MA, 1993. IEEE Press, Piscataway, NJ.
304. P.P. Vaidyanathan. A unified approach to orthogonal digital filters and wave digital filters, based on lbr two-pair extraction. *IEEE Transactions on Circuits and Systems*, 32(7):673–686, 1985.
305. P.J.M. Van den Hof and R.J.P. Schrama. Identification and control–closed loop issues. *Automatica*, 31:1751–1770, 1995.
306. P.M.J. Van den Hof, P.S.C. Heuberger, and J. Bokor. System identification with generalized orthonormal basis functions. *Automatica*, 31(12):1821–1834, 1995.
307. J. Van Deun and A. Bultheel. The computation of orthogonal rational functions on an interval. *Journal of Computational and Applied Mathematics*, 2004. Accepted.

308. E.T. van Donkelaar. *Improvement of Efficiency in Identification and Model Predictive Control of Industrial Processes—A Flexible Linear Parametrization Approach.* PhD thesis, Delft University of Technology, Delft, The Netherlands, November 2000.
309. E.T. van Donkelaar, P.S.C. Heuberger, and P.M.J. Van den Hof. Identification of a fluidized catalytic cracking unit: an orthonormal basis functions approach. In *Proc. 1998 American Control Conf.*, pages 1814–1917, Philadelphia, PA, 1998. IEEE Press, Piscataway, NJ.
310. P. Van gucht and A. Bultheel. State space representation for arbitrary orthogonal rational functions. *Systems & Control Letters*, 49:91–98, 2003.
311. S.M. Veres and J.P. Norton. Predictive self tuning control by parametric bounding and worst–case design. *Automatica*, 29:911–928, 1993.
312. T. Von Schroeter. Frequency warping with arbitrary allpass maps. *IEEE Signal Processing Letters*, 6(5):116–118, 1999.
313. B. Wahlberg. System identification using Laguerre models. *IEEE Transactions on Automatic Control*, 36(5):551–562, 1991.
314. B. Wahlberg. Laguerre and Kautz models. In *Proc. 10th IFAC Symp. on System Identification (SYSID'94)*, volume Number 3, pages 1–12, Copenhagen, 1994. Elsevier Science Ltd., Oxford, UK.
315. B. Wahlberg. System identification using Kautz models. *IEEE Transactions on Automatic Control*, 39(6):1276–1282, 1994.
316. B. Wahlberg. On spectral analysis using models with pre-specified zeros. In A. Rantzer and C.I. Byrnes, editors, *Directions in Mathematical Systems Theory and Optimization*, volume 286 of *Lecture Notes in Control and Information Science*, pages 333–343. Springer, Berlin Heidelberg New York, 2003.
317. B. Wahlberg. Orthonormal basis functions models: A transformation analysis. *SIAM Review*, 45(4):689–705, 2003.
318. B. Wahlberg and E.J. Hannan. Parametric signal modelling using Laguerre filters. *The Annals of Applied Probability*, 3(2):467–496, 1993.
319. B. Wahlberg and P.M. Mäkilä. On approximation of stable linear dynamical systems using Laguerre and Kautz functions. *Automatica*, 32(5):693–708, 1996.
320. J.L. Walsh. *Interpolation and Approximation by Rational Functions in the Complex Domain*, volume XX of *American Mathematical Society Colloquium Publications*. American Mathematical Society, Providence, Rhode Island, fifth edition, 1975. First edition in 1935.
321. L. Wang. Discrete time model predictive control design using Laguerre functions. In *Proc. 2001 American Control Conf.*, volume 3, pages 2430–2435, Arlington, VA, 2001. IEEE Press, Piscataway, NJ.
322. L. Wang and W.R. Cluett. Optimal choice of time-scaling factor for linear system approximations using Laguerre models. *IEEE Transactions on Automatic Control*, 39(7):1463–1467, 1994.
323. L.P. Wang. Discrete model predictive controller design using Laguerre functions. *Journal of Process Control*, 14:131–142, 2004.
324. N.F. Dudley Ward and J.R. Partington. Rational wavelet decomposition of transfer functions in Hardy–Sobolev classes. Technical report, University of Leeds, Dept. of Pure Mathematics, 1995.
325. N.F. Dudley Ward and J.R. Partington. Robust identification in the disc algebra using rational wavelets and orthonormal basis functions. *International Journal of Control*, 96(3):409–423, 1996.

326. A.H. Whitfield. Asymptotic behavior of transfer function synthesis methods. *International Journal of Control*, 45:1083–1092, 1987.
327. H. Widom. *Studies in Real and Complex Analysis*. MAA Studies in Mathematics. Prentice-Hall, Englewood Cliffs, NJ, 1965.
328. N. Wiener. *The Fourier Integral and Certain of its Applications*. Cambridge University Press, Cambridge, UK, 1933.
329. N. Wiener. *Extrapolation, Interpolation and Smoothing of Stationary Time Series*. M.I.T.-Press, Cambridge, MA, 1949.
330. R.A. Wiggins and E.A. Robinson. Recursive solution to the multichannel filtering problem. *Journal of Geophysical Research*, 70(8):1885–1891, 1965.
331. G.A. Williamson. Tracking random walk systems with vector space adaptive filters. *IEEE Transactions on Circuits and Systems II*, 42(8):543–547, 1995.
332. L.L. Xie and L. Ljung. Asymptotic variance expressions for estimated frequency functions. *IEEE Transactions on Automatic Control*, 46:1887–1899, 2001.
333. G. Yen and S.B. Lee. Multiple model approach by orthonormal bases for controller design. In *Proc. 2000 American Control Conf.*, pages 2321–2325, Chicago, IL, 2000. IEEE Press, Piscataway, NJ.
334. A. Yonernoto, T. Hisakado, and K. Okumura. Expression of transient phenomena at faults on ideal transmission lines by Laguerre functions. *IEE Proceedings of Circuits, Devices and Systems*, 150:141–147, 2003.
335. Z.D. Yuan and L. Ljung. Black box identification of multivariable transfer functions: Asymptotic properties and optimal input design. *International Journal of Control*, 40:233–256, 1984.
336. Z. Zang, B.N. Vo, A. Cantoni, and K.L. Teo. Iterative algorithms for envelope constrained recursive filter design via Laguerre functions. *IEEE Transactions on Circuits and Systems I*, 46:1342–1348, 1999.
337. H.P. Zeiger and J. McEwen. Approximate linear realizations of given dimension via Ho's algorithm. *IEEE Transactions on Automatic Control*, 19:153, 1974.
338. C. Zervos, P.R. Belanger, and G.A. Dumont. Controller tuning using orthonormal series identification. *Automatica*, 24:165–175, 1988.
339. C. Zervos and G.A. Dumont. Deterministic adaptive control based on Laguerre series representation. *International Journal of Control*, 48:2333–2359, 1988.
340. K. Zhou, C. Doyle, and K. Glover. *Robust and Optimal Control*. Prentice-Hall, Upper Saddle River, NJ, 1996.
341. Q.G. Zhou and E.J Davison. Balanced realization of Laguerre network models. In *Proc. 2000 American Control Conf.*, pages 2874–2878, Chicago, IL, June 2000. IEEE Press, Piscataway, NJ.
342. Q.G. Zhou and E.J. Davison. A simplified algorithm for balanced realization of Laguerre network models. In *Proc. 39th IEEE Conf. on Decision and Control*, pages 4336–4640. IEEE Press, Piscataway, NJ, December 2000.
343. T. Zhou and H. Kimura. Time domain identification for robust control. *Systems & Control Letters*, 20(3):167–178, 1993.
344. A. Zygmund. *Trigonometric Series*. Cambridge University Press, Cambridge, UK, 1959.

Index

CPSIA information can be obtained at www.ICGtesting.com
Printed in the USA
LVOW10s1408160214

373889LV00003B/81/P

9 781849 969765